EARTH, EMPIRE AND SACRED TEXT

Comparative Islamic Studies
Series Editor: Brannon Wheeler, US Naval Academy

This series, like its companion journal of the same title, publishes work that integrates Islamic studies into the contemporary study of religion, thus providing an opportunity for expert scholars of Islam to demonstrate the more general significance of their research both to comparativists and to specialists working in other areas. Attention to Islamic materials from outside the central Arabic lands is of special interest, as are comparisons which stress the diversity of Islam as it interacts with changing human conditions.

Notes from the Fortune-Telling Parrot:
Islam and the Struggle for Religious Pluralism in Pakistan
David Pinault

Ibn Arabi and the Contemporary West:
Beshara and the Ibn Arabi Society
Isobel Jeffery-Street

Prolegomena to a History of Islamicate Manichaeism
John C. Reeves

Orientalists, Islamists and the Global Public Sphere:
A Historical Genealogy of the Modern Image of Islam
Dietrich Jung

Prophecy and Power:
Muhammad and the Qur an in the Light of Comparison
Marilyn Robinson Waldman
Edited by Bruce B. Lawrence

The Qur'ān:
A New Annotated Translation
Edited by A. J. Droge

East by Mid-East:
Studies in Cultural, Historical and Strategic Connectivities
Edited by Brannon Wheeler and Anchi Hoh

EARTH, EMPIRE AND SACRED TEXT

MUSLIMS AND CHRISTIANS AS TRUSTEES OF CREATION

DAVID L. JOHNSTON

SHEFFIELD UK   BRISTOL CT

Published by Equinox Publishing Ltd.

UK: Kelham House, 3 Lancaster Street, Sheffield, S3 8AF
USA: ISD, 70 Enterprise Drive, Bristol, CT 06010

www.equinoxpub.com

Paperback edition published 2013.

British Library Cataloguing-in-Publication Data
A catalogue record for this book is available from the British Library.

Library of Congress Cataloging-in-Publication Data
A catalogue record for this book is available from the Library of Congress

ISBN: 978 1 84553 225 3   (hardback)
ISBN: 978 1 78179 035 9   (paperback)

Typeset by Forthcoming Publications Ltd. (www.forthpub.com)
Printed and bound in the UK by Lightning Source UK Ltd., Milton Keynes and Lightning Source Inc., La Vergne, TN

*To J. Dudley Woodberry and Bishop Kenneth Cragg,
my mentors in Muslim-Christian dialogue—
gracious, scholarly and persevering*

CONTENTS

Part I
OUR POSTMODERN SITUATION

Part III
HUMAN TRUSTEESHIP:
MUSLIM AND CHRISTIAN INTERPRETATIONS

FOREWORD BY BISHOP KENNETH CRAGG

'Good wine needs no bush', says an old English proverb, referring to the spray of leaves which vintners and wine-shops used to hang outside their premises to suggest their vineyard. 'Good writing needs no gush'—the 'gush' publishers sometimes offer on jackets and fly-leaves as to what is 'widely acclaimed' well before it has reached any sober readership. 'Forewords' in such terms are likewise suspect.

Not this one, which is happy to salute an assiduous exploration of a vital theme between the Semitic faiths, namely, our custody of a good earth, an earth responsive to our will and skill and itself derived from an otherwise incredible sequence of awesome 'chances' that might well have doomed it in the very precariousness of the proceeding from which it at length eventuated. A secular humanism like that of the Oxford novelist-philosopher Iris Murdoch may opine 'We are simply here' and recognise in human life no external cause or reason.

Others, perhaps more accurately, find our very 'here-ness'—as it were—'internalising external cause' precisely in the evident mandate to occupy and possess which all human awe and reverence must discern in our experience of mortal place and time. Such is the Biblical/Qur'anic theme of *khilafah*/'dominion'—what, forty years ago, I ventured to call 'The Privilege of Man' and tried three years ago to focus again via Surah 7:172's question: 'Am I Not Your Lord?' Addressed to an imagined concourse of all human generations, tribes and haunts, the question in the negative was thereby plainly seeking the answer 'Yes!, One Allah', as it were, by dint of willing recognition (and not, by non-concurrence otherwise). All 'knowing God as God', His 'letting be', turn on a response of ours to His delegacy into our hand, our mind and will, of the natural order, whence all arts, sciences, economies and politics derive their quality as 'trusts', not perquisites, as tenancies not lordships.

It is this shared perception of creation for creaturehood in both Islam and Christianity which David Johnston's study examines and explores with meticulous attention to its exposition through writers down the centuries. He casts a large and fine net. There are few facets that escape him. I have learned much from the ample detail of his findings.

His study is an exemplary contribution to the territory between faiths and between faith and ecology. What gives its scholarly merits such current value is its accent on human custody of the rare planet and the urgency of subduing our passions to its tutelage of gratitude, humility and a due *islam* or honest handling of the 'entrustedness' which is its clarion call for justice and compassion as from an Allah who took the enormous 'risk' of these in the enterprise of radical divine decision of which Surah 2:30 dramatically tells. It was a decision which necessitated the entire corollary of prophethood, for guidance, direction, education and motivation in its discharge, as so magnanimously deputed into human care.

Perhaps in that context, a Foreword of eager commendation may venture three ever concurrent points about that whole enterprise of 'God through us and we for God'. The first would be what one of our faiths understands as 'the originality of sin'. What is striking about Surah 2:30 and cognate passages is that Allah is—as it were—forewarned about the risk present in the perversity of humans. 'Is He going to entrust all that is precious to those who will shed blood and corrupt the earth?' Allah over-rides this strong (shall we call it?) warning, replying mysteriously: 'I know what to you is unknown'. 'I have my own mind—my resources, my capacities: risk with these mortals is what I will incur.'

There is thus no neglecting or denying 'the originality of sin', some-thing there at the heart of things, not heredity, nor fate, nor primal deed of one that all the rest must inherit, but something about us all of self-will, something law will only excite when aiming to restrain, an impul-sive 'self-demand' which 'corrupts' a pure selfhood. Hence the perennial problem of being 'rightly me' inside my own egoism—the very burden at the core of Buddhism. How apposite the cry: 'Who shall deliver me?'

The second point—reached in Dr. Johnston's last section—would be the issue of secularity and outright secularisation (unless the terms are carefully distinguished). This bears heavily on the theme of separation, or junction, between religion and politics. It is urgent to appreciate what deeply theological reasons there are for 'secular statehood'. For only a secular state ensures to the private self a citizen's aspects of his/her 'caliphate' in respect of how the state is governed and kept sane. A secular state is not imperiously dominated by any one religion whose Shari'ah, or political regime, precludes that personal and citizen dignity only present under secular auspices or those of a religion ready to co-exist with others.

This is highly significant now for that quarter of the world's Muslims who are in diaspora and whose *Dar al-Islam* can no longer be its long,

Medinan thing of power and control. It can only be the *Dar al-Islam* of a personal *taqwa* (devotion) and *sabr* (patience) inside the moral *jihad* and *ikhlas* (sincerity), with the pillars of religion. All this can also have a great bearing on the vocation of those Muslim states that have mixed citizenships or almost exclusive Muslim character, as to the civil duty.

Sadly, twentieth-century Islam had a crudely brutal experience (or rather the Turkish segment of it) in the violent abolition of the old Caliphate under Atatürk and his mentor, Ziya Gölkap, who made Islam merely the cultural aspect of Turkism. It is surely possible now for the Muslim mind to engage more openly and sanely with the secular nature of the state. For that theme only returns it to Islam's original quality in its defining origins in Mecca, where, without the benefit of power, it was simply a *balagh* informing a community in obedience to Allah. The priority of that Meccan version is surely indubitable for the contemporary self-expression of Islam.

Thirdly, this Foreword would make a plea, in the overall range of David Johnston's work, for the theme and concept of the 'sacramental'. Islam my find the term uncongenial. But it has a ring of urgency. As did the 'pledge' or *sacramentum* of the Latin world. It binds and recruits the heart and will. To talk of 'holism' and the 'holistic' is abstract and cerebral. For the Christian the 'table of the Lord' that is the landscape under human hand becomes again the table in the sanctuary, the centre-piece of a worship that responds alike to a created order in trust and a redemptive grace in action. 'The bread and the wine', the nurture and the hospitality, the labour and the joy, which are the themes of our being human become in turn the home of our imagination and the summons to our wills. Such is the text which ultimately we here encounter. Read it well.

Kenneth Cragg
Oxford, 2007

ACKNOWLEDGMENTS

This nine-year project would not have been completed without the help of many people. First, my wife Charlotte, a skillful editor and great encourager. Then, those who stood by me in my doctoral program at Fuller Theological Seminary: my mentor, J. Dudley Woodberry, always encouraging, knowledgeable and wise in things Islamic; R. Daniel Shaw, who guided me in anthropology, both in research theory and in analyzing the data I collected from two trips to the West Bank and Gaza Strip (1998, 1999); and theologian Charles Van Engen, who taught me how theology is always pieced together in context.

I owe my five years of research at Yale University to Professor Gerhard Böwering's generosity and friendship. He also provided useful comments on Chapter 7 of this book. Steven Davis kindly commented on my dissertation in my first year at Yale. It was his astute questioning that led me to correct, rewrite and expand what has now become Chapters 2–5. Many thanks also to colleagues there (especially Frank Griffel and Adel Allouche), who were kind enough to critique articles I wrote along the way. In writing those I began to grasp the necessary intermingling of Islamic law, theology and ethics, and realized the crucial role played by decisions people make (usually unconscious) in matters of epistemology and hermeneutics.

Also important was the financial grant I received from Fuller Seminary in 2003–2004 (and again in 2006) for the revision of my dissertation in view of publication. I have been truly blessed by Professor David Augsburger and the Conflict Transformation Grant committee's vote of confidence and tangible support for this project.

Finally, I wish to acknowledge the inspiration and stimulation my Muslim friends have given me over the years, from Algeria, where I first settled in 1978, to Egypt, Israel-Palestine, Iraq, and more recently in United States. May this book widen the scope of present cooperative ventures in many places and spark the beginning of many more!

LIST OF FIGURES

INTRODUCTION

Muslims, Jews and Christians construct their vision of reality—as it is and how it *ought* to be—based on their understanding, sifting out and prioritizing of the material in their sacred texts. Naturally, this theological effort is fed and often constrained by the weight of inherited tradition, yet it can sometimes break out of the received mold to exhibit the radical tones and buoyant colors of the kingdom of God in fresh and compelling ways. Nevertheless, as Omid Safi indicates, any new burst of "progressive" thinking must also represent "a serious engagement with tradition."[1]

The book before you takes these strategies seriously with the conviction that the task of theology has to be undertaken anew in each age and in each locale. Core religious beliefs and values remain, but the vision that results from their application to specific historical configurations necessarily changes from period to period and from context to context. Without discounting the necessity of local interpretations, the following pages seek to interpret and reflect on the ethical imperatives arising out of the creation accounts in the Qur'an and Bible against the horizon of humanity's global reality at the beginning of the twenty-first century CE. By building on the idea of God's empowering the human person as His trustee on earth, the common theological ground between Muslims and Christians on the nature and calling of humanity and its relationship with creation is expanded. It must be said from the outset that even though the main focus here is on Islam and Christianity, the assumption throughout is that since the Genesis account of creation (indeed the whole Hebrew Bible) is shared by Christians and Jews, what is argued about Christians

---

[1] Omid Safi, "Introduction: The Times they Are a-Changin'—A Muslim Quest for Justice, Gender Equality, and Pluralism," in *Progressive Muslims: On Justice, Gender and Pluralism*, ed. Omid Safi (Oxford: Oneworld, 2003), 7. For an entirely female chorus of Muslims asserting their faith and identity in the American context, see *Living Islam Out Loud: American Muslim Women Speak*, ed. Saleemah Adul-Ghafur (Boston: Beacon Press, 2005). As the editor puts it in her Introduction, "We are women who understand that following disempowering interpretations of sacred text isn't for us… This book reflects the diversity of American Muslim women in all our complexity" (p. 6).

also applies to Jews. There will be references here and there to Judaism, but again, the purpose of this project is to move beyond the useful Muslim–Christian dialogue in process since the 1970s between theologians and scholars[2] by seeking to widen and motivate the circle of potential activists on both sides to cooperate for the sake of building a world that is more just, peaceful and respectful of the great diversity of humankind.

A Personal Note

In a postmodern context in which suspicion of unconscious or ulterior motives and inescapable issues of power in human discourse, in speech or writing, are taken for granted, it is appropriate for authors to first signal their own perspectives. In my case, besides being an Islamicist, a Christian minister (Congregationalist) and evangelical theologian (the Bible is my authoritative guide), I must also indicate that this study is borne out of three personal traits. The first is a passion for observing and comparing human cultures. I was raised as an American in France, with a fully French education, complete with Baccalauréat. My university years in the USA were spent in trying to become American—at least enough so that one would not notice I was still quite French inside. Then I spent nearly sixteen years in three Arabic-speaking countries (Algeria, Egypt and Palestine [West Bank]), striving hard to speak the language well and to understand the culture from within. Though one never fully succeeds at this, relationships that are developed bring joy and fulfillment. That process has continued in the multi-cultural setting of Southern California and the international atmosphere of Yale University and the University of Pennsylvania.

The second trait that lies behind this project is my fascination with the study of religion. Always and in every location I have enjoyed studying and teaching the Bible. At the same time, I have been enriched by discussing issues of faith with my Muslim friends—a more natural (and inevitable) pastime in the Middle East. There, people live, walk and breathe their faith. So, I have naturally turned my attention to Islam, to its sacred texts and to the much-needed task of establishing meaningful Muslim–Christian dialogue.

The final trait that urges me forward in this venture is a passion for peace and reconciliation. Though part of my personality, this drive also

---

[2] The best summary of this effort from a Muslim perspective is Ataullah Siddiqi's work, *Christian–Muslim Dialogue in the Twentieth Century* (London: Macmillan; New York: St. Martin's Press, 1997).

finds inspiration in the words of Jesus to his followers, "Blessed are the peacemakers, for they will be called children of God" (Mt. 5:9), and his command to love even one's enemy. I have thus been grieved in my friendship with Muslims to witness the gaping wounds left in their psyche by a long history of Crusades, colonialism and "neo-colonialism," as one takes into account the current hegemony of the West in matters of economics and politics. Moreover, having lived with my family in the Occupied Territories during the end of the first intifada and then through two and a half years of the Oslo Peace Process, I could not help but internalize some of the anger, frustration and helplessness of my Palestinian friends. At the same time, I was heartened and inspired by the courageous peace activism undertaken in Israel/Palestine by a coalition of Jews from Israel and abroad, Palestinians, both Muslims and Christians, and internationals.

To be even more specific, I am a Christian theologian wrestling with questions of interfaith dialogue and pluralism with mostly Muslim partners. I have been privileged to participate of the scholarly end of an ongoing evangelical Muslim dialogue co-sponsored by Fuller Theological Seminary and the Salam Institute for Peace and Justice of Washington, DC.[3] After two conferences facilitated by a grant made by the US Justice Department in 2003,[4] a joint publication by the participating Muslim and Christian scholars has been published.[5] Besides the obvious intellectual enrichment attested by all participants, many of us noted with delight that important barriers had come down. In particular, the second morning of the first conference was almost entirely devoted to the thorny issue of Christian mission, which has been particularly active worldwide through evangelical agencies in recent years. I anticipate that, on the basis of the groundwork achieved in those conversations, many fresh initiatives for peacebuilding, human rights advocacy and development will be taken, affecting various regions of the world.

In that spirit, therefore, this book proposes a common theology of creation as one possible means to break the deadlock of ancient Muslim–Christian animosities—from the Muslim conquests of the seventh century

---

[3] This was also in cooperation with the Islamic Society of North America (ISNA), the Graduate School of Islamic Social Sciences (GSISS) and the Mohamed Said Farsi Chair of Islamic Peace at American University (the present holder of that chair, Mohammed Abu-Nimer, was also co-chair of the conference).

[4] Conflict Transformation Grant: Creating Cooperation and Reducing Conflict between Muslims and Christians.

[5] Mohammed Abu-Nimer and David Augsburger, eds., *Peace by, between, and beyond Muslims and Evangelical Christians* (Lanham, MD: Lexington Books, 2009).

to the Iraq War of 2003. Only such a vision of solidarity and mutual cooperation can defuse the current post-Cold War tensions that some have called "a clash of civilizations."[6]

Part I:
Our Postmodern Condition

In the first part of the book, I present the interpretive framework that brings the vision into focus. If theology necessarily grows out of specific contexts, then it is imperative to understand our present-day context, which I label here "postmodern"—in two different senses. The first sense has to do with the process of globalization with its socioeconomic structures, its cultural and political ramifications. This is what I am calling "postmodernity."[7] "Postmodern" in its second sense evokes the interpretive mood and specific strategies in philosophy and literary criticism, architecture and the arts, cultural/religious pluralism and social protest that many have called "postmodernism."

I first came to this distinction through the work of Cambridge sociologist Bryan S. Turner, who has also studied Islamic societies. Turner defines postmodernity "as a state of affairs standing in opposition to [Max] Weber's view of modernization or as a state of affairs coming after rationalist modernization."[8] He then adds such key factors as globalization and its resistance by "postmodern" groups, which call for a greater emphasis on local economies and cultures and attention to the human impact on the environment. Yet on my view, he fails to offer a viable definition and description of postmodernism.

University of Virginia theologian John Milbank has also contributed to this discussion. Like many others, his use of "postmodernity" plainly

---

[6] The leading exponent of this view has been Samuel Huntington; see his *The Clash of Civilizations and the Remaking of the World Order* (New York: Simon & Schuster, 1996). See Akbar S. Ahmed's essay, "Ibn Khaldun's Understanding of Civilizations and the Dilemmas of Islam and the West Today," *Middle East Journal* 56.1 (Winter 2002): 20-45, for a thoughtful Muslim response to this paradigm.

[7] This is by no means original. The classic study in this sense of "postmodernity" is the urban specialist David Harvey's *The Condition of Postmodernity* (Malden, MA: Blackwell, 1990). He saw the early 1970s as a turning point, not only in the arts, architecture and philosophy (mostly imported from France), but also in the long tendency of Fordist capitalism to "overaccumulate." With time, this process led to the 1973 recession, and then to "a revival of entrepreneurialism and neo-conservatism" (p. 124).

[8] Bryan S. Turner, *Orientalism, Postmodernism and Globalism* (London: Routledge, 1994), 198.

includes "postmodernism," but his analysis serves as a useful introduction to the way I will be using the terms. For Milbank, "postmodernity means the obliteration of boundaries… Everything is made to run into everything else; everything gets blended, undone, and then re-blended."[9] Interestingly, only the first of his four main areas of blurring fits into my category of "postmodernism," and at that, not in a tidy fashion: the confusion of the boundaries between nature and culture. Both the idea of a single human essence (primarily considered rational since the Enlightenment) and the idea of fixed "natural laws" in nature seem to have been abolished of late, while artistic production and cultural artifacts increasingly blur traditional categories of childhood, adulthood and old age, or even sexual distinctions. On the other hand (here we move into my view of postmodernity), human interventions in nature—from the genetic manipulation of plants and crops to the disastrous affect of greenhouse gases on our global climate—opens up a horizon of anxiety for us as we look into the future.

Milbank's other three categories of blended distinctions fall more neatly into the sociological, economic and political dimensions of what I will argue should be labeled postmodernity: the blurring of interior and exterior (the private self is invaded by the media, redefined in the "virtual spaciality" of the Internet and chopped up by the utopian promises of a consumerist society with its shopping malls and marketing campaigns); the confusion of traditional economic categories (the postindustrial age is in fact the age of information, in which classical boundaries between production, marketing and financing, and between labor and management are often slippery); and finally, the process of globalization. As a social critic this is where, I shall argue, the central feature of postmodernity lies: "The times of postmodernity are in no sense post-capitalist times, but rather times of capital writ still larger. Indeed, capital has always been a force of abstraction; today it reaps even larger material benefits from increased abstraction."[10] This calls for a brief look at the West's recent past.

Following on the heels of the Renaissance and its cosmopolitan outlook, the era of Western colonialism in effect coincided with the fall of the last Muslim enclave of Andalusia, the proud city of Granada (1492). One of the ironies of history is that it was the pluralistic, largely tolerant and knowledge-thirsty civilization of Muslim Spain that ignited the Renaissance in the first place. Sadly, this inclusive impulse of Andalusia

---

[9] John Milbank, "The Gospel of Affinity," in *The Future of Hope: Christian Tradition amid Modernity and Postmodernity*," ed. Miroslav Volf and William Katerberg (Grand Rapids/Cambridge: Eerdmans, 2004), 149.
[10] Ibid., 154.

was overpowered and destroyed by the twin engines of religious impe-rialism and the greed for power and money.[11] What is more, the discov-ery of the Americas helped to fuel the growing European appetite for trade, and, in the wake of the Industrial Revolution, its search for new markets to absorb a surplus of goods.

But Christendom did not survive these radical shifts unchanged. In fact, the Protestant Reformation helped to break the monopoly of Rome on the interpretation of the Christian tradition in the west and provided a new impetus to the expanding reality of nation-states: Max Weber's intuition about the connection between the Protestant ethic and the birth of capitalism—a process that added to the competition of European nations to exploit the natural resources of other continents for the benefit of their own industries.

The inauguration of the "modern" era thus coincided with a new system of industrial production in parallel with the economic system we now call capitalism. Traditionally, as with the cottage industries, for example, the privatization of the means of production was already assumed. But with the rise of the nation-state and especially the pheno-menal growth of industry caused by the invention of machines, capital-ism took on a life of its own. Above all, it was the idea that financial power could be consolidated and used to create new projects and indus-tries. According to William Greider, "capitalism ingeniously strives to produce more output from less output—the efficiency that demands (according to its own definitions) the most effective allocation of resources and capital."[12] It is this kind of efficiency and productivity that enabled the imperial powers of the nineteenth century to develop the kind of economic domination and lethal weaponry to maintain their grip on lands they had previously pillaged.

The twentieth century, however, proved to be a mixed blessing to these European-American powers. For one, they were brought to their knees through acts of unprecedented mutual slaughter: two "world" wars, and the subsequent unraveling of the colonial enterprise. What is more, they achieved two more revolutions in the economic and social realm.

---

[11] Akbar Ahmed informs us that the rehabilitation of Andalusia in contemporary Spain has led the king of Spain to apologize to the Jews for the atrocities committed to them in the Reconquista, but has yet to do so toward the Muslims; see Ahmed's *Islam Today: A Short Introduction to the Muslim World* (London: I. B. Tauris, 1999), 220.

[12] William Greider, *The Soul of Capitalism* (New York: Simon & Schuster, 2003), 19.

The first had to with mass production, often dubbed "Fordist," after the innovations set up by Henry Ford in his automobile plants, whose primary objective was efficiency of production. The second revolution came with the affluence of the post-World War II period: the focus turned from the producer to the consumer, who in turn became at once the engine of economic health and its main victim.

The nation-state ideal—shaped as it was in the American and French revolutions—was articulated in the idiom of the European Enlightenment: individual agents with rights to freedom and the pursuit of happiness, protected by the state through the principles spelled out in a constitution and/or a bill of rights, within a collective political structure dubbed a "democracy" because the people, not the king, are sovereign. But the logic of the market—if left unchecked—is inherently inimical to democratic ideals, and thus it had to be tamed, for instance by passing laws forbidding child labor or limiting the working week to forty hours, and so on.[13]

By way of illustration, let us take the United States of America as a case in point. Now the richest and most powerful among a handful of nations that have largely surpassed the satisfaction of their basic needs, it finds itself coming to a crossroads. Americans are born and bred on the notion that each citizen is entitled to "more"—more cars and SUVs, more gadgets in bigger homes, more products with designer labels, more movies and cable channels, more advanced drugs, more exercise and outdoor activities, more and more software and hardware upgrades. Yet in the midst of this overwhelming abundance (unthinkable even three generations ago), Greider argues that the current trend of mass consumption is threatening the very fabric of society. The rising tide of stock prices in the 1990s upheld the myth that small investors could store away enough cash for the lean days ahead, but when the bubble popped "American households swiftly lost $3 or $4 trillion in putative wealth."[14] Perhaps the best way to gauge the growing unease from the poor up to the higher middle classes is to use the definition of poverty proposed by Cambridge economist and Nobel Prize-laureate Amartya Sen: poverty is "when people lack the means to appear in public without shame."[15] Consumption is not just competition between neighbors; more importantly it

---

[13] This is one of Benjamin Barber's main arguments in his *Jihad Vs. McWorld: Terrorism's Challenge to Democracy* (New York: Ballantine Books, 1995 [with a new introduction, 2001]).

[14] Greider, *The Soul of Capitalism*, 13.

[15] Ibid., 17.

is whatever is required to be seen as prosperous, healthy and attractive according to the fluctuating definitions set by corporate marketing executives. This, in fact, is a treadmill—socially enforced—the allure of which few can resist.

The truth is that the working class has seen its income steadily eroded over the last couple of decades.[16] In order to keep up, often both spouses have worked harder, longer, thus dangerously reducing time spent with their children. Many have overspent their credit cards while often borrowing from their home equity to pay for more products. These dead-end strategies do not bode well for the future. At the end of the collision course lies the stark reality of homelessness for some, drastically downsized consumer profiles for others, and social shame for all involved. If one adds to this the outsourcing of white-collar jobs to South Asia and elsewhere, one begins to see how high up the economic ladder the prevailing sense of insecurity has reached. Yet these are only the signs of malaise provoked by "postmodernity" in the most affluent country of the globe. In order to understand the global implications of this process, another look back is needed.

After the horrendous massacres of the First and Second World Wars and the revelation of the West's bankrupt leadership, the consensus of the emerging powers was that some international governance must be put in place in order to forestall any repeat of this madness. The United Nations were born, a proclamation of "universal human rights" was decided upon, right after the victorious powers had created a structure of international financial institutions. The World Bank (WB) and the International Monetary Fund (IMF) were meant to oversee the reconstruction of war-devastated Europe and Asia. They soon became tools in the hands

---

[16] The 2005 US Census Bureau found that while jobs had been created from 2003 to 2004, the rate of poverty had grown from 12.5 to 12.7 percent, that is, 1.1 million people joined the ranks of the "poor." Poverty in the US has now risen for the fourth year in a row. Among these, only Asians experienced a relative decrease in poverty, and while other ethnic groups remained stable (with poverty on the rise in the same degree as before), only non-Hispanic whites showed an increase. Duke sociology professor David Brady indicated that the definition of poverty was part of the problem. Many people living above the official poverty line are no longer able to make ends meet. Yet politicians from both parties will not tackle this issue of accountability, though even with the current definition, poverty has been on the rise for a generation. "We don't care enough about the poor," he quipped (quoted by Sara B. Miller and Amanda Paulson, "Despite More Jobs, US Poverty Rises," Christian Science Monitor Online [August 31, 2005], http://www.csmonitor.com/2005/0831/p02s01-usec.html?s=itm).

of the free-market side of the Cold War to assist in the development of those nations newly emerging from the night of colonial rule. Originally they were "designed to guide and regulate private-sector interests in the name of public sector reconstruction," yet over the decades "they became the instruments of the very private-sector interests they were meant to channel and keep in check."[17]

With the neoliberal turn in the US and UK economies of the 1980s (think of Ronald Reagan and Margaret Thatcher), the logic of the market was given much greater latitude in the international sphere, a trend that only increased after the self-implosion of the Soviet Union in 1989. Paradoxically, this ensured less American influence in the flow of global capital and markets. The key players were the transnational corporations, now made more powerful than ever. It is not surprising therefore that the WB and the IMF were "left at the mercy of the true multinationals of our epoch, the transnational corporations and thousands upon thousands of nongovernmental interest groups and associations that constitute the international market."[18]

In other words, capitalism has gone global. Indeed, there is an inescapable interconnectedness of exchange rates, interest rates, fluctuations of the stock market around the world. As multi-billionaire George Soros put it recently, "[g]iven the decisive role the international financial capital plays in the fortunes of individual countries, it is appropriate to speak of a global capitalism system."[19] Since capital is the necessary engine of production, weaker countries desperately need to attract it; but when they succeed in doing so, they are not free to tax it or use it in ways they deem necessary to benefit their own people, because that capital might suddenly move to another location. Inevitably, then, international finance overshadows local finance. When the system is in good working order, "[g]lobal financial markets work like a gigantic circulatory system, sucking up capital into the financial institutions and markets at the center, then pumping it out to the periphery either directly, in the form of credits and portfolio investments, or indirectly through multinational corporations."[20] However, the system does run amuck, and whereas the center

---

[17] Barber, *Jihad Vs. McWorld*, xxix. See also the classic work of French development thinker, Serge Latouche, on this theme: *In the Wake of the Affluent Society: An Exploration of Post-Development*, intro. and trans. Martin O'Connor and Rosemary Arnoux (London/New York: Zed, 1993).

[18] Barber, *Jihad Vs. McWorld*, 240.

[19] George Soros, *The Bubble of American Supremacy: Correcting the Misuse of American Power* (New York: PublicAffairs, 2004), 83.

[20] Ibid., 85.

benefits from protective measures put into place for decades, the periphery has no such protection and thus pays a heavy price in human suffering.

Postmodernity as defined in this book includes this tension between weaker states attempting to manage their own affairs and the enormous pressure exerted on them by the combined forces of international finance, self-interest of powerful nations, transnational corporations and the WB, IMF and the World Trade Organization (WTO, which emerged from GATT, General Agreement on Tariffs and Trade, in 1995). As such, postmodernity is a mass consumption regime imposed on a global scale, with predesigned technological tracks to follow, and predetermined labels from which we are forced to choose.[21] As historian J. G. A. Pocock quipped, "[today we find] ourselves in a post-industrial and post-modern world in which more and more of us were [*sic*] consumers of information and fewer and fewer of us producers or possessors of anything, including our own identities."[22]

Postmodernity, then, is more than the growing uniformization of culture and social psychology. Its current configuration reveals political and economic power structures that have dramatically increased the gap between haves and have-nots in the last twenty to thirty years.[23] This is an important fact, especially in the eyes of Muslims whose main representative body, the Organization of the Islamic Conference (OIC), can only deplore that most of its fifty-four members are considered "developing economies," and most of those "are located in the no-growth or very slow-growth regions of the world, with a high degree of trade dependence on extra-regional sources."[24] The chairman of the Institute of

---

[21] On this, see also Fredric Jameson's *Postmodernism, or, The Cultural Logic of Late Capitalism* (Durham: Duke University Press, 1991); *The Cultural Turn: Selected Writings on the Postmodern 1983–1998* (London/New York: Verso, 1998).

[22] J. G. A. Pocock, "The Ideal of Citizenship Since Classical Times," *Queen's Quarterly* (Spring 1992): 55, quoted in Barber, *Jihad Vs. McWorld*, 274.

[23] This is also one of the main themes of Barber's *Jihad Vs. McWorld*: what the world needs now in the wake of September 11 is a Declaration of Interdependence, for the chaos and suffering caused by the reign of "wild capitalism" is the same space that causes global terrorism to thrive. No nation can be shielded from a determined martyr. "Nor is any nation ever again likely to experience untroubled prosperity and plenty unless others are given the same opportunity. Suffering too has been democratized, and those most likely to experience it will find a way to compel those most remote from it to share in the pain. If there cannot be equity of justice, there will be equity of injustice" (ibid., "2001 Introduction," xxiv).

[24] Syed Nawab Haider Naqvi, "Exogenous Shocks and Islamic Economic Response," in *The Economic and Financial Imperatives of Globalization: An Islamic Response*,

Islamic Understanding Malaysia (IKIM) expresses the frustration of most Muslim economists, scholars and thinkers when he writes,

> The precise mechanisms of international trade and economic relationships, in which the West has retained its economic position, are characterized by the exploitation of the many poor countries by the rich. The exploitation is carried-out [*sic*] through two simple mechanisms i.e. [*sic*] terms of trade and keeping the many poor nations disunited. The terms of trade are designed in such a way that many poor commodity producers have to sell cheap but are forced to buy dearer industrial products, including technology. These hegemonic powers have control over raw materials, sources of capital, markets, and competitive advantages in the production of highly valued goods.[25]

Hence, in the spirit of Latin American liberation theology,[26] the first chapter of this book seeks to uncover the structures of injustice at work in the global institutions of contemporary capitalism and their collusion with the giant transnational corporations that continue to undermine and reconfigure the reigning twentieth-century ideology of nationalism. This constellation of powerful socioeconomic and political forces in a post-Cold War era is a fact of life people around the globe are forced to confront in their daily lives. Yet, if Americans suffer from social dislocation and economic deprivation for a growing minority, what about the third of humanity living below the absolute poverty line, and worse yet, those tens of thousands dying daily of malnutrition, of HIV-AIDS, or the millions of refugees, victims of regional violence, particularly on the African continent?

What is called for is a new vision of the world. The conditions of postmodernity can be changed, no doubt, and yet part of the solution

---

ed. Nik Mustapha Nik Hassan and Mazilan Musa (Kuala Lumpur: Institute of Islamic Understanding Malaysia, 2000), 2.

[25] Tan Sri Dato Seri (Dr) Ahmad Sarij bin Abdul Hamid, "Preface," in *The Economic and Financial Imperatives of Globalization*, xi-xii.

[26] "Liberation theology" designates a Christian (Roman Catholic at first) theological movement that began in Latin America in the 1960s and spread elsewhere. Though later condemned by Rome, many of its tenets and practices have been incorporated into mainstream Catholicism and strongly influenced Protestant theology as well. One of the contributors in *Progressive Muslims* is the South African Farid Esack, who, in the matrix of a shared struggle against apartheid in his country, made common cause with Christian liberation theologians. As opposed to the Muslim and Christian leaders who advocated the status quo ("accommodation theology"), he practiced a "liberation theology." See Esack's *Qur'an, Liberation & Pluralism: An Islamic Perspective on Interreligious Solidarity against Oppression* (Oxford: Oneworld, 1997).

could come from another concurrent phenomenon, this time in the realm of thought: postmodernism. Here is a way of thinking that is primarily secular, but which has profound implications for the kind of religious venture promoted in this book.

In Europe and North America, postmodernism is a much contested phenomenon, and, as such, it is articulated and described in various ways.[27] Here is a brief summary of the salient points of a complex movement:

1. A radical questioning (and often rejection) of the Enlightenment's view of history as a unitary human march toward progress. The inherent optimism of modernity's technological feats is dampened and subverted from the start by the imperialistic designs of colonialism. Jean-François Lyotard is typical of this trend, as he resolutely turns his back on any scheme of totality—"metanarratives" as he calls them (whether religion, Marxism, socialism, capitalism, nationalism, etc.)[28]

2. The Enlightenment dogma of "objective scientific truth," incorporated into Western consciousness since Francis Bacon, is left behind with the recognition of the social, economic and political dimensions of scientific research, and, equally, that science advances by succeeding paradigms displacing one another.

3. More generally, postmodernism, by rejecting Enlightenment rationalism (now there is no universal human *reason*), raises afresh the fundamental questions of *epistemology* (what do we know and how do we come to know it?), *ontology* (what is "being," and how does it relate to "truth," somehow standing outside of a human subject or humankind as a whole?) and *hermeneutics* (how do persons apprehend the meaning of a text, or interpret phenomena which they perceive through their five senses, whether the data of scientific research or the past events of history?).

---

[27] Thomas Docherty ("Postmodernism: An Introduction," in *Postmodernism: A Reader*, ed. and intro. Thomas Docherty [New York: Columbia University Press, 1993], 1) explains that although postmodernism as a current of thought has impacted just about every academic discipline in the last twenty years, "the debate around the postmodern has never been seriously engaged. The term itself hovers uncertainly in most current writings between—on one hand—extremely complex and difficult philosophical senses, and—one the other hand—an extremely simplistic mediation as a nihilistic, cynical tendency in contemporary culture."

[28] Ibid., 25. Lyotard's classic text is *The Postmodern Condition: A Report on Knowledge*, trans. G. Bennington and B. Massumi (Manchester: Manchester University Press, 1984).

4. As implied in the first point, postmodernism often takes on radical political stances. A case in point is Michel Foucault's denouncing of the power dimension behind discourse,[29] which led the Palestinian–American literary critic Edward Said to analyze Western writing on the Middle East and Islam as a discourse that from the eighteenth century onward—"Orientalism"—became "a Western style for dominating, restructuring, and having authority over the Orient."[30] Said's analysis of how one conceives and depicts the Other in turn stimulated and strongly impacted feminist, black and postcolonial studies.

5. Postmodernism is also a hodge-podge of social movements and currents of thought in revolt against the certainties of the Western Enlightenment project. American philosopher Fredric Jameson proposes that we rather speak of "postmodernisms," with an emphasis on the reaction of architecture in the 1970s against "the high modernism of the so-called International Style (Frank Lloyd Wright, Le Corbusier, Mies)," a style "credited with the destruction of the fabric of the traditional city and its older neighborhood culture."[31] More broadly, this movement sought to destroy the border between high culture and mass (or commercial) culture: "The postmodernisms have in fact been fascinated precisely by this whole 'degraded' landscape of schlock and kitsch, of TV series and *Reader's Digest* culture, of advertising and motels, of the late show and the grade-B Hollywood film, of so-called paraliterature with its airport paperback categories of the gothic and the romance, the popular biography, the murder mystery and science-fiction or fantasy novels."[32]

6. Along with its great diversity, secular postmodernism inevitably contains contradictory tendencies and impulses. Among them, I signal two in particular that impinge on this project: (a) how does one reconcile the conscious shift from truth to narrative knowledge and a commitment to "progressive" values of social and gender justice, pluralism, democracy and human rights?[33]

[29] Michel Foucault, *The Archeology of Knowledge*, trans. A. M. Sheridan Smith (London: Tavistock, 1974).

[30] Edward Said, *Orientalism* (New York: Pantheon, 1978).

[31] Fredric Jameson, "Postmodernism, or the Cultural Logic of Late Capitalism," in Docherty, ed., *Postmodernism*, 63. Urban architecture was also David Harvey's starting point in his classic *The Condition of Postmodernity*.

[32] Ibid.

[33] Feminist writer Sabina Lovibond argues that the main weakness of the thoroughgoing postmodernist critique of universal reason is that it fails to explain how, in a

(b) As the anthropologist (and former Pakistani High Commissioner to the UK) Akbar S. Ahmed notes, postmodernism is primarily a movement driven by the urban elites of the West.[34] How then can an elitist movement bred and nourished by the culture of consumerism provide advocacy for humankind's majority of disinherited and downtrodden peoples? The answer is found in another branch of the amorphous postmodern collage: the Marxists-turned-green, or the anti-globalization alliances that came together in Seattle (November 1999), Washington, DC (April 2000), and constituted the World Social Forum in Porto Allegre (2001, 2002 and 2003) and in Mumbai (2004). To this grassroots global coalition one must add the mainline, often UN-connected, NGO development theorists in the 1960s and 1970s, who gradually came to question the very notion of "development" as a Western, modernist and hegemonic project. Thus one encounters David C. Korten's "sustainable development,"[35] Serge Latouche's concept of "post-development,"[36] Gustavo Esteva and Madhu Suri Prakash's "radical democracy,"[37] along with many others.

This, then, in summary fashion, is a taste for the scope of secular post-modernist thought and practice. My reasons for building on some of these strategies in an interfaith project are threefold. First, postmodernism confirms the general Muslim sense of feeling oppressed by a Western-led, neoliberal, corporate-run globalization process. Here postmodernity is effectively critiqued by the international protest side of postmodernism. Indeed, Roger Burbach uses the term "postmodern politics," and his explanation is useful to this study:

world of abysmal wealth and resource disparities, environmental destruction and military madness, humankind might reclaim "this kind of classically humanist agenda" ("Feminism and Postmodernism," in Docherty, ed., *Postmodernism*, 392; originally in *Postmodernism and Society*, ed. R. Boyne and A. Rattansi [Basingstoke: Macmillan Foundation; New York: St Martin's Press, 1990], 403).

[34] Akbar S. Ahmed, *Postmodernism and Islam: Predicament and Promise* (London: Routledge, 1992), 23.

[35] David C. Korten, *Getting to the 21st Century: Voluntary Action and the Corporate Agenda* (West Hartford, CT: Kumerian, 1990); *When Corporations Rule the World* (West Hartford, CT: Kumerian, 1995); and *The Post-Corporate World: Life After Capitalism* (San Francisco, CA: Berret-Koehler/West Hartford, CT: Kumerian, 1999).

[36] Latouche, *In the Wake of the Affluent Society*.

[37] Gustavo Esteva and Madhu Suri Prakash, *Grassroots Post-Modernism: Remaking the Soil of Cultures* (London/New York: Zed Books, 1998).

Sustained and continual opposition to the hegemony of the new global order comes from rebellious factions that appear anywhere and everywhere, as the Battle of Seattle [1999] demonstrated. The opposition is postmodern in the sense that it has no clear rationale or logic to its activities while it instinctively recognizes that it cannot be effective by working through a "modern" political party, or by taking state power. It functions from below as an almost permanent rebellion, placing continuous demands on all the powers that be.[38]

Second, postmodernists have rehabilitated the values of local culture, Other-oriented thought, pluralism and multiculturalism. This way of perceiving the world provides fodder against a "clash of civilization" mentality.

Finally, the postmodernist hermeneutic subverts the literalism and rigidity of both Muslim and Christian conservatives in their approach to the Qur'an and Bible, thereby increasing their chances of finding common ground without watering down core theological commitments.

Part II:

Human Trusteeship

Part II explores past Muslim and Christian interpretations of humanity's role in the world as spelled out by their Creator. Starting with a brief overview of the classical commentaries on the qur'anic concept of Adam as God's "caliph" (*khalīfa*), or "representative," "trustee," or "vicegerent,"[39] I examine the modern period in which the human caliphate functions as a divine mandate to shape the created order—both nature and fellow human beings—according to God's design.

According to Q. 2:30, speaking of Adam, God addresses the heavenly council in these words, "Behold, I am about to place my trustee (*khalīfa*) on earth." For a variety of reasons, this verse received scant attention among the classical Muslim commentators. From the eighteenth century on, however, starting with the Indian scholar/jurist Shah Wali Allah, the idea that Adam's call was in reality intended for humanity as a whole began to take root and spread. This divine call of Adam and his descendants involved more than the politically charged word "caliphate" had meant up until then. There had been hints in the more rationalist side

---

[38] Roger Burbach, *Globalization and Postmodern Politics: From Zapatistas to High Tech Robber Barons* (London/Sterling, VA: Pluto; Kingston, Jamaica: Arawak, 2001), 11.

[39] I will be exploring this in depth in a subsequent book project, *Adam as God's Caliph: Qurʾanic Anthropology in Early and Classical Islam*. Whereas Part I is mostly new material, Part II is a slightly updated version of my dissertation material.

of medieval Islamic thought that humankind had been endowed with intelligence, moral sensitivity and free will, yet I will argue that it was not until Muslim scholars had been confronted with a new, technologically advanced and aggressive civilization that they began to entertain the theme of human mastery over nature and rethink the traditional Sunni consensus on reason vs. revelation,[40] with its implications of new forms of sociopolitical formations. The concept of the human trusteeship became one of the central pieces in the new configuration of core Islamic beliefs.

The modernist movement found its most articulate and influential spokesman in the Egyptian Grand Mufti Muhammad Abduh (1849–1905). In his attempt to synthesize traditional Islamic thought with the Western tenets that he deemed most useful in helping the Muslim umma (worldwide community) catch up with the West's advances, Abduh singled out Q. 2:30 as a key to God's will for his human creatures: "It is because of their ability [human beings] to harness the resources of nature to their own advantage, to subdue the forces that might well destroy them, and to invent constantly new and more imaginative technologies that God has placed them as his deputies on earth."[41]

Modernists found difficulty in coping theologically with the fact that politically and economically Muslims were under the thumb of Western colonialism, especially with the abolition of the Ottoman Caliphate in 1924. While Abduh had remained mostly apolitical in his borrowing from Western thought, his disciple and collaborator, the Syrian Rashid Rida (1865–1935), saw matters differently. For him, despite much lip service paid to the priority of rethinking the Islamic heritage, the traditional institution of the caliphate was tied to God's sacred law, the Shari'a, and was not something that could be interpreted away. In practice, he was much more of a literalist than he admitted. Thus, around

---

[40] The emergence of a theological consensus in the ninth and tenth centuries soon became known as "Sunni" (a triumph of the more conservative specialists of the Prophet's traditions—the Sunna, or the "people of the Sunna," over their more rationalist opponents, the Mu'tazilites). Though these major theological rivalries are mostly limited to the Sunni branch of Islam, this distinction should not be confused with the earlier division between Sunnis and Shi'ites, mainly over a political issue—whence the leadership of the Muslim community was to come. This majority theological position among Sunnis was named "Ash'arism," after the great theologian Abu al-Hasan al-Ash'ari (d. 935). On this, see W. Montgomery Watt's *The Formative Period of Islamic Thought* (Edinburgh: Edinburgh University Press, 1973); for the changes in the modern period, see Albert Hourani, *Arabic Thought in the Liberal Age, 1798–1939*, 2d ed. (Cambridge: Cambridge University Press, 1983).

[41] David L. Johnston, "The Human *Khalīfa*: A Growing Overlap of Reformism and Islamism on Human Rights Discourse?," *Islamochristiana* 28 (2002): 42.

1930, when the newly founded revivalist network of young Hasan al-Banna was spreading like wildfire in Egypt, al-Banna paid regular visits to Rida, considering him to be a kindred spirit. It soon became apparent to Rida that the Society of the Muslim Brothers represented a logical progression of his own ideological leanings. After his death in 1935, al-Banna's organization officially took over the latter's prestigious journal *al-Manār* (*The Lighthouse*), alongside its other publications.

Al-Banna's Muslim Brotherhood (founded in 1928) was the first Islamic socio-political movement built on a modern activist model. A parallel movement appeared in India (1941), and soon in Pakistan, founded by the charismatic Abu al-A'la al-Mawdudi (1903–1979).[42] As Abduh, both he and the great ideologue for the Muslim Brotherhood, Sayyid Qutb (1906–1966), made use of the idea of humanity's deputy-ship, but narrowing it to the covenant contracted before the dawn of time in which God called together all the souls of future humanity and make them promise to recognize Him as their Lord (Q. 7:172). Qutb's inference is that human vicegerency is limited to those who fulfill God's covenant, and since the Jews have reneged on their part of the covenant and the Christians have missed it altogether, only Muslims are left to pick up the torch and act as God's vanguard for a new humanity.[43]

Though a number of splinter groups have embraced Qutb's ideology of exclusion since he was executed by president Gamal Abd al-Nasser—many of them reinterpreting the traditional term *jihad* to mean the necessary and violent overthrow of corrupt and "un-Islamic" Muslim regimes—the caliphate of humanity has been appropriated as a key theological concept by a variety of groups, from the moderate islamists[44] like the Tunisian Rachid Ghannouchi[45] and traditionalist qur'anic commentators, to more liberal thinkers as well.

---

[42] This movement, called Jamaat-i Islami, as is the Muslim Brotherhood, is still active. Yet whereas the latter has gradually moderated its positions, particularly on the use of violence for achieving political gains, the Jamaat-i Islami has always engaged in local and parliamentary elections with the goal of reforming Muslim society from the bottom-up.

[43] Johnston, "The Human *Khalīfa*," 44-46.

[44] In my writing I have followed the French lead and use a lower case "i" for islamism. This is especially to highlight the fact that as an ideology that fuses certain traditional features of the Muslim tradition to form a modern political package it should be distinguished from Islam or the adjective "Islamic." Further, in contemporary discourse, it functions in the same way as the adjectives "conservative," "fundamentalist," and "communist."

[45] That is the French spelling he himself uses. See the recent monograph devoted to his thought, Azzam Tamimi, *Rachid Ghannouchi: A Democrat within Islamism* (Oxford/New York: Oxford University Press, 2001).

The second half of this section explores Christian interpretations of the Genesis account, and in particular the injunctions, "Be fruitful and multiply, and fill the earth and subdue it; and have dominion over... every living thing" (1:28, RSV), and "The Lord God took the man and put him in the Garden of Eden to till it and keep it" (2:15, RSV). Besides laying down some groundwork for the interpretation of these passages, I will introduce some of the findings of historian Jeremy Cohen, who traces the "career" of Gen. 1:28 in medieval literature, both Jewish and Christian.[46] Such material not only informs us today about the content of past interpretations, but also sheds light on the process of articulating theology—an eminently situated process.

Part III:
Toward an Applied Muslim–Christian Theology of Trusteeship

In the third part of this book, attention is again focused on the vision that would enable Muslims and Christians to engage together in refashioning our "globalized" planet in a form that more closely resembles what God intended in the first place. By unabashedly advocating theology in this way, I am following R. Scott Appleby who has recently drawn attention to the role of religions in the current push and pull of global politics. Appleby, the co-editor of the Fundamentalism Project, dismisses the common notion that religion inevitably veers toward fundamentalism and therefore lies at the heart of most of today's conflicts worldwide.

While accepting this as a partial truth, Appleby's research points in a different direction. To those who assume that religion cannot be counted on to help resolve deadly conflicts, he counters "that a new form of conflict transformation— 'religious peacebuilding'—is taking shape on the ground, in and across local communities plagued by violence."[47] In order to encourage this as of yet "inchoate and fragile" development, Appleby seeks to highlight positive case studies in various regions of the world. In addition, he hopes to "bring [religious peacebuilding's] funda-mental concepts and methods together in a coherent presentation that orients trainees to a core set of skills and concepts that can be applied with sensitivity to specific cultural situations."[48]

---

[46] Jeremy Cohen, *"Be Fertile and Increase, Fill the Earth and Master It": The Ancient and Medieval Career of a Biblical Text* (Ithaca, NY: Cornell University Press, 1989).

[47] R. Scott Appleby, *The Ambivalence of the Sacred: Religion, Violence and Reconciliation* (Lanham, MD: Rowman & Littlefield, 2000), 6.

[48] Ibid., 7.

This is precisely the context into which I wish to insert the present research. In the post-Cold War era of proliferating ethnic and religious struggles, writes Appleby, it becomes all the more urgent to search for a "second-order religious language—a common cross-cultural vocabulary that facilitates dialogue while remaining true to the primary theological claims of each participating community."[49]

By assuming a prior convergence on the notion of a Creator God—in the monotheistic faiths of Muslims, Christians and Jews—I propose that crucial theological reflection lies in one's view of humankind as created by God. Following the text of the Bible and Qur'an, might we not come to a compatible view of God's intended purpose for his creation of man and woman and his placing of them on the earth? And if so, might this common mission not only break down barriers between Muslims and Christians, but also allow them to contribute effectively to the remedy of some of the thorniest problems of our planet in the twenty-first century?

As I embark on the development of a common theology of creation, I am taking Appleby's proposal one step further: instead of a "second-order" religious language, I am seeking to develop a "first-order" theological framework, one which will stimulate and enhance cooperative efforts between Muslims and Christians to care proactively for the creation they believe God has graciously granted to all. By this I mean that it will be theological discourse that touches at the heart of each tradition, while only incorporating the common ground between the two faiths. Naturally, shaping the contours of this common core calls for sustained dialogue.

Books like the present one are only one part of the solution. As Muslims and Christians roll up their sleeves and embark on projects together, they will extend their conversation about God's good creation and expand each in their own way on the meaning of the human vocation on earth.[50] In my view, this is the mark of "good" theology.

Here, then, is my working definition of theology: an ever-growing and evolving reflection—in the light of sacred texts and in interaction with a specific religious tradition—that leads people to better articulate their relationship to God, provides answers to the ultimate questions of human

---

[49] Ibid., 151.

[50] I am assuming in this project that people act for the sake of justice and poverty reduction in a given society on the basis of a theological (with God or without—secularism is at least a form of counter-theology) conception of humankind. Why are we here? How are we related? Are people qualitatively different from animals? This naturally leads to a consideration of ethics. Thus a theology, or an ethical vision of humanity necessarily undergirds any attempt to better human society.

existence and gives shape to a life-style in the world as community that
best reflects that understanding. As such, theology must always be constructed within a particular socio-political, historical and cultural situation. Thus, one commonly speaks of "African theologies," or "feminist
theologies." Yet another reason for this cultural rootedness of theology is
that the reader of a text can only extract meaning in reference to a specific perspective—one's own as an individual and as a member of a faith
community, one's social status and role in society, one's socio-political
context and the like.[51] In fact, this entire study is an exercise in "doing
theology in context," to borrow a phrase from one of my mentors, theologian Charles Van Engen.[52] For this reason my methodology and
sources draw from a variety of disciplines: biblical and qur'anic studies,
Christian and Muslim theology, religious studies, philosophy, history,
sociology and anthropology, development theory, critical theory and
ethics.

In the light of the above, the objective of this book—to reinvision a
better world through a common Muslim–Christian theological effort—is
not as outlandish as it may sound to some, but it will require a closer
look at the concept of "worldview." Somehow, culturally patterned ways
of viewing the world must feed into a more universal picture of what all
people might like to see our world become. Ninian Smart, the British
scholar of comparative religion, is the one who likely influenced more
than anyone else the course of religious studies in universities worldwide
in the past half century. For him, to study religions is to study the worldview of particular peoples, or blocks of peoples. It is to explore the way
people articulate their communal understanding of the physical world
(which he calls "cosmos") around common myths, rituals, ethical norms
and practices that help to shape their experiences and give meaning to
their existence, past, present and future. This comes rather close to my
own definition of theology. As such, all religions point to a reality
beyond the cosmos, a transcendence that takes the form of theism for the
monotheistic religions, and simply a "Beyond" in the case of Eastern
spiritualities. But the Enlightenment project too projected a particular
worldview, which Smart calls "secular humanism": knowledge of the
world comes to us through science and there is nothing beyond human

---

[51] The fourth and fifth chapters deal with these questions of hermeneutics.
[52] This is the title of one of his courses ("Doing Theology in Context," Fuller Theological Seminary, Fall 1998), and a reoccurring theme of his work. Like others, he
has been influenced in this regard by the liberation theologians. See, for instance,
José Miguez-Bonimo, *Doing Theology in a Revolutionary Situation* (Philadelphia:
Westminster Press, 1975).

beings from which they might derive meaning. The ideologies of Marxism and nationalism, besides being spin-offs from secular humanism, share many common characteristics with religion.[53]

In this age of the "emerging global civilization," argues Smart, worldview analysis becomes a vital tool for crosscultural communications within our increasingly pluralistic societies. This means that we must combine loyalty to our own tradition with an ability to respect the dignity of the Other, and in particular in being willing to listen with empathy and learn from her or him.[54]

Admittedly, this is more difficult for the more conservative elements of these faiths. Part of the fundamentalist instinct is to make abstraction of centuries of theological efforts and reconnect directly with the origins of one's faith. Yet, besides the obvious disadvantage of not being able to learn from past achievements and mistakes, there lurks another pitfall: that of idealizing the origins, or of refashioning them in the image of the present, and thus reading our own values into them. The Christian fundamentalist movement in the 1920s, for instance, sought to draw a hedge against what it considered the destructive critical methods of Protestant liberalism around what it considered the "fundamentals" of the faith. On the other hand, what began as an impulse of conservative self-defense became a launching pad for a large movement of revitalization in the American Protestant landscape. Though few today would accept the label "fundamentalist," the evangelical movement it has spawned has grown much faster than the traditional mainstream, and with a considerable spectrum of views.[55]

In Islam, it was Muhammad Abduh, the reformer, who founded the *Salafi* movement—the return to the "righteous forbears," the Companions of the Prophet and their Successors. This was a useful strategy for Egypt's Grand Mufti, a leader responsible for steering the legislation of

---

[53] The state demands a strong loyalty on the part of the citizen, including the payment of taxes and the readiness to die in its defense. Its ideology is "clothed in religious garments. There is a national anthem (solemn, it is hoped, and tear-jerking), a national flag and other emblems, and pomp and circumstance surrounding state events... war is the great sacrament of the modern state" (*Worldviews: Crosscultural Explorations of Human Beliefs*, 3d ed. [Upper Saddle River, NJ: Prentice-Hall, 2000], 43). To this must be added the founding myths of the nation, reproduced and inculcated in the young through the educational system.

[54] Ibid., 156-57.

[55] See, for instance, Christian Smith, *American Evangelicalism: Embattled and Thriving*, with Michael Emerson et al. (Chicago/London: University of Chicago Press, 1998); *Christian America? What Evangelicals Really Want* (Berkeley: University of California Press, 2000).

his country in a direction that was more compatible with the legal code put into place by the British. Rashid Rida and all the revivalist movements that had sprung up in the wake of the Muslim Brotherhood used the *Salafi* label in a much more restrictive way, but the instinct was the same. Even a rapid survey of Islamic history would convince one that the ideals projected by the Qur'an and Sunna were embodied in a variety of ways—politically, socially and culturally—over time.

In a recent book, *Islam Under Siege*, Akbar S. Ahmed remarks that Islamic civilization represented for many in the past an experience analogous to our situation today: "societies living within different ethnic, geographic, and political boundaries, but speaking a language understood throughout, enjoying a common cultural sensibility, and recognizing the same overarching ethos in the world-view."[56] Even after the waning of the Abbasid Caliphate and the sacking of Baghdad in 1258, the explorer Ibn Battuta spent a lifetime roaming over the Muslim world and thereby handing down to posterity his famous *Travelogue*. Besides serving as a Maliki judge[57] in Dehli and the Maldive Islands, the Moorish traveler left us with detailed notes about his observations on life in Niger, the Arabian Gulf, Central Asia, India and Sumatra.

Ahmed is even more impressed by another fourteenth-century figure, Ibn Khaldun, "the founder of sociology."[58] A high-ranking civil servant in a variety of political regimes across North Africa, Ibn Khaldun engaged in cross-cultural and international comparative sociopolitical research. At the heart of his theory was the Arabic word ʿaṣabiyya: social cohesion in tribal settings, but now applied more widely to traditional cultures, both peasant and urban. "With asabiyya, society fulfills its primary purpose to transmit with integrity its values and ideas to the next generation."[59] This process, however, tends to break down in urban settings, observes Ibn Khaldun. How much more so in the twenty-first century, exclaims Ahmed, with the onslaught of the following elements in today's mostly traditional Muslim societies:

> Massive urbanization, dramatic demographic changes, a population explosion, large-scale migrations to the West, the gap between rich and poor (which is growing ominously), the widespread corruption coupled with the low premium on education, the crisis of identity, and, perhaps

---

[56] Akbar S. Ahmed, *Islam Under Siege: Living Dangerously in a Post-Honor World* (Cambridge: Polity, 2003).

[57] One of four main legal schools in Sunni Islam, especially predominant in North Africa.

[58] Ibid., 74-80.

[59] Ibid., 78.

most significantly, new and often alien ideas and images, at once seduc-
tive and repellent, and instantly communicated from the West, ideas and
images which challenge traditional values and customs.[60]

In essence, the present book has much in common with *Islam Under
Siege*. Ahmed's keen anthropologist eye is trained on the transcultural
theme of honor and its perversion into "hyper-*asabiyya*" in the hands of
militant fundamentalists of all stripes, which leads the reader to the last
chapter, "Toward a Global Paradigm." Further, his Introduction is entitled
"God's Gamble," which attempts to show how the Creator placed
humans on this earth, endowed with free will and commissioned to be
His deputies. "Seeing the state of affairs in the early 21ˢᵗ century—the
widespread poverty, the lack of justice and compassion, the willful
depletion of the resources of the planet, the senseless and widespread
violence—God may well be regretting human creation now."[61] Yet there
is hope, Ahmed reminds us. Abraham and other prophets are common to
Judaism, Christianity and Islam. The qur'anic assertion, "There is no
compulsion in religion" (Q. 2:256) should put to rest the exclusivist
tendencies of some believers, or at least their justifications for violence.

It is unlikely, however, that most Muslims, still imbued with the red-
hot fervor of religious revivalism of the 1980s and 1990s, will follow
Ahmed's deliberate recourse to the values of Sufism (Islam's mystical
tradition). For instance, he weaves the ideas of God's oneness, and consequently the oneness of all being (*tawhid*), together with humankind's
deputyship and imitation of God's qualities:

> Everything was from God and the more man was in the image of God the
> closer he was to fulfilling his destiny and becoming God's "deputy." It is
> in this context that the idea of merging with God, which finds historic
> expression in the Sufi dictum *ana al-haqq* or "I am God," to be under-
> stood. Although the idea was sacrilegious to the orthodox it was not as
> far-fetched as it seemed. By God wishing to create divine qualities in
> man, He was emphasizing the unity and integration of creation itself.[62]

By the same token, Ahmed's articulation of Islamic theology is by no
means unique. In fact, a group of Muslim academics teaching mostly in
the West have issued a sort of manifesto of late. The result is Omid Safi's
edited book, *Progressive Muslims: On Justice, Gender and Pluralism*.[63]

---

[60] Ibid., 81.

[61] Ibid., 2-3.

[62] Ibid., 4. What is more, speaking of violence undertaken against God's messengers
on the next page, Ahmed casually states, "Jesus was tortured and crucified"—hardly
an orthodox view, though not so uncommon today.

[63] See n. 1.

What do these writers have in common? They are postmodern in that they critique the sheer arrogance of Western modernity and leverage the feminist concept of "multiple critique"[64] in order to accomplish two tasks: first, to refute the dangerous ideology of "the Wahhabi and neo-Wahhabi" groups,"[65] and second, to resist "increasingly hegemonic Western political, economic, and intellectual structures that perpetuate an unequal distribution of resources around the world."[66] All of this is premised on the values taught by the sacred text, and, above all, "the common humanity of all of God's creation" and the inherent dignity of all human persons, "because, as the Qur'an reminds us, each of us has the breath of God breathed into our being."[67]

My argument in this work is that these scholars are consciously "doing theology," intentionally reinvisioning the world according to the ethical and spiritual vision they perceive in the Qur'an and Sunna. They and more conservative colleagues are calling for a rethinking of the Islamic tradition that will provide guidance and empowerment to Muslims on today's globalized planet. My conviction here is that their focus on God's act and purpose of creation—the physical world, the animals and, supremely, humankind—can help to frame a fruitful interfaith conversation and to articulate a common theology that refashions the world in a more just and peaceful way.

Finally, a piece of practical advice to my readers: Chapter 2, 3 and 4 deal with the philosophical issues swirling around the transition from modernism to postmodernism. These impact the way believers read their sacred text and think theologically about the world around them. I argue that, though these discussions happened in the West, through globalization they affect everyone to some extent. These chapters, then, feed into Chapter 5, which directly focuses on reading our scriptures. However, some readers will prefer to skip over Chapters 2–4 and pick up the flow in Chapter 5. Not everyone cares about the theoretical issues—nor does everyone have the background to do so.

---

[64] "Introduction," in Safi, ed., *Progressive Muslims*, 2: "a multi-headed approach based on a simultaneous critique of the many communities and discourses that we find ourselves positioned in."

[65] Wahhabism refers to the body of teachings passed down from the eighteenth-century Arabian scholar Ibn Abd al-Wahhab, and to its conservative ideology that remains the keystone of contemporary Saudi Arabia.

[66] Ibid., 2-3.

[67] Ibid., 3.

Part I

OUR POSTMODERN SITUATION

Chapter 1

POSTMODERNITY AND THE DOUBLE WALL

In this chapter I describe postmodernity   the current interconnected, global, neoliberal system of political and economic instruments, institutions and alliances—as the logical outcome of a chain of events in Western civilization: (1) the Western imperial drive from 1492 on to exploit foreign territories resulting in the dispossession of native populations; (2) the rise of capitalism, and especially its insistence on the commodification of public goods, which started, arguably, with the acquisition of land belonging to native peoples by Anglo-American settlers in the early eighteenth century; (3) and, much more recently, the post-Fordist, postindustrial and consumerist era arose out of the ashes of the post-World War II Bretton Woods arrangements that collapsed around 1970.

The "double war" idea is from Arthur Mitzman, who argues that the current model of global relations is headed for a crisis on two fronts, the imminent security risk caused by accelerating disparities between rich and poor all over the globe—especially between the most wealthy and the most impoverished nations—and the risk of impending ecological disaster. The crisis is in fact so imminent that it is rather like a car careering toward two walls, one immediately behind the other. These alarming conditions call for an urgent rethinking of the world order. In light of this, "the enormous risks we face at the beginning of the twenty-first century have more to do with the ideological fundamentalism of neoliberal capitalism than with that of Islamic terror networks."[1] What is needed is a holistic vision that jettisons the fetishism of growth inherited from modernity and encompasses the aspirations of Third World

---

[1] Arthur Mitzman, *Prometheus Revisited*, xvi. In saying this, he was applauding Ulrich Beck's article, "Globalization's Chernobyl," *Financial Times* (November 6, 2001), that was arguing just that: 9/11 was the equivalent of Chernobyl, the tragedy that exposed "the false promise of neoliberalism." Beck is also the author of *Risk Society: Toward a New Modernity* (London: Sage, 1992).

peasants, native peoples and the urban poor, as well as the majority of working and middle class people in other countries. This vision will have to focus on a sustainable *modus vivendi* for all people in harmony with the earth for which they share a common responsibility.

Postmodernity and the Wall of Social Injustice

By way of clarification, I am using "postmodernity" more as a heuristic device than as a hard and fast historical category. The respected British sociologist Anthony Giddens, for example, notes the radical discontinuity between the project of modernity since the Enlightenment and "high-modernity" of the last quarter of the twentieth century. Yet he refuses to use the word "postmodernity." Modernity for him is inherently globalizing, prone to creating discontinuity in social configurations and doubt in people's minds (think of Descartes' starting point),[2] but not until recently has its dizzying rate of change engulfed the whole planet. The best metaphor, contends Giddens, is that of the juggernaut, "a runaway engine of enormous power which, collectively as human beings, we can drive to some extent but which also threatens to rush out of our control and which could rend itself asunder."[3] To resist it is to risk being crushed, and to follow it is always disconcerting, because however steady its path may usually seem, it can so easily bolt off in unexpected directions. It can be an exhilarating ride at times and instill great hope. Yet security will always remain elusive, if only because for the first time ecological collapse and nuclear war now stand at our doorstep.

Giddens's picture of the juggernaut of modernity certainly has descriptive power for the middle classes (whether of the "blue" or "white" collar variety) struggling to keep up across the globe. For the majority of humanity, however, the "wild ride" has long left them in the dust, in the gutter of no return, with no hope of betterment for their children, if indeed they survive into adulthood. There is no hope of cultural production here, and little chance of creating art for the sake of art while life is squeezed out of people in the midst of the ubiquitous symbols of mass-produced and mass-marketed artifacts exported by the west—only a tattered Nike T-shirt here, and a Coke bottle on the ground over there. It is difficult to escape the pervasive impact of economic systems and their political or geopolitical supports.

[2] Radical doubt was the foundation of rationalism and the pathway to certainty: *dubito, ergo sum* ("I think, therefore I am").
[3] Anthony Giddens, *The Consequences of Modernity* (Stanford, CA: Stanford University Press, 1990), 139.

In fact, few would deny today the connection between the economic power of multinational corporations and the nefarious effect of this monocultural steamroller of consumerism on local cultures and traditions. Behind all of this, argues Moroccan academic Anouar Majid, is the "free-market fundamentalist ideology."[4] This is precisely the negative side of globalization, driven by the neoliberal ideology of free markets and leading the world community into a dead-end—the wall of social injustice, as Mitzman has it, or the growing disparity between the few rich and the multiplying poor.[5] An influential member of Uganda's parliament, Norbert Mao, recently wrote, "Globalization could benefit Africa, but in its current raw form, it will only paralyze the poverty-struck continent by turning it into a cluster of wagon economies whose engines are in the Western world."[6]

Postmodernity as Post-Fordist Flexible Accumulation
Postmodernity is driven by economic forces in the hands of a handful of powerful countries. For this reason I have identified David Harvey's landmark book, *The Condition of Postmodernity*, as a first guide for explaining how postmodernity came to be and how it should be defined.[7] A British geographer who came to the United States in 1969,[8] Harvey concluded early in his career in Baltimore that Marxian historical materialism (history explained primarily by material conditions) provided the

---

[4] Anouar Majid, *Unveiling Traditions: Postcolonial Islam in a Polycentric World* (Durham, NC/London: Duke University Press, 2000), 15.
[5] Mitzman, *Prometheus Revisited*, see especially his Preface: "The Double Wall before the Future" (pp. xv-xxiii).
[6] Norbert Mao, "Unevenly Yoked: Has Globalization Dealt Africa a Bad Hand?," *YaleGlobal* 3 (November 2003), http://yaleglobal.yale.edu/display.article?id=2721. Moa, though strongly critical of the neoliberal status quo, offers a constructive view: "If Africa is to help itself, developed countries must give it the tools to do so; they must open their markets to African goods, offer debt relief, and, above all, provide focused developmental assistance based on agreed social goals."
[7] David Harvey, *The Condition of Postmodernity: An Inquiry into the Origins of Social Change* (Oxford/Cambridge, MA: Blackwell, 1989). I could also have chosen Fredric Jameson, whom Harvey quotes and draws from repeatedly. See his *Postmodernism, or the Logic of Late Capitalism* (Durham, NC: Duke University Press, 1991). But I have found Harvey's economic and political analysis more detailed, especially from an historical perspective.
[8] Harvey went back to England from 1987 to 1993 where he held the Halford Mackinder Chair of Geography at Oxford University. For a useful summary of his intellectual odyssey, see the interview of Harvey held at the University of California at Berkeley's Institute for International Studies in 2004, a transcript of which appears online: http://globetrotter.berkeley.edu/people4/Harvey/ harvey-con1.html.

best theoretical framework to account for the graphic social disparities in
US cities. He progressively developed an analysis that identified the
cultural, intellectual and aesthetic practices of postmodernism in terms of
capitalist cycles of accumulation and overaccumulation. In the third part
of *The Condition of Postmodernity*, "The Experience of Space and
Time," Harvey links the growing "compression" of time and space to
three successive crises of capitalist overaccumulation, starting with the
European depression of 1846–47, which sent shockwaves worldwide.[9]

This first crisis not only sparked political upheavals in Europe, he
explains, but it created "a radical readjustment in the sense of time and
space in economic, political and cultural life. Before 1848, progressive
elements within the bourgeoisie could reasonably hold to the Enlightenment sense of time… But after 1848, that progressive sense of time was
called into question in many important respects."[10] Harvey argues that
part of this legacy was to be seen in the literature of James Joyce who
"began his quest to capture the sense of the simultaneity in space and
time," and of Césanne who initiated experimentation in the arts that
"tried to represent time through a fragmentation of space."[11]

Ironically, the second crisis, the "great depression" of 1929, is only
incidental in the fine-tuning of the archetypal modern economic innovation: Fordism. Harvey sees 1914 as the symbolic launching of Fordism,
"when Henry Ford introduced his five-dollar, eight hour day as recom-
pense for workers manning the automated car-assembly line he had
established the year before at Dearborn, Michigan."[12] Yet Fordism was
more than just a jump ahead in efficiency. Ford had a vision which he
was able to pass on to his generation: "his explicit recognition that mass
production meant mass consumption, a new system of the reproduction
of labour power, a new politics of labour control and management, a new
aesthetics and psychology, in short, a new kind of rationalized, modernist, and populist democratic society."[13]

Two major obstacles stood in the way of those who attempted to apply
Fordism and Taylorism on a national scale in the US and in Europe

---

[9] I deal with the concept of "compression of time" in the next chapter.

[10] Harvey, *The Condition of Postmodernity*, 260-61.

[11] Ibid., 267.

[12] Ibid., 125.

[13] Ibid., 125-26. This was also the observation made by the leader of the Italian communist leader, Antonio Gramsci, while in prison under Mussolini. This American
method of production should be seen as "the biggest collective effort to date to
create, with unprecedented speed, and with a consciousness of purpose unmatched in
history, a new type of worker and a new type of man" (Gramsci, from his *Prison
Notebooks*, quoted by Harvey, ibid., 126).

before World War II. The first was a class issue: most of Ford's labor force was recent immigrants, and willing to put up with difficult conditions, but in general American workers remained hostile to this standardization of production. The modern state represented the second obstacle. With an overaccumulation of products that could not be absorbed, and a frenzied, overheated speculatory stock market, the entire economy collapsed overnight in 1929. This is when state-sponsored solutions came to the fore: Roosevelt's New Deal in America and national socialist programs initiated in Germany, Italy and Japan, with their strong militaristic impulses.

Thus it was only after the war that Fordism entered its own, as "a fully-fledged and distinctive regime of accumulation," a regime that survived unchanged until 1973. Capitalist countries expanded their economies and began to draw in their wake a number of recently independent countries. For one thing, the industries related to the war effort had "been pushed to new extremes of rationalization." For another, the reconstruction effort initiated by the allies in Europe helped to spark an unprecedented investment in transport, urban and communication infrastructure everywhere. This was made possible by establishing the financial nerve center of the global economy in the West, attracting "massive supplies of raw materials from the rest of the non-communist world," and intentionally aiming "to dominate an increasingly homogenous mass world market with their products."[14]

It is this international dimension of postwar Fordist capitalism and its distinctive configuration of capital that most impacts the present situation. For Harvey, the 1944 war agreement of Bretton Woods adopted the dollar as the international exchange and reserve currency and "tied the world's economic development firmly into US fiscal and monetary policy." With this move US economic hegemony was sealed:

> The United States acted as the world's banker in return for an opening up of the world's commodity and capital markets to the power of the large corporations. Under this umbrella, Fordism spread unevenly as each state sought its own mode of management of labour relations, monetary and fiscal policy, welfare and public investment strategies, limited internally only by the state of class relations and externally by its hierarchical position in the world economy and by the fixed exchange rate against the dollar.[15]

Moreover, on the American stage, the postwar economic bonanza was made possible in part by a muzzling of the labor unions through the

[14] Harvey, *The Condition of Postmodernity*, 132.
[15] Ibid., 137.

Taft–Hartley Act of 1952 at the height of the anti-communist McCarthy-ite period. Nevertheless, a compromise was progressively hammered out to different degrees in different places during this period. As Mitzman avers, "War economies under advanced capitalism amount to a kind of socialism for the rich, with a significant trickle-down effect for the rest of the population."[16] This indeed was the period of the welfare state, in which governments dramatically widened benefit packages for all citizens and intervened directly to regulate the cycles of recession and inflation. In other words, a truce had been worked out with the working classes and the economy remained on a fairly stable course.

These social-democratic ideals and practices, however, were not destined to endure and the third crisis was not far behind. Already in the 1960s, the Japanese and West-European economies had completely recovered and, with their internal markets saturated, were seeking other markets to absorb their surplus output. Further, the International Monetary Fund (IMF) and World Bank (WB) were just beginning to impose import-substitution in many "developing" countries, particularly in Latin America. This, at least initially, resulted in a number of industrial plants opening up and competing with American products. What is more, US industries were beginning their first off-shore experiments and Southeast Asia in particular posted spectacular growth in industrialization. These factors combined put a great deal of strain on the production capacity of the American economy, and the credit crunch of 1966–67 revealed its growing inability to control inflation. The rigidity of the whole Bretton Woods system was severely taxed and, finally, in 1973, ended up in a crisis caused by a crash of property markets, OPEC's decision to raise the price of crude oil, and the Arab embargo as a result of the 1973 Israeli–Arab War. This combination of acute inflation and sudden rise in energy costs forced Western industries dramatically to reduce expenditures and find new ways to recycle surplus petro-dollars.

The momentous pressures that mounted against the Fordist superstructure forced a difficult two decades of "economic restructuring and social and political readjustment." Harvey, along with many others, sees 1973 as the transition to "an entirely new regime of accumulation."[17] This third

---

[16] Mitzman, *Prometheus Revisited*, 79.

[17] Harvey, *The Condition of Postmodernity*, 145. George Soros sees the break in 1980, "because globalization was a market fundamentalist undertaking." Reagan and Thatcher, he notes, succeeded in reducing the state's ability to interfere with the markets, in coaxing other states to follow suit and, as they permanently fixed London and New York as the world's banking capitals, they were able to assure their own competitive advantage in the game (Soros, *The Bubble of American Supremacy*, 90).

crisis produced what he calls "flexible accumulation," that post-Fordist arrangement which was largely responsible for the growing economic disparities, both locally and globally:

> It rests on flexibility with respect to labour processes, labour markets, products, and patterns of consumption. It is characterized by the emergence of entirely news sectors of production, new ways of providing financial services, new markets, and, above all, greatly intensified rates of commercial, technological, and organizational innovation. It has entrained rapid shifts in the patterning of uneven development, both between sectors and between geographical regions, giving rise, for example, to a vast surge in so-called "service sector" employment as well as to entirely new industrial ensembles in hitherto underdeveloped regions.[18]

The year 1973 is also crucial because this is when "the whole financial architecture of the Bretton Woods system collapsed," notes Harvey in a more recent work.[19] "The collusion now documented between the Nixon Administration and the Saudis and Iranians to push oil prices sky-high in 1973" created a surplus of dollars that flooded the world financial institutions. The United States, now losing its former leadership in the area of production (the South Asian countries having matched Europe's industrial production capabilities), was eager to monopolize the recycling of excess petro-dollars and assert its hegemony through finance.[20] Gold, therefore, was now abandoned as the standard against which economies measured their value, and, as Greider put it, capital itself became "abstracted and etherealized, mystified by dense mathematical calculation and accounting definitions, invested with unknowable intangible qualities like corporate 'goodwill'."[21] So, from then on, the dollar became the international currency of choice, and particularly in the buying and selling of oil.

This combination of factors led to the present state of capitalism, in which even in the First World the middle classes desperately struggle to stay on top of the consumerist treadmill, while the swelling ranks of the poor, marginalized and homeless no longer have the protection of the pre-1980 welfare provisions. Harvey, for his part, documents the close connection between national and international arrangements of capital, the resulting space–time compression, and the cultural production of postmodernism in architecture, the humanities and the arts. In the next

---

[18] Harvey, *The Condition of Postmodernity*, 147.
[19] David Harvey, *The New Imperialism* (Oxford/New York: Oxford University Press, 2003), 62.
[20] Ibid.
[21] Greider, *The Soul of Capitalism*, 94-95.

section I emphasize a point made by many (including Harvey) that post-modernity is also the logical outworking of modernity's urge to privatize everything, from the land seized from native populations beginning in 1492, to the selling of industrial patents based on biological life forms today. Postmodernity, then, like modernity, has a cultural identity—a point often made by Muslim critics: the European (formerly) Christian nations, and today supremely, the United States of America.

Postmodernity: Commodification of the Commons
Though he does not call it postmodernity, Benjamin Barber's analysis of the current tension between McWorld and "Jihad" (the reaffirmation of identity politics, ethnically and religiously) is useful here. For Barber, McWorld is about a culture of consumption that is being forced onto poorer countries through the relentless advertising campaigns of multina-tional corporations, the "soft imperialism" of the American "infotainment industry" (Hollywood, MTV, CNN and the like) and, more deviously, the conditions of IMF and WB debt servicing tied to recipients opening their territory to "free trade"—meaning, allowing free reign to multinationals.[22]

On one level, McWorld seems rather innocuous. It is a theme park culture, notes Barber. Yet beyond Disneyland, Six Flags and Universal Studios, this culture is displayed in the ubiquitous commercial strips, malls and fast food chains. McDonald's is undoubtedly the most recog-nized brand worldwide, opening about one thousand new franchises a year and spawning entertainment and charity ventures. Like other brand names, it sells an experience, "and the experience becomes the defining attribute of a food marketplace that is also a theater of consumption and a theme park of lifestyles… The McDonald's way of eating is a way of life: an ideology as theme park more intrusive (if more subtle) than any Marx or Mao ever contrived."[23]

---

[22] See also the penetrating analysis along these lines in the work of British sociolo-gist, Leslie Sklair: *Sociology of the Global System*, 2d ed. (Baltimore, MD: The Johns Hopkins University Press, 1995), and *Transnational Capitalist Class* (Oxford/Malden, MA: Blackwell, 2001).

[23] Barber, *Jihad Vs. McWorld*, 129. The veteran neo-Marxist and "third-worldist" Samir Amin defines cultural life as "the mode of organization of use values." The way in which Western-dominated capitalism deploys its forces of production is through a "generalized exchange value"—currently determined by US banks and securities. This has the effect of standardization. What is needed, he argues, is for whole ethnic and cultural blocks to "delink" from the system, much as precapitalist societies held "use value" independently of "exchange value" and thus were free to follow a diversity of paths toward development (Samir Amin, *Eurocentrism*, trans.

The US phenomenon of suburban sprawl since the 1960s also coincided with a philosophy of development that aimed to "contain the entire world within the shopping plaza," and to configure the internal architecture of the mall so that consuming will become addictive. This has far-reaching implications for urban or suburban communities and in effect reduces "every other human activity into a variation of buying and selling."[24]

This privatization of public space and its dedication by the powers-that-be to the cult of consumption is but one manifestation of what Barber calls the "market ideology," which manifests itself as a "commitment to the privatization of all things public and the commercialization of all things private."[25] One extreme—which so far has not been adequately discussed from an ethical viewpoint, let alone regulated—is the way in which the mapping of the human genome has been patented off to individuals and corporations for billions of dollars. This is a deplorable development, because the commercialization of scientific research is not only a hindrance for the necessary flow of information among scientists globally,[26] but, thanks to the emphasis in the 1990s on Intellectual Property Rights (IRPs), the World Trade Organization has sought to enforce the dominion of transnational corporations over local indigenous knowledge systems and the rule of monocultures over traditional "economies based on diversity and decentralization."[27] The rich knowledge pool of indigenous plant and animal life and how best to

Russell Moore [New York: Monthly Review Press, 1989], 137-38). Amin, an Egyptian with an academic career in France, is the author of over thirty books. He now heads up the Third World Forum based in Dakar, Senegal.

[24] Ibid., 130.

[25] Ibid., xvii. Soros adds a needed warning in this regard: "It is dangerous, however, to place excessive reliance on the market mechanism. Markets are designed to facilitate the free exchange of goods and services among willing participants, but are not capable, on their own, of taking care of collective needs. Nor are they competent to ensure social justice. These 'public goods' can only be provided by a political process" (*The Bubble of American Supremacy*, 91).

[26] Kenneth Martin highlighted almost twenty years ago the dangers of research on university campuses being tied to the financial interests of large corporations. He quotes Martin Kenny, "the fear of being scooped or of seeing one's work transformed into a commodity can silence those who presumably are colleagues. To see a thing that one produced turned into a product for sale by someone over whom one has no control can leave a person feeling violated" (*Biotechnology: The University-Industrial Complex* [New Haven, CT: Yale University Press, 1986], 110, quoted in Vandana Shiva, *Biopiracy: The Plunder of Nature and Knowledge* [Boston, MA: South End, 1997], 15).

[27] Ibid., 72.

interact with it from a human standpoint is in danger of being lost through the arrogant intervention of Western (monocultural) corporations.

Vandana Shiva, the prolific Indian author, physicist and ecologist, explains what this policy of IPRs entails: "While both kinds of economies use biodiversity as an input, only economies based on diversity produce diversity. Monoculture economies produce monocultures."[28] The highly industrialized nations are situated in zones that are poor in biodiversity while most Third World countries find themselves in the richest zones of biodiversity, the tropics. Beyond the obvious ecological dimension relative to the dramatic loss of this biodiversity due to various forms of pollution (discussed in the next section), this unbalance has of late translated into what Shiva and others have called "biopiracy." A delegate from Malaysia to the UN General Assembly in November 1990 put it this way:

> There are various instances where transnational corporations have exploited the rich genetic diversity of developing countries as a free resource for research and development. The products of such research are then patented and sold back to the developing countries at excessively high prices. This must cease. We must formulate mechanisms for effective cooperation with reciprocal benefits between biotechnologically-rich developed countries and gene-rich developing countries.[29]

The result is increased tensions between the IPR regime and local populations. In India, for instance, the native Neem tree "has been used for centuries as a biopesticide and a medicine... Communities have invested centuries of care, respect, and knowledge in propagating, protecting, and using neem in fields, field bunds, homesteads, and common lands."[30] Western ignorance of these natural properties has suddenly given way, however, to a "new discovery" and a rush to patent new products based on the neem tree—and this in spite of neem toothpaste produced by Calcutta Chemicals for over 30 years. Because it was considered common knowledge, the Indian Central Insecticides Act of 1968 did not even list the neem tree.

I agree with Shiva that this is a case of two opposing conceptions of value. The indigenous perception of biodiversity is that it is valuable

---

[28] Ibid.

[29] David Cooper, "Genes for Sustainable Development: Overcoming the Obstacles to a Global Agreement on Conservation and Sustainable Use of Biodiversity," in *Biodiversity: Social & Ecological Perspectives*, ed. Vandana Shiva et al. (London/New Jersey: Zed Books; Penang: World Rainforest Movement, 1991), 110-11.

[30] Shiva, *Biopiracy*, 69.

because it contributes to meeting human needs. In the IPR conception of biodiversity, biodiversity is valuable because it can turn a profit. Thus "tinkering becomes necessary to add value." Yet the arrogance of this way of thinking is manifest: without the local knowledge this property of the neem tree would not have been "discovered," and the genetic engineering in most cases adds little value to the traditional uses of the plant. Shiva concludes, "IPRs allow for the privatization of biodiversity and the intellectual commons."[31]

Imperialism, Privatization and Dispossession
In this section I argue that the issue of the "commodification of the commons" is the guiding thread of modernity. And, as mentioned above, it raises the plight of the world's indigenous populations—an issue that should gravely concern both Muslims and Christians from the standpoint of creation. Here I find Anthony J. Hall's analysis particularly useful. For Hall, the issue of biodiversity and the privatization of what in traditional societies was considered a common legacy to be used by all (nature, folk knowledge and the like) goes back to the West's colonial project born in 1492—"the Columbian conquests," as he puts it.[32] Equally, the imperialist roots of modernity can plausibly be traced to the Christian conviction of that time that Christian princes had the God-given right to appropriate the lands of the "savages" (read "non-Christians"). This belief was enshrined by Pope Alexander VI in his "Bull of Donation," which granted all territories "discovered and to be discovered, one hundred leagues to the West and South of the Azores towards India" to the Catholic monarchs Isabel of Castille and Ferdinand of Aragon. Just the year before, these monarchs had granted Columbus the charter to discover and conquer new territory. Indeed, the pope saw himself as God's "caliph on earth," much as the "rightly guided caliphs" who succeeded the Prophet Muhammad and who presided over an equally impressive campaign of military and imperial expansion some seven centuries before: "The pope as the vicar of God commanded the world, as if it were a tool in his hands; the pope, supported by the canonists [experts in canon law, or church jurisprudence], considered the world as his property to be disposed according to his will."[33]

---

[31] Ibid., 72.
[32] Anthony J. Hall, *The American Empire and the Fourth World: The Bowl with One Spoon*, vol. 1 (Montreal/Kingston: McGill-Queen's University Press, 2003).
[33] Walter Ullmann, *Medieval Papalism*, quoted in Shiva, *Biopiracy*, 1. Also cited in Hall, *The American Empire*, xxviii-ix, 303.

The Christians, however, were not as tolerant as their imperial Muslim counterparts.[34] Yet, any attempt to quantify the human toll resulting from the Western powers' conquest of South and North America is a delicate and contested terrain. One useful indication, perhaps, is provided by Russell Thornton's careful investigations: from 72 million indigenous inhabitants at the beginning of the sixteenth century, the number shrank to about five million in 1900.[35] Hall offers another telling description:

> The lethal impact of the New World expansionism initiated with the Columbian conquests has been manifest in the early and violent demise of the vast multitudes of Aboriginal individuals, together with the extermination of more than a thousand distinct Aboriginal societies. The extent of this cataclysm in the Americas, the primary frontier of Europe's most aggressive episodes of expansionary zeal, has been monumental. According to Todorov, the founding of New Spain and Portuguese Brazil in the "sixteenth century perpetuated the greatest genocide in human history."[36]

Hall's ambitious project centers around the emergence of a structure of "unregulated, superpower hegemony that currently defines the main outlines of world order" that came to fill the vacuum left by the unfulfilled promises of the de-colonization movement after WWII. Indigenous peoples, possibly three-quarters of the world's population in Hall's reckoning, resent the fact that "the formal structure of empires, colonies, and subject peoples has not been replaced with a fairer means of

---

[34] At the beginning of the second Islamic dynasty, the Abbasids, who moved their capital from Damascus to Baghdad, the great majority of the empire's population from Spain to the Indus River was still non-Muslim. Marshall G. S. Hodgson, for example, emphasizes the effect of the new high culture of the Abbasid imperial court and the accompanying economic expansion of the cities as the main reasons for the accelerated rate of conversion during the eighth and ninth centuries CE (*The Venture of Islam: Conscience and History in a World Civilization*, vol. 1, The Classical Age of Islam [Chicago: University of Chicago Press, 1974, paperback ed. 1977], 303-5). For Ira M. Lapidus, the traditional Western belief that Islam was "spread by the sword" is no longer tenable: "It is now apparent that conversion by force, while not unknown in Muslim countries, was, in fact, rare. Muslim conquerors ordinarily wished to dominate rather than convert, and most conversions to Islam were voluntary" (*A History of Islamic Societies*, 2d ed. [Cambridge: Cambridge University Press, 2002], 198).

[35] Russell Thornton, *American Indian Holocaust and Survival* (Norman, OK: University of Oklahoma Press, 1987), 42-43, quoted in Hall, *The American Empire*, 27.

[36] Hall, *The American Empire*, 26. Hall is quoting Tzvetan Todorov, *The Conquest of America: The Question of the Other*, trans. Richard Howard (New York: Harper & Row, 1984), 5. Hall adds that the "genocide" is even better understood in light of the "extinction of more than three-quarters of the 2,200 or so Aboriginal languages and dialects spoken in the Americas in 1492" (p. 27).

organizing human relationships."[37] Hall highlights the close attitudinal and ideological impulses in the early American messianic concept of "Manifest Destiny," which gradually subjugated and destroyed most of the native population and culture in its path. Here too I agree with him that the Bush Administration's declaration of war on terrorism after September 11, 2003 and the American and British invasion of Iraq in April 2003 partake of the same imperial spirit.

The irony of the tanks rolling into the heart of ancient Mesopotamia in April 2003, notes Hall, is that the United States itself was a society born of a violent revolution against the old powers of Europe.[38] The founding documents of this newly emerging state were written by the likes of Thomas Jefferson, men steeped in the Enlightenment writings of John Locke and the French *philosophes*: Montesquieu, Diderot and Voltaire. Yet, from the beginning, the proclamation of human equality and dignity did not hold for all people. The last in a long list of grievances presented to King George III in the Declaration of Independence reads, "He has excited domestic insurrections amongst us, and has endeavoured to bring on the inhabitants of our frontiers, *the merciless Indian Savages* whose known rule of warfare, is an undistinguished destruction of all ages, sexes and conditions."[39] The Anglo-American colonists, much as the European powers they were opposing, conceived of themselves as the recipients of a God-given civilizing mission to push back the frontiers of barbarism and savagery. What is more, as the above grievance shows, the Anglo-American colonists castigated the British attempts at establishing large Aboriginal territories free from colonist incursions and conquests.

In this regard, the Royal Proclamation of 1763 provided a constitutional framework for British America that sparked revolutionary action in the colonial communities. Among its provisions, King George outlawed the westward expansion of the colonists in order to keep Canada

---

[37] Ibid., xxix.

[38] Another irony in the Iraq war is the naming of key weaponry after the native freedom fighters who best resisted American military genocidal policies. Hall cites the agile "Apache" and "Black Hawk" helicopters. Black Hawk was an Aboriginal leader who spearheaded a resistance movement in 1832 in the Illinois area. His capture and incarceration marked "the final pre-emption by the United States of the sovereign aspirations of the Indian Confederacy" (ibid., xvi).

[39] Retrieved online at http://www.ushistory.org/declaration/document/index.htm (emphasis added). Hall draws the parallel between the parading of the defeated warrior Geronimo by Theodore Roosevelt in his inaugural procession down Pennsylvania Avenue in Washington (and the denial of rights generally to the native population) and the refusal of the Bush Administration to allow the Guantanamo Bay detainees the right to a fair trial in a US court.

as an Indian hunting ground and as a source of fur for Montreal. This prospect of shared jurisdiction between indigenous peoples and British colonists smacked as "provocative and tyrannical to many Protestant colonists, who regarded Catholic–Aboriginal Canada as a conquered realm and as their rightful inheritance by natural law."[40]

The fundamental issue at stake was a choice between two opposing conceptions of British imperialism in North America. King George III had opted for what we call today "Aboriginal and treaty rights," or a "regime of recognition" of the rights of the indigenous population within the colonized territories. In an effort to atone for the dispossession of the natives and the mounting inequalities that resulted from the founding charters and patents granted by his predecessors, King George, along with an influential minority of voices, issued his 1763 proclamation in order to "add to the legal edifice of America some basic guarantees of the inherent human right of Indigenous peoples," who then would no longer be " robbed of their lives, their lands, or their collective capacities for self-determination."[41]

Over the decades, the movement of opposition to the Royal Proclamation—the first attempt at nailing down some elements of international law—had sharpened its arguments and, in so doing, grew into what was to become the mainstream of the American Revolution. In Hall's view, this revolutionary movement, which was gradually to claim the whole continent as their rightful domain, came from two sources, one religious and one secular. Part of its inspiration drew from the Puritan heritage, a powerful source of identity and direction for many of the original colonists, especially in New England. Here we touch on the central theological theme of the present book: already in 1630, the Puritan writer John Cotton argued that the take-over of Indian country was justified on the basis of the creation mandate: "in a vacant soyle, hee that taketh possession of it, and bestoweth culture and husbandry on it, his Right it is."[42] Further, the general mood in Protestant circles was that the Indian religion was idolatrous—a worship of nature instead of God. Ignoring the fact that their worship of the Great Spirit was tantalizingly close to monotheism, what seemed to provoke their ire more than anything was the fact that Aboriginal thinking did not accept the European conviction that human beings are to exert mastery over creation.

---

[40] Hall, *The American Empire*, 310.
[41] Ibid., 315.
[42] Patricia Seed, *Ceremonies of Possession in Europe's Conquest of the New World, 1492–1640* (Cambridge: Cambridge University Press, 1995), 30, cited in Hall, *The American Empire*, 309.

This linkage of dominion over the earth to private property and commercial contracts represents the second stream that flowed into the American Revolutionary ethos. Indeed, the ideas of John Locke held particular sway in the thinking of Thomas Jefferson. Thus for Hall, the entrenchment of the worldview of "possessive individualism" emanates from a belief that natural law configures society as relationships between proprietors and that "political society becomes a calculated device for the protection of this property and for the maintenance of an orderly relation of exchange."[43] It was not by chance that Adam Smith wrote *The Wealth of the Nations* during this period (it was published in 1776). The charters that sanctioned British colonization in North America were steeped in capitalist ideology.

Lutheran theologian Larry Rasmussen shows that the key institution of modernity was not the nation-state (though it is crucial) but the corporation.[44] Furthermore, it is intimately linked to the vast European enterprise of the last 500 years, the "industrial revolution."[45] Indeed, its roots are traced to England's exploration project of the "New World." The central concept of "limited liability" enabled these chartered corporations to accomplish two goals at once: "absorb risk-taking and promote exploration and settlement...a dazzling innovation unwittingly tailor-made for the transition from agricultural societies to globalizing conquest, commerce and industrialism."[46] The process continues today: corporations go on extracting the world's resources, globalizing means of production, technology and finance, and the promotion of values inherent in these processes. The figures are astounding:

[43] C. B. Macpherson, *The Political Theory of Possessive Individualism: Hobbes to Locke* (Oxford: Oxford University Press, 1979), 3, cited in Hall, *The American Empire*, 308.

[44] Larry Rasmussen, *Earth Community, Earth Ethics* (Maryknoll, NY: Orbis, 1996). See also David C. Korten, *When Corporations Rule the World* (Hartford, CT: Kumerian, 1995). A lay theologian of the Evangelical Lutheran Church in America, Rasmussen held the Reinhold Niebuhr Chair of Social Ethics at Union Theological Seminary in Manhattan, New York, until he retired in July 2004.

[45] The "agricultural revolution" started with the neolithic villages of the Middle East about 10,000 years ago. It was a multi-ethnic enterprise. The Industrial Revolution, and its heir, the "information revolution," originated in Europe and expanded with the power of transformed raw materials, growing scientific knowledge and therefore, superior weaponry. Colonialism ensured that this domination by one race/culture would be perpetuated—thus the mushrooming of "neo-Europes" all over the globe (ibid., 73-74).

[46] Ibid., 63.

> In the 1990s the one hundred largest corporations in the world had more
> economic power than 80 percent of the world's people. In 1991 the
> aggregate sales of the world's ten largest corporations totaled more than
> the aggregate GNP of the one hundred smallest countries of the world.
> The world's five hundred largest industrial enterprises, employing only
> 0.05 of 1 percent of the planet's population, nonetheless controls 25
> percent of the planet's economic output.[47]

A week after the September 2001 attacks on New York and Washington,
Rasmussen (who lives and works in Manhattan) commented that the
Statue of Liberty that stood so close to the rubble of the Twin Towers
had come to represent for Americans "not the Statue of Liberty so much
as the Statue of Wealth." He then adds that both Presidents Bush, a
decade apart, declared that "the American way of life is not up for
negotiation." This assumes a definition of democracy "in market terms":
"The democratic society is one with virtually unrestricted liberty to
acquire and enjoy wealth. This vision renders the right to property and its
uses more basic than the right to use government as an equalizing
force."[48] His theological vision of "earth community" (to which we shall
return) leads us in quite another direction. A democratic society, in this
understanding, is one that stands in solidarity with all others on the
planet—including the non-people (living beings and physical environment)—particularly those most impoverished and vulnerable, because
our prosperity, and indeed survival, depend on everyone being able to
thrive in a sustainable lifestyle.

Globalization and the Fourth World
The wall of social injustice that we as a global community are now
careering toward was graphically portrayed by the former president of
the World Bank, James Wolfensohn, who noted that, out of the two
billion people to be born in the next twenty-five years, only fifty million
will be born in a rich country. He went on, "The vast majority will be in
the poorer nations: born with the prospect of growing up into poverty
and unemployment and disillusioned with a world that they will inevitably view as inequitable and unjust."[49] This is why I believe the concept of

---

[47] Ibid.

[48] Larry Rasmussen's contribution to the "Union Theological Seminary (New York)
Faculty Panel of 9/20/01 on the Twin Towers Disaster," *The Missionary Society
of Connecticut* (UCC) web page (last updated January 11, 2003), http://www
.ctconfucc.org/resources/fromgroundzero/rasmussen.html.

[49] David Fickling, "World Bank Condemns Defense Spending," in *The Guardian*
(February 14, 2004), http://www.guardian.co.uk/globalisation/story/ 0,7369,1147888
,00.html. The article ends with a damning statistic: "One sixth of the world's six

"Fourth World" is so important to the equation of a sustainable political, economic and cultural agenda for our planet. George Manuel was the Canadian Shuswap activist who became the leading philosopher of the North American Indigenous revival of the 1960s and 1970s. His vision of the Fourth World emerged out of his own contacts, first in Tanzania, then with a growing coalition of Aboriginal peoples in New Zealand, Australia and the Saami in Scandinavia. The pioneering efforts of President Julius Nyerere of Tanzania within the larger African decolonization enterprise provided Manuel with a theoretical framework for North America: "Nyerere personified the animating impetus of egalitarian sharing and cooperation, …one starkly in contrast to that of 'destruction, conquest and suppression' which had been integral to the European colonization of the New World."[50] He elaborates on this theme in a speech, quoted by Hall: "We have only to watch Julius Nyerere taking time from his executive duties to work alongside the day labourers in a small village to know that the traditions of our grandfathers have a place in the modern, technological world."[51]

Manuel identified the ideological enemy as Social Darwinism, the credo of the social scientists and politicians who had cast indigenous peoples into the role of "savages," "natives," "backward," and "tribal," in contrast to societies of Europe and beyond that had embraced the ideology of the nation-state and its attendant paradigms of modernity and social "progress." In a world divided by two superpowers, whether of the capitalist or the socialist kind, the Third World was inevitably caught between the two, fought over, thoroughly patronized and subjugated, with no room left for the dignity, value, or "contemporary applicability of Indigenous knowledge and philosophy." Hall goes on to explain:

> Fourth World thinking is necessarily antagonistic to the bias of Third World thinking, a mode of conceptualization that promotes external models of change for most of humanity to mimic or duplicate. Unlike Third World thinking, with its emphasis on imposing standardized, mono-cultural moulds of growth and development on different societies, Fourth World thinking emphasizes the freedom of diverse peoples to chart their own distinct courses of social, legal, economic, technological, and political change. The object of this change is to ameliorate the pluralistic ecology of human relationships in ways that reflect and project forward the Aboriginal inspiration and dynamics of First Nation cultures.[52]

billion people owned 80% of its wealth, while another sixth earned less than a dollar a day."
[50] Hall, *The American Empire*, 239.
[51] Ibid.
[52] Ibid., 240-41.

Manuel's dream of Canada becoming the first Western nation-state to enter into a meaningful relationship with its indigenous population was partially realized posthumously through the Canadian Patriation Act of 1982, which recognized the Aboriginal and Treaty rights inherent to the 1763 Royal Proclamation. Yet, in three subsequent rounds of constitutional amendments (with First Nations representatives participating in the discussions), these rights were watered down and the only winner to emerge in the process was the Province of Quebec, whose special status (ironically, considering its colonial past) was recognized. That struggle is still underway and in spite of high-profile actions by coalitions of Indians and non-Indians in America, the bid to revisit broken treaties of the past faces greater obstacles in the US.[53] Indeed, the clash between what Hall calls "the American Empire of private property" and the paradigm of "the Indian Country of Canada" has become all the more poignant since the neoliberal push to spread free trade throughout the Americas under the umbrella of NAFTA and the WTO has resulted in the dispossession of many native peoples throughout the continent.

Nevertheless, Manuel's legacy lives on. Indeed, it was Manuel who could simultaneously spearhead a cultural and political revival all across the spectrum of indigenous north-American nations and elicit the support of influential anthropologists and ethnographers, who, partly through his inspiration, had founded Survival International (based at Harvard University), the Scandinavian Association and the International Work Group on Indigenous Affairs. Manuel attended a Scandinavian-based conference in 1973, one also attended by Inuit, Dene, Cree and Saami representatives. Two years later, these relationships formed the network Manuel used in order to found the World Council of Indigenous Peoples.

---

[53] Indian activism picked up after WWII, though it was not until the 1960s, with the birth of the civil rights movement for African Americans, opposition to the Vietnam War and the spread of ecological consciousness that "Red Power" came to the foreground. Some of the defining moments in the movement, according to Hall, were: (1) the founding of the National Indian Youth Council at the University of Chicago in 1961; (2) the fish-ins of the mid-sixties in Washington state, given a high profile through actor Marlon Brando's involvement; (3) the 1969 occupation of Alcatraz Island in the San Francisco Bay which drew activists from all over North America; (4) the 1969 Pulitzer Prize awarded to the Kiowa author N. Scott Momaday for his *House Made of Dawn*; (5) the 1972 cross-country pilgrimage, Trail of Broken Treaties, which ended up with partial destruction of the Bureau of Indian Affairs in Washington, DC; (6) the sometimes violent conflict at Wounded Knee on the Pine Ridge reservation (South Dakota) in the summer of 1973 (ibid., 238-99).

To sum up, the vision of the emerging Fourth World finds its inspiration in that of Tecumseh, the leader of the Indian Confederacy who lost militarily to the Americans in the War of 1812 (but thereby secured the existence of Canada, "apart from the realm of the Stars and Stripes"), and the legacies of George Manuel and other native leaders worldwide—the philosophy of the bowl with one spoon. Hall explains the central symbol of his project:

> The bowl with one spoon is an Aboriginal pictorial representation of the principle that certain hunting territories are to be held in common. In northeastern North America the image appeared frequently in the design of many wampum belts used to signify the terms of treaty agreements. In the era of Tecumseh the image came to signify the need for federal unity among Indigenous peoples if the shared Indian Country was ever to achieve sovereign recognition in international law.[54]

The contemporary nature of Tecumseh's fated struggle was poignantly illustrated by the Zapatista uprising of 1994 in Chiapas, Mexico. The Zapatista Liberation Army took control of several localities in Chiapas, deliberately timing their simultaneous attacks to coincide with the signing of the North American Free Trade Agreement (NAFTA). But far from being a fundamentalist exercise in communitarian politics, the masked leader ("Subcommandante Marcos") "initiated an unorthodox campaign for political change that was simultaneously locally rooted and globally oriented."[55] By reclaiming the revolutionary aura of Emiliano Zapata, Mexico's founder, the Zapatistas sparked passionate interest around the globe. Particularly in North American and European cities some 45,000 sympathetic websites were created to support the movement and/or to extend it to other venues and issues.

Suddenly, people everywhere were connecting the dispossession of poor Indian farmers by NAFTA's policies of industrialization to the wave of impoverishment created by the neoliberal machine globally. In Ecuador, the rural indigenous population revolted when the government announced the conversion of the local currency to the dollar, while in Bolivia Indians rioted against "major increases in the cost of water, as well as against high fuel prices, unemployment, and pauperization" and forced the government to abandon its IMF-imposed water privatization campaign.[56] Journalist Greg Palast did some investigating into these riots,

---

[54] Ibid., 1.

[55] Ibid., 143.

[56] Mitzman, *Prometheus Revisited*, 158. The scandal in this case is that the privatization imposed by the IMF had brought in a British firm to do the job, International

which were not in the least covered in the US media. In fact, they claimed six lives, with "175 injured including two children blinded after the military fired tear gas, then bullets, at demonstrators opposing the 35 per cent hike in water prices imposed on the city of Cochabamba by the new owners of the water system, International Waters (IWL) of London."[57]

Fourth World politics, invigorated in the 1990s by the Zapatista rebellion, were not the plaything of a few leftist dreamers. Rather, they were propelled by a groundswell of anger on the part of dispossessed and marginalized people (in part Aboriginals) everywhere. Palast counters the optimism of both Thomas Friedman[58] and Anthony Giddens[59] on the topic of globalization. Palast, who managed to get a hold of scores of "confidential" IMF and WB documents, recognizes that these institutions have attempted to reform some of the excesses of their earlier neoliberal zeal, as shown in the change of vocabulary—from the deadly "Structural Assistance Plans" to the more innocuous-sounding "Poverty Reduction Strategies." This is an urgent move—though he remains skeptical about the actual substance of the change, in view of the fact that the IMF admitted in its April 2000 World Report that, "in the recent decades, nearly one fifth of the world population have regressed. This is arguably one of the greatest economic failures of the 20th Century."[60]

Tellingly, Joseph Stiglitz, the former chief economist for the World Bank (and chairman of President Clinton's Council of Economic Advisers) was fired from his job in 1999 and spent several days in April of 2001 explaining to Palast why he regreted the ideology that had driven him and the WB for so many years.[61] Confirmed by the documents in hand, Stiglitz reveals here that every poor country in need of funds is given the same four steps to follow:

Waters Ltd, controlled in turn by US giant Bechtel. The riots forced the state to cancel their contract.

[57] Greg Palast, *The Best Democracy Money Can Buy: An Investigative Reporter Exposes the Truth about Globalization* (London/Sterling, VA: Pluto, 2002), 54.

[58] Thomas Friedman, *The Lexus and the Olive Tree* (New York: Farrar, Straus & Giroux, 1999). Palast calls it "a long, deep kiss to globalization" (*The Best Democracy*, 44).

[59] Palast heard Giddens's lecture "an earnest crowd of the London School of Economics," and quotes him as declaring, "Globalization is a *fact*, and it is driven by the communication revolution" (ibid., 46, emphasis original).

[60] Ibid., 50.

[61] Palast explains, "He was not allowed quiet retirement; US Treasury Secretary Larry Summers, I'm told, demanded a public excommunication for Stiglitz having expressed his first mild dissent from globalization World Bank-style" (ibid.). Stiglitz teaches at Columbia University and won the Nobel Prize for Economics in 2001.

1. Privatization: state industries have to become private—in a process Stiglitz calls "briberization," due to the high incidence of bribes offered to elites in power to coax them into selling off water and electricity companies.

2. Capital market liberalization: this insures that investment capital flows unhindered in and out of the country.[62] Unfortunately, as it happened to Indonesia and Brazil, money flows out much more readily than it flows in. As a nation's reserves begin to dry up, the IMF seeks to draw speculators back to the country by demanding a hike in interest rates "to 30 per cent, 50 per cent and 80 per cent"—with predictable results, adds Stiglitz: "demolished property values, savaged industrial production and drained national treasuries."[63]

3. Market-based pricing: all subsidies on basic foodstuffs, water, fuel and domestic gas must be lifted. This of course, says Stiglitz, leads to step three and a half: "the IMF riot." Thus the riots of Indonesia (1998), of Bolivia (2000 and 2001), of Ecuador and so on. Every country report announces coldly that "social unrest" will result in the short term. What is more, a secret report on Ecuador reveals that the plan to make the dollar the national currency "has pushed 51 per cent of the population below the poverty line." As noted with regard to the Bolivian riots, these are usually dispersed with the aid of "bullets, tanks and tear gas," further driving away investment capital and allowing foreign corporations to "pick off remaining assets, such as the odd mining concession or port, at fire sale prices."[64] Who are the biggest winners? "The Western banks and US Treasury," who are able to skim off large profits from all the turmoil.

4. Poverty reduction strategy, that is, free trade—according to rules laid out by the IMF and WB:[65] just like the Opium Wars of the

---

[62] Mao points to the abundance of quality mangoes in Uganda. The fruit sector, then, deserves to be diversified, that is, juice-processing plants should be built. Without foreign investment, however, this could never happen. "Yet foreign investors have neglected this productive sector, preferring instead to invest in services like mobile phones, cheap electronics, and gigantic supermarkets. Consequently, Uganda still imports mango juice from the Middle East" ("Unevenly Yoked," 2).

[63] Stiglitz, *The Best Democracy*, 51.

[64] Ibid., 52.

[65] On the connection between free trade and ethnic violence, see Yale law professor Amy Chua's *World on Fire: How Exporting Free-Market Democracy Breeds Ethnic Hatred and Global Instability* (New York: Anchor Books, 2004). Note how she con-

nineteenth century, the West has many ways to force down barriers to their products in Asia, Africa and Latin America, while
at the same time "barricading their own markets against Third
World agriculture." And here Stiglitz (with great emotion,
recalls Palast) rejoins Shiva in decrying the effect of the WTO's
intellectual property rights regime (TRIPS)—"the new global
order has 'condemned people to death' by imposing impossible
tariffs and tributes to pay to pharmaceutical companies for
branded medicines."[66]

The first of the double walls blocking humanity's future, then, is this
"social-economic malaise"[67] that was resolutely resisted in the Zapatista-
related movements of the 1990s; in the general strikes in France at the
end of 1995; in the anti-globalization protests in Seattle (November
1999), Washington and Sydney (both April 2000);[68] in the climax of the
country-long Zapatista coalition march that ended in a mass gathering in
the capital's Zócalo Plaza;[69] and finally in the millions who marched in
January 2003 to protest the upcoming US war in Iraq. Yet our crashing
into the wall of global injustice is still avoidable. Commenting on the
breakdown of the WTO talks in September 2003 (the poor countries
walked out), Benjamin Mkapa, president of Tanzania, still believes free
trade could benefit the poorest regions of the world:

> In our increasingly interconnected world, global stability is of interest to
> everyone. But it can only be assured where global governance is mani-
> festly just, where it is premised on a value system that recognises all
> players as equal stakeholders, worthy of a place at the negotiating table

nects this to Iraq: "Our Most Dangerous Export: Imposing Free-Market Democracy
on Iraq Has Unleashed Ethnic Hatred," *The Guardian* (February 28, 2004), http://
www.guardian.co.uk/Iraq/Story/0,2763,1158215,00.html.
[66] Palast, *The Best Democracy*, 53. Palast asks him if any country had escaped this
fate. He answered that Botswana had, but only via rejection of the IMF's conditions.
[67] Mitzman, *Prometheus Revisited*, xix.
[68] Hall says that the quarter-million march on Sydney Harbor Bridge represented
"one of the many significant rituals in the worldwide movement emphasizing the
need for major initiatives, both domestically and internationally, to reverse the
destructive course of the ongoing Columbia conquests through various forms of
reconciliation with Indigenous peoples" (*The American Empire*, 143).
[69] Hall (ibid.) notes that this gathering was attended by such Hollywood icons
as Robert Redford and Oliver Stone, as well as leading intellectuals from Spain,
Portugal and Canada. Naomi Klein likened the event to Martin Luther King Jr's
march on Washington—the dispossessed on the move in order to speak truth and
power.

and the dining table. It also helps when the whole world is seen as being committed, in practical terms, to the war on poverty.[70]

This will not happen until power is redistributed, not only among countries, but within them as well. The struggle of the Fourth World, then, is to promote a democracy, as Majid sees it, which gives people "the right to determine their own cultural and economic agendas."[71] This in turn will require some serious rethinking of the foundations of contemporary capitalism, and particularly the central notion of corporate identity. Jeffrey Kaplan shows that until about 1840 US state legislators ensured that corporations were severely limited: in time and in the scope of their activities (only for projects of public utility). But this quickly changed in the following decades. De Tocqueville expressed concern that the emergence of "an industrial aristocracy" in America—more destructive than its European counterpart—would crush democracy:

> In 1886, without comment, the United States Supreme Court ruled for corporate owners in *Santa Clara v. Southern Pacific Railroad*, allowing corporations to be considered "persons," thereby opening the door to free speech and other civil rights under the Bill of Rights. By the early 1890s, states had largely eliminated restrictions on corporations owning each other, and by 1904, 318 corporations owned forty percent of all manufacturing assets.[72]

However, as Kaplan shows, there is now a groundswell of resistance to this concept of corporations with "human" rights, both on the part of states that are trying to overturn these legal definitions and on the part of townships on a more local level.[73] If this kind of reversal is possible in the US—though the movement is yet at its infancy—it is all the more urgent that global corporate capitalism be challenged where it is causing the most intense human suffering. However, this challenge will have to be tackled together with its twin—the second wall Mitzman highlights—the severe degradation of our collective human environment.

---

[70] Benjamin Mkapa, "Giving Everyone a Place at Global Dining Table," *The Guardian* (February 16, 2004), http://www.guardian.co.uk/business/story/0,,1148792,00.html.

[71] Majid, *Unveiling Traditions*, 124.

[72] Jeffrey Kaplan, "Consent of the Governed: The Corporate Usurpation of Democracy and the Valiant Struggle to Win it Back," *Orion* (November–December 2003): 54.

[73] Kaplan illustrates this by pointing to the movement of small farmers in Pennsylvania who are successfully tackling the pressures of large agribusiness corporations on their territory (ibid.).

Postmodernity and the Destruction of Earth

It comes as no surprise that the corporate greed that plundered whole continents and shamelessly exploited its peoples is now causing irreparable damage to our common home. According to James Gustave Speth, Dean of the Yale School of Forestry and Environmental Studies and veteran international advisor on ecological issues, "[a] global crisis has unfolded quickly, and, as in classic Greek tragedy, we have been told what the future may hold, but so far we seem unable to step from the path to disaster that has been mapped out for us. The last act is about to begin."[74]

The 2002 Living Planet Report of the World Wildlife Fund concluded from its careful analysis of our biological environment, which measured the human factor as an "ecological footprint," by asking how much land was needed to sustain a human being at the current rate of consumption. The average per capita "footprint" is now about 2.3 hectares per person. The problem with this finding is twofold, however. First, it is not sustainable, since "the 'biological capacity' of the earth is equal to just 1.9 hectares per person."[75] Around 1980, humankind began to consume more of the earth's resources than was being replenished. Today, says the report, this consumption is 20 percent higher than the planet can afford, and if the present trends hold, that percentage will rise to 50 percent by 2050. But the second problem may be even more worrisome, writes Mitzman: "[t]he report indicates that the ecological footprint is much deeper in North America and Europe—9.6 and 5 hectares per person respectively—than in Asia and Africa, where the use of resources is estimated at 1.4 hectares per person."[76]

The *National Geographic* magazine recently featured a cover story, "Global Warming: Bulletins from a Warmer World."[77] In an article gleaning from the work of scientists from many nationalities, universities and organizations, the consensus is emerging that our planet has warmed up by one degree in the last century, and that this temperature is projected to climb between 3 to 10 degrees Fahrenheit in this century, with disastrous consequences for humanity, and especially the poor. Among

---

[74] James Gustave Speth, *Red Sky at Morning: America and the Crisis of the Global Environment* (New Haven, CT/London: Yale University Press, 2004), 1.

[75] Mitzman, *Prometheus Revisited*, xix.

[76] Ibid. This again shows the interrelatedness of the "two walls," and why the term "ecojustice"—cf. below—is so appropriate.

[77] *National Geographic* (September 2004): 2-75.

the ominous signs of a brisk and brutal shift in global climate is upon us with unforeseeable costs on life and ecosystems are the following:[78]

- The Arctic perennial ice cover has decreased by 9 percent per decade since 1979.[79]
- The average winter temperatures of Antarctica have risen nearly 9° F since 1900, with several species now threatened in their ability to survive.[80]
- Alaska has warmed up over 3° F over the last three decades, with much of its permafrost now melting.[81]
- The five hottest years on record are since 1998.
- Oceans, which are important centers of carbon dioxide (CO_2) absorption, are warming up dangerously, with CO_2 levels rising fastest in the deepest waters.
- The rapid melting of both polar caps has altered the salinity of the oceans, and with rising temperatures of the waters this portends major disruptions in the thermohaline circulation system (the pattern of ocean currents, which in turn are crucial to global climate control).[82]
- One barometer of warming seas is its coral reefs: the hottest year, 1998, bleached 16% of the world reefs, most of which will never recover.

---

[78] These data are all taken from the *National Geographic* article.

[79] See also recent reports on melting glaciers in Greenland: Steve Connor, "Melting Greenland Glacier May Hasten Rise in Sea Level," *The Independent Online* (July 25, 2005), http://news.independent.co.uk/world/environment/article301493.ece; and Richard Hollingham, "Icy Greenland Turns Green," *BBC News Online* (August 14, 2005), http://news.bbc.co.uk/1/hi/programmes/from_our_own_correspondent/4145034.stm.

[80] The more frightening scenario at the other end is the possible melting of the grounded ice of the mammoth West Antarctic Ice Sheet (WAIS). According to estimates by British and Norwegian scientists, there is a five percent chance that the WAIS would melt to such an extent "that sea levels will rise three to six feet over the next two hundred years" (Speth, *Red Sky at Morning*, 60).

[81] Melting permafrost is occurring on a large scale, with at least two drawbacks: whole mountainsides can crumble; frozen soils that melt release more greenhouse gases (ibid., 59.)

[82] Speth tells us that a 2002 National Academy of Sciences report "predicts that we are likely to see surprises, sudden shifts, and even drastic upheavals in global climate and its impacts." Among these potential surprises: "Fossil evidence shows that the Gulf Stream has shut down in the past, quickly, and plunged the North Atlantic region (not just Europe) into a dramatically cooler area. Today's computer models suggest that a shutdown of the Gulf Stream would produce winters twice as cold as the worst winters on record in the eastern United States" (ibid.).

- In the cold regions of the world, winter starts measurably later each fall/autumn and ends earlier each spring.
- The world's glaciers are melting dramatically, at all latitudes, portending two ominous scenarios in the not too distant future: loss of drinking water for many populations, and rising sea levels that will decimate low-lying poor countries like Bangladesh.

Contemporary studies based on ice cores, fossil pollen, fossil marine organisms and sediments from oceans and lakes, show that the earth has experienced several cycles of ice ages (with the last global meltdown around 19,000 years ago), suggesting that we are now in an extended interglacial period. This would seem to refute the contention of most scientists that the rising levels of CO_2 and global temperatures are the main causes of the climate changes observed today, mainly by increased burning of fossil fuels, the clearing of forests and the higher emissions of heat-trapping gases such as carbon dioxide, nitrous oxide and methane. Though we now know that these rhythms are also caused by astronomical phenomena, the sharp rise in CO_2 levels since 1950 has led the majority of scientists to impugn humanity's footprint. In the words of the 2002 report by the National Academy of Sciences, "Recent scientific evidence shows that major and widespread climate changes have occurred with startling speed… [G]reenhouse warming and other human alterations of the earth system may increase the possibility of large, abrupt, and unwelcome regional or global climactic events."[83]

And yet, dramatic climate change is only part of the alarming ailments of our planet. Butting against the headlong rush of modernity to madly produce and consume is the hard fact of environmental deterioration. Its two major trends are "pollution and biological impoverishment."[84] Pollution, first of all, should be defined as the presence of excessive quantities of elements that disrupt ecosystems. Plant nutrients like phosphates are essential for healthy rivers and streams, but beyond a certain quantity, they rob the water of the oxygen that aquatic organisms need. Equally, carbon dioxide keeps the atmosphere around the earth warm enough for its inhabitants to live, but in greater quantities, it acts like a greenhouse and traps the rising heat. Many pollutants, such as mercury, dioxin and PCBs, are toxic even in small quantities, causing cancer and reproductive

---

[83] National Research Council, *Abrupt Climate Change: Inevitable Surprises* (Washington, DC: National Academy Press, 2002), 1, cited in Speth, *Red Sky at Morning*, 60.

[84] Ibid., 23.

disorders. Speth estimates that "pollution is occurring on a vast and unprecedented scale worldwide. It is pervasive, quite literally, affecting in some way virtually everyone and everything."[85]

Speth discerns four trends in the spread of pollution: (1) the quantities of pollutants are growing exponentially, notably the levels of greenhouse gases, sulfur dioxide and nitrogen oxide (causing smog and acid rain), hazardous wastes (2.5 billion pounds released annually in the US alone); (2) pollutants vary in nature from visible agents to microscope ones, mainly resulting from chemical ("pesticides, plastics, industrial chemicals, medical products, detergents, food additives"[86]) and nuclear industries; (3) pollutants are everywhere from First World to Third World: the developing world is host to the worst water and air pollution today, with higher exposure to toxic chemicals, and to many devastating industrial accidents;[87] (4) the effects of pollutants are both local and global: acid rain and smog destroy the quality of people's lives in many regions of the world, and "the depletion of the stratosphere's ozone layer, despite dramatic reductions in CFC use.[88] Finally, the greatest threat to the survival of our planet's life forms—including our own—is global warming and the inevitability of severe climate changes up ahead.

The equally troubling fact of biodiversity loss also surfaced in the early 1980s. Biologist E. O. Wilson gave this now famous defense of the Endangered Species Act of 1982 before Congress:

---

[85] Ibid., 44.

[86] Ibid., 46. It is estimated that about 80,000 chemicals are circulating today. What is frightening is that "few toxicity data are publicly available for most of these chemicals... The U.S. Environmental Protection Agency has reviewed the data available on 2,863 commercial-scale synthetic chemicals. For 43 percent there was a complete absence of toxicity data; full testing and data were available for only 7 percent" (ibid., 47). The European Commission, on the other hand, has ordered extensive testing on a scale unknown in the US.

[87] Arguably the worst chemical disaster occurred in Bhopal, India, in 1984. For a harrowing account of how the people of India (and its poor, primarily) have suffered from free-market deregulation, see Vandana Shiva et al., eds., *Licence to Kill: How the Holy Trinity—the World Bank, the International Monetary Fund and the World Trade Organisation—Are Killing Livelihoods, Environment and Democracy in India* (New Delhi: Research Foundation for Science, Technology and Ecology, 2000).

[88] CFCs, or cholofluorocarbons, present in industrial solvents, refrigerators and in the release of aerosol cans, were blamed in a 1974 report for damaging the earth's ozone shield. Speth cites this case as the greatest success story of the environmental movement. Indeed, many countries took energetic measure to ban these products and by the late 1970s world production of CFCs had dramatically decreased (*Red Sky at Morning*, 54).

> The worst thing that can happen during the 1980s is not energy depletion, economic collapse, limited nuclear war, or conquest by a totalitarian government. As terrible as these catastrophes would be for us, they can be repaired within a few generations. The one process ongoing in the 1980s that will take millions of years to correct is the loss of genetic and species diversity by the destruction of natural habitats. This is the folly our descendants are least likely to forgive us.[89]

Over twenty years later, the global scientific community is not so much concerned about specific endangered species than it is about conserving ecosystems—which is far more serious in the long run. The biologist at the head of the international Millenium Ecosystem Assessment, Walter Reid, warns that "ecosystem change…[and loss] of the ability of these systems to meet human needs"—much more than the problem of species extinction—has far-reaching "practical consequences for human livelihoods and U.S. interests."[90] As humans we draw our very life from the ecosystems that nourish us. They are "the productive engines of the planet…from the water we drink to the food we eat, from the sea that gives up its wealth of its products, to the land on which we build our homes."[91] Ecosystems are what provide us with pure water and air, supervise the process of decomposition and recycling of nutrients and protect the biodiversity that is vital to their sustainability. Yet life is seeping away from the system as we speak:

> At this moment, in all nations—rich and poor—people are experiencing the effects of ecosystem decline in one guise or another: water shortages in the Punjab, India; soil erosion in Tuva, Russia; fish kills off the coast of North Carolina in the United States; landslides on the deforested slopes of Honduras; fires in the disturbed forests of Borneo and Sumatra in Indonesia.[92]

Who or what is responsible for this? Undoubtedly, Speth is right in saying that behind the answer lies "a reality of immense complexity."[93] Quoting others, he opines that the Seattle demonstrators were not objecting to globalization as such, but the neoliberal, market-driven kind. Then

---

[89] Ibid., 24.

[90] Walter Reid, "Biodiversity, Ecosystem Change, and International Development," *Environment* 43.3 (2001): 22, cited in Speth, *Red Sky at Morning*, 26.

[91] World Resources Institute, *World Resources, 2000–2001* (Washington, DC: WRI, 2000), 3-4, cited in Speth, *Red Sky at Morning*, 26.

[92] World Resources Institute, *World Resources, 2000–2001*, viii, cited in Speth, *Red Sky at Morning*, 26.

[93] Speth, *Red Sky at Morning*, 141.

he discards for a moment his characteristic caution and cites approvingly Martin Khor, director of Third World Network, in his assessment of the world community's failure to implement the clear guidelines emanating from the 1992 Rio Earth Summit. Khor does not see this failure coming from the paradigm of sustainable development itself. Rather, "intense competition came from a rival—the countervailing paradigm of globalization, driven by the industrialized North and its corporations, that has swept the world in recent years."[94] In fact, in several places of his book, Speth blames the Bush administration for its dismal record on ecological integrity, and especially its reneging on the 1997 Kyoto Agreement.

Beyond the pointing of fingers, at stake is a vision of the world and the possibility of creating a common perception of who we are as humans in relation to our physical environment. The modern paradigms of unlimited economic expansion and the commodification of the commons are, when all is said and done, at the root of our current predicament.

Beyond Modernity: A Holistic Vision of Eco-Justice

As might be expected, Vandana Shiva is not as sanguine as some about the chances of the present world system's ability to reform itself. In a recent book, she argues that "[t]he water crisis is the most pervasive, most severe, and most invisible dimension of the ecological devastation of the earth."[95] Already a prominent physicist in her native India while in her thirties, Shiva was shocked by the devastation inflicted by the greed of multinationals and their local allies, and particularly the foresting industry in the Himalayas. She recounts:

> Cherapunji in northeast India is the wettest region on earth, with 11 meters of rainfall a year. Today, its forests are gone and Cherapunji has a drinking-water problem. My own transition from physics to ecology was spurred by the disappearance of Himalayan streams in which I played as a child. The Chipko movement was launched to stop the destruction of water resources through logging in the area.[96]

---

[94] Martin Khor, "Globalization and Sustainable Development: The Choices Before Rio + 10," *International Review for Environmental Strategies* 2.2 (2001): 210, cited in Speth, *Red Sky at Morning*, 142.

[95] Vandana Shiva, *Water Wars: Privatization, Pollution and Profit* (Cambridge, MA: South End, 2002), 1. Since 1970, she reports, "the global per capita water supply has declined by 33 percent." It is not just a result of population growth, she contends, but "it is exacerbated by excessive water use as well" (ibid., 2).

[96] Ibid., 3.

The systematic elimination of the forests triggered a chain of negative results, some more predictable than others: soil erosion, mud slides, flooding of the plains, the unsustainability of the ecosystem due to the firs planted in place of the original oaks, and the beginning of more extreme storms. Indeed, deforestation, industrial agriculture, overmining and aquaculture have unleashed an era of ruthless climate change. In the state of Orissa, Shiva describes the havoc wreaked by the 1999 cyclone: nearly two million houses destroyed; extensive destruction of paddy crops in twelve coastal districts; all of the banana and papaya plantations destroyed; 80 percent of coconut trees uprooted or cut in two, and 15,000 ponds either salinated or contaminated. In addition, the cyclone killed more than 300,000 cattle and, by some estimates, over 20,000 people. Two years later, Orissa experienced its worst drought on record, followed by its worst flood, severely affecting more than six million people.

The most recent report of the Intergovernmental Panel on Climate Change (IPCC) involved the collaboration of over one thousand scientists. According to the "Climate Change 2001" report, the climbing temperatures of the earth "will lead to crop failures, water shortages, increased disease, flooding, landslides, and cyclones." Insurance companies are now greatly concerned about the issue: "The Global Commons Institute has assessed that damages due to climate change could amount to $200 billion by 2005" and that by 2050 "the property damage could reach $20 trillion."[97]

Much of this can be attributed to the avarice of unregulated business and commerce. The multiplication of shrimp ponds (destined for the enjoyment of the rich Westerners), for instance, along the coast of India and Bangladesh, account for the systematic destruction of the mangroves that once stood between ocean and land, forming a natural barrier against tides and storms and absorbing the nitrates and phosphates of waters flowing into the ocean. Yet besides industrial greed, one would also have to indict the Western drive to subdue nature in the form of dams and large-scale irrigation. Already in the western United States specialists deplore the building of the great dams. In these states, "irrigation accounts for 90 percent of total water consumption. Irrigated land increased from four million acres in 1890 to nearly 60 million in 1977... These areas are also affected by soil salinity because of salts dumped into rivers when irrigation waters drain." The rising salinity of the soils decreases the fertility of the soil, and that problem compounds with time.

---

[97] Ibid., 42.

In California's artificial "green belt," the San Joaquin Valley, "crop yields have declined by 10 percent since 1970, an estimated loss of $312 million annually."[98]

As always, the result of reckless technologies in the hands of corporations and governments that have privatized that which from time immemorial belonged to all people has created the greatest suffering among the world's poor. Yet, as Shiva shows, a revival of indigenous technologies and community management of water resources is noticeable, and it spells hope for the future. She explains: "cultures that waste water or destroy the fragile web of the water cycle create scarcity even under conditions of abundance. Those that save every drop can create abundance out of scarcity. Indigenous cultures and local communities have excelled in water conservation technologies."[99] A vision urgently needed today is a mainstay of India's Hindu culture. For Indians, every river is sacred.

Recall that at the heart of the modern (and Western) expansionist paradigm launched in 1492 was the idea of collective ownership of the world (due to the superior rights God had granted to Christian kings) and a nascent capitalist ideology—expressed in the initial charters and patents and in the preference of private property over that of community management of the commons—progressively gave birth to the corporations. These, in turn, propelled the Industrial Revolution that empowered the European Empires to establish and exploit their far-flung empires. As colonial independence movements gathered momentum in the early twentieth century, the inherently expansionist tendencies of capitalist accumulation—coupled with growing nationalism in the wealthy states of Europe—created a tension that eventually exploded in 1914, dragging the whole world into Europe's civil war, and then into a second one in 1939. After World War II, however, what was supposed to have been a process of decolonization quickly gave way to a new kind of political and economic colonization of the so-called Third World—the raw powers of modernity unleashed in two different modes, both equally voracious when it comes to devouring natural resources and polluting the commons of humanity—water and air.

When the Second World collapsed in 1989, the neoliberal, free-market fundamentalist brand of capitalism unleashed in the 1970s now became the ruling ideology of the United States, Japan and their European allies, and the transnational corporations merged back and forth, growing into

---

[98] Marq De Villiers, *Water: The Fate of Our Most Precious Resource* (New York: Houghton Mifflin, 2000), 143, cited in Shiva, *Water Wars*, 114.
[99] Ibid., 119.

behemoths and reaching everywhere. Speth in his book, *Red Sky at Morning*, is cautiously critical of this system, if only, I surmise, because he is an insider who wants to convince American opinion leaders and politicians to change their ways. Indeed, back in 1977, President Jimmy Carter asked the State Department and the Council on Environmental Quality (CEQ) to study the "probable changes in the world's population, natural resources and environment through the end of the century." Speth, as one of the three members of CEQ, was soon to become its chair. The first part of the report was presented to Carter in 1979 and a separate report by the National Academy of Sciences, the "Charney Report," bolstered their conclusions. From then on, Speth and his colleagues focused their attention on climate change, producing in 1981 a report that detailed the potentially disastrous effect of the global production of greenhouse gases and made detailed recommendations for an international effort to curb this trend. Significantly, this report contains a vision of the world that borders on the theological:

> Whatever the consequences of the carbon dioxide experiment for humanity over the long term, our duty to exercise a conserving and protecting restraint extends as well to the community of life—animal and plant— that evolved around us. There are limits beyond which we should not go in disrupting or changing this community of life, which, after all, we did not create. Although our dominion over earth may be nearly absolute, our right to exercise it is not.[100]

With the knowledge we now have of the past, as human occupants of this earth and as a species embedded in it and totally dependent on its well-being, we dare not ignore the tell-tale signs of devastation ahead. This is the message that scientists from the International Geosphere-Biosphere Program want to pass on to all of us today:

> The evidence is now overwhelming that [rising temperatures] are a consequence of human activities... [W]e are now pushing the planet beyond anything experienced naturally for many thousands of years. The records of the past show that climate shifts can appear abruptly and be global in extent, while archaeological and other data emphasize that such shifts have had devastating consequences for human societies. In the past, therefore, lies a lesson.[101]

---

[100] U.S. Council on Environmental Quality, *Global Energy Futures and the Carbon Dioxide Problem* (Washington, DC: Government Printing Office, 1981), cited in Speth, *Red Sky at Morning*, 5.

[101] Keith Alverson et al., *Environmental Variability and Climate Change* (International Geosphere-Biosphere Program Science Series 3, 2001 [http://www.igbp.net/documents/resources/science-3.pdf]), cited in Speth, *Red Sky at Morning*, 60.

Unfortunately, it is still business as usual, and in the corporate and finance centers of the world the modern worship of economic growth continues unabated. We seem oblivious to the fact that humankind has now passed the "historical transitional point in the evolutionary development of our species from living in a world of open frontiers to living in a full world."[102] Speth tabulates the growth statistics of the last three decades:

- Global population   up 35 percent
- World economic output   up 75 percent
- Global energy use   up 40 percent
- Global meat consumption   up 70 percent
- World auto production   up 45 percent
- Global paper use   up 90 percent
- Advertising globally   up 100 percent[103]

The science of ecology reminds us that as human beings we share the same planet. This fact takes on greater meaning in the twenty-first century as we realize that not only do the rich of today have a moral obligation to the poor, but also to all the future generations who will inherit a more toxic, infertile and inhospitable planet. Already in 1968, Kenneth Boulding was chiding his fellow Americans that they considered the earth a vast cowboy prairie. Manifest destiny—beyond its racist implications—assumes the modern paradigm of limitless frontiers and resources to exploit. No, says Boulding, the earth is more like a spaceship. David Korten extends Boulding's analogy: "Life on a spaceship can be sustained only through the cooperation of all of the spaceship's inhabitants… No increase in economic output in the spaceship can be counted as an advance unless it is based on sustainable processes *and* translates into justly distributed benefits for the spaceship's inhabitants."[104]

In a similar vein, Boutros Boutros-Ghali, then General Secretary of the UN, spoke at the 1992 Rio Earth Summit about a needed paradigm shift in the self-consciousness of the world community: "Every new triumph over nature will in fact be a triumph over ourselves. Progress, then, is not necessarily compatible with life; we may no longer take the logic of the infinite for granted." He then went on to define "sustainable development": "Development that meets the needs of the present as long as resources are renewed, or, in other words, that does not compromise the

---

[102] Korten, *When Corporations Rule the World*, 28.
[103] Speth, *Red Sky at Morning*, 20-21.
[104] Korten, *Getting to the 21st Century*, 37 (emphasis original).

development of future generations."[105] This implies shifting to a con-servation mode. Among other implications, this means ensuring that indigenous people, especially those who have not yet been forced into the urban centers, conserve their natural resources so as to allow them to "preserve and enhance their quality of life."[106]

Before crashing into the double wall of unsustainability of the present world—social injustice and ecological collapse—we can make a choice, as people from many cultures and perspectives, and especially as Muslims and Christians, and opt for dialogue.

I end here simply with the theological questions that will preoccupy us for the rest of the book: what is the connection between the human trusteeship and our current predicament? How do we interpret our human mastery over the rest of creation? And what of our "dominion" of the earth, which under the modern paradigm has meant tearing down, destroying and subverting its God-given balance and fertility?

Leaving behind the modern paradigm that led us to colonialism, neo-colonialism and ecological devastation, we turn to a new, more holistic one, which in a postmodern world must seek to include both "community responsibility and ecological balance." Rasmussen believes this has the merit of pointing "to webs of social relationships that define human community, together with ecosystem, webs and the regenerative capaci-ties of both human and ecosystems communities."[107] Earth community, as the purpose and goal of creation, includes humankind and "otherkind." Issues of social justice necessarily involve ecological considerations.[108]

Maurice Strong, the Secretary-General for the UNCED, the United Nations branch that hosted the non-governmental consultations at the Rio Summit of 1992, is also a man of faith. In his foreword to one of the best evangelical books on faith and ecology, he writes that the mandate of Genesis 1 was not only to "subdue" the earth but also to "replenish" it. He then expresses the hope that

---

[105] Cited in Wesley Grandberg-Michaelson, *Redeeming the Creation: The Rio Summit: Challenge for the Churches* (Geneva: WCC, 1992), 7.

[106] As Korten notes, "Too often the exploitation of natural resources for export deprives local people of their land and livelihoods—in order to repay loans that benefited only the rich by catering to the overconsumption of wealthy foreign consumers" (*Getting to the 21st Century*, 220).

[107] Rasmussen, *Earth Community, Earth Ethics*, 131.

[108] See Rasmussen (ibid., 75-78) on global patterns (and in the US) of systematically dumping toxic waste in the "neighborhoods" of the poor and people of color.

the great religions of the world—especially Islam, Judaism and Christianity, which believe in the God of creation—may rise to their responsibilities and empower this movement with the commitment and support of the global communities they represent. I remain convinced that religious conviction, and the disciplined behavior that flows from it, have the unique potential to produce the resolve and motivation required to "replenish the earth."[109]

It is this overall purpose of "eco-justice"—caring for the poorest and weakest in the family of humanity as we care for the earth itself—that must fill in the contents of "the human caliphate" and emphasize harmony rather than mastery in our mandate to manage the earth.

[109] Dayton W. Roberts, *Patching God's Garment: Environment and Mission in the 21st Century* (Monrovia, CA: MARC, 1994), iv.

Chapter 2

BEYOND MODERNISM:
TIME, SPACE AND THE SELF

Whether by design or by default, theology is always pieced together in context. In order to define a theology of human deputyship in a world increasingly shrunk by the forces of postmodernity, one must take into account the current intellectual climate. The previous chapter argued that the modern paradigm of human domination and exploitation of the earth through technological prowess and unlimited economic growth is now morally bankrupt, to the point where if this planet is to go on sustaining the life of its inhabitants, some drastic changes must be implemented. In particular, the process of commodification of all aspects of human and non-human existence, begun with the Western colonial conquests and carried to its extreme with the present neoliberal capitalist system writ on a global scale, must be reversed. This is especially urgent from a Muslim and Judeo-Christian perspective,[1] for all human beings are held accountable by their Creator to manage the wealth of creation among all in a just manner.

This chapter and the next aim to show how the modern paradigm of growth and possession has identifiable philosophical roots in the Western Enlightenment project. Only by identifying those roots can we see why and how the seeds of a new tree must be planted. In the biblical book of Proverbs—that collection of wisdom sayings common to much of the ancient Near East—we read, "Where there is no vision the people perish: but he that keepeth the law, happy is he."[2] The overall theme of the

---

[1] See Richard W. Bulliet, *The Case for Islamo-Christian Civilization* (New York: Columbia University Press, 2004), for an excellent historical argument for the affinity of Islamic and Christian theological, sociological, intellectual and political manifestations over the centuries. In fact, the term "Judeo-Christian" came into vogue only in the post-World War II (post-holocaust) period.
[2] Prov. 29:18 (King James Version).

chapter in which this saying occurs, ch. 29, is the contrast between a society with wicked rulers ("the people groan under an intolerable burden as injustice and violence flourish unchecked") and one with godly rulers.[3] Striking a familiar theme in the Torah, v. 7 emphasizes "the rights of the poor," which here means, "to actively promote justice for the poor."[4] The above quoted v. 18 uses the Hebrew word *chazon*, literally a "prophetic vision."[5] No doubt "vision" here encompasses all aspects of revelation, prophecy and law, which are "essential to the harmony and well being of society and the individuals within it."[6]

Appropriately, the word vision sums up nicely what Jews, Muslims and Christians understand as the role prophets are to play—whether Isaiah, Jeremiah, John the Baptist or Muhammad[7]—calling people to repentance by boldly highlighting individual and corporate sins. This also explains my reliance on "critical theory" in this chapter.[8] What is more, prophets go beyond the stark diagnoses of social ills by offering visions of a better world governed by the wise guidelines God provides.

In a strikingly similar way to Isaiah, Martin Luther King Jr proclaimed "I have a dream" to the crowd drawn to Washington by its unquenchable thirst for justice and dignity. Almost five years later, just the day before he was assassinated, the weary yet ever combative King boarded a plane in Atlanta, Memphis-bound, in order to carry out the Poor People's Campaign, a march that was to take place on Monday of Holy Week. A bomb threat—which, according to the pilot was directed at him—grounded the plane for over an hour on the runway tarmac. Nonetheless, the day proceeded uneventfully, though there was talk about a violent storm expected that evening. Three thousand people gathered for a rally that night at a black Pentecostal church. At first, King sent Ralph Abernathy, because he was exhausted. Shortly after, King received a phone

---

[3] John Barton and John Muddiman, eds., *The Oxford Bible Commentary* (Oxford/ New York: Oxford University Press, 2001), 421.

[4] Ibid.

[5] See, for instance, how it is used repeatedly in Ezek. 12 and 13.

[6] *The Oxford Bible Commentary*, 421.

[7] As a "warner" (*nadhīr*) who proclaimed the judgment of the Creator God—the One and the Almighty—on the idolatrous and socially oppressive society of his day, Muhammad, for me, roughly fulfills the role of the Hebrew prophets he so often refers to in the Qur'an. By this I do not mean to say that he is in all respects their spiritual heir. As a follower of Jesus and in the light of the New Testament writings, there arises for me as well a necessary dissonance in this comparison. See Appendix D for a fuller statement on how I conceive Muslim–Christian dialogue.

[8] See below for an explanation of this term.

call from Abernathy. People had risked the heavy rains, winds and thunderclaps, for only one reason, he pleaded. King obliged. This turned out to be one of his most powerful orations. I offer the following excerpt to show how King instinctively dons the mantle of the prophet—identifying here with Moses in his last hour:

> Well, I don't know what will happen now. We've got some difficult days ahead. But it doesn't matter with me now. Because I've been to the mountaintop. And I don't mind. Like anybody, I would like to live a long life. Longevity has its place. But I'm not concerned about that now. I just want to do God's will.
>
> And He's allowed me to go up to the mountain. And I've looked over. And I've *seen* the promised land. I may not go there with you. But I want you to know tonight, that we, as a people will get to the promised land. And I'm happy, tonight. I'm not worried about anything. I'm not fearing any man. Mine eyes have seen the glory of the coming of the Lord.[9]

That "promised land" was an America in which racial and class walls would come crashing down and all would work together for the common good while tearing down the fiendish idols of racism, materialism, poverty and militarism. That vision, that burning sense of "utopia" or hope for which one is willing to sacrifice, rejoins an aspiration people the world over, including Muslims and Christians, find welling up within their souls.

This book argues that such a vision, which for many has no religious overtones, can be reinforced by the contribution of Muslims and Christians through a common reflection on humankind's vocation at creation— the vision of the inherent dignity of each and every person, and a mandate to manage together the earth's bounty. While in the previous chapter I focused on the bitter fruit of modernity, in this chapter and the next two I inspect its philosophical underpinnings, which could lead in one of two directions.

According to the more extreme versions of postmodernism, the human person is mired in a murky pool fed by the cultural construction of language, thought and practices. Personal agency was a modern chimera, we are told, and social criticism—if possible at all—will have to content itself with local narratives that articulate their own moral standards, which can only be validated for that time and place. Cultures and communities may attempt to communicate with one another, but since there is no

---

[9] Stewart Burns, *To the Mountaintop: Martin Luther King Jr.'s Mission to Save America: 1955–1968* (New York: HarperSanFrancisco/HarperCollins, 2004), 443-44, emphasis original.

more "grand narrative" such as the modern myth of unlimited progress, we are left with a polycentric world in which difference remains the norm and any rational coordination and conversation becomes unthinkable.

Yet the ethos of the Enlightenment can be recast in another mold. While leaving behind the arrogant certainties of the modern disembedded and autonomous self that seeks to control both nature and other peoples in light of its own interest, we could also take a humbler approach that still kept alive the hope of human emancipation and the possibility of rational interaction among the world's vastly divergent cultures and groupings.

My contention here is that this is a necessary theoretical task to be undertaken. What is the human self? How is it related to other selves? Does reason construct language, or does language determine reason? How do I know there is a world outside of me? If there is—and this has always been a human assumption—then how can I know it? These are some of the questions raised and discussed in this and the next chapter.

The preceding chapter ended with the profoundly spiritual vision of eco-justice. Any theological vision, however, must be articulated within a particular worldview, at the deepest level of one's culture and core beliefs.[10] At the same time, if it has any chance of transcending one's own community's context, its presuppositions will have to be laid out in some kind of rational discourse. Though much has been written in the last three decades about "the end of philosophy," I will contend that any ethical reflection that seeks to encompass humanity as a whole and provoke a meaningful conversation at that level will have to enlist the help of philosophy. Here, then, is a simple working definition of that discipline by philosopher E. J. Ashworth:

> Philosophy aims at intellectually responsible accounts of the most basic and general aspects of reality. Part of what it is to provide an *intellectually responsible* account, clearly, is for us to make sense of our own place in reality—as, among other things, beings who conceive and formulate descriptions and explanations of it.[11]

---

[10] Evangelical philosopher David K. Naugle traces the intellectual history of the concept of worldview (*Weltanschauung*) to nineteenth-century German idealism, and shows how it has been transformed and reworked more recently across the disciplines of philosophy, psychology, sociology and cultural anthropology. He also raises the philosophical questions that are germane to the present work in a way that, for the most part, I find helpful. In Chapter 4 I will be quoting from his book, *Worldview: The History of a Concept* (Grand Rapids/Cambridge: Eerdmans, 2002).
[11] "Philosophy of Language," in *Routledge Encyclopedia of Philosophy*, gen. ed. Edward Craig (London/New York: Routledge, 1998), 6:409, emphasis added.

It would be "irresponsible," then, for Christians or Muslims to think theologically about action in a "postmodern" world without interacting with the philosophical currents that guide and inform its thinking. But why, the reader might ask, go down this road, since the Enlightenment and its aftermath are all Western phenomena? My answer is that for good or for ill we are all interconnected and that philosophical ideas have a way of filtering down into our worldwide mass culture. Put differently, McWorld and the identity politics of ethnic or class-based groups are often intertwined, particularly as their conversations are shaped through the Internet. Add to this my own belief that an effective advocacy for a more just world must deal with issues of international law that in turn necessarily grow out of the complex interaction between local cultures, specific religious convictions, national interests and ethical values capable of sustaining intercommunal and international dialogue. As I see it, a Muslim–Christian dialogue that will impact the world in a fruitful way will have to confront some of the basic philosophical questions I raise in this and the next two chapters.

Ashworth's definition of philosophy rightly showcases the central importance language has come to have in contemporary philosophical debates about humankind. Thus the necessary triangle of key elements to be examined, each in turn and also in relation to one another: self, language and world. Further, Ashworth's use of the term "reality" takes on a crucial role in the next chapter, and is related to the human ability to apprehend what is outside its own consciousness. The answer to this question determines to a large extent how we understand science (from the natural or physical sciences to the social sciences), organize society, understand ethics while applying them to politics, and read our sacred texts.

Undeniably, our twenty-first century international context has been shaped by certain emancipatory ideals of modernity, notably those of freedom, equality, democracy and human rights. The theology of hope adopted from the start of this project compels us to explore to what extent these utopian themes can be retrieved from a postmodern critique of the "Enlightenment project." To this end, I return to David Harvey for some important insights on the impact of economic realities on human thought in culture, and in the next section I designate Jürgen Habermas and Seyla Benhabib as particularly useful allies in our attempt to think through a Muslim–Christian theology of creation in the predominantly secular mindset of international law and society today.

The main focus of this chapter, then, is the plausibility of the modern view of the autonomous, rational self. The next chapter considers the concepts of history, metaphysics and language, as they have been singled out for critique by postmodern scholars. In the end, all four elements will appear as deeply interrelated. My contention is that this seeming detour into postmodernism leads us (both Muslims and Christians) into a more fruitful theology of creation.

The postmodernist critique of the autonomous self has been leveled on a variety of fronts, depending on the discipline through which a writer comes to the issue. As the discussion proceeds, I shall present several perspectives on the modern project from a variety of disciplines. My aim is to offer the reader diverse approaches to this topic and connect them so that their reoccurring themes reinforce each other progressively, much like the web a spider spins. This admittedly postmodern way of proceeding intentionally reveals an epistemic conviction that runs throughout this book and that I present in the next chapter: critical realism within a postmodern framework of holism, a position with clear affinities with the original Semitic worldview in which Judaism, Christianity and Islam were shaped, yet also more in tune with the post-Enlightenment world in which we now live.

Critical realism involves the idea that true knowledge of the world around us, including human society, is possible, but will always remain tentative and incomplete; and further, that reality cannot be compartmentalized without imposing serious distortions in our understanding of it. Thus, to look at several different theories will inevitably widen and enrich our perspective, even if only some aspects are retained for further use. This approach leads us, as Stephen Toulmin perceptively demonstrated, to adopt a more modest and tolerant view of human knowing, much like the humanist convictions of such sixteenth-century Renaissance thinkers as Erasmus, Montaigne and Shakespeare.[12] In the end, it may be that only the dogmatic, ultra-rationalist side of Descartes's seventeenth-century modernism and its goal of human exploitive domination of nature need to be jettisoned.

Modernity, Capitalism, and the Compression of Time and Space

In this first section I look at the self in society and ponder how its self-reflection might have been transformed by its socioeconomic environment. Anouar Majid has complained that many social critics—Muslim and non-Muslim alike—ignore the economic and political ramifications

---

[12] Stephen Toulmin, *Cosmopolis: The Hidden Agenda of Modernity* (Chicago: University of Chicago Press, 1992).

of today's global corporate capitalism. Yet because "the dictates of profit" have overtaken just about all other social values, "the dream for a more enlightened human civilization is diminishing," and the global stage seems to be set for another wave of violence similar to that of the first world wars "amid a widespread historical amnesia." He explains, "It is this ominous prospect that justifies a reevaluation of the world's indigenous traditions, including Islam."[13] One of the causes for this "historical amnesia," argues David Harvey, is the compression of time and space induced by capitalism under the conditions of modernity.

As we have seen, Giddens's sociological analysis captures well how it "feels" to be caught on the wild ride of modernity's juggernaut. One might classify his approach as psycho-social. The key element that drives it onward is the "emptying of time and space"—the inevitable dislocation of people's rootnedness in localities within traditional societies. Though modernity's roots are characteristically Western, this runaway engine, now in its "high modernity" stage, is unstoppable and "inherently globalising." Has it affected people's thinking? Indeed, for Giddens, human thought has "cut loose from its moorings in the reassurance of tradition," with the only dominant "vantage point" left being the "dominance of the West."[14] High modernity is precisely this era in which human society's profoundly reflexive self contemplates the future with a certain "utopian realism" against the backdrop of high-consequence risk—including nuclear self-annihilation. Despite the atmosphere of doom in some quarters, asserts Giddens, we see the stirrings everywhere of an "emancipatory politics linked to 'politics of self-actualisation' which seek to balance the welfare of the individual with the welfare of the world community."[15]

Giddens's cautiously optimistic view is, manifestly, politically conservative—the risks may be high, but people will find a way to work together. The system is basically sound. Though he does offer some analysis of the contributing causes of capitalism since the industrial revolution in the West, Giddens has no interest in pointing out the growing disparities in the recent expansion of neoliberal economics. Lacking in his optimistic picture is any analysis of the macro dimensions of global capitalism and its likely impact in the near future, either on the centers of capital wealth, or on its peripheries. The same too can be said of sociologist Bryan Turner, who differentiates postmodernism as "an alternate set of social theories," and postmodernity as a global configuration that erases the modern differentiation of spheres (as defined by

---

[13] Majid, *Unveiling Traditions*, 152.
[14] Giddens, *The Consequences of Modernity*, 176.
[15] Ibid., 173.

Weber), together with "an erosion of certainty in the value of economic capitalism and a growing awareness of the importance of environmental and green issues."[16]

David Harvey, by contrast, finds himself in the tradition of critical theory initiated by the Frankfurt School of the 1930s. Because of the rise of fascism and their distaste for the institutionalization of Marxism in the Soviet Union, these Marxian thinkers moved to the Institute for Social Research at Columbia University. Several relocated to Frankfurt in the 1950s, with Theodor Adorno as the leader of their movement. Thomas Docherty finds in the intellectual production of the Frankfurt School "a major source for the contemporary debates around the postmodern." In fact, long before Jean-François Lyotard, it was Adorno and Max Hork-heimer who in their 1944 work, *Dialectic of Enlightenment*, underscored the totalitarian nature of the modern project.[17] Knowledge for the Enlightenment philosophers was to enable humankind to master nature. It was irreversibly tied to power. Though the belief of mastery through technology is merely an illusion, it nonetheless manifested itself in tangible acts of oppression and slavery with regard to those with less power, that is, the working classes and colonized peoples of the world.

In the same line of reasoning, Harvey sharpens his critique through the impact of material elements on people's perception of time and space. His working assumption is that ideas are shaped by people's physical environment more than the other way around. As a case in point, the maps of the Renaissance are radically different from previous maps drawn in medieval times. The latter strike us as emphasizing human interests and preferences in often sensuous ways. The former, however, informed by the work of travelers to new continents, revealed a world that was at the same time finite and knowable. Thus, for the first time, maps aimed to be rational and objective. "Geographical knowledge became a valued commodity in a society that was becoming more and more profit-conscious."[18] On the one hand, as wealth and power increased, capital became associated with personal knowledge, which in turn yielded a new control over space. On the other hand, with new information about competing trade and military powers in the wider world, a certain sense of vulnerability crept into the picture.

---

[16] Turner, *Orientalism, Postmodernism and Globalism*, 198. The last sentence presents a thesis that I will pick up again: green activism is indeed a "postmodern" phenomenon.

[17] Docherty, "Postmodernism: An Introduction," 5.

[18] Harvey, *The Condition of Postmodernity*, 244.

This revolution in people's understanding of time and space began to gain momentum. Harvey offers a dramatic illustration: a funnel-shaped figure includes four black and white world maps, each one smaller than the preceding one as one looks at the funnel from top to bottom. The first—the large one at the top—is dated "1500–1840," when the "best average speed of horse-drawn coaches and sailing ships was ten miles per hour. The second world map is considerably smaller. The dates are "1850–1930" and the scale is controlled by the speed of locomotives and ships powered by steam engines (36 miles per hour). The second map is followed by a much smaller one representing the 1950s: propeller aircraft further reduced the world's size at speeds of 300 to 400 miles per hour. Finally, at the bottom of the funnel sits the smallest world map. The globe has now shrunk to the dimensions of a jet passenger aircraft of the 1960s that travels at speeds of between 500 and 700 miles per hour.[19]

But it was not only that space was being compressed; it was being rationalized in a way that was bound to have theological consequences. The baroque penchant for "twisting perspectives and intense force fields constructed to the glory of God" in architecture began to give way to a new landscape brashly ordered and rationalized by men proclaiming human reason's imperium over creation—the essence of the Enlightenment project. In Harvey's description that follows, a theological question emerges as to how to assess critically humankind's mastery of nature in the modern mindset:

> What many now look upon as the first great surge of modernist thinking, took the domination of nature as a necessary condition of human emancipation. Since space is a "fact" of nature, this meant that the conquest and rational ordering of space became an integral part of the modernizing project. The difference this time was that space and time had to be organized not to reflect the glory of God, but to celebrate and facilitate the liberation of "Man" as a free and active individual, endowed with consciousness and will… Enlightenment thinkers similarly looked to command over the future through powers of scientific prediction, through social engineering and rational planning, and the institutionalization of rational systems of social regulation and control.[20]

---

[19] Ibid., 241.

[20] Ibid., 249. This is precisely Toulmin's image of "cosmopolis" (cosmos + polis, the realm of nature understood in Newtonian terms as a machine, and human society operating under different laws). To understand some of the powerful economic forces involved in this race to master nature, see Dava Sobel, *Longitude: The True Story of a Lone Genius Who Solved the Greatest Scientific Problem of His Time* (New York: Walker, 1995). Until the early eighteenth century, scores of ships sunk and thousands of sailors perished because there was yet no way of measuring

Even before the deist proclamations of the eighteenth-century *philosophes*, many began to realize that human reason might, if not supplant revelation, at least carve out a much greater sphere of autonomous human activity. There is no doubt that the scientific revolution—later spilling over into the Industrial Revolution—was leading many to rethink their theological postulates. Equally, as the autocratic politics of monarchs was seen to stand in collusion with the Church's agenda, the bourgeois class was more and more inclined to imagine "a new, more democratic, healthier, and more affluent society." Again, technology seemed to point the way: "Accurate maps and chronometers were essential tools within Enlightenment vision of how the world should be organized."[21]

If the abstraction of space through the deployment of scientific tools could be carried out in order to rethink the world and impose order on the vast diversity of populations, cultures and their physical environments (with the Western powers as the masters), the same kind of operation could be applied to time.[22] By means of the mechanical division of time provided by the swing of the clock's pendulum, Westerners thought increasingly of time as a linear projection, whether looking backwards or forwards. History was becoming a scientific discipline for the first time, with experts suggesting to their rulers that past patterns of events could provide the means rationally to project, predict and manipulate the future. In the economic sphere as well, one could calculate capital's rate of return, hourly wages and interests rates, and thereby make "enlightened" micro- and macroeconomic decisions in a capitalist context. "What all of this adds up to," muses Harvey, "is the by now well accepted fact that

latitude. The kings of Europe all set a bounty for the one who would solve this riddle. The highest prize was fixed by the British Parliament in 1714, "several million dollars in today's currency," for a "Practical and Useful" method of calculating longitude (ibid., 8). The brilliant English clockmaker John Harrison, after forty years of intrigue and fiendish opposition by the aristocracy, finally claimed his prize in 1773.

[21] Ibid.

[22] As reported in the first chapter, this rationalization of space had strong political implications in the conquest of indigenous populations. As Frederic Jameson wrote recently, "the gradual colonization of space" is an ongoing process today in Eastern Europe and Israel, and its genesis can be pinpointed rather precisely: "it is the moment in which a Western system of private property in real estate displaces the various systems of land tenure it confronts in the course of its successive enlargements (or, in the European situation itself, from which it gradually emerges for the first time in its own right)" (*The Cultural Turn: Selected Writings on the Postmodern 1983–1998* [London/New York: Verso, 1998, 65]).

Enlightenment thought operated within the confines of a rather mechanical 'Newtonian' vision of the universe, in which the presumed absolutes of homogenous time and space formed limiting containers to thought and action." Not only was the world shrinking, but the human mastery of nature could now be measured, quantified, right down to the economic issues of trade and finance. This is precisely what brings us to his thesis of time–space compression—the process that unfolded in the late nineteenth and early twentieth-century modern thought.

The Newtonian picture of the world as quantifiable mechanics that gave the West its scientific revolution, nevertheless proved untenable. Here, economics were slowly putting pressure on a society's worldview, and indeed, on its very social make-up. According to Harvey, the "absolute" reference points of time and space that had anchored modern thought growing out of the Renaissance began to break down with the economic depression that paralyzed Britain in 1846–47 and engulfed the rest of the Western world thereafter. On the surface, the crisis was easily explainable: many of the banks and early financial institutions crashed all over Europe as the tension between credit (mostly for railroad construction) and the commodities that upheld its value (gold, jewelry, and so on) exploded. This was coupled with rising unemployment of the traditional classes of artisans whose trades were now seriously threatened by the rise of industrial production. Finally, a series of social revolts (later called "revolutions") spread to nearly all the major European cities.

Yet, at a deeper level, these traumatic shifts "created a crisis of representation."[23] The year 1848, proffers Harvey, was the first instance in which philosophers began to ask, "What time are we in?" Traditional and trusted reference points seemed to be drifting away with the tide. "Have we entered a new era?" Physical and social time had begun to part ways, after the Enlightenment had wedded them. New classes were emerging with more money, education and power. What of the old elites? What about the newly formed working classes, whose livelihood increasingly seemed to erode? Other uncertainties surfaced as well. The synchronic eruption of revolutions across the continent contrasted with the diachronic time of capitalist investment. "The certainty of absolute time and space gave way to the insecurities of a shifting relative space, in which events in one place could have immediate and ramifying effects in several other places."[24] For one thing, the universalizing theory of Marx's *Communist Manifesto* made sense to people struggling to explain the growing chaos. For another, the radical reorganization of the financial

---

[23] Harvey, *The Condition of Postmodernity*, 262.
[24] Ibid., 261.

and stock markets after 1850 never solved the inner contradiction between money functioning as credit and money tethered to commodities, between "money as a lubricant of exchange and investment" and "money as a measure and store of value."[25]

The new disjointedness in the economic and social spheres soon made its impact on cultural productions. From the 1850s on, modernism in the arts started to reflect this crisis of representation. Flaubert in his novels explored the problematic nature of language and its ability to denote time and space as experienced so differently by people. Manet launched impressionism—brushstrokes "that began to decompose the traditional space of painting and to alter its frame," seeking to uncover "the fragmentations of light and colour."[26] But this was only the beginning of that tight reactivity between time and space compression and evolving modes of representation in cultural production. The first Western crisis of capitalist overaccumulation of 1847 was resolved in part by a series of innovations in "temporal and spatial displacement": new systems of finance and corporate organization and new methods of production (increased specialization, fragmenting and de-skilling of the work force) and distribution (large department stores). More importantly, this was the colossal growth spurt of Western imperialism: "[t]he expansion of the railway network, accompanied by the advent of the telegraph, the growth of steam shipping, and the building of the Suez Canal, the beginnings of radio communication and bicycle and automobile travel at the end of the century, all changed the sense of time and space in radical ways."[27] Moreover, the rapid expansion of foreign trade and investment set the colonial powers on a competitive course that proved fatal in the long run.

The First World War completed a growing process of deterritorializing of global space. The form in which international space had been configured by the imperial powers was now reshaped by war at a time when other earth-shattering changes were taking place: Einstein's special theory of relativity was published in 1905 (his general theory in 1916); Ford set up his historic assembly line in 1913;[28] in the same year "the first radio signal was beamed around the world from the Eiffel Tower, thus emphasizing the capacity to collapse space into the simultaneity of

---

[25] Ibid., 262.

[26] Ibid., 263.

[27] Ibid., 264.

[28] By fragmenting tasks and distributing them in space so as to maximize the efficiency of production, Ford was in effect deploying "a certain form of spatial organization to accelerate the time of capital in production"—or, an acceleration of time "by virtue of the control established through organizing and fragmenting the spatial order of production" (ibid., 266).

an instant in universal public time";[29] in 1914 it was calculated that 38 billion telephone calls were made in the USA. The cultural repercussions of these innovations were not long in coming, opines Harvey. James Joyce and Marcel Proust wrote novels that played with concepts of space and time in surprising ways. Picasso, Delaunay and Braque began to fragment space in their paintings, thus setting up cubism as a movement for a deconstruction of traditional linear perspectives.

In philosophy as well, this is the time when Nietzsche put forward some bold theses, sending shock waves that reverberated far beyond the world of philosophy. He considered Hegel's grandiose project of Enlightenment reason as self-reconciling knowledge totally misguided. By contrast, "Nietzsche uses the ladder of historical reason in order to cast it away at the end and to gain a foothold in myth as the other of reason."[30] Ortega y Gasset follows Nietzsche's nihilism in propounding in 1910 a new theory of perspectivism: "there are as many spaces in reality as there are perspectives on it" and "there are as many realities as points of view."[31] It was in 1912 that Durkheim, generally considered the founder of modern sociology, published his *Elementary Forms of the Religious Life*. Though not embracing the radical subjectivism of Gasset, Durkheim nevertheless proposes that time is constructed socially and that, as a result, there must always be a multiplicity of spatial visions within and among different societies. Despite its coming to fruition in sync with the imperial conquests, this relativization of space could also serve a renewed Enlightenment ideal of democracy and emancipation. Certainly this theme had inspired at least part of Jefferson's homesteading system of land distribution in early American history, and it contributed to the popularity of the many World Exhibitions, from the Crystal Palace in 1851 to the grandiose Columbian Exhibition of Chicago in 1893. For Harvey, modernism from the start contained these contradictory polarities of internationalism and nationalism, parochialism and universalism.[32]

What is constant, then, is the gradually increasing compression of people's notions of time and space. For Harvey, this was a major factor behind the variegated cultural expressions of modernism in the first half of the twentieth century. It led as well to the cultural revolution of the 1970s—the postmodern "rise of aesthetic populism" as Jameson would

---

[29] Ibid.

[30] Jürgen Habermas, *The Philosophical Discourse of Modernity: Twelve Lectures*, trans. Frederick G. Lawrence, Studies in Contemporary German Social Thought (Cambridge, MA: MIT Press, 1987), 86. See also Harvey's commentary on Nietzsche's *The Will to Power*, in *The Condition of Postmodernity*, 15-20, 273-74.

[31] Harvey, *The Condition of Postmodernity*, 268.

[32] Ibid., 275-76.

have it.[33] What I find particularly fruitful in Harvey's theory—beyond his reading of the contemporary production of postmodern art and culture—is the link between the experience of time and space compression and the sociopolitical nature of today's global neoliberal capitalism. More specifically, it is the element of power emanating from its capital, Washington DC, that helps to make sense of where "we" are, as a world community.

This is the theme of Harvey's recent book, *The New Imperialism*, which opens with a reflection on the American/Coalition forces' invasion of Iraq in April 2003. The consensus of the many millions of people who protested this war in the months leading up to it was that it was less about democracy and more about Middle Eastern oil and the US bid to control it more effectively. But for Harvey oil is still at the surface of the issues at stake. One has to back up to WWII, which decisively marked the ebbing of the British empire and the corresponding flow of the American empire, even if it was never overtly discussed in those terms. As mentioned in the first chapter, the two significant shifts that consolidated US hegemony were (1) the 1973 financial crash that allowed the US to repatriate the Eurodollars that were used to run the global oil trade and enabled US banks to monopolize world finance now set loose from its traditional gold standard; and (2) the fall of the "iron curtain" in 1989. New York then became the center of the global economy.

Into this context Harvey inserts a theory about empire that owes some of its insights to Hannah Arendt, a scholar well within the orbit of the Frankfurt School, though she chose to spend the rest of her life in the US. Arendt noted that late nineteenth-century imperialism was "the first stage in the political rule of the bourgeoisie rather than the last stage of capitalism."[34] In fact, the first capitalist accumulation disaster of 1847–48 had set off bourgeois revolutionary fervor all over Europe until 1850—with bold rebellions sometimes sparking bourgeois/working-class solidarities.[35] One of the results of these widespread movements was to leverage more bourgeois participation in the European state structures. Also, from an economic angle, the crisis was solved in two ways: (1) "long-term infrastructural investments" (e.g. Haussmann's gigantic public works in Paris); (2) expansion of trade overseas (chiefly with the US).[36]

---

[33] I quoted from this passage by Jameson in the "Introduction," under the fifth characteristic of postmodernism (cf. n. 30).

[34] Hannah Arendt, *Imperialism* (New York: Harcourt Brace Janovitch, 1968), 18, cited in Harvey, *The New Imperialism*, 42.

[35] Harvey defines "overaccumulation" as "a surplus of capital lacking profitable means of employment" (ibid.).

[36] Ibid., 43.

But soon this absorption of surplus capital was cut short, in part by the American Civil War, and in part by civil unrest again spreading through Europe (e.g. the Paris Commune of 1871). In a pattern that was to be duplicated many times over, the capitalist classes in power resisted local investments in social infrastructure (such as housing, education, transport and the like) and pushed for colossal investments and speculation in other parts of the world—seeking, in Harvey's words, "spatio-temporal fixes" for an efficient use of excess capital.[37] As Arendt argued nearly four decades ago, "endless accumulation requires the endless accumulation of political power."[38] The result is what Harvey calls "accumulation by dispossession":

> Overaccumulation, recall, is a condition where surpluses of capital (perhaps accompanied by surpluses of labour) lie idle with no profitable outlets in sight. The operative term here, however, is capital surplus. What accumulation by dispossession does is to release a set of assets (including labour power) at very low (and in some instances zero) cost. Overaccumulated capital can seize hold of such assets and immediately turn them to profitable use… Privatization (of social housing, telecommunications, transportation, water, etc. in Britain, for example) has, in recent years, opened up vast fields for overaccumulated capital to seize upon.[39]

Since 1973, global capitalism has experienced chronic problems of surplus capital needing to be absorbed.[40] In some cases too, political changes have been fortuitous: "The collapse of the Soviet Union and then the opening up of China entailed a massive release of hitherto unavailable assets into the mainstream of capital accumulation."[41] But then also, the neoliberal project of privatizing public assets on a grandiose scale fits into this pattern, as do IMF policies in debt-ridden countries mentioned in the previous chapter. Another way of mopping up surplus capital is to feed cheap raw materials such as oil into the world market,[42]

---

[37] Ibid.

[38] Ibid., 140.

[39] Ibid., 149.

[40] Relate this thought to Shiva's observation about transnational corporations dominating and undermining traditional local economies.

[41] Ibid., 150.

[42] The fact that the price of a barrel of oil climbed over $140 in 2008 was just one of the many flagrant signs of overaccumulation once again. Arendt remarked that wars turn out to be the best tools for mopping up surpluses. The Iraq and Afghanistan wars certainly fall into this category. As Harvey observes, there is no firm evidence that Roosevelt's New Deal turned the Great Depression around: "It took the travails of war between the capitalist states to bring territorial strategies back into line so as to put the economy back on a stable path of continuous and widespread capital accumulation" (ibid., 76).

or by devaluating existing capital assets (think of the Asian crisis of 1997–98)[43] and muzzling the power of labor unions while streamlining companies or resorting to "out-sourcing."

What is the "new imperialism" then? Harvey explains his analytic tool in these words:

> I here define that special brand of [imperialism] called "capitalist impe-
> rialism" as a contradictory fusion of "the politics of state and empire"
> (imperialism as a distinctively political project on the part of actors whose
> power is based in command of a territory and a capacity to mobilize its
> human and natural resources towards political, economic, and military
> ends) and "the molecular processes of capital accumulation in space and
> time" (imperialism as a diffuse political-economic process in space and
> time in which command over and use of capital is primary).[44]

I have no space here to follow his arguments in detail, but only to show how his dialectical theory of the two logics of imperialism impinges on US external policies today.[45] On the one hand, the war in Iraq makes sense as a long-term strategy to cut off oil supplies to its greatest potential foes—Europe, and especially China, which with its booming economy of production is increasingly dependent on Middle Eastern and Caspian Sea oil. Equally, the Chinese–US relationship reflects the tensions of a competitive race on two fronts: economic and political spheres of influence from Central Asia to East Asia. Miffed for instance by being forced to withdraw its bid to acquire the American oil company Unocal in August 2005, China will also have to find a balance between its age-old imperialist drives and its current market-driven economic ambitions.[46] Though the US is undoubtedly the greatest military power in the early twenty-first century, it may be already extending itself too far as it reels

---

[43] The crisis was particularly acute for Thailand and Indonesia who were forced by the IMF to devaluate massively their currencies, liberalize their economies, which resulted in a huge transfer of ownership and power to Western banks (ibid., 150).

[44] Ibid., 26.

[45] Harvey notes that since 2002 the mainstream US media have openly discussed the issue of American imperialism. Michael Ignatieff of Harvard University, for instance, published a cover piece in the *New York Times Sunday Magazine* ("American Empire: Get Used to It" [January 5, 2003]: 22-54), in which he describes the "war on terror" as "an exercise in imperialism."

[46] David Barboza and Andrew Ross Sorkin, "Chinese Company Drops Bid to Buy U.S. Oil Concern," *New York Times* (August 3, 2005): A1 and C4. This was particularly infuriating for the Chinese since their bid of $18.5 billion was far higher than the US firm that acquired Unocal for $17 billion, Chevron. On the other hand, the bid by Cnooc (70% owned by the Chinese state) raised "a political firestorm in Washington" over what many consider the acquisition of a "strategic asset," and thus constituting a threat for US national security.

under the fallout of the greatest recession since the Great Depression and struggles to disengage with Iraq and Afghanistan.[47] On the other hand, the logic of global capitalism, while it enabled the US to gain control over the levers of international finance in the 1970s, dramatically worked at cross-purposes with the territorial imperialism of George W. Bush's neo-conservative administration. President Obama has from the start adopted multilateralism, apparently putting behind US imperial designs. Yet, his administration is saddled with seemingly impossible challenges: two wars, a deep recession, a health care system in shambles, a soaring foreign debt and the ominous threats related to climate change.

What is even more worrisome is the prospect of foreign investors pulling out the gigantic sums that still keep the US economy afloat.[48] Before the Asian financial crisis of 1997–98, the "Asian Tiger" economies imported capital in order to feed their economic growth. After the crisis, in a move forever to avert a similar crisis, they embarked on a frenzied campaign of saving capital, importing huge quantities of foreign assets. This overaccumulation (China now own over a trillion dollars of US treasury bonds) in turn translated into "a world awash in cheap money, looking for somewhere to go."[49] This capital flooded not only the US but also some of the emerging European economies—which all imploded with the event of the 2008 financial crisis: Iceland, Ireland and Bosnia, among others.

This horizontal rush of capital to the US in the 2000s has only multiplied American debt, putting increasing pressure on the dollar. In May of 2009 it was clear that international confidence in the greenback was waning. The US economy is no longer the safe-haven of investment it used to be. If this loss of confidence truly materializes, among foreign investors and American citizens alike, a flight from the US dollar would have dire consequences for the US economy.[50] Either way, this crisis has

---

[47] Harvey notes that no one seems to have paid much attention to Paul Kennedy's book of 1990, *The Rise and Fall of the Great Powers: Economic Change and Military Conflict from 1500 to 2000* (New York: Fontana Press). There he pointedly warned about the danger of "overextension and overreach" that has proved to be "the Achilles' heel of hegemonic states and empires (Rome, Venice, Holland, Britain)" (ibid., 35).

[48] Harvey explains: "Foreigners now own over a third of US government debt and 18 per cent of corporate debt (more than double the ratios in around 1980), and the US now depends on over $2 billion a day of net foreign investment inflow to cover its continuously rising current account deficit with the rest of the world" (ibid., 206).

[49] Paul Krugman, "Revenge of the Glut," *New York Times* (March 1, 2009): http://www.nytimes.com/2009/03/02/opinion/02krugman.html.

[50] Harvey, *The New Imperialism*, 207-8.

forced American leaders to favor the logic of global capitalism over that of territoriality.

As Harvey points out, it is not difficult to imagine nationalism and racism starting to rise in countries or blocks that withdraw "into regional configurations of capital circulation and accumulation," with the potential for new conflicts on a much greater scale. His solution, at least as a first step, would be some form of "New Deal" between the US and Europe (add today: the G8 and G20) that "might, by adequate pursuit of some long-term spatio-temporal fix, actually assuage the problems of overaccumulation for at least a few years and diminish the need to accumulate by dispossession." Further, this "might encourage democratic, progressive, and human forces to align themselves behind it and turn it into some kind of practical reality."[51] Though such a move still represents a form of "benevolent imperialism," it is better than "the raw militaristic imperialism currently offered by the neo-conservative movement in the United States."[52]

Without subscribing wholesale to Harvey's (or Jameson's) Marxian thesis of historical-geographical materialism,[53] I am pointing to the explanatory power of their theses, and particularly Harvey's theory of time and space compression. Meanwhile, both Harvey and Jameson sink their intellectual roots into the soil of Western Marxism,[54] and though they incorporate insights on the nature of culture and aesthetics from other contemporary currents and thus reinvigorate Marxist theory in today's context, they should still be seen as committed to the "grand narrative" of human liberation from the alienating effects of global finance capitalism—a purpose that certainly overlaps with the concerns of the present book.

---

[51] Ibid., 210-11.

[52] Ibid., 211.

[53] Both of them have taken the postmodern turn away from traditional Marxist analysis, which was properly modern and tended toward reductionism. From a foundationalist perspective, such a theory would imply that cultural production is solely caused by social and economic phenomena rising out of late capitalism. Rather, they tacitly endorse the kind of critical realism I will be advocating (i.e. this is a working theory that for now best accounts for the data at hand). Further, I do not read them as levelling all human agency and creativity in the midst of these processes, a position to which system theorists are almost inevitably led.

[54] Western Marxism is associated with the pioneering works of Georg Lukacs and Ernst Bloch in the 1920s, as well as the critical theory proponed by the Frankfurt School. In general, it has distanced itself from Leninism and has deemphasized the role of class struggle in its sociopolitical analysis.

These passing remarks on Marxian analysis also serve as a useful transition into the next section: the modern utopia of Marx is no less modern than that of Rousseau, Voltaire or de Tocqueville in France and Thomas Jefferson in the United States. Enlightenment thinking starts with a transcendent subject, one who can accurately apprehend the world as it is and therefore shape it in its image. Adorno and Horkheimer argued vehemently in the 1940s that this could not be so—thus the title of the next section. For Harvey and Jameson too, human rationality is constructed within the dynamic interplay between political and economic forces and persons caught up in the maelstrom of shifting ideas in society. A scholar reflecting on cultural artifacts of the 1990s, for instance, would have sought to understand and theorize on the basis of observations (an empirical approach), while at the same time confessing to the necessary distortion of any such theory that emerges. One might object, however, that both Harvey and Jameson seem to advocate their theories with a great deal of self-assurance. In my reading, however, both thinkers contend that any careful observer will discern a clear linkage between (1) the commodification of nature and human life that reached its zenith in the early 1970s, and (2) the cultural production of a global elite (should I say "bourgeoisie"?) since then that has been dubbed "post-modern" by a good number of social critics. The resulting picture is more that of limpidity than of opaqueness. For William McPheron, "Jameson's guiding premise is that cultural artifacts are oblique representations of their historical circumstances, whose concrete social contradictions they variously distort, repress, and transform through the abstractions of aesthetic form." The critic's role is not so much to comment on a work's aesthetic virtues as it is "to lay bare its roots in political and economic conditions and to explain how and why these roots have been obscured."[55]

Critical theorists such as Adorno and Horkheimer adopted this stance, laying at the feet of modernity's instrumental reason the brunt of responsibility for the "iron cage" of late capitalist society (to use Max Weber's expression), without, however, offering an alternative. At least, this is the indictment of their work that their heir, Jürgen Habermas, would pronounce already in the 1960s.[56] Critical analysis should be constructive, he wrote—an objective shared by the likes of Harvey and Jameson.

---

[55] From the Stanford University Libraries website, "Frederic Jameson pages edited by William McPheron, William Saroyan Curator for American and British Literature (1999)," http://prelectur.stanford.edu/lecturers/jameson/.

[56] Peter Dews, "Communicative Rationality," in *Routledge Encyclopedia of Philosophy*, vol. 2:459. I continue with Habermas below.

I contend, then, that in order to support a utopian vision of greater human freedom and social justice,[57] the more radical claims of post-modern deconstruction must at least be softened. After all, is social criticism even possible without a subject able to make some meaningful pronouncements about society "out there"? In what follows, I look more closely at the question of the self.

Death of the Autonomous Subject?

In this section and the next chapter, I turn to Yale's political philosopher, Seyla Benhabib, as our chief guide to the philosophical underpinnings of the modern and postmodern projects—with German philosopher Jürgen Habermas close at hand. The discussion revolves around three assertions made by postmodernists, which Benhabib borrows from Jane Flax, who, like her, is concerned to redefine the contentious borders between post-modernism and feminism.[58] Flax argues that postmodernism announces the death of man, the death of history and the death of metaphysics, using that threefold thesis as a canvas upon which to paint her own picture of postmodernism. I intend to show (agreeing with Benhabib) that while the philosophical landscape has irreversibly turned "postmo-dernist," one need not subscribe to the extreme version Flax herself has chosen. While the remainder of this chapter concerns "the death of man," the next chapter takes up the question of history and metaphysics and wraps up the discussion by returning to the notion of rationality.

The Epistemic Crisis of Representation

Benhabib defines postmodernism primarily in terms of an epistemologi-cal crisis, that is, what does it mean to "know"? Turning to Lyotard's landmark book, *The Postmodern Condition*, the modernist "episteme" (epistemological paradigm) impels us to take either one of two directions in our thinking and imagination. Either:

---

[57] Again, I use "utopian" in a positive sense, following Seyla Benhabib: while the Enlightenment ideal that seeks to restructure "our social and political universe according to some rationally worked out plan" has "ceased to convince," "the long-ing for the 'wholly other' (*das ganz Andere*), for that which is not yet" is a needed ingredient of "utopian thinking," and in our present context, "a practical-moral imperative" ("Feminism and the Question of Postmodernism," in *Situating the Self: Gender, Community and Postmodernism in Contemporary Ethics* [New York: Routledge, 1992], 229).

[58] See Benhabib's "Feminism," 203-41. The book by Jane Flax is *Psychoanalysis, Feminism, and Postmodernism* (Berkeley: University of California Press, 1990).

1. society is seen as a whole in which knowledge is appropriated through "performativity" (knowledge is associated with the power of scientific knowledge, which increases our efficiency and control through technology—thus knowledge is power, and power opens the way to more knowledge while legitimating it);
2. or society is bifurcated into two parts, with a subject recruiting this critical knowledge for the sake of self-empowerment—"It seeks not to enhance the efficiency of the apparatus but to further the self-formation of humanity; not to reduce complexity but to create a world in which a reconciled humanity recognizes itself."[59]

For Lyotard, that emancipatory ideal (2), borne of nineteenth-century romanticism, propels the work of contemporary German philosopher, Jürgen Habermas, whose attempt to construct "a metadiscourse which is 'universally valid for language games,'" seeks not to build a nation but rather international "consensus, transparency and reconciliation." Meanwhile, Lyotard remains unconvinced: "Consensus has become an outmoded and suspect value… We must…arrive at an idea and practice of justice that is not linked to that of consensus."[60]

Naturally, Benhabib's question is, "Can Lyotard convince?" This is a critical question in view of the recent alliance between feminism and postmodernism, she asserts. The nub of this rapprochement is the common recognition of "the demise of the episteme of representation." Indeed, the crucial epistemological moment of modernism should be seen in René Descartes's double move "to take nothing and no authority for granted whose content and strictures had not been subjected to rigorous examination, and that had not withstood the test of 'clarity and distinctness,'" and thus to withdraw from the world of sense perception.[61] On Benhabib's view, this withdrawal was both ethical (the Stoic suspension of the self from worldly involvement) and religious (a Jesuit form of meditation). Those who followed, however, retrieved only the epistemological implication of Descartes's double movement: "The corporeal, ethico-moral self was *reduced* to a pure subject of knowledge, to consciousness or to mind."[62]

Fuller Seminary philosopher Nancey Murphy rightly calls this a dualist strategy, the conviction "that essential humanness is associated with the

---

[59] Benhabib, "Feminism," 205.

[60] Ibid.

[61] Ibid., 205-6.

[62] Ibid., 206, emphasis added. I will come back to this dualistic vision of humankind and nature, with its reductionistic tendencies.

mind and thus is quite independent of the workings of mechanistic nature."[63] But what is this "thinking substance" Descartes has now posited, and how do mind and body interact? Historically, Murphy argues, this dilemma led to the sharp distinction between the natural sciences and the humanities, and, with the reigning paradigm of the logical positivists in the first half of last century,[64] to a movement of extreme reductionism: the social sciences can, in the end, be reduced to physics and chemistry.

In fact, this kind of progression could unfold because, from Descartes and Locke to Hume—whether one places "true" knowledge in ideas (rationalism) or in sense perception (empiricism)—the agenda of philosophy was to figure out how a subject could connect the pictures of the world it entertained in the mind with the world "out there." David Hume showed that if all we have as humans is the world of our five senses, our knowledge is dubious at best. For Descartes and many others until (and including) Immanuel Kant, it is the mind of the Creator that guarantees the integrity of human reason and its ability to apprehend the world as it is.[65] In the final analysis, concludes Benhabib, be they empiricists or rationalists, "modern epistemologists agreed that the task of knowledge, whatever its origins, was to build an adequate representation of things."[66] The modern episteme of representation, then, consists of three distinct elements: (1) the ideas and sensations represented in the human consciousness (raising the ontological status of the subject of knowing itself); (2) the language that bridged between individual and societal experiences of the world; and (3) the origin of the representations themselves, that is, the world outside the human subject—back to our familiar triangle of self (or subject), language and world.

Here Murphy would helpfully interject a gloss on the first element: since Descartes, the dominant picture of knowledge is that of a building and the idea that a building is only as strong as its foundations—what is known as foundationalism. As he recalls in his *Discourse on Method*, Descartes was looking at surrounding buildings from the room in Germany where he was staying at the time. He reflected on the fact that buildings built upon shaky foundations are in danger of crumbling, and thus must be rebuilt from the bottom up. In the same way, he thought

---

[63] Murphy, *Anglo-American Postmodernity*, 17.

[64] I take up logical positivism in the next chapter.

[65] But as I show below, Kant was a watershed thinker in this regard. Though he remained optimistic thanks to his theistic perspective, he drove a crucial wedge between what the mind can know and the world as it "really" is.

[66] Benhabib, "Feminism," 206.

that once in his lifetime he should reject all the ideas he inherited, either to adopt them again or to discard them for good, depending on whether they could fit into a "rational scheme" or not. He wrote, "And I firmly believed that by this means I would succeed in conducting my life much better than if I built only upon the old foundation and gave credence to the principles which I had acquired in my childhood without ever having examined them to see whether they were true or not."[67] This kind of foundationalism, Murphy notes, has given rise to two basic questions in modern epistemology: (1) what is the foundation of knowledge—clear and distinct ideas (rationalism), or sense data (empiricism)? And (2), what kind of reasoning will be used to build on this foundation— deductive, inductive, constructive, and so on?

And yet, as Benhabib is apt to remind us, all of these issues still presuppose a knowing subject, somehow able to adopt a critical distance from the data it attempts to process. Already in the last two centuries, however, three broad movements began to assail the very notion of this Cartesian subject as objective observer. First, a line of critique was launched by German idealism, followed by Marx and Freud, then in 1937 by Horkheimer[68] and finally by Habermas in his *Knowledge and Human Interests* (1968). The Marxists and Freudians denied the validity of an autonomous ego's "clear and distinct" ideas somehow unaffected by the determinations of history on the one hand, or by the swirling currents of the unconscious on the other. A second critical assault came from Nietzsche, who drew Kant's distinction between the phenomenal realm (as represented in the human mind) and the noumenal realm (the world as it is, *das Ding an sich*) to its logical conclusion. Thus, through Nietzsche's pen, Descartes's initial doubt that was to lead to certainty now leads to universal doubt, and the mind–body dualism is turned into an irreversible axiom.

---

[67] René Descartes, *Discourse on Method and Meditations*, trans. Laurence J. Lafleur (Indianapolis: Bobbs-Merrill, 1960 [1637]), 12, quoted in Murphy, *Anglo-American Postmodernity*, 9-10.

[68] Docherty ("Postmodernism, An Introduction," 8) helpfully summarizes the central argument of Adorno and Hokheimer's *Dialetic of Enlightenment* in these words: "Enlightenment itself is not the great demystifying force which will reveal and unmask ideology; rather, it is precisely the locus of ideology, thoroughly contaminated internally by the ideological assumption that the world can match—indeed, can be encompassed by—our reasoning about it, or that the human is not alienated by the very process of consciousness itself from the material world of which it desires knowledge in the first place. Enlightenment, postulated upon reason, is— potentially, at least—undone by the form that such reason takes."

Then Martin Heidegger picked up where Nietzsche left off, though rejecting the nihilism to which the latter's thought seemed to lead. Taking stock of the dead end of Enlightenment's two rival solutions— rationalism and empiricism—Heidegger defined the multiplicity of ideas and manifestations of the world in the subject's mind as the very core of the modern conception of being (being as *presence to* human conscious-ness), with the ever present temptation to impose its own order on the world as it is perceived so that it might better control it. Finally, Adorno and Horkheimer concluded that it is this modern idea of knowledge as representation that "culminates in the technical triumph of the western *ratio*, which can only know things in that it comes to dominate them."[69]

The idea that human thoughts and categories take center stage in philosophy and the social sciences was initiated in the work of Edmund Husserl in the 1890s, spawning a movement that is still influential today: phenomenology. This broadly flowing stream emphasized human con-sciousness as shaped by culture and context, and though much of it in the early decades espoused realism and transcendentalism in metaphysics ("seeing and describing universal essences"),[70] it gradually embraced the earlier skepticism of Nietzsche about the Enlightenment's subject-centered reason, and later, that of Danish theologian Søren Kierkegaard. As a result, phenomenology branched out into existential phenomenol-ogy (Hannah Arendt, Gabriel Marcel, Jean Paul Sartre and Simone de Beauvoir) and hermeneutical phenomenology (Hans-Georg Gadamer and Paul Ricœur).[71]

Throughout its long career, thinkers associated with all strands of phenomenology have consciously rejected the idea of the world as fully explainable through the categories of natural science (naturalism) and the notion of human behavior as reduced to biochemical reactions (behavior-ism). This partly explains why the mindset and methods of the phenome-nology of religion have come to dominate the academic study of religion in the last few decades. The fuller explanation, however, lies in the per-ceived failure of the traditional "normative" approaches to explore the validity of religious truth claims. The phenomenological turn began with

---

[69] Benhabib, "Feminism," 208.

[70] Lester Embree, "Phenomenological Movement," in *Routledge Encyclopedia of Philosophy*, vol. 6:334.

[71] Martin Heidegger is considered to have anticipated either or both the existential and hermeneutical strands of phenomenology, depending on how one interprets his 1927 *Sein und Zeit* (*Being and Time*). Embree, for instance, writes that the exis-tentialists misconstrued Heidegger: "This incomplete masterpiece is actually not devoted to human existence but rather 'fundamental ontology'" (ibid., 337).

a bracketing of the epistemological and metaphysical questions and the focusing on the task of describing varieties of beliefs, rites and practices, and finding new ways to classify religious phenomena.[72] I take this up more fully in the next chapter, because it affects our understanding of all religious discourse. In this sense, phenomenology is a post-Enlightenment movement, though not "postmodern" in the sense used by Lyotard.

Finally, Benhabib recounts a third front of criticism on the "knowing subject," one that emerged from the tradition of linguistics launched in the 1920s, from Ferdinand de Saussure and Charles Pierce to Gottlob Frege and Wittgenstein's philosophy of language. What they hold in common is the contention that human language functions as a socially determined and thus arbitrary system, the meaning of which has nothing to do with the images or concepts of an individual subject. Admittedly, these thinkers disagreed as to where meaning was located—whether in the linguistic signs themselves (in words or in sentences), or in the infinite number of social contexts in which these signs are put to use. Just the same, this is the tradition with which Lyotard is at home. In fact, some of the most influential philosophical traditions of the twentieth century fall into the orbit of the linguistic paradigm. Thus Heidegger's interest in being as the center of human knowledge shifted to language for the proponents of analytical philosophy, hermeneutical phenomenology and French poststructuralism.[73]

According to this paradigm, the modern episteme of representation no longer holds, because knowledge is mediated by language, which itself results from the linguistic traditions of a given society. The focus of epistemology is no longer on a private subject but rather on the public sign-system created by a community of subjects, whose individuality is now greatly reduced. Within this framework, however, one can still choose to emphasize the interpreting process of the individual within a particular community. In this sense, the modern episteme widens its horizons from individual to community. This is the option Benhabib herself adopts, based on what she calls a "social pragmatic conception of language."[74] On the other hand, European structuralism and poststructu-

---

[72] Merold Westphal, "Phenomenology of Religion," in *Routledge Encyclopedia of Philosophy*, vol. 6:334.

[73] Heidegger's thought and influence goes beyond this, however. He was intensely interested in hermeneutics—how people come to understand a text—and, in that sense, the fourth movement of phenomenology associated with Gadamer and Ricœur does have some loose connections to Anglo-American philosophy of language (from Wittgenstein to Searle and Ayers). More will be said on this in Chapter 4.

[74] Benhabib, "Feminism," 209.

ralism have chosen to emphasize the all-encompassing role of language itself in the production of meaning, with a dramatic de-emphasis on the subject of knowing. From their standpoint, as language takes on a life of its own, a text's meaning is either constructed by its reader or is to be found in the text itself. For others, meaning is fundamentally absent— along with the text's author. For Lyotard, the nature of language-games is such that, though knowledge and meaning do break through, they only appear within local traditions. Consensus, therefore, is an unrealistic aim, because the criteria of validity and truth can only be agreed upon locally—thus the term the "agonistics" of language: "to speak is to fight, in the sense of playing, and speech-acts fall within the domain of a general agonistics."[75] In this sense, the modern subject, who articulates his or her own understanding of the world (including the meaning of texts), is dead. Language now totally dominates "mind" and "world."

Communicative Action and Discourse Ethics
While Lyotard argues that the postmodern proclamation of the subject's death leads to a "polytheism of values," Benhabib demurs, anxious to recover some level of discourse that could apply to humanity as a whole (a prerequisite for the agenda of feminism and discourse in general). Sure enough, a "monotheism" of values is no longer possible, she concedes. Yet the demise of the episteme of representation need not imply "the death of the subject," as Lyotard and the French postmodernists would have it. Hegel's critique of Kant's universal imperative, with its moral focus on the individual, can also lead to a concept of rationality that is negotiated through social actors entering a moral conversation as equals with a commitment to reciprocity.[76] Hence, Benhabib proposes the theory of "communicative action" (pioneered by Habermas)[77] as the rational condition for a universalizable discourse of human rights. It is above all a pragmatic moral theory that potentially protects and enhances the rights of both individuals and collectivities within the framework of a demo-cratic nation-state. As such, it is still situated within the horizon of modernity; at the same time, it transcends the foundationalism and individualism of the Cartesian episteme.

---

[75] Lyotard, *The Postmodern Condition*, 10, cited in Benhabib, "Feminism," 209.
[76] As Benhabib shows, contemporary social ethics still revolve around Hegel's criticism of Kant's formula, "Act only on that maxim through which you can at the same time will that it should become a universal law" (ibid., 26-38).
[77] See his magnum opus in two volumes, *The Theory of Communicative Action*, trans. Thomas McCarthy (Boston: Beacon Press, 1984).

It is clear, then, that when Flax writes the following, it is the French deconstructionist version of postmodernism (following Jacques Derrida) that she has in mind:

> Postmodernists wish to destroy all essentialist conceptions of human being and nature... In fact Man is a social, historical or linguistic artifact, not a noumenal or transcendental Being... Man is forever caught in the web of fictive meaning, in chains of signification, in which the subject is merely another position in language.[78]

By subscribing to an intersubjective moral validity of social action, Benhabib convincingly argues for a weaker version of postmodernism. While it is true that from a secular perspective today—an understandable stance (at the very least) in our multicultural, multireligious Western democratic societies[79]—arguments about the common good cannot be based on an "essentialist conception of human being and nature," this need not imply that there is no justification for the equal respect and dignity of all people across nations and cultures. In fact, one purpose of the present work is to reinforce the public conception of human dignity as enshrined in the Universal Declaration of Human Rights (UDHR, 1948) with a joint Muslim–Christian theological reflection on the issue.

This philosophical and quasi-theological statement of human dignity, which is almost taken for granted in much of the world today, requires some further elaboration. Indeed, the idea of human rights points to a link between the triangle of subject (or mind)–language–world and the weighty matters of ethics and political theory. At the risk of over-simplifying the historical progression, I will suggest four steps that led to the fifth, namely, the codification of human rights in international law (the UDHR):

---

[78] Flax, *Thinking Fragments*, 32, cited in Benhabib, "Feminism," 211.

[79] I am not saying here that the secular assumptions of Western society are necessary to the functioning of democracy—only that this is an understandable position, given the brutal wars of religion that until the first half of the seventeenth century wreaked havoc in Europe. The principal framer of the US Declaration of Independence, Thomas Jefferson, made it clear in many of his letters that his leanings towards deism (belief in a Supreme Being without any particular religious framework—though he officially remained Anglican) and/or Unitarianism was in part dictated by the extreme intolerance of religion that he witnessed first-hand as ambassador for the new republic in several European capitals (see, for instance, Charles B. Sanford, *The Religious Life of Thomas Jefferson* [Charlottesville: University Press of Virginia, 1984]).

1. The emergence of natural law theory: partly a product of Greek thought, biblical notions of human dignity and creation, and Roman jurisprudence, the idea that humans, by virtue of their reasoning ability, could discern universal moral laws and use them to foster a just political and social order.[80]

2. By the seventeenth century this notion was becoming secularized, with the theory of Dutch legal philosopher Hugo Grotius (d. 1645) that established natural rights for all human persons by virtue of their reason, independently of God.[81]

3. The idea that human rationality could be the grounds for human rights in general gained ground in the early modern period. Natural law had given way to "natural rights"—a theory developed especially by Thomas Hobbes (d. 1679) and John Locke (d. 1704). These writings led to political activism, culminating in the Declaration of Independence of the United States of America (1776), the French Declaration of the Rights of Man and of the Citizen (1789) and the Bill of Rights of the United States of America (1791). Whereas the English Bill of Rights (1689) had only spoken of the "indubitable rights and liberties of the people of this kingdom," the proponents of the French and American revolutions proclaimed these universal and inalienable rights as, respectively, "liberty, property, security, and resistance to oppression," and "Life, Liberty and the pursuit of Happiness."[82] Here is a civil religion or a secular social order—depending on one's viewpoint—that still leans on a natural law perception: because humans are "created equal" they have rights, which *ought* to be respected.

---

[80] The greatest medieval Christian philosopher, Thomas Aquinas (d. 1274), wrote in his *Summa Theologica* a response to the question "Is there a natural law?" that humanity as "the rational creature" is not only "subject to Divine providence," but also "has a share of the Eternal Reason, whereby it has a natural inclination to its proper act and end: and this participation of the eternal law in the rational creature is called the natural law." This ability to discern right from wrong in harmony with the natural law "is nothing else than an imprint on us of the Divine Light" ("second article" or "Question 91: Of the Various Kinds of Law," quoted in Patrick Hayden, ed., *The Philosophy of Human Rights*, Paragon Issues in Philosophy [St Paul, MN: Paragon House, 2001], 45).

[81] Hayden, ed., *The Philosophy of Human Rights*, 4.

[82] All four documents are found in ibid., 339-52. The French document places itself under the "auspices of the Supreme Being" (a clearly deist stance) and the US Declaration of Independence refers to the "Laws of Nature and Nature's God" (a compromise between the deists and Christians).

4. Philosophically, one could assert that the UDHR participates in this mindset, with one important difference: neither God nor any Supreme Being is mentioned. This is understandable, given the circumstances of its promulgation: a broad sweep of nations at the end of the twentieth century's second horrific "world war" wanted to set up new standards for international law that would somehow prevent such devastating conflicts in the future. Since then, the International Bill of Human Rights has been ratified by most UN member states, consisting of the UDHR, the International Covenant on Economic, Social, and Cultural Rights (ICESCR) and the International Covenant on Civil and Political Rights (ICCPR) (the last two coming into force in 1976).[83]

Yet this apparent unanimity is misleading. For one, several cultural blocks have earnestly argued that the UDHR is a Western document in need of retooling in order to adapt it more widely.[84] More importantly perhaps, the philosophical issues swirling around cultural relativism threaten the very concept of "universal" rights. This line of reasoning began with Kant in the nineteenth century, as he raised the epistemological question—how does the human mind connect to the world? Recall Kant's fundamental distinction between the two worlds of reality: the world as it is, exterior to human consciousness and sense experience (noumenal world), and the world as shaped by human categories of thought (phenomenal world). This represented simultaneously the high water mark of Enlightenment rationalism and the beginning of its demise. Indeed, with the help of the empiricists and fuel provided by anthropological findings, it was not long before people concluded that human reason was neither universal nor trustworthy, but rather a product of subjectivity and cultural conditioning. This led to the second question: what role does language play between subject and world? One answer came from the empiricists of the late nineteenth century: language consisted of mere symbols that referred to sense data, which were bandied

---

[83] Ann Elizabeth Mayer, *Islam and Human Rights: Tradition and Politics*, 3d ed. (Boulder, CO: Westview, 1999), 18. Some important newer instruments of human rights include the Convention on the Elimination of All Forms of Discrimination against Women (1979), the Vienna Declaration and Programme of Action (1993) and the Declaration on the Rights of Persons Belonging to National, Ethnic, Religious or Linguistic Minorities (1993).

[84] For a discussion of Asian exceptionalism, see Xiaorong Li, "'Asian Values' and the Universality of Human Rights," *Report from the Institute for Philosophy and Public Policy* 16.2 (1996): 18-23, reprinted in Hayden, *The Philosophy of Human Rights*, 397-408.

about by people in their minds. This move led to a kind of empirical psychology, one which sought to provide all the answers for mathematics, logic and grammar—a kind of reductionism usually referred to as "psychologism."

Habermas shows that this reasoning not only turned out to be an impasse, but it provoked a reaction, which in turn set the course for philosophy in the twentieth century. Language was the key battleground, and what happened was, in his words, "a linguistic turn." The work of Edmund Husserl and the above-mentioned movement of phenomenology were part of this reaction to strict empiricism. In the course of events, Gottlob Frede turned out to be the pioneer of the linguistic turn. Each individual, Frede contended, has specific mental representations that are necessarily conditioned by a specific time and place. Yet, through the use of language, thoughts "overstep the boundaries of an individual consciousness," and these thoughts contain elements that are shared by people everywhere.[85] Mental representation tied to sense data represents only a small part of what takes place, as language is used in a person's thoughts or in conversation with others. When the reference to a particular object is made the subject of a sentence, for example, the sentence itself can either "express a proposition or report a state of affairs," which can be shown to be true (then it is a "fact") or false. Habermas notes, "With this critique, Frege takes the first step in the linguistic turn. From this point on, thoughts and facts can no longer be located immediately in the world of perceived or imagined objects; they are accessible only as linguistically 'represented' (*dargestellt*), that is, as states of affairs expressed in sentences."[86]

So, the epistemological turn was made around language, which from then on became the focus of much philosophical reflection. But beyond the question of language, a third issue was raised: the discourse of human rights points to the problematic relation between reason and ethics. Kant had distinguished between two kinds of "reason" at humanity's disposal: theoretical reason, which arrives at conclusions and principles in philosophy that are certain, but is in fact incapable of attaining any certain knowledge of the world; and practical reason, which is endowed with the double ability to make correct moral judgments and to decide to act on them. Having said that, however, Kant asserted that ethical values cannot be derived from any religious or metaphysical concept of the good, thereby attempting to rule out means–ends calculations based on some

---

[85] Jürgen Habermas, *Between Facts and Norms: Contributions to a Discourse Theory of Law and Democracy*, trans. William Rehg (Cambridge: Polity Press, 1996), 10.
[86] Ibid., 11.

putative good. Put otherwise, Kant (and his followers today) must reject both consequentialist and teleological moral theories.[87] By contrast, Kant proposed a deontological theory of ethical values (the right precedes the good, not vice versa). Specifically, practical reason establishes a universal principle to which all human subjects may agree: "act only on that maxim through which you can at the same time will that it become a universal law." A corollary to this is the following, "treat humanity... never simply as a means, but always at the same time as an end."[88] The imperative of treating every other human being as a person (and not a "thing") with the same rights of autonomy and freedom as oneself is merely, in my view, a secular reformulation of the Golden Rule—"Do to others what you would have them to do to you" (Mt. 7:12).

Ironically, it was the Protestant Reformation, as Weber once theorized, that brought to the fore the power and dignity of the individual conscience. By defending the right of persons to interpret God's word for themselves and by harnessing the newly invented printing press for this purpose, the Reformers not only successfully challenged the political absolutism of the Roman Church, but they also "laid the basis for subsequent intellectual breakthroughs made by Galileo, Copernicus, Kepler, and Newton in interpreting the divinely ordered natural cosmos."[89] As mentioned above, the dignity of the human person was justified in theo-

---

[87] Right actions can be discerned on the basis of good results (consequentialist view) or good ends (teleological view).

[88] Both quotes are from Kant's short introduction to ethics, *Groundwork of the Metaphysics of Morals* (originally *The Moral Law*), trans. with notes by H. J. Paton (London: Hutchinson, 1948; repr. New York: Harper & Row, 1964 [1785]), vol. 4:421, 429, cited by Onora O'Neill, "Kantian Ethics," in *Routledge Encyclopedia of Philosophy*, vol. 5:201. O'Neill concedes that the linkage between Kant's two maxims is not an obvious one and proposes that Kant's critique of consequentialist ethics holds the key to this query: the consequentialist "not merely permits but requires that persons be used as mere means if this will produce good results." This would indicate that, for Kant, to treat others as persons means that one allows them to exercise the same capacity to make choices that we ourselves take for granted (ibid.).

[89] Mitzman, *Prometheus Revisited*, 16-17. One of Mitzman's early contributions to social theory was *The Iron Cage: An Historical Interpretation of Max Weber* (New York: The Universal Library, 1971), in which he demonstrated the link between Weber's "disenchantment" with modernity and his own mental breakdown between 1898 and 1902 (triggered by his father's death shortly after he—the son—had cut off his relationship with him). Weber's later productivity can also be tied to his anguished determination to rid himself of the demons of his past. Above all, he (consciously or unconsciously) linked the strict authoritarianism of his Protestant father and the political hegemony of the Junkers over the Prussian land workers.

logical terms, but with the impact of eighteenth-century Enlightenment thinking, the grounds for justification moved from religious to secular modes.

Today, for more than six decades, the world's states have lived under the umbrella of the Universal Declaration of Human Rights. Though several important covenants have been drafted and ratified by the UN since then, one crucial document (in light of Chapter 1) has yet to be finalized and ratified: the Draft United Nations Declaration on the Rights of the Indigenous Peoples. Here, in the spirit of Benhabib's project of defining a universalizable moral discourse, I want to ferret out two philosophical problems that demand urgent attention, and one political conundrum as a case study.

Human Rights and Moral Validation

First, regarding the universalizing project of Kantian ethics, what kind of validation should be invoked for human rights and the specific international covenants that seek to protect them? Habermas offers additional historical background to this discussion. At the time, Kant was reacting to Hobbes's consequentialist theory, hitherto influential, that sought to justify individual liberties on the basis of enlightened self-interest—the archetypal liberal concern that sustains the towering work of John Rawls, for example.[90] For Kant, Hobbes was the "theoretician of a bourgeois rule of law without democracy" who failed to show how "an ordered system of egoism that is favored by all the participants" could be maintained in practice.[91] In fact, there could be no ground for legitimacy, Kant argued, and this in itself would lead the system to self-destruction. One cannot negotiate a transition from private law modeled on commercial contracts to a general law that coerces people—and therefore limits their freedoms—without invoking some principle of morality to ensure its legitimacy. The leap from private contract to social contract is a huge one, asserts Kant, and an empiricist framework will not be up to the task. Thus the enacting of public laws can only happen on the basis of some rational framework each and every citizen shares. I quote Habermas:

> Kant sees this primordial human right as grounded in the autonomous will of individuals who, as moral persons, have at their prior disposal the social perspective of a practical reason that tests laws. On the basis of this reason, they have *moral*—and not just prudential—grounds for their move out of the condition of unprotected freedom. At the same time,

[90] See his classic work on distributive justice, *A Theory of Justice* (Cambridge, MA: Harvard University Press, 1972).
[91] Habermas, *Between Facts and Norms*, 90.

> Kant sees that the "single human right" must differentiate itself into a
> *system of rights* through which both "the freedom of every member of
> society as a human being" as well as "the equality of each member with
> every other as a subject" assume a positive shape.[92]

According to Habermas, then, we should not entirely jettison this subject-
centered moral perspective that we inherited from the Enlightenment
project. Nonetheless, Habermas's famous expression, "the unfinished
project of modernity,"[93] remains controversial. Though the expression
does not appear as such in *The Philosophical Discourse of Modernity*,[94]
there is no doubt that the publication of these lectures (1985) aimed to
retrieve a modified conception of rationality (communicative vs. subject-
centered reason) from the Enlightenment tradition in the face of its most
vociferous critics, from Nietzsche to Heidegger, and from Georges
Bataille to Jacques Derrida and Michel Foucault. The thread running
through these last writings is a critique of subject-centered reason that
targets "the self-assertive and self-aggrandizing notion of reason that
underlies Western 'logocentrism'"—and thus ushering a smug declaration of the end of philosophy.[95]

Habermas agrees to some extent with these critics: we can no longer
entertain the idea of a sovereign and autonomous subject, with its
thoughts free from the influence of society, culture and events, and
indeed, its own unconscious. What stood for rationality now is seen to
include large doses of irrationality, mainly in its will to dominate others
while fulfilling its own desires. Yet, just as Foucault, to take one example,
sought to unmask the dimensions of power inherent in any "rational
discourse" and provide an illuminating "genealogy" or "archeology" of
knowledge, his analysis in the end remains trapped in the very aporias of

---

[92] Ibid., 93, emphasis original. The quotes from Kant are taken from "On the
Proverb: That May Be True in Theory, But Is of No Practical Use," in *Perpetual
Peace and Other Essays*, trans. T. Humphrey (Indianapolis: Hackett, 1983), 72.
[93] This is taken from his essay, "Modernity—An Incomplete Project," *New German
Critique* 22 (Winter 1981): 3-15, which was based on his acceptance speech upon
receiving the Adorno Prize in 1980 in Frankfurt. See also Maurizio Passerin
D'Entrèves and Seyla Benhabib, eds., *Habermas and the Unfinished Project of
Modernity: Critical Essays on The Philosophical Discourse of Modernity* (Cambridge: Polity Press; Oxford: in association with Blackwell, 1996).
[94] He begins his Preface to *The Philosophical Discourse* by alluding to this speech.
He writes, "This theme, disputed and multifaceted as it is, never lost its hold on me"
(p. xix). The lectures as a whole seek to confront the critique of reason as it is
mobilized by French "neostructuralism."
[95] Thomas McCarthy, "Introduction," in Habermas, *The Philosophical Discourse*,
viii.

the subject-centered philosophy he had tried to escape. What objective stance, asks Habermas, could a Foucaultian critique of power in scientific discourses adopt to make that judgment in the first place? He writes,

> To the objectivism of self-mastery on the part of the human sciences there corresponds a subjectivism of self-forgetfulness on Foucault's part. Presentatism, relativism, and cryptonormativism are the consequences of his attempt to preserve the transcendental moment proper to generative performances in the basic concept of power while driving from it every trace of subjectivity. This concept of power does not free the genealogist from contradictory self-thematizations.[96]

Recall that this was also the tension I pointed out in Jameson's work: he wants to maintain a progressive vision of society (critical of the neo-imperialism of the international structures of late capitalism) while at the same time ostensibly adopting the radical epistemology of postmodernism. That is precisely the position of Foucault, argues Habermas. If language, and in a wider sense, discourse, is indissolubly wedded to a power that wipes out any contribution by individuals (subjectivity is banned), then this leaves no room for personal initiatives.

I believe Habermas is right on both counts: the self is to some extent "embedded" in culture and language; and it is in the very nature of language to make validity claims that expect to be redeemed in conversation with others. Thus, if rationality is considered to be the product of people interacting and communicating with one another within their social context (there is no "pure" reason), then everyday language turns out to be the tool through which a democratic politics can be worked out. As Habermas puts it, "[a]greement arrived at through communication, which is measured by the intersubjective recognition of validity claims, makes possible a networking of social interaction and lifeworld contexts."[97] It is true that in this process the nature of autonomous, or "pure" reason as conceived in Cartesian–Kantian terms has been scaled down considerably. We are now dealing with "postmetaphysical" theory, contends Habermas. Admittedly, any contemporary theory of human rights projected on a universal screen is by definition a post-Enlightenment conversation.[98] There can be no essentialized, reified conception of

---

[96] Ibid., 294-95.

[97] Ibid., 322.

[98] See also Michael Freeman, *Human Rights: An Interdisciplinary Approach* (Cambridge: Polity Press, 2002), especially his Chapter 6, "Universality, Diversity and Difference: Culture and Human Rights"; and Michael Ignatieff, *Human Rights as Politics and Idolatry*, ed. and intro. Amy Gutmann (Princeton, NJ: Princeton University Press, 2001).

human rights and democracy, but rather an ongoing negotiation carried out in practice by interlocutors on an international scale (as in the UN, for example). A certain transcendence breaks through—as a universalization of ethical thinking—but at the same time, its rationality is constructed ad hoc, through an immanent kind of practice.[99]

This tension between truth as locally constructed through the practice of persons involved in validating or invalidating their mutual truth claims aimed toward specific actions (let us say, how to eliminate toxic waste from their community) and the higher claim that this whole process is what morality is all about is embedded into the very nature of communicative ethics. This "discourse ethics" is also organically related to law—the human rights concept is, after all, behind the idea of international law. Habermas is also correct, then, when he highlights the parallel tension between facticity and validity in law (see his book, *Between Facts and Norms*). On the facticity side, "Modern law is formed by a system of norms that are coercive, positive, and, so it is claimed, freedom-guaranteeing." On the validity side,

> The formal properties of coercion and positivity are associated with the claim to legitimacy: the fact that norms backed by the threat of state sanction stem from the changeable decisions of a political lawgiver is linked with the expectation that these norms guarantee the autonomy of all legal persons equally. This expectation of legitimacy is intertwined with the facticity of making and enforcing law.[100]

Yet this legitimacy, unlike the one assumed by Hobbes or Rousseau (as social contract), is grounded in a discursive form of rationality in which the human self finds its own identity in conversation with others. While "strategic action"—the purposive behavior of those trying to reach their own ends, even at the expense of others—may dominate in a society where the market rules almighty, still, "communicative action" represents the ethical ideal of a democratic ethics and politics. Indeed, communicative action presupposes actors with an intentional orientation toward the common good. Thus the legitimacy of civil rights must go beyond their being written in a state's constitution. That provides only facticity: "the

---

[99] Katarina Dalacoura also wrestles with these issues as a social scientist studying the political cultures and Egypt and Tunisia (*Liberalism and Human Rights: Implications for International Relations*, rev. ed. (London/New York: I. B. Tauris, 2003). For her, the human rights concept has a metaphysical dimension to it, one which cannot be negotiated on the basis of reason. Whether from a secular or religious perspective, the idea of human dignity requires a step of faith—hence the hospitality of human rights to religious perspectives.

[100] Habermas, *Between Facts and Norms*, 447.

self-constitution of a community of free and equal persons."[101] Legitimacy can only come through the moral intuition of the democratic procedure in itself:

> In the form of individual rights, the energies of free choice, strategic action, and self-realization are simultaneously released and channeled by compelling norms, about which citizens must reach an understanding by following democratic procedures and publicly exercising their legally guaranteed communicative liberties... A force that otherwise stands opposed to the socially integrating force of communication is, in the form of legitimate coercion, thus converted into the means of social integration itself.[102]

This position is anything but morally neutral, argues Benhabib. Discourse ethics, or communicative ethics, require of us two basic moral principles—both deontological, in a Kantian fashion:

> (1) that we recognize the right of all beings capable of speech and action to be participants in the moral conversation—I will call this *the principle of universal moral respect*; (2) these conditions further stipulate that within such conversations each has the same symmetrical rights to various speech acts, to initiate new topics, to ask for reflection about the presuppositions of the conversation, etc. I call this *the principle of egalitarian reciprocity*. The very presuppositions of the argumentation situation then have a normative content that precedes the moral argument itself.[103]

As mentioned earlier, John Rawls expounded his post-Enlightenment theory of justice from a traditional liberal perspective. Besides the postmodernists who questioned the very idea of progress and thereby castigated the liberal agenda, widely blamed for two world wars and the growing inequalities between haves and have-nots in more recent years, criticism has arisen in communitarian circles along similar lines. Communitarian critics such as Alasdair MacIntyre, Michael Sandel, Charles Taylor and Michael Walzer, allege that the modern project of a social contract between free and autonomous subjects is incoherent and indefensible. Deontological ethics lead nowhere, because the good life can only be defined within a particular worldview, which, by definition, is culturally determined.[104] Virtue arises only in particular communities

---

[101] Ibid., 462.

[102] Ibid.

[103] Benhabib, *Situating the Self*, 29, emphasis original.

[104] Naugle shows how Wilhelm Dilthey (1833–1911) became the "father of world-view theory" (*Worldview*, 82-98). Though better known for his theories on the human sciences (*Geiteswissenschaften*) and his contributions to hermeneutics, he

and its definition varies, depending on the tradition of each community. In other words, there can be no "unencumbered" self, as the modern project would have us believe.

To some extent, Benhabib agrees, but retorts that a situated self does not rule out the benefit of modern reflexivity, that is, the ability of the self to stand at a distance from the ethical norms taken for granted by their community and make moral judgments. Yes, she concedes, the self does forge its identity within "constitutive communities," but again this is different from a "conventionalist or role-conformist" attitude in a pre-modern setting. "The specifically modern achievement of being able to criticize, challenge and question the content of these constitutive identities and the 'prima facie' duties and obligations they impose upon us should not be rejected." She goes on to target the communitarians, who, she argues, "are hard put to distinguish their emphasis upon constitutive communities from an endorsement of social conformism, authoritarianism and, from the standpoint of women, of patriarchalism."[105]

Habermas is correct, continues Benhabib, in pointing to the Weberian differentiation of value spheres as a result of modernity. One consequence of the detachment of religion and morality from science, law and aesthetics is that we no longer have one definition of the good life. And this is all the more true today, with the radical pluralization of worldviews, particularly in the West. Yet this is precisely where communicative ethics becomes so important: "Under conditions of value differentiation, we have to conceive of the unity of reason not in the image of a homogeneous, transparent glass sphere into which we can fit all our cognitive and value commitments, but more as bits and pieces of dispersed crystals whose contours shine out from under the rubble." Visions of the good life must be hashed out and continually negotiated within the rational, discursive practice of democracy. In this sense, the right will always precede the good. This is an important point in light of my overall purpose here. Though I inject a theological element absent in both Habermas and Benhabib, I believe that Muslims and Christians—precisely because

---

was the first to develop a comprehensive theory of how people envision the world and ascribe meaning to their lives in that context (*Weltanschauunglehre*). This thinking begins with people's shared view of life's meaning (rooted in and patterned by culture) and this he calls the "lifeworld" (*Lebenswelt*). In turn the lifeworld "begets a worldview," the attempt by individuals in community to solve "the riddle of life" (ibid., 86). Habermas builds into this definition of "lifeworld" his own categories, taking into account the political and economic institutions that constrain the democratic ideals.

[105] Benhabib, *Situating the Self*, 74.

they treasure the dignity of the human person as empowered by its Creator—will always demand a democratic form of government as a means to preserve and promote their vision of the good life in human society. Despotism and autocracy are ruled out.

The Problem of the Self and Otherness. The second philosophical problem is the following: how can individual rights be balanced against collective ones—a question made more urgent by the problem of minority rights in some Muslim states and the larger issue of indigenous rights for countries like Australia, Canada and the United States? And what about issues of gender? These issues are all related to the question of the self and otherness. We saw the weakness of the communitarian critique in this respect—are we forever condemned to remain imprisoned in the ethical framework of our own community? To concede the solid embeddedness of the subject as seen from both the communitarian and postmodern viewpoints, however, still does not force us to posit the subject "as merely another position in language," as Flax put it. Rather, according to Benhabib's reasoning, a kind of reformulation of the modern ideal is needed:

> The traditional attributes of the philosophical subject of the West, like self-reflexivity, the capacity for acting on principles, rational accountability for one's actions and the ability to project a life-plan into the future, in short, some form of autonomy and rationality, could then be reformulated by taking account of the radical situatedness of the subject.[106]

The fact is that self-identity is always constructed through the use of language, understood within the deep symbolic structures of one's culture-shaped worldview, and interpreted in narrative categories. For instance, "I was born into a Tamil family in Southern India, my grandparents emigrated from Sri Lanka, I am one of three daughters and my father is an imam in the local mosque." Nevertheless, contends Benhabib, "we must still argue that we are not merely extensions of our histories, that vis-à-vis our own stories we are in the position of author and character at once."[107] In fact, we expect to find in every person a normal tension between the cultural conditioning and gendering and the desire to break out of that mold in autonomous ways. There could be no feminist movement, notes Benhabib, without "such a regulative ideal of enhancing the agency, autonomy and selfhood of women." For her, feminists such as Judith Butler adopt so totally the premises of post-

---

[106] Benhabib, "Feminism," 214.
[107] Ibid.

modernism that they are forced to posit a subject that "is merely a bland slate upon whom are inscribed the codes of a culture, a kind of Lockean tabula rasa in latter-day Foucaultian garb!"[108]

To be sure, one can profitably study how the individual's subjectivity is constituted within the specific parameters of cultural symbols, but this does not answer the following question: "what mechanisms and dynamics are involved in the developmental process through which the human infant, a vulnerable and dependent body, becomes a distinct self with the ability to speak its language and the ability to participate in the complex social processes which define its world?"[109] Indeed, there are "structural processes and dynamics of socialization and individuation" that must be taken into account. As the same time, we must not neglect to study the chronological process by which persons come to define their own identity and life-meaning. No doubt, we are linguistic beings. By the same token, "we have to explain how every human infant can become the initiator of a unique life-story, of a meaningful tale—which certainly is only meaningful if we know the cultural codes under which it is constructed—but which we cannot predict even if we knew these cultural codes."[110]

That individuals are still agents who can be both authors and characters in the narrative they are in the process of piecing together is a key insight behind the theory of discourse ethics and communicative action. Indeed, it carries with it strong moral overtones that directly feed into a politics of human dignity. This idea of "an enlarged mentality" or of "representative thinking" is the insight that Hannah Arendt had gleaned from Kant, developed so forcefully and which led her to consider judgment "a political rather than a moral faculty."[111] Such a position not

---

[108] Ibid., 217. The book Benhabib is referring to is Butler's *Gender Trouble: Feminism and the Subversion of Identity* (New York: Routledge, 1990).

[109] Ibid.

[110] Ibid., 217-18. For a similar argument by Nobel-prize-winning economist Amartya K. Sen, see his *Identity and Violence: The Illusion of Destiny* (New York: Norton, 2006). Sen's particular life quest to understand identity, culture and community is particularly striking: born in Bengal to a Hindu family in 1933, his homeland became Pakistan in 1947 and Bangladesh in 1971. We are all torn apart by overlapping and often conflicting identities and loyalties, he writes. In the end, we have to make choices. This is where the democratic process engages everyone.

[111] Benhabib, *Situating the Self*, 9. The bringing together of the private and public spheres is important to Benhabib, since feminist theory has shown that "traditional modes of drawing this distinction have been part of a discourse of domination which legitimizes women's oppression and exploitation in the private realm." In this sense, both discourse ethics and social movements such as feminism "project the extension of a postconventional and egalitarian morality into spheres of life which were hitherto controlled by tradition, custom, rigid role expectations and outright inegali-

only privileges the public sphere, which she connects to "the macro-political institutions of a democratic polity" in order to allow a multiplicity of views the chance to coalesce into shared action for the common good, but also rehabilitates Kantian universalist ethics "along the model of a moral conversation in which the capacity to reverse perspectives, that is, the willingness to hear from the others' point of view, and the sensitivity to hear their voice is paramount."[112]

Recall that Benhabib's discourse ethics required two substantive moral norms: (1) the principle of universal respect; and (2) the principle of egalitarian reciprocity. When this is combined with communicative action, the result is a six-step process. In her words:

1. A philosophical theory of morality must show wherein the justifiability of moral judgments and/or normative assertions reside.

2. To justify means to show that if you and I argued about a particular moral judgment ("It was wrong not to help the refugees and to let them die on the wide sea") and a set of assertions ("Education should be free for all for the first eighteen years of their lives") that we could in principle come to a reasonable agreement (*rationales Einverständnis*).

3. A "reasonable agreement" must be arrived at under conditions which correspond to our idea of a fair debate.

4. These rules of fair debate can be formulated as the "universal-pragmatic" presuppositions of argumentative speech and these can be stated as a set of procedural rules.

5. These rules reflect the moral ideal that we ought to *respect* each other as beings whose standpoint is worthy of equal consideration (the principle of universal moral respect) and that furthermore,

6. We ought to treat each other as concrete human beings whose capacity to express this standpoint we ought to enhance by creating, whenever possible, social practices embodying the discursive ideal (the principle of egalitarian reciprocity).[113]

What is significant here is Benhabib's method of rational procedure. Instead of being "a single *deductive chain* of reasoning," this kind of principle is established by a "weak justification strategy," namely, "there

---

tarian exploitation of women and their work" (ibid., 110). Hence, such roles as "husband" and "wife" must be constantly questioned and renegotiated through the process of communicative action.

[112] Ibid., 8.

[113] Ibid., 30-31, emphasis original.

is a family of arguments and considerations each supporting the centrality of this principle as a basic moral norm."[114] The progression between
rules of debate and the norms of respect and reciprocity is at least
plausible, and at most pragmatically necessary. In her words, "All
communicative action entails symmetry and reciprocity of normative
expectations among group members."[115] In terms of social action theory,
to be part of a human group implies this kind of respectful attitude in
order for the group to function in a healthy manner. Conversely, in
conditions of conflict and war, self-esteem and appreciation for others
quickly vanish, "leading to the breakdown of mutuality" and an atmosphere of "extreme indifference and forms of atomized individualism."
Again, this approach leaves much room for those who would argue for a
theological justification behind universal respect and reciprocity.

Put differently, are we simply back to an intuitive grasp of the Golden
Rule? Yes, to some extent, concedes Benhabib. The deontological ethic
of universalizability "enjoins us to reverse perspectives among members
of a 'moral community' and judge from the point of view of other(s)."[116]
The difference between premodern and modern ethical theories lies in
their scope: instead of reciprocity within the boundaries of the in-group
(defined religiously or ethnically), the modern period proposed the
extension of this respect to all human beings.[117] This becomes all the more
relevant today when societies have become pluralistic in the extreme,
and when groups with a premodern view, such as religious traditionalists
or fundamentalists, enter the moral conversation on a national or international level. The presuppositions attached to the conversation are not
a case of circular reasoning, since they can be questioned from within the

---

[114] Ibid., emphasis original. Remember: "weak" here is a compliment for a sound
argument that is not based on a modern-like foundation (that in the end is unjustifiable). This is the kind of postmodern holism I will argue for in the next chapters.
[115] Ibid., 32.
[116] Ibid.
[117] Naturally, I disagree with her that it was the modern and secular perspective that
granted universal respect, a disagreement shared by Jews, Muslims and Christians,
notwithstanding all the wars fought in the name of religion. After all, this is the point
of the creation narrative in both Bible and Qur'an. The fact that all religious groups
have at times interpreted their call narrowly does not invalidate the message of their
sacred texts. I have argued this much on the basis of Jesus' parable of the Good
Samaritan (which answered the question, "who is my neighbor?") relative to the
position of US Christians and US policies in the Middle East. See my essay in a
book by evangelicals who reflect on the "Bush Doctrine": "Loving Neighbors in a
Globalized World: US Christians and Muslims in the Mideast," in *Anxious about
Empire: Theological Essays on the New Global Realities*, ed. Wes D. Avram (Grand
Rapids: Brazos, 2004).

debate. Yet, if they are jettisoned, violence, coercion and oppression might well follow. Sadly, in today's world, it is the latter that predominates over the peaceful moral conversations. Still, if every side is able to perceive the advantages of a discourse ethic that unfolds in a respectful exchange of views with the purpose of finding common ground and practical solutions to particular conflicts, then the ideal of discourse ethics has served humanity in a precious way.

We have seen that Benhabib, together with Habermas, subscribes to a deontological theory of ethics—right precedes good and is defined by the pragmatic-rational procedure of interactive, communicative action. This is an intentional neo-Kantian attempt to find a universal moral discourse that can sustain a democratic politics in an increasingly pluralistic global society. At the same time, she does not dismiss the communitarian concern for the fashioning and strengthening of virtue as a vision of the good life within communities. In effect, she is crafting a political ethic that combines the right and the good. Whereas Kant posited a rupture "between the public virtue of impersonal justice and the private virtue of goodness," Benhabib claims that they can be joined. As she puts it, "The gap between the demands of justice, as it articulates the morally right, and the demands of virtue, as it defines the quality of our relations to others in the everyday lifeworld, can be bridged by cultivating qualities of civic friendship and solidarity."[118] Also, departing from Hannah Arendt who differentiated between moral judgment and political practice, Benhabib, through the "enlarged thought" of discourse ethics, advocates "a fundamental link between a civic culture of public participation" and the moral quality of trying to see from the other's perspective.[119]

This stance also represents a departure from the Kantian assumption that ethics start with a solitary thinker who engages in a purely rational procedure. I already pointed to the problematic of his two-world metaphysics (noumenal vs. phenomenal). But Benhabib takes Habermas to task as well for his Kantian penchant for an ultra-rational conception of morality, which places justice above all other virtues. By taking into account "the perspectives of natality, plurality and the narrativity of action," we are able "to think from the standpoint of everyone else," and thus contribute to a public culture in which more and more room is made for the perspectives of others, and especially the oppressed—women and minorities of all sorts.[120] This view of the public sphere also has the

[118] Benhabib, *Situating the Self*, 140.
[119] Ibid.
[120] Ibid., 141.

advantage of being holistic: morality (including the principles of respect and reciprocity) permeates all levels of society and discourse ethics becomes the glue that allows a democratic society to cohere, from the private spheres of family and friends to the civil society of churches, mosques and associations of all kinds, and to the institutions of the state.

Coming back to the international norms of human rights as spelled out through the various covenants passed under the aegis of the United Nations, we can see how hospitable discourse ethics turns out to be for this mindset. Though secular in nature, religious people of all stripes can use this platform and come to the table of discussion—whether on the local, national or international levels—in order to participate in the ongoing debates about democracy and human rights.[121] Further, discourse ethics allow us to subscribe to the "weak version" of postmodernism and communitarianism, for although the individual is "constituted by discourse," he or she is "not determined by it."[122] As persons embedded in culture and history, we nevertheless fulfill the double role of authors and characters in the narrative that we help to write, whether on a page or in daily life.

Indigenous Peoples as Case Study. This principle also has the consequence that at the macro level of politics majority rule should never allow states to oppress the life and culture of minority groups. Nevertheless, the task of adjudicating such norms is no simple one, particularly for developing countries that came into existence by fiat of the colonial powers, often with borders that arbitrarily divided ethnic or religious groups down the middle. A further quandary is that represented by indigenous peoples. I offer one simple question to illustrate what I have said: how can the non-controversial principle of self-determination of

---

[121] Here I anticipate a later chapter by referring indirectly to Catholic theologian Hans Küng, who helped to draft the Declaration (ratified quasi-unanimously) of the Parliament of the World's Religions. Küng finds the common ground of all religions to be in the concept of humanity as intrinsically worthy of respect. See Hans Küng and Karl-Josef Kushel, eds., *A Global Ethic: The Declaration of the Parliament of the World's Religions* (New York: Continuum, 1993). See also *Islam and the Secular State: Negotiating the Future of Shari'a* (Cambridge, MA: Harvard University Press, 2008) by Sudanese-American scholar Abdullahi Ahmed An-Na'im. In this landmark book, the culmination of several decades of research, An-Na'im argues persuasively that Islam today can only flourish in a secular state in which all factions can equally participate in a democratic conversation.

[122] This is Benhabib quoting Butler (*Gender Trouble*, no page given), in "Feminism," 218.

peoples (read "nation-states") be extended to indigenous peoples within existing nation-state structures? Though I have no space seriously to answer this question, I will use discourse ethics to point in a specific direction.

Australian academic Paul Keal explores the concept of "group rights" concerning the claims made in recent years by indigenous peoples.[123] No doubt, the issue of self-determination of a minority group within the confines of a nation-state is a problematic one. Moreover, it is tied to the notion of culture: it is not individuals who claim rights, but rather a cultural group. In order to clarify the difficulty, Keal uses the term "international society" as a tool for discussing collective rights. He defines it as "a society of states."[124] He then explains:

> This means that in crucial respects indigenous peoples, in common with non-indigenous individuals, generally have had a place in international society only as citizens of states. But one of the complaints of indigenous peoples is precisely that the states of which they are a part have deprived, and continue to deprive, them of political, cultural, and property rights. Consequently many indigenous peoples seek recognition of an international personality that will support their claims against states over issues not already covered in existing human rights instruments. To bring peoples into international society in this sense would be to give them a distinct international personality and ensure their group rights.[125]

A little background is called for here. Positive international law grew out of the last century's world wars. In one sense, it was a constructive effort to protect the world from the devastating destruction it had just experienced. In another sense, it contributed to an artificial separation between the ethical status of individual states (which could be represented in the UN) and the ethical status of international society. One of Keal's central arguments is that the moral legitimacy of international society is directly linked to the moral legitimacy of its member states. Historically, international society expanded through conquest and dispossession. From a moral standpoint, it could be argued that the present

---

[123] Paul Keal, *European Conquest and the Rights of Indigenous Peoples: The Moral Backwardness of International Society*, Cambridge Studies in International Relations 92 (Cambridge: Cambridge University Press, 2003).

[124] One of his main references for this concept (from a liberal perspective) is Chris Brown, *International Relations Theory: New Normative Approaches* (New York: Columbia University Press, 1992). Keal is more in line, however, with the critical approach taken in Timothy Dunne's study, *Inventing International Society: A History of the English School* (London: Macmillan, 1998).

[125] Keal, *European Conquest*, 53.

structure of hierarchical power with the former imperial states at the top lacks legitimacy. In addition, other peoples were excluded from membership until they conformed to Western ways of political organization—adding a second layer of injurious Western hegemony. Some argue that as long as world stability is maintained, there is no reason to interfere with domestic policies within individual states.[126] Yet this is to ignore the moral legitimacy of the system itself. One has to consider the past—the colonial legacy of conquest and the Western imposition of norms—but, equally, one should evaluate the present: the moral legitimacy of the whole (international society) can only be derived from the moral status of its composing states. If any state routinely tortures, deprives or suppresses the rights of any of its minorities, then the moral legitimacy of both that state and international society as a whole should be called into question.

All this is to say that a moral conversation must be initiated on the international level, along with the parallel debates on the reformation of the UN, and, in particular, the need to democratize the security council. Though Keal strongly advocates a "rational" discussion in order to delink the norms of sovereignty and self-determination from the state, he is skeptical that the rationalist tradition in international relations is able to contribute to this effort. After all, "Rationalism draws on the very classical theory that was implicated in the denial of indigenous rights and codified difference." Critical theory and other more postmodern approaches to international relations offer more promise in this regard—especially as they deal with the following concerns: "accepting and dealing with difference, the meaning of autonomy and what is needed to achieve it, and, extension of the boundaries of moral community."[127] Specifically, Keal cites Andrew Linklater who has consistently worked with discourse ethics in this area. For Linklater, discourse ethics "argues that human beings need to be reflective about the ways in which they include or exclude others from dialogue." This means that boundaries of all kinds should be "problematized" and "that the legitimacy of practices is questionable if they have failed to take into account the interests of

---

[126] This is the position argued by French literary theorist-turned ethicist Tzvetan Todorov (born in Bulgaria in 1939) in his lecture, "Right to Intervene or Duty to Assist?" (in *Human Rights, Human Wrongs: The Oxford Amnesty Lectures 2001*, ed. Nicholas Owen [Oxford/New York: Oxford University Press, 2003], 28-48). Harvard scholar Michael Ignatieff in "Human Rights, Sovereignty and Intervention" (pp. 52-87 of the same volume) argues for intervention in specific cases, though he shares Todorov's reservations about the use of outside military force to coerce states to apply international norms of human rights.

[127] Keal, *European Conquest*, 23.

outsiders." As a rule, "Discourse ethics argues that norms cannot be valid unless they can command the consent of everyone whose interests stand to be affected by them."[128]

The conclusion of Keal's book, not surprisingly, is a plea for an international adoption of the UN's Draft Declaration on the Rights of the Indigenous Peoples. Drafted in 1993 with an unprecedented participation of indigenous peoples worldwide, the Draft UN Declaration is comprised of nine parts and 45 articles altogether. Paradoxically, Keal notes, it is a document that draws inspiration from the Western liberal tradition of rights discourse and yet at the same time subverts that very tradition. The crucial issue here is the role of states in international society. Again, the legitimacy of the world order depends to a great extent on the possibility of holding individual states accountable for mistreating their populations. Hence, it is the concept of "human community" that, in the name of individual human rights, must trump "the society of nations" and the notion of "norms of world order" that must set the tone for a universalist moral discourse that at the same time preserves diversity. In one sense, to grant self-determination to indigenous populations in and out of existing state borders, is to affirm a plurality of definitions of the good life, and, in some cases, definitions that run counter to the current notion of human rights. In effect, it is a modification of the present conception of human rights that allows indigenous peoples "communitarian" rights.[129] On the other hand, it is an expansion of classical liberal theory in a more democratic direction, much in the vein of what Benhabib is advocating in a more recent work.[130]

The irony of the present time, writes Benhabib, is that "global integration is proceeding alongside sociocultural disintegration, the resurgence of various separatisms, and international terrorism."[131] Global capitalism and regional unification movements such as the European Union have redefined the modern conception of citizenship in radical ways.[132]

[128] A. Linklater, "Citizenship and Sovereignty in the Post-Westphalian State," *European Journal of International Relations* 2.1 (1996): 85-86, cited in Keal, *European Conquest*, 194.

[129] Article 34 states it clearly: "Indigenous peoples have the collective right to determine the responsibility of individuals to their communities" (Appendix in ibid., 233).

[130] Seyla Benhabib, *The Claims of Culture: Equality and Diversity in the Global Era* (Princeton, NJ/Oxford: Princeton University Press, 2002).

[131] Ibid., viii.

[132] She refers to Weber's unitary definition of citizenship, based on four elements: residency, administrative subjection, democratic participation, and cultural membership. This provided "'ideal typical' model of citizenship in the modern nation-state of the West" (ibid., 180-81).

Significantly, her concluding chapter is entitled, "What Lies Beyond the Nation State?" What direction should we take? In answering this question, she disagrees with the "multicultural theorists," who, in their promotion of cultural diversity, tacitly accept an incommensurability of cultural dialogue.[133] These thinkers arrive at such a conclusion because of their confused epistemology—they have in effect reified cultures, transforming them into unchangeable essences. In reality, just as the self is constituted within the horizon of a particular culture and uses its agency to weave a personal narrative amid "the macronarrative of collective identity,"[134] so the cultural narratives themselves, in constant interaction with other cultures and dominant political and economic forces from outside, are always in the process of redefining their own norms and practices. Whether referred to by conservatives or progressives, this "reductionist sociology of culture" assumes three "faulty epistemic premises:"

> (1) that cultures are clearly delineable wholes; (2) that cultures are congruent with population groups and that a noncontroversial description of the culture of a human group is possible; and (3) that even if cultures and groups do not stand in a one-to-one correspondence...this poses no important problem for politics or policy.[135]

By contrast, leaving behind the multiculturalist theory, Benhabib proffers a "democratic theory" that is concerned with the contesting of cultural identities in civic spaces.[136] In fact, a certain amount of legal pluralism within countries is acceptable under three conditions that follow logically from her commitment to communicative ethics.[137] The first, "egalitarian reciprocity," stipulates that minorities should enjoy the same civil, political, cultural and economic rights as the majorities. The second, "voluntary self-ascription," is the right of individuals to choose

---

[133] This is the ethical relativist thesis, long in vogue in anthropological circles, now somewhat modified to suit social theory in a more multidisciplinary way.

[134] Ibid., 15.

[135] Ibid., 4.

[136] For a good example of how democratic procedures can potentially redress injustices done to indigenous peoples, see Euan Denholm's "Congo's Marginalized Pygmies See Hope in Polls," *Reuters Online* (August 21, 2006): http://today.reuters .com/news/articlenews.aspx?type=worldNews&storyID=2006-08-22T010531Z_01 _L15576369_RTRUKOC_0_US-CONGO-DEMOCRATIC-PYGMIES.xml& archived=False (accessed August 22, 2006).

[137] Particularly with respect to gender issues, Benhabib considers several case studies, including the "scarf affair" in France and the necessities for alternate family law regimes for minorities—already a reality in countries such as India, Egypt, Australia, the United Kingdom, and Israel.

to determine continuing membership in whatever group to which the state may assume (or impose) their belonging. Finally, the principle of "freedom of exit and association" expands the preceding principle in allowing individuals to change their membership from one community to another—a case especially important for spouses marrying outside their native communities.

Though she does not treat Aboriginal issues, Benhabib does briefly reflect upon this question. In promoting her "vision of cultural plurality and democratic contestation," she sees two main obstacles to its actual implementation today.[138] The first category is composed of those—like the indigenous populations of the Amazon or the Saami of Norway— whose cultural differences are tied to a particular territory. They rightly seek to preserve their traditional ways of life, opines Benhabib, particularly in view of the past history of conquest and subjugation. Her solution goes far in the direction advocated by Keal: "While being greatly skeptical about the chances for survival of many of these cultural groups, I think that from the standpoint of deliberative democracy, we need to create institutions through which members of these communities can negotiate and debate the future of their own conditions of existence."[139]

Meanwhile, the second greatest obstacle to a global intercultural dialogue for Benhabib is the rise of fundamentalism across a wide spectrum of religions and cultures. While some fundamentalisms are able to discover modes of adaptation to and compromise with the pluralist ethos of democratic societies, one category apparently lacks this ability: "the rejectionist fundamentalists who find it most difficult to live in a globalized world of uncertainty, hybridity, fluidity, and contestation."[140] Aside from the racist ideologies of exclusion still active in the traditionally Christian territories of the United States and Europe and the reassertion of Hindu supremacy in India in the last two decades, the most worrisome for the Western-dominated global order of this new century is undoubtedly, she notes, the Islamic version of fundamentalism.

Admittedly, these issues of Aboriginal rights and current intercultural confrontations that threaten to spin out of control will not find simple solutions. Yet I would argue with Benhabib and others that discourse ethics—in combination with what people of various faiths can contribute in order to bolster a universal norm of human dignity—does point to a way forward that is both hopeful and pragmatic. And, as I argued

---

[138] Benhabib, *The Claims of Culture*, 184.
[139] Ibid., 185.
[140] Ibid., 186. See Chapter 9 for a fuller discussion of these issues in a Muslim context.

previously, respect for the inalienable rights of the human person can still be salvaged from the wreckage of the modernist paradigm of the autonomous subject who ponders the world from a distance, the better to subjugate and exploit it. The communitarian and postmodernist paradigms, notwithstanding their valuable contribution of a situated subject, move too far in the opposite direction: critical distance as well as personal agency can become obliterated in the process. The individual subject cannot die, unless one gives up the hope of improving the lot of abused individuals and oppressed minorities in our shared world. I suggest therefore that in the company of Habermas (with Harvey and the many others quoted in this chapter) we ought to reflect on how to continue the project of modernity in a more constructive fashion. Part of that task will force us to take another look at history.

Interestingly, the ethical questions raised by today's global democratic challenges—issues of Turkey's potential admittance into the European Union, for example, or the Security Council responses to Iran's self-affirmed right to uranium enrichment—touch upon how one conceives the human person in a pluralistic context. Modern definitions, shaped as they were in the Western crucible of the Renaissance, empire-building and the industrial revolution, are increasingly contested, both from secular quarters (be they liberals, communitarians or poststructuralists) and from religious quarters of various stripes. As Stephen K. White argued in his seminal book *Sustaining Affirmation*, what we desperately need is a "weak ontology."[141] These are the questions taken up in the next chapter—specifically, the "death of history" and the "death of metaphysics."

---

[141] Stephen K. White, *Sustaining Affirmation: The Strengths of Weak Ontology in Political Theory* (Princeton, NJ/Oxford: Princeton University Press, 2000). I have no space in this project to comment on this important work. Yet this is also by choice: White's arguments draw on the work of theorists George Kateb, Charles Taylor, Judith Butler and William Connolly to argue these "weak ontologists" are effectively countering the strong ontologies of many liberals and yet paying attention to the necessary work of ontology (what is the human being—the self?), as opposed to the communitarians, postmodernists and poststructuralists who mostly sidestep the issue altogether. His plea for a more open, kinder and compassionate democratic conversation in the public square is very similar to Benhabib's use of Hannah Arendt and her own ethic inspired by Habermas. Benhabib appears nowhere in White's book, however. Apparently, critical theorists are not part of the conversation.

Chapter 3

BEYOND MODERNISM:
FROM THEORY TO ACTIVISM

Malcolm X and Martin Luther King Jr may not have agreed on the principle of nonviolence, but they both contributed to the 1960s civil rights movements in crucial ways. Both were launched into mission in 1948—Malcolm X as a prison convert and King as an ordained Baptist minister. Strikingly, both in the last year or so of their lives broke out of the confines of black emancipation to pursue a global quest for human liberation.

In March 1964 Malcolm severed his ties with the racist ideology of the Nation of Islam, and upon founding The Mosque, Inc., he embraced a prophetic mission to unite and empower America's Black Muslims. His design was to "expand the civil-rights struggle to a higher level—to the level of human rights" and to bring US racial sins before the tribunal of international opinion.[1] The true turning point, however, came during a five-week trek the next month through a number of Muslim countries, culminating in his performance of the Hajj. From Lagos, Nigeria, he wrote of the impressions that had begun to shatter his previous world-view:

> Never have I witnessed such sincere hospitality and the overwhelming spirit of true brotherhood, as is practiced by people *of all colors and races* here in this ancient holy land, the home of Abraham, Muhammad and all the other prophets of the Holy Scriptures... There were tens of thousands of pilgrims from all over the world. They were *of all colors*, from blue-eyed blonds to black-skinned Africans, but were all participating in the same ritual, displaying a spirit of unity and brotherhood that my experience in America had led me to believe could never exist between the white and non-white.[2]

[1] Malcolm X, "The Ballot and the Bullet," in *Malcolm X Speaks* (New York: Grove Press, 1965), 35, cited by Burns, *To the Mountaintop*, 231.
[2] Malcolm X, "Letters from Abroad," in *Malcolm X Speaks: Selected Speeches and Statements*, ed. George Breitman, 5th print (New York: Pathfinder, 2002), 74-75, emphasis original.

A couple of days after his return, Malcolm X confessed the following to a crowd in Chicago: "I have permitted myself to be used to make sweeping indictments of all white people, and these generalizations have caused injuries to some white people who did not deserve them."[3] He went on to condemn all forms of racism, wishing "nothing but freedom, justice and equality: life, liberty and the pursuit of happiness—for all people." Sadly, these God-given, "inalienable rights" were being kept from his people, the "Afro-Americans." No doubt, the speech in Cleveland, uttered just a couple of months before ("The Ballot or the Bullet"), had laid out a strategy of expanding "the civil-rights struggle to the level of human rights," and taking it "into the United Nations, where our African brothers can throw their weight on our side," along with all other people of color, from Latin America to Asia.[4] But now, he was speaking after his return from the Hajj, and his program of human rights came on the heels of his "spiritual rebirth." At this stage, he was willing to work with a wider spectrum of people, including American whites, those at least whom he considered willing to discuss the issues in good faith.

At the end of June, Malcolm founded the Organization of Afro-American Unity and was permitted to attend as observer the second meeting of the Organization of African Unity in Cairo. In an eight-page memorandum handed to the African heads of state, Malcolm recounted the brutal history of his people's slavery in America, "your long-lost brothers and sisters," and what they now suffer—"the most inhuman forms of physical and psychological tortures imaginable."[5] He also reported how several visiting Africans had recently been beaten by the New York police, mistaking them for "American Negroes." Pleading with these African leaders on the basis that "our problem is your problem," Malcolm declared that this injustice concerns all of humanity: "It is not a problem of civil rights but a problem of human rights."[6]

In February of 1965, during one of King's longer stays in the Selma jail, two of his leading activists drove an hour east to hear Malcolm X speak. They went up to him after his speech and invited him to visit Selma. To the shock of the top Southern Christian Leadership Conference (SNLC) cadres, James Bevel and Andrew Young, Malcolm X appeared the next morning at an SNLC youth rally at the Selma Brown

---

[3] Ibid., 73.

[4] Ibid., 48. The editor, George Breitman, informs us that this speech was given to the Congress of Racial Equality meeting in a United Methodist church in Cleveland on April 3, 1964, just ten days before this trip abroad.

[5] Malcolm X, "Appeal to African Heads of State," in ibid., 90.

[6] Ibid., 91.

Chapel. Bevel and Young cautiously invited him to speak in the pre-rally brief, but with the understanding that he should measure his words. Malcolm's retort is now famous, "Nobody puts words in my mouth."[7] They later apologized to him, for his speech was extremely supportive of King's effort and he had been at pains throughout to emphasize common ground. His only practical exhortation—beyond doing what they were already doing—was to air their grievances at the United Nations.

Though he was not able to visit King in jail, Malcolm told Coretta King to greet her husband warmly for him. Tragically, only two weeks later, Malcolm was assassinated by members of the Nation of Islam, who considered him to be an apostate. What adds to the tragedy is the strong presumption that Malcolm X and King would have cooperated in the future, adding more strength and legitimacy to a movement that was now expanding to a global scale.

Stewart Burns observes that "by 1967 King seemed to be following the example of Malcolm X." In a SNLC staff retreat, King recounted to his followers how the signing of the Bill of Civil Rights had in fact not brought about the end of racism and poverty for US blacks—in fact, the resulting backlash had put them further behind. He had come to realize that "civil rights" carries with it too much of the Western baggage of "individualism, and not enough counterweight from a tradition of communitarian impulses, collective striving, and common good." This legacy had been handed down to them by their forbears, by the likes of Gabriel Prosser, Frederick Douglas, John Mercer Langston and A. Jay Du Bois. For these black leaders, rights were always viewed through the prism of their black spirituality. Burns writes, "African Americans had perceived their human rights, no matter how poorly fulfilled, as a covenant with their personal God intervening in history on the side of justice."[8] Hence, King's words to his staff in May 1967: "It is necessary to realize that we have moved from the era of civil rights to the era of human rights. When you deal with human rights you are not dealing with something clearly defined in the Constitution. They are rights that are clearly defined by the mandates of a humanitarian concern."[9]

The parallel pilgrimage of Malcolm X and Martin Luther King Jr provides a graphic illustration of a Muslim–Christian theological reflection in a particular community context and in the furnace of gutsy social activism. This chapter continues the argument that the Enlightenment subject, bearer of individual rights at the expense of communal ones and

---

[7] Burns, *To the Mountaintop*, 269.
[8] Ibid., 323.
[9] Ibid., 322.

capable of objectively surmising the world, is better left behind. This point is admirably driven home by Burns as he reflects on King's view that "rights were more than individual possessions." His was a more holistic and ethically informed view of human rights:

> They were a moral imperative that transcended individual needs. He was rehabilitating the old preindustrial meaning of *right*: something that was right or just (righteous), that one therefore had a "right" to. Rights rightly understood were not whatever a person claimed as his or her due, with no boundaries; but what was required for all people, and thus for each, by the higher laws of justice and love. They were those entitlements that constituted the moral foundation of the beloved community.[10]

King's understanding of rights, as noted before, was also theologically informed. The moral laws were written into the universe by the Creator, who formed human beings in his image. These teachings of the Hebrew Scriptures project a holistic and communitarian worldview that is shared by traditional societies in general. Accordingly, the rights of the individual impel him or her to make sure others enjoy the same rights. The bonds of solidarity from family to clan, from village to nation, are reinforced by God-given rights, which are clearly portrayed in the Torah as "shared resources." Burns contrasts this with modernism: "This perspective diverged sharply from the classic liberal ideology of unbounded rights, owned by isolated, unencumbered selves devoid of community ties." He adds, tellingly: "King came to have hardly more affinity for such individualistic rights than he had for unbounded freedom or democracy, coins of the same realm."[11]

As noted previously, theology, philosophy, politics and economics are mutually dependent and, when we compartmentalize them, we end up distorting reality. I argued in the previous chapter that the modernist view of the autonomous self was progressively rejected by a flurry of movements from the late nineteenth century on, and that the human "I" can only understand itself in a movement of self-reflection over time—the subject as perceived through narrative in a community of "Thous." But with its emphasis on narratives spun out by individuals and communities, postmodernism has, most famously in the words of Jean-François Lyotard, expressed its "incredulity towards metanarratives."[12] This observation led Jane Flax to weigh the postmodernist claim of "the death of history." I continue here Leyla Benhabib's discussion of these

[10] Ibid., 323.
[11] Ibid.
[12] Lyotard, *The Postmodern Condition*, xxiv, quoted in Naugle, *Worldview*, 174; Docherty, *Postmodernism*, 11.

issues, first with the problem of history and then with the larger question of metaphysics—though, admittedly, they overlap at points. The reason for starting this chapter with Malcolm X and King, however, was also to highlight the issue of activism. Thus I later move from theory to the practice of a "postmodern" kind of activism, a kind of holistic perspective that is rooted in specific communities, but which breaks out onto the global scene, much like the Mexican Mayans accomplished with their Zapatista movement.

The Death of History?

Arguably, it was Heidegger, more than any other twentieth-century thinker, who initiated the discussion about being and history, and tilted it toward postmodernism. Against his mentor, Edmund Husserl, Heidegger stated that there could be no "transcendental phenomenology," no timeless philosophy that might claim objective rigor. This is so, he argued, because the basic philosophical category, *Dasein* ("being there," i.e., in the world)[13] relates to being, and, as the human person is immersed in history and culture, ontology cannot be divorced from time—thus his magnum opus, *Being and Time*.[14] Traditional philosophy, from the Greeks onward, imagined that it was possible to build on common human values such as goodness, truth and beauty, and articulate a comprehensive picture of reality, not realizing that such attempts are all caught up in cultural constructions. Worldview, in fact, is always framed and determined by a particular cultural perspective. The modern positivists, in the spirit of the Enlightenment's "prejudice against prejudice,"[15] sought to escape the limitation of human knowledge by confining science to the realm of empirical investigation and thus arrive at reliable scientific results. "In the context of modern critical consciousness, therefore, scientific philosophy becomes the foundation for and culminates in a scientific worldview."[16] Both of these positions, counters Heidegger, obscure the fact that philosophy and worldview are two distinct human phenomena.

Precisely because Heidegger places the human subject (embedded in culture) at the center of philosophical inquiry, he ends up constructing a

---

[13] Also denoting "life" or "existence" in German. For Heidegger, it was human existence as described from a human vantage point.

[14] Naugle, *Worldview*, 128.

[15] This is Naugle's wording (ibid., 253), but he attributes the idea to Hans-Georg Gadamer in his seminal work, *Truth and Method*, 2d rev. ed., translation rev. Joel Weinsheimer and Donald G. Marshall (New York: Continuum, 1993), 269-77.

[16] Naugle, *Worldview*, 134.

hyper-modernism. Ironically, his whole career had been devoted to the critique of the subject-centered philosophy of Descartes that eventually gave way to Nietzsche's nihilism.[17] In a lecture delivered in 1938, "The Establishing by Metaphysics of the Modern World Picture," Heidegger asserted that when people see themselves as subjects in the world (without a god in the background, of course), they can then relate to the world as an object, which presents itself to them as a picture. This objectification of the world also implies that humans must master and dominate it. Thus human beings, since the time of the Enlightenment, have drawn up their own guidelines for the world that they see represented in front of them. This does not mean, however, that the world-views that emerge are all the same or even compatible. In a way that seems to contradict his previous assertions about the necessary distinctions about philosophy and worldview, Heiddegger maintains that this is the age of worldviews, and that, leaning on science, humans now compete in their conflicting worldviews to establish their mastery over the world. For Habermas, this clear-cut alliance of worldview and power is a confusion of philosophy and ideology—a mistake that will have regrettable consequences.

The early Heidegger, recently turned atheist,[18] still found some inspiration in the work of Danish theologian Søren Kierkegaard, the pioneer existentialist: the individual, wrapped up in time and culture, must stake an authentic claim on its existence by acting in history. This was undoubtedly the original motive and insight of his philosophy of Dasein—the self-awareness of the human person as finite and moving toward death. Personal conversion is the resolute embrace of that finality, and "fallenness" is now redefined as forgetting one's condition.

---

[17] Habermas puts it this way: "Heidegger tries to break out of the enchanted circle of the philosophy of the subject by setting its foundations aflow temporarily. The super-foundationalism of a history of Being abstracted from all concrete history shows that he remains fixated by the categories he negates" (*The Philosophical Discourse of Modernity*, 104).

[18] Heidegger had wanted to become a Catholic priest and was known as a Thomist philosopher (in the tradition of Thomas Aquinas) in 1915 when he began lecturing. He was married as a Catholic two years later, just before he was drafted as a weatherman into the German forces in World War I. Returning to his lectureship at Freiburg in 1919, he suddenly broke with Catholic philosophy and theology and, becoming Husserl's assistant, embarked on the philosophical project that led to the first half of *Being and Time* in 1927 (cf. Thomas Sheehan, "Martin Heidegger," in *Routledge Encyclopedia of Philosophy*, vol. 4:308-9). One would have to wonder whether the war had anything to do with his rejection of religious faith; whatever the reasons, the impact of this decision profoundly affected his philosophy.

From 1933 on, however, Heidegger hitched his star to Hitler's and "now he substitutes for this 'in-each-case-mine' Dasein the collective Dasein of a fatefully existing and 'in-each-case-our' people [*Volk*]."[19] The title of his address to German scholars in Leipzig (11 November 1933) on the occasion of being named "Rector" of its university was "Profession of Faith in Adolf Hitler and the National Socialist Movement."[20] The ideological turn explains how his philosophy fell into the very trap it had originally set out to avoid. Following Husserl's critique of the self-delusional claim of scientific objectivity by the positivists, Heidegger had re-immersed humanity into being and rehabilitated the notion of a situated subject. However, with the advent of a powerful sociopolitical movement that was turning the German economy around and endowing its people with a new sense of empowerment, he began to stress the collective dimension of Being and the necessary struggle of worldviews in their bid to master the world. What is particularly "irritating" for his compatriot Habermas, "is the unwillingness and the inability of this philosopher, after the end of the Nazi regime, to admit his error with so much as *one* sentence."[21]

The fact is, to use philosophy to justify an ideology of domination and violence is both to condemn that philosophy to oblivion (at least in that form) and to repeat the imperial mistakes of all the powers that emerged from the Enlightenment age. Morally, it is wrong to condone acts of aggression against fellow human beings, from the Holocaust to the more than forty million souls who perished as a result of the war against Nazi Germany. This could be a moral judgment based on the religiously inspired respect for the sanctity of the human person, or it could be the Habermasian verdict of discourse ethics. On both counts, history condemned Heidegger.

What, then, with the postmodernist claim about "the death of history"? Considering all the disillusionment with the Enlightenment's myth of progress we have surveyed (starting in the nineteenth century), this common assertion by several postmodernist thinkers seems initially more plausible than the preceding one ("death of the subject"). Max Weber's

---

[19] Habermas, *The Philosophical Discourse of Modernity*, 157. Sheehan nuances this allegiance somewhat: "Heidegger supported Hitler's policies with great enthusiasm for at least one year, and with quieter conviction for some ten years thereafter" ("Martin Heidegger," 309). I am aware of the fact that Habermas' interpretation is contested. Sheehan recognizes that the link or lack thereof between Heidegger's philosophy and his Nazi involvement "has been the subject of heated debate." He, for one, sees no link and no change in Heidegger's thought.

[20] Ibid., 155.

[21] Ibid., emphasis original.

labeling of modernity's consequence as "disenchantment" is, after all, a thesis to which (at least in some version) all critical theorists, feminists, communitarians and liberals could subscribe. The strong version of postmodernism, however, might spell the end of all rational reflection altogether. It likely means, as Benhabib puts it, "a prima facie rejection of any historical narrative that concerns itself with the long durée and that focuses on macro- rather than micro-social practices."[22] But then there could be no vantage point for passing an ethical judgment on Heidegger's later thought. This was precisely the point of Habermas's critique of Lyotard, Foucault and Jacques Derrida in *The Philosophical Discourse of Modernity*. This is also why Benhabib devotes a whole chapter to cautioning feminist scholars about their alliance with post-modernism.[23] The "death of history" thesis precludes any interest in historical research and the art of writing history. This is bad news for oppressed groups who struggle "to interpret history in light of a moral-political imperative, namely, the imperative of the future interest in emancipation." If history holds no possibility of meaning for those on the margins of economic development or political participation, then activism on their behalf also becomes meaningless. Benhabib offers this illustration:

> Think for a moment not only of the way in which feminist historians in the last two decades have discovered women and their hitherto invisible lives and work, but of the manner in which they have also revalorized and taught us to see with different eyes such traditionally female and pre-viously denigrated activities like gossip, quilt-making, and even forms of typically female sicknesses like headaches, hysteria and taking to bed during menstruation. In this process of the "feminist transvaluation of values" our *present* interest in women's strategies of survival and histori-cal resistance has led us to imbue these *past* activities, which were wholly uninteresting from the standpoint of the traditional historian, with new meaning and significance.[24]

---

[22] Benhabib, *Situating the Self*, 219-20.

[23] In parallel fashion, feminist writer Sabina Lovibond sees the postmodernist turn as a move away from "Platonism," "the idea that truth goes beyond, or 'transcends', our current criteria of truth. A recurrent feature of postmodernist theory is the claim that Platonism in this sense is obsolete—that is, that it is no longer possible to believe in a transcendent truth against which the whole intellectual achievement of the human race to date could be measured and found wanting" (Lovibond, "Femi-nism and Postmodernism," in Docherty, *Postmodernism*, 392; originally, in *Post-modernism and Society*, ed. R. Boyne and A. Rattansi [Basingstoke: Macmillan Foundation/New York: St Martin's Press, 1990], 154-86).

[24] Ibid., 20, emphasis original.

On the other hand, asserts Benhabib, postmodernism has taught us a great deal, and in particular has reminded us that "it is no longer possible or desirable to produce 'grand narratives' of history."[25] The very reason the Enlightenment sought to elevate reason as the infallible instrument of humanity's emancipation became in Derrida's analysis a "logocentrism" in need of legitimation and a dangerous tool in the hands of totalitarians and imperialists. By virtue of being human, each person must construct his or her own identity on the basis of difference—whether rooted in gender, class, degrees of power, religion or ethnicity. Yet, for Derrida, difference is irreducible and ultimately impossible to overcome.

Early on, in a text he was to present in October 1968 at an international colloquium on "Philosophy and Anthropology," Derrida introduced himself as the disciple of Heidegger who, on the basis of the latter's critique of the West's anthropocentric tradition, could now announce that "a radical trembling" was at the door. "This trembling is played [out] in the violent relationship of the whole of the West to its other, whether a 'linguistic' relationship...or ethnological, economic, political, military, relationships, etc."[26] In a long preamble he had pointed out that an international gathering of this type presupposed a "form of democracy." By that he means, "*democracy* must be the *form* of the political organization of society,"[27] and it guarantees, at the very least, that philosophers who participate do so without having to endorse the politics of their respective governments. He had come to this conference assured that he could publicly affirm his solidarity "with those, in this country, who were fighting against what was then their country's official policy in certain parts of the world, notably in Vietnam."[28]

But Derrida remains, nonetheless, difficult to decipher. On the one hand, the "linguistic violence" of the West against its Other (which boils down to a definition of Being, he says) is likely to be in "structural solidarity" with its military and economic violence—a clear political statement. On the other hand, "the 'logic' of every relation to the outside is very complex and surprising."[29] We find no elucidation of this "complexity" in the closing paragraphs, except perhaps in his mention of the binary opposition of "superior man" and "superman" in Nietzsche. We

[25] Ibid.
[26] Jacques Derrida, "The Ends of Man," in Derrida, *Margins of Philosophy*, trans. with additional notes by Alan Bass (Chicago: University of Chicago Press, 1982), 134-35.
[27] Ibid., 135, emphasis original.
[28] Ibid., 113.
[29] Ibid., 135.

might interpret the former as the Enlightenment man, who in the words of Heidegger, "must first learn to exist in the nameless."[30] In that case perhaps, we should "take the question of the truth of Being as the last sleeping shudder of the superior man" and conclude that both Nietzsche and Heidegger were pointing in the direction of his own deconstructive agenda by announcing the end of metaphysics.[31] Where does that leave the political question?

In an essay published three years later (1971), "White Mythology: Metaphor in the Text of Philosophy," Derrida broadly canvasses the perennial texts of philosophy in order to "deconstruct" the language on the basis of which metaphysics (the metaphysics of presence, on his reading) has been erected—and Western rationalism by the same token. The entire article centers around an abstract notion of language and economy that Derrida picks up elsewhere. Here the use of metaphor in philosophy "promises more than it gives" and this fact is "in our interest," that is, it brings us more profit than loss. But the term "white mythology" only appears in one small passage, out of character with the rest of the essay, and it is the only place that refers to the West. Yet it substantiates the claim by Benhabib that for Derrida "the logocentrism of the West" is connected to "the imperialist gesture with which the West 'appropriates' its other(s)":[32]

> Metaphysics—the white mythology which reassembles and reflects the culture of the West: the white man takes his own mythology, Indo-European mythology, his own *logos*, that is, the *mythos* of his idiom, for the

[30] The sentence is taken from Heidegger's "Letter on Humanism," in *Basic Writings*, ed. David Farrell Krell (New York: Harper & Row, 1977 [1947]), 199. The full sentence reads, "But if man is to find his way once again into the nearness of Being (*in die Nähe des Seins*) he must first learn to exist in the nameless (*im Namenlosen*)." For Habermas, Derrida has interpreted Heidegger in such a way as to justify his own philosophical project, namely, "the self-overcoming of metaphysics"; the destruction of humanity's access to Being (or reality, one might say) feeds into his project of deconstruction. Commenting on this essay, Habermas identifies the two "ends of man" that Derrida has in mind: "The human being as the being toward death has always lived in relation to its natural end. But now it is a matter of the end of its humanistic self-understanding: In [*sic*] the homelessness of nihilism it is not the human being but the essence of the human that wanders blindly about. And the end is supposed to be disclosed in the thinking about Being initiated by Heidegger" (Habermas, *The Philosophical Discourse of Modernity*, 161).

[31] Derrida, "The Ends of Man," p. 136.

[32] Benhabib, *Situating the Self*, 15. She explains: "The Orient is there to enable the Occident, Africa is there to enable western civilization to fulfill its mission, the woman is there to help man actualize himself in her womb, etc.... The logic of binary oppositions is also a logic of subordination and domination" (ibid.).

universal form of that he must still wish to call Reason. Which does not go uncontested…

White mythology—metaphysics has erased within itself the fabulous scene that has produced it, the scene that nevertheless remains active and stirring, inscribed in white ink, an invisible design covered over in palimpsest.[33]

This is no doubt an exaggeration, Benhabib contends, or rather a distortion of the Western philosophical tradition. By claiming on the heels of Heidegger that "Western metaphysics has been under the spell of the 'metaphysics of presence' at least since Plato," Derrida schematizes this tradition in such a way that his own "metaphysics of absence" (the result of deconstruction) can easily knock down any rival system. Having said that, for Benhabib there is much credibility in the postmodern contention that discourse often hides within it structures of power that are easily unleashed in oppressive ways, both socially and politically.[34] For example, she highlights Michel Foucault's contribution in the denouncing of Western reason's hegemonic ambitions: "In his impressive genealogies of reason Foucault uncovers the discursive practices which have drawn the line between madness and civilization, mental health and sickness, criminality and normality, sexual deviance and sexual conformism. Foucault shows that the other of reason comes to haunt this very reason."[35]

The unmasking of the hegemonic potential within language and its connection to social practices (both participating in the "discursive" mode for postmodernists) already presupposes the Enlightenment ideal of human autonomy and the revolutionary right to oppose autocratic regimes—violently if necessary, as in the French Revolution. For Benhabib, this is the irony of the postmodern stance. The positing of "a fractured, opaque self" hardly allows battered individuals or groups to

---

[33] Jacques Derrida, "White Mythology: Metaphor in the Text of Philosophy," in *Margins of Philosophy*, 213.

[34] Andrew Cutrofello defends Derrida on the issue of whether, as some critics maintain, "deconstruction undermines the very possibility of making ethical and political determinations." He answers, "Such interpretations overlook the constant concern with justice that informs Derrida's works. Working both with and against the Kantian conception of regulative ideals, Derrida seeks to delineate a conception of justice as an *aporia* which both calls for, and is called forth by, the work of deconstruction" (Cutrofello, "Jacques Derrida," in *Routledge Encyclopedia of Philosophy*, vol. 2:308-9). Without doubt, there is enough fluidity and ambiguity in Derrida's work (as in Heidegger before him) to allow for a spectrum of judgments on its meaning and implications.

[35] Benhabib, *Situating the Self*, 14.

struggle for social justice.[36] Equally, the suspicion of all master narratives leads Foucault to see "every act of resistance" as "but another manifestation of an omnipresent discourse-power complex."[37]

Yet neither the human self nor human history is as opaque as some postmodernist thought has made them out to be. As I have argued before, plausible connections can be found between economic policies and cultural phenomena, between economic and political conditions (recall the European revolutions of 1848) and the evolution of philosophy and the arts. Nor is it difficult to imagine how the late Heidegger's seduction under the spell of a particularly noxious kind of master narrative created ripples in post-World War II philosophy. Other events only reinforced the sense that the West's foundations had been irreversibly damaged in that war.

Thomas Docherty cites the year 1968 as a turning point, when spontaneous uprisings occurred all over Europe, with workers and intellectuals hand in hand. Without doubt, this eruption of anger was fueled by leftist students and pro-communist trade unions. But in fact, what Arthur Mitzman calls the French "almost revolution" of 1968 turned out to be in retrospect the seedbed of continental postmodernism.[38] The essay Derrida presented in October 1968, which spoke of the "radical trembling" intellectuals were sensing in the world, was written in Paris during the momentous May riots that shook France to the core.[39] Docherty acquiesces: the fact that expectations were not met—European political and social structures remained on a capitalist, top-down, technocratic track—led many Marxists and left-leaning people to rethink seriously the tenets of classical Marxism.[40] Class was no longer the only issue; there

---

[36] Ibid., 16.

[37] Ibid., 222.

[38] Mitzman offers an enlightening case study of French oppositional politics and their metamorphosis from the "events" of 1968 into the general strike of 1995 as a microcosm of the global movement now gaining ground against corporate capitalism (*Prometheus Revisited*, 161-65). I can personally testify that in May–June 1968, as a lycéen (high school student) in Lyon, revolution was indeed in the air—just about all secondary students were also on strike for that period. However, the goals of the movement were still local, not yet global.

[39] Derrida mentions them in the long preamble, as well as the assassination of Martin Luther King Jr.

[40] One irony of history is that student protesters decided to occupy the German premises of the Frankfurt School in January 1969. Here was a movement that had tried to integrate sociology and philosophy, economics and history in order to promote a critique of society that would express the true ideals of human emancipation, and yet now it was opposed by a younger generation in the name of the same

was a general dissatisfaction among the youth with "the Fordist welfare state at its high point." This translated in a three-pronged revolt:

1. Young people protested the rigid mores of an older generation that had not grown up with the material ease and security they had experienced in the sixties; the "hippie" culture of American youth also reflected this new way of being young and free.

2. The young rebelled against the prospect of being forced into a dull, bureaucratic and alienating work environment, whether in offices or factories.

3. The fallout out of the brutal wars of Indochina and Algeria for the French served to enhance the general repulsion of European youth by the imperial project of the leader of the so-called free countries in Vietnam; behind the mask of humanism and democracy, the US was cruelly beating down on a local liberation movement.[41]

Indeed, the social revolts of 1968 led in short order to the European green parties—one of the political manifestations of postmodernism in its activist mode. In switching from a "red" revolution to a "green" one, many young intellectuals and activists latched on to a new paradigm that connected the anti-colonial movements of days past with a new concept of international solidarity based on ecology and local autonomy rather than class. Along with the post-Marxist gropings of Lyotard[42] and Jean Baudrillard,[43] the period was also giving birth to postcolonial thought. Frantz Fanon's explosive work, *The Wretched of the Earth*, came out in 1968.[44]

---

ideals ("Frankfurt School," in *The Penguin Dictionary of Critical Theory*, ed. David Macey [London/New York: Penguin Books, 2000], 139).

[41] Mitzman, *Prometheus Revisited*, 162.

[42] Jean-François Lyotard, *L'Economie Libidinale* (Paris: Minuit, 1974).

[43] Jean Baudrillard, *Le Système des Objets* (Paris: Gallimard, 1968); *La Société de Consommation* (Paris: Gallimard, 1970); *Le Miroir de la Production* (Paris: Casterman, Tournail, 1973).

[44] Frantz Fanon, *The Wretched of the Earth*, trans. Constance Farrington (New York: Grove Weidenfeld, 1968). Ato Quayson emphasizes the eclectic nature of postcolonialism as a movement. Above all, it is a "critical practice" that "involves a studied engagement with the experience of colonialism and its past and present effects at the levels of material culture and of representation." Both the experience of past colonies and the life of people living in diaspora today "are frequently linked to the continuing power and authority of the West in the global political, economic and symbolic spheres and the ways in which resistance to, appropriation of and negotiation with the West's order are prosecuted" (Quayson, "Postcolonialism," in *Routledge Encyclopedia of Philosophy*, vol. 7:578). This present work, undoubtedly, has affinities with postcolonialist thought.

Among the "postcolonial" writers who found their voice at the time, Edward Said (1929–2003), a Palestinian literary critic at Columbia University, merits some attention here. He not only provided articulate analyses of his people's plight with a consistent activist stance, but he internalized the postmodern critique of language and deployed it in ways that energized and inspired scholars in many fields. Further, his work has influenced many Muslim writers.

In the introduction to his seminal work, *Orientalism*, Said explains the three interrelated meanings of that word. "Orientalism" is most commonly understood as that field of academic endeavor which takes the "Orient" as its object of study and to which anthropologists, sociologists, historians and philologists bring their expertise. It is also a style of thought based on the alleged epistemological distinction between the "Occident" and the "Orient." A great many writers (he mentions Aeschylus, Victor Hugo, Dante, and Marx) "have accepted the basic distinction between East and West as the starting point for elaborate theories, epics, novels, social descriptions, and political accounts concerning the Orient, its people, customs, 'mind,' destiny, and so on."[45] Of course, he goes on to say, there has been a lot of crossover between the two genres. Yet it is the third definition that claims his energies in this volume—a phenomenon more precisely defined in history:

> Taking the late eighteenth century as a very roughly defined starting point Orientalism can be discussed and analyzed as the corporate institution for dealing with the Orient—dealing with it by making statements about it, authorizing views about it, describing it, by teaching it, settling it, ruling over it: in short, Orientalism as a Western style for dominating, restructuring, and having authority over the Orient.[46]

From the beginning, Said acknowledges his intellectual debt: "I have found helpful to employ Michel Foucault's notion of discourse, as described by him in *The Archaeology of Knowledge*...to identify Orientalism."[47] Said taps into Foucault's revelation of the power dimension of discourse in order to lay bare the West's political, cultural

---

[45] Edward Said, *Orientalism* (New York: Pantheon, 1978), 2-3.

[46] Ibid., 3. An intriguing perspective comes from Richard C. Martin and Mark R. Woodward (with Dwi S. Atmaja). They note that in providing critical editions of Hindu scriptures, which had been hitherto fragmented and dispersed, the Western Sanskrit scholars helped to create "a 'Hindu' historical and textual tradition. This in turn became an (unintended) scriptural focus for Hindu (and Muslim) communalism based on religious nationalism" (*Defenders of Reason in Islam: Mu'tazilism from Medieval School to Modern Symbol* [Oxford: Oneworld, 1997], 3).

[47] Said, *Orientalism*, 3.

and ideological manipulation of the Orient. The Orientalist discourse in many cases reveals more about the anxieties of the European self-identity than about the actual culture and thinking of the "Other":

> This is not to say that Orientalism unilaterally determines what can be said about the Orient, but that it is the whole network of interests inevitably brought to bear on (and therefore always involved in) any occasion when that particular entity "Orient" is in question. How this happens is what this book tries to demonstrate. It also tries to show that European culture gained strength and identity by setting itself off against the Orient as a sort of surrogate and even underground self.[48]

There is no need here to discuss his thesis in detail—Said has been critiqued at great length and his work is now somewhat dated.[49] What occupies us here, again, is the method. It is Said's epistemology that highlights the subjective character of text production and strategies of readers as they encounter the text from within their own context.[50] Though he is careful not to denigrate the massive work of collecting, translating, and annotating ancient manuscripts and the intellectual prowess in general done by western islamicists, he nonetheless exposes their prejudices and the connection between their work and the colonial

---

[48] Ibid.

[49] I would have to agree with Hanan Ashrawi, however, that Orientalism is alive and well in current US policies and discourse about the Israeli–Palestinian conflict. Deborah Sontag reported that the former Palestinian Authority Minister of Higher Education lamented Washington's one-sided praise for Prime Minister Barak's stance at the Camp David II peace talks. The assumption was that Mr Arafat could persuade his people to accept any peace deal he made. She goes on, "The American State Department functions from an Orientalist textbook... Arab people are sheep, and all you have to do is push buttons and they can be herded. I have been repeatedly cautioning them. There is public opinion in our society" (Sontag, "As Arafat Embraces Revolt, His Sagging Popularity Rises," *The New York Times* [December 8, 2000, Section A], 1, 14).

[50] The powerful influence of Said's book goes back to his epistemology, coupled with his own activist stance for the Palestinian cause. Bryan S. Turner is worth quoting here: "An exciting and important challenge of Said's work was what one may call 'the methodology of the text', that is, Said was able to apply the more advanced aspects of American literary studies to the analysis of history and the social sciences; and through what is popularly called deconstructionism, Said was able to provide new directions for the analysis of historical and social phenomena. Certainly Said's approach was very attractive at the time because he provided a model of what we might call the intellectual hero. Said was not simply someone who sat on the margins of literary studies and analytical research, he was actually seen to be at the forefront of Palestinian politics and Middle Eastern politics" (Turner, *Orientalism, Postmodernism and Globalism*, 4).

expansion engineered by their political leaders. On the one hand, admirable research; on the other, "Orientalism overrode the Orient." By way of illustration:

> [A]n observation about a tenth-century Arab poet multiplied itself into a policy towards (and about) the Oriental mentality in Egypt, Iraq, or Arabia. Similarly a verse from the Koran would be considered the best evidence of an eradicable Muslim sensuality. Orientalism assumed an unchanging Orient, absolutely different (the reasons change from epoch to epoch) from the West. And Orientalism, in its post-eighteenth-century form, could never revise itself. All this makes Cromer and Balfour, as observers and administrators of the Orient, inevitable.[51]

Thus the appeal of Said's work lay more in his hermeneutical method. In order to extricate the meaning of a text, one has to ferret out the sociopolitical and cultural elements behind its production. Meaning heavily depends on social structure. In a fashion similar to that of Talal Asad,[52] Said squarely raises the issue of any group trying to analyze another, the inevitable "Other." In all cases there will be philosophical difficulties in vouching for the veracity of the translation. To what extent is it possible at all? At least by making explicit the extra-textual influences, the task is made easier. But humans naturally tend to highlight the differences between "us" and "them," and then to simplify the discrepancies as much as possible. As Bryan Turner explains, "We understand other cultures by slotting them into a pre-existing code or discourse which renders their oddity intelligible."[53] By drawing attention to this problem, Said opened the way for a whole spate of new directions in the social sciences:

---

[51] Ibid., 96. He also illustrates the ambivalence of Islamicists, as in the case of H. A. R. Gibb. In a 1945 speech, Gibb characterizes the "oriental mind" as incapable of abstract reasoning, "lacking a sense of law," and as rejecting rationalist modes of thought because of the "atomism and discreetness of the Arab imagination." Naturally, we would have to concur: "this is pure Orientalism." Eighteen years later, speaking to an English audience and now as director of the Harvard Center for Middle Eastern Studies, Gibb stresses that to prepare students for careers "in public life and business" what is needed is a combination of "the traditional Orientalist plus a good social scientist working together." Yet the orientalist mindset does not die easily, remarks Said, as shown in Gibb's next sentence: "History, politics, and economics do not matter. Islam is Islam, the Orient is the Orient, and please take all your ideas about a left and a right wing, revolutions, and change back to Disneyland" (ibid., 106-7).
[52] Edward Said, *Genealogies of Religion: Discipline and Reasons of Power in Christianity and Islam* (Baltimore, MD: The Johns Hopkins University Press, 1993).
[53] Said, *Orientalism, Postmodernism and Globalism*, 37.

Said's work was significant in showing how discourses, values and patterns of knowledge actually constructed the "facts" which scholars were attempting to study, apparently independently. Over the years the classical approach to orientalism has largely shaped what people understand by the notion of 'Otherness,' and the problem of the 'Other' in human cultures has been taken up first of all by feminism, by black studies and more recently by postmodernism.[54]

By way of summary, therefore, Said was a literary critic engaged in both cultural and historical studies, whose hermeneutical strategy in *Orientalism* sparked enormous interest for other fields. Though I take up the topic of hermeneutics in the next chapter, it must be said here that a philosophy of discourse heavily influences historical studies—in fact, there can be no "pure" historical research. By necessity, it involves the discriminating hand of researchers, who are guided by choices largely dictated by their own sociocultural context (and who is paying them), and who then organize the material collected around themes of their choosing with the emphases of their choosing. More than ever, historians are acutely aware of the inescapability of hermeneutics.[55] They also realize that they must contend with the research of colleagues in other fields.

In this section, I have attempted to show that, starting with Heidegger's focus on human existence as described and interpreted by a subject situated in time and culture, a current of social critique developed, one which called into question the cornerstone of Enlightenment philoso-

---

[54] Ibid., 4. Specifically, "The point of orientalism, according to Said, was to orientalise the Orient and it did so in the context of fundamental colonial inequalities. Orientalism was based on the fact that we know or talk about the Orientals, while they neither know themselves adequately nor talk about us. According to Said, there is no comparable discourse of occidentalism" (ibid.). The nature of orientalism has changed over the years but "much of the underlying politics of power remains" (Said, *Orientalism*, 44-45).

[55] This is recognized by sociologist Roland Robertson, whose main concern is to establish the theoretical boundaries of the term "globalization." It is a phenomenon, he argues, which urgently calls for interdisciplinary treatment—and in particular historical and comparative issues in sociology and anthropology: "While there have been attempts to carve out a new discipline for the study of the world as a whole, including the long-historical making of the contemporary 'world system'…my position is that it is not so much that we need a new discipline in order to study the world as a whole but that social theory in the broadest sense—as a perspective which stretches across the social sciences and humanities…and even the natural sciences—should be refocused and expanded so as to make concern with 'the world' a central hermeneutic, and in such a way as to constrain empirical and comparative-historical research in the same direction" (Robertson, *Globalization, Social Theory and Global Culture* [London: Sage, 1992], 52).

phy—the use of a "universal reason" to foster a more just and peaceful world. Following Benhabib, I distinguished between the strong postmodern "death of history" thesis and its weaker one. Ironically, the leading proponents of the stronger view, including Derrida and Foucault, engaged in spirited sociopolitical debates, and their connection to the May 1968 revolts reveals the ambiguities of their epistemic positions. I also argued that one must take to heart Said's simultaneous unpacking of orientalism as a discourse of power, his call for a listening posture toward the other and his lifelong advocacy for Palestinian rights.

In light of all this, I submit that the postmodernist skepticism of metanarratives and its highlighting of the tentativeness and vulnerability of all historical narrative can actually become useful perspectives for Muslims and Christians attempting to create a common vision of a better world ahead. On the basis of a theology of creation, both Malcolm X and King fought for universal human rights that were at the same time rooted in community identities and bonding. In this sense, they were anticipating the current bubbling of activism all around the world, which may properly be dubbed "postmodernist."[56] Yet it is a sad fact of history that both Muslims and Christians have resorted to violence to impose religious and political domination. What is more, the sacred texts of Jews, Christians and Muslims bear the marks of the patriarchal societies in which they were revealed. If there is any consensus in the "postmodern" intellectual atmosphere of today, it is that peacebuilding cannot proceed without progress in the area of human rights, and among them, gender equality.[57] On the Christian side, the fact that the idea of female clergy is gaining acceptance in many American evangelical circles demonstrates two things: (1) how firmly entrenched the literalistic hermeneutical tradition has been in conservative circles; (2) and that, even in those circles, changing gender norms and social practices eventually lead people to reinterpret the texts.[58]

---

[56] I will explain below, in the final section ("Postmodern Activism," 152).

[57] The recent edited volume quoted above and below, *Progressive Muslims*, amply demonstrates this point.

[58] A critical work that paved the way for this opening was Paul K. Jewett's *Man as Male and Female: A Study in Sexual Relationships from a Theological Point of View* (Grand Rapids, MI: Eerdmans, 1975). One of the most controversial claims he made therein was that the apostle Paul was simply reflecting his rabbinic training when he argued for the headship of the male in the family and in the Church on the basis that Eve was created after Adam and from one of his ribs. In reality, this was inconsistent with his own theological rule: "There is neither Jew nor Greek, slave nor free, male nor female, for you are all one in Christ Jesus" (Gal 3:28), and in contrast to Jesus' teaching and practice.

In the same way, Ebrahim Moosa, a Duke scholar from South Africa, complains about the reification of Islamic traditions in contemporary Muslim discourse. Complex ideas and practices are reduced to a simplified form, declared to be "the Islamic teaching on the matter," and then history is distorted in order to score ideological points against one's adversaries. Hence, to prove that the "spirit of Islam" is equality and justice, writers adduce the fact that the caliph 'Umar prohibited the sale of women slaves who had children through their masters. This issue, along with that of inheritance, is more than problematic, comments Moosa:

> It is not very clear whether 'Umar was actuated by concerns of freedom in limiting the sale of female slaves who had offspring or whether he wanted to prevent the proliferation of incest. For there were real concerns that a young female slave separated from her offspring when sold off could years later unknowingly be sold as a concubine to her wealthy offspring. It is also uncertain whether the inheritance system intended to further justice. However, there are clear indications that the new system of intergenerational succession attempted to further a specific form and system of kinship based on patriarchy.[59]

In the same context, Moosa asserts that "reason is not a self-evident faculty but a socially constructed one."[60] Though his statements may sound like the "strong version" of postmodernism, in practice he subscribes to the weaker version deployed by Benhabib. Thus he writes that, while there is one "Islam," if we mean certain historical events that took place in seventh-century CE Arabia, we may also profitably speak of a multiplicity of "Islams": "For, whatever Islam *is*, the closest we can come to what 'it' is or is not, is through its embodiment in concrete forms, practices, beliefs, traditions, values, prejudices, tastes, forms of power that emanate from human beings who profess and claim to be Muslim or profess belonging to a community that calls itself Muslim."[61] Naturally, the same can be said about Christianity. Both faiths make assertions about history, but ultimately the meaning of those events or artifacts are tied to statements about a metaphysical reality beyond history. This meaning is then unpacked in the form of traditions that develop in specific contexts and later migrate to other cultural situations, and so on. The metanarrative remains at the core, but its interpretation, ritual and discursive embodiment will always vary according to time and place.

[59] Ebrahim Moosa, "The Debts and Burdens of Critical Islam," in *Progressive Muslims*, 121.
[60] Ibid., 118.
[61] Ibid., 114.

History, then, retains its importance, whether for religious faiths or for social and political causes. Moreover, to assert that, for all the fragility of historical analysis, persons can nevertheless make some sense of history and learn from it, is to return to the beginning of these two chapters on philosophy. Like the conclusions I drew in the previous chapter about the located self, we are now in a position, I would suggest, to make some wider metaphysical claims about the triangle of subject, language and world. The next section tackles the problem of how a person can know and manipulate the world of which she or he is a part. This turns out to be an epistemological question relating to science, language and society, which, if faced head on, helps to clarify the notion of social activism as well.

The Death of Metaphysics?

As mentioned above, Benhabib takes issue with the postmodern assertion that the history of Western philosophy is the ill-fated attempt to define a metaphysics of presence, and in particular, a construing of the "Real" in terms of domination and subjugation.[62] Certainly Heidegger had come close to this when he defined human consciousness (Dasein) as picturing the world the better to master it. He thus ended up with competing worldviews (representing their respective nations), which all sought to exert their technological, economic and political dominion more efficiently and ruthlessly than the others. One would have to admit that the history of the modern period—whether as embodiment of the scientific revolution of the Renaissance or the political aspirations of the Enlightenment—is closely identified with the imperialist ventures of the Western nations and ended in the horrific butchery of two world wars.

But is metaphysical realism (Descartes's positing of the world from the vantage point of the thinking subject) inherently linked to the exploitation of nature and weaker peoples? To assert that would qualify as the advance of another equally "unprovable" metanarrative. This is why I turned to Harvey, Mitzman and Jameson, in order to show that colonial conquests were also tied to the development of capitalism in Western Europe—new ideas and practices of economic exchange, personal property, and the progressive commodification of the commons. Yet to reduce the complexity of historical phenomena to material conditions would be to fall back on another master narrative or metaphysical construction, which is the very reflex contemporary Marxian thinkers

---

[62] Benhabib, "Feminism," 223-24.

have opposed. Rather, as Nancey Murphy has asserted, we must resist all the reductive tendencies of modernism, embrace a form of holism, both in epistemology and metaphysics, while refusing the dead-end road of relativism.[63] Starting with Murphy, and then continuing with theologian, physicist and philosopher of science Ian G. Barbour, I will now seek to map out what this position looks like, and explain how it may inform a Muslim–Christian theology of creation that engages with the "double wall" challenges of the twenty-first century world.

One could argue that the postmodern turn away from the rationalistic dogma of modernity since Francis Bacon began in 1958, the year Michael Polanyi, chemist and philosopher, published his critique of scientific detachment, *Personal Knowledge: Towards a Post-Critical Philosophy.*[64] Starting with the findings of Gestalt psychology, he showed that knowledge in the "exact" sciences is a process of self-involvement, which requires the use of tools, a growing skill in using them, and above all, an art that has been passed on to the disciple by a master within a particular tradition. Students in chemistry, biology and medicine spend long hours in laboratories or with patients learning skills and "connoisseurship" as apprentices to their masters. This fact alone "offers an impressive demonstration of the extent to which the art of knowing has remained unspecifiable at the very heart of science."[65]

On the other hand, though research is tied to tradition and the artful use of accepted tools in the field, it is not merely subjective. The idea of truth has its importance, claims Polanyi: "Intellectual commitment is a responsible decision, in submission to the compelling claims of what in good conscience I conceive to be true."[66] I believe I can connect with truth, he says. But that belief is not in itself verifiable. It is a faith commitment, much in the same way religious belief is.[67] This, naturally,

---

[63] Choosing to explore the realism–antirealism debate (and especially as it relates to science) is to part ways temporarily with Benhabib. She pursues the questions relating to "the death of metaphysics" in the field of ethics and social theory, to which I will return subsequently. Because this work is concerned about humanity's vocation with respect to creation, it is imperative that the metaphysical and epistemological issue be resolved first. Philosophically, how do we believers in a Creator interpret our relationship with the created order around us?

[64] Michael Polanyi, *Personal Knowledge: Towards a Post-Critical Philosophy* (Chicago: University of Chicago Press, 1958).

[65] Ibid., 52.

[66] Ibid., 65.

[67] He calls both science and religious faith "heuristic vision," which a person has chosen to "in-dwell." For instance, reading the Bible (which could mean something quite different to the non-believing person) can supply clues to the Christian's

raise the question of epistemology—one's theory of knowledge. Recall our discussion of the modern episteme of representation that came under assault by a variety of movements, long before the advent of postmodernism in the 1970s. Descartes had wanted to build a solid philosophical superstructure on the foundation of one certitude: namely, that he as a thinking subject could reflect on the world outside of himself. Thus he posited two substances, human thinking with its own rational methods, and the world, which, as Newton had shown, behaved in predictable mechanistic patterns. The modern worldview involves both a metaphysical theory (the world—including other people—exists as an entity outside my mind) and an epistemological one (I can apprehend the truth about the laws that govern it).

The theological worldview of the Middle Ages, whether among the Muslim scientists and philosophers in Baghdad or Cordova or the Christians who borrowed from their insights in polemic form like Thomas Aquinas, taught the reality of the world perceived by human senses and explained by rational human powers. Creation for them was a hierarchical structure, one which postulated God (with planets next in Ptolemaic theory), angels, humans, animals and plants in descending order. Yet all was orderly, law-abiding, essentially static and moral rather than mechanical. In that sense, the world was seen in the picture of "a Kingdom—a fixed, hierarchical, ordered society under a sovereign Lord."[68]

Galileo Galilei's (d. 1642) combined use of experimentation and mathematical equations enabled him to confirm the Copernican model put forward in the previous century.[69] Further, he applied the notions of time, velocity and mass to a variety of physical phenomena and represented them by mathematical symbols. For instance, he calculated the equations of accelerated motion. As in many other instances, he moved

knowledge about God and strengthen his or her faith. Knowledge can come either by direct experience or from books (as in the sciences): "Both kinds of comprehension establish their own heuristic vision which asserts no specific fact. They are forms of highly personal knowledge which subsidiarily comprise a set of relatively impersonal experiences. This relation of factual clues to a heuristic vision is similar to the relation of factual experience to mathematics and to works of art" (ibid., 283).

[68] Ian G. Barbour, *Religion and Science: Historical and Contemporary Issues*, A Revised and Expanded Edition of Religion in an Age of Science [1990] (New York: HarperSanFrancisco, an imprint of HarperCollins Publishers, 1997), 6. This is the landmark book I will be using as my guide in this section.

[69] In addition, by making use of the newly invented telescope, Galileo could point to the mountains on the moon's surface and prove that it was not a perfectly shaped "celestial sphere"; his discovery of Jupiter's moons equally displaced the idea of the earth being at the center of all motion (ibid., 10).

from theoretical assumptions to mathematical calculations, and from experimental testing of these assumptions to a modified theory. Yet, as Ian Barbour cautions, it would be misguided to pinpoint observation as the key to the "new science," as Sir Francis Bacon did in Galileo's lifetime, followed by David Hume and, more recently, the logical positivists. That view contended that only observation and induction were needed, which then led to summarizing and generalizing. The assumption was that nature is mechanistic and that people can unlock its mysteries by simply describing what is "out there."[70] What Bacon's account fails to explain is the role of theory in science. Take Galileo's idea of motion without air resistance, Barbour suggests. This was a counterintuitive idea, "a conception of the world we do *not* experience," which, as it turned out, was crucial in order to discover the principle of inertia:

> He imagined observed motion to arise from two sources, neither of which could be observed alone: a continuing uniform inertial motion and a frictional retarding force. Aristotle's view had been closer to everyday observations; a cart left to itself does come to its "natural" state of rest if there is no horse to keep it moving. Galileo imagined an idealized frictionless case that, left to itself, would continue to move uniformly. Starting from such an "ideal case," he could argue that the cart comes to rest not because of any natural tendency to do so but because friction hinders its uniform motion.[71]

In essence, Galileo was crafting a double-pronged theory *behind* the data everyone can observe. Furthermore, a medieval scientist was trained to ask *why* objects move, not *how* they move. By contrast, only efficient causes commanded the modern scientist's attention; no longer questions of teleology (questions about causes directed to the future) or metaphysical ones (causes in the essence of objects). Hence, Galileo distinguished between the two "primary qualities" of objects, mass and motion, and their "secondary qualities," like color or temperature, pain or taste. The latter result entirely from human sense organs. As we know, his contemporary Descartes (d. 1650), took this budding mind/matter dualism to an extreme, calling the external world the realm of bodies in motion or that

---

[70] Born the year Galileo died, Sir Isaac Newton, who fully developed the image of "the world as an intricate machine," must be seen as the pre-eminent spokesman of the Western scientific revolution (ibid., 17). As did Galileo, however, he considered only those properties that could be understood mathematically (mass and motion) belonged to the "real world." Any other properties were relegated to the subjective realm of the human mind. Still, Newton took for granted the existence of a Creator God and the freedom of the human spirit. These assumptions would soon come under much scrutiny.

[71] Ibid., 11.

of matter stretching into space, while mind was an entirely different realm with only "thinking substance."[72]

We saw in the last chapter that this metaphysical dualism was progressively attacked from several quarters. Yet the scientific picture of reality as an agglomeration of small units that can be broken down into smaller ones became the hallmark of the modern worldview until very recently. Nancey Murphy argues that this mechanistic view of the external world not only reinforced the classical realism it had inherited (the world is just as I see it and describe it), but also extended an atomistic and reductionist methodology to all other branches of sciences, including psychology and the social sciences. Reductionism "is the strategy not only of analyzing a thing into its parts, but also of explaining the properties or behavior of the thing in terms of the properties and behavior of the parts."[73] Therefore, the reasoning goes, if biology can be reduced to the interaction of atoms and molecules, so too one should be able to understand human behavior in terms of biological processes, family upbringing and social conditioning. In its most extreme form, one could argue that human behavior could ultimately be reduced to molecular physics.

For Murphy, this strictly "modern" way of thinking has been extended to the philosophy of language, ethics and political theory. Above all, it is the reflex of seeing parts as determinative of the whole, no matter the discipline of inquiry: "Thus the common good is a summation of the goods for individuals; psychological variables explain social phenomena; atomic facts provide the justifying foundation for more general knowledge claims; the meaning of a text is a function of the meaning of its parts."[74] A case in point is the atomism behind Thomas Hobbes's social contract theory. Individuals and their rights come before the commonwealth, which in fact is only an artificial body. As individuals strive for self-preservation, they find a way to register their collective property rights and moral obligations in a social contract.

This is, so far, the story of modern thought. But that story is now being challenged (though not yet supplanted) by a postmodern story, Murphy contends. As with the modern worldview, the initiative has come from the sciences. As many scientists have noted the failure of reductionist strategies, they have increasingly turned to new theories of causal explanations. Murphy notes three interrelated features of the natural order, as one moves from one level of analysis to another:

---

[72] Ibid., 12.
[73] Murphy, *Anglo-American Postmodernity*, 12.
[74] Ibid., 14.

1. Emergence: certain properties and processes at one level, say, of the biological order, can only be explained by bringing in concepts that apply to a higher level. A cell "behaves" in a way not reducible to its simply being a collection of molecules.

2. Decoupling: in the hierarchy of the sciences each level has a kind of autonomy. Changes in the macrolevel often do not affect the microlevel. Conversely, as causal laws from below are loosened, scientists discover that higher levels display emergent laws not seen at lower levels.

3. Top-down or whole-part causation: following from the two previous points, the opportunity arises for a more thorough critique of reductionism:

> It is now recognized in a variety of sciences that interactions at the lower levels cannot be predicted by a look at the structure of those levels alone. Higher-level variables, some of which cannot be reduced to lower-level properties or processes, have genuine causal impact. Biochemists were among the first to notice this: chemical reactions do not work the same in a flask as they do within a living organism. The science of ecology is based on the recognition that organisms behave differently in different environments. Thus, in general, the higher-level system, which is constituted by the entity and its environment, needs to be considered in a complete causal account.[75]

This thinking becomes useful when the development of the philosophy of language is taken into consideration. The bottom-up reductionism of Bertrand Russell's logical positivism (post-World War I) is completely challenged by Wittgenstein (in his later career) and J. L. Austin. For them the meaning of a term is derived from usage, and this can only be understood in a social context. Language is a game people play in a variety of settings, but always according to rules that are culturally determined. This top-down reasoning is what accounts for speech act theory—a topic to be explored in more depth in the next chapter. A word does not primarily mean something because it refers to "something out there." Expressions find their meaning in their actual use as a "game" in a particular social context. As in the sciences, one recognizes a hierarchy of levels and the truth that each level brings with it new forms of meaning not found at the lower levels. Murphy calls this semantic holism (cf. Fig. 1).

---

[75] Ibid., 21.

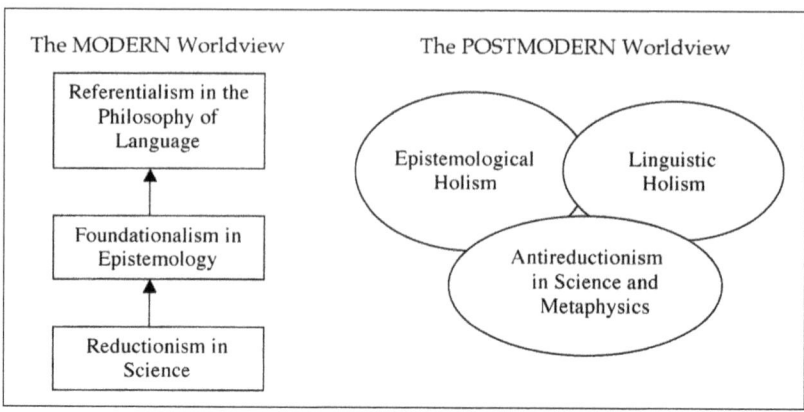

Figure 1. *Nancey Murphy's Postmodern Shift*[76]

In the end, Murphy, as philosopher and theologian, would like to see the foundationalism of Cartesian epistemology replaced by postmodern holism. The main reason for this is that the foundations in the modern paradigm proved to be either very wobbly or irrelevant. It turned out that ideas that were clear and indubitable to Descartes could be equally vague and uncertain to others. The other foundation had been empirical "facts"; yet from the beginning, David Hume showed them to be useless, as observation can only yield immediate experience and no certain knowledge.[77] A further difficulty, as Polanyi and Thomas Kuhn have shown, is that all facts are theory-laden. The modern attempt to separate them by

---

[76] I do not wish this figure to be misread. Two comments are in order: First, the bottom-up causation in evidence through the arrows on the modern side does not imply that all of these perspectives are not interrelated. Plainly, they are. I am trying, however, to do justice to Murphy's contention that the scientific perspective led to a similar outlook in the philosophy of language and metaphysics, while at the same time illustrating the bottom-up causation so characteristic of this perspective. When Murphy herself looked at this figure; however, she was not sure which to put at the bottom: the reductionism in science, or the foundationalism in epistemology. Second, the postmodern worldview is pictured more holistically, not exactly as a web but at least as circles touching one another. At the same time, I do not see Murphy claiming that there was any kind of order in the relative formation of these positions. They all interact with one another, back and forth.

[77] Not without a touch of dry humor, Murphy introduces here her own "Murphy's law": "whenever the foundations are suitably undubitable, they will turn out to be useless for justifying any interesting claims; when we do find beliefs that are useful for justifying the rest of the structure, they always turn out to be questionable" (ibid., 26).

claiming some royal objectivity on the part of the subject failed miserably. And finally, Descartes's ambition to start philosophy with a clean slate was doomed from the start, because any assertion involves language, which is largely conditioned by one's cultural worldview. Any reason to doubt a belief must appeal to another reason that itself cannot be called into question. The real problem is with the picture of foundationalism.[78]

I have leaned on Murphy to bring out the implications of the modern scientific method—atomism and reductionism, and how these impulses were carried out within a wide range of disciplines. Ian Barbour takes us further on our pilgrimage toward a Muslim–Christian theology of creation in a postmodern setting, even though he does not raise directly the issue of postmodernism. In the next chapter I will present David Naugle's view of the human person as essentially "semiotic"—a bold evangelical interaction with postmodern thought. Here, however, I offer Barbour's rapprochement of theology and science as a contribution to an ongoing dialogue between Muslims and Christians.

Western science grew out of the work that preceded it within the Muslims states of al-Andalus (Muslim Spain), the Abbasid caliphate and their various successors to the east. It is high time to return the favor, and particularly in a two-way, give and take sort of conversation. Just as Christians and Jews advanced the sciences with Muslims in al-Andalus of old, many Muslim scholars today conduct research and publish in the West with Christian, Jewish and agnostic colleagues. Yet it is my opinion that we can no longer hold premodern conceptions of "science." It would behoove both sides to consider the wider picture—of which I offer little windows here—of what actually takes place in writing texts and furthering scientific research. On my view, Barbour's account of "critical realism" would be a useful theoretical ally for all sides to the conversation.

With a Ph.D. in physics (University of Chicago, 1949), Barbour early on sought to relate science and theology. In 1953 he enrolled in the Yale Divinity School to study theology, philosophy and ethics. Two years later he was hired by Carlton College in Minnesota, where he stayed for

---

[78] The mental picture inherited from the Enlightenment is difficult to set aside. W. V. O. Quine offers a quintessentially postmodern picture to illustrate this different epistemological perspective: the web, or net. Murphy explains: "There are no sharp distinctions among the kinds of beliefs in the web, and so there is no distinction between basic (foundational) beliefs and nonbasic beliefs. Beliefs differ only in their 'distance' from experience, which provides the 'boundary conditions' for knowledge. The requirement of consistency transmits experiential control throughout the web" (ibid., 27).

the rest of his career.[79] He first attracted the attention of scholars from several disciplines in the mid-sixties with his book, *Issues in Science and Religion*.[80] He published another seminal book in 1974, *Myths, Models and Paradigms*, which analyzed the philosophy of science of Kuhn, Imre Lakatos and others, while going beyond.[81] The following section is based on his 1997 work, which summarizes his lifelong research on the intersection of science and religion: *Religion and Science: Historical and Contemporary Issues*.[82] This is the book that won for him the 1999 Templeton Prize for Progress in Religion.

Since the "scientific revolution" engineered by Galileo and Newton contained two main elements—experimentation on the data resulting from observation and formulation of theories—Barbour proposes to examine in detail how these two relate in practice. When in fact one searches for a link between the data of observation and the forming of theories, that link is only indirect. On the one hand, new concepts often crop up as analogies to situations already familiar to the researcher. For example, Newton's atomism helped to conceptualize how a gas remains enclosed in a container. Let us imagine, says Barbour, that the gas is composed of minute elastic balls bouncing off each other, much in the same way as billiard balls collide. This theory advances a "conceptual model" that posits entities that cannot be observed (or properties—recall Galileo's idea of resistance to motion). The theoretical properties of mass and velocity could be expressed mathematically as they interacted with the energy and momentum of these hypothetical spheres. This model, then, allowed the formulation of the kinetic theory of gases. In one application, Boyle's law states that "if the volume (V) of a gas is reduced by 50% (by compressing the air in a bicycle pump, for example) then the pressure (P) of the gas will double."[83] Here we see a model, based on an analogy, leading to a theory that attempted to explain observed patterns in the state of gases.

---

[79] I gleaned the biographical information from the Templeton Prize website: http://www.templeton.org/archives/IB-bio.asp.

[80] Ian G. Barbour, *Issues in Science and Religion* (Englewood Cliffs, NJ: Prentice-Hall, 1966).

[81] Ian G. Barbour, *Myths, Models and Paradigms: The Nature of Scientific and Religious Language* (London: SCM Press, 1974).

[82] Ian G. Barbour, *Religion and Science: Historical and Contemporary Issues*, rev. and expd. ed. of *Religion in an Age of Science* (New York: Harper SanFrancisco, 1997 [1990]).

[83] Barbour, *Myths, Models and Paradigms*, 31.

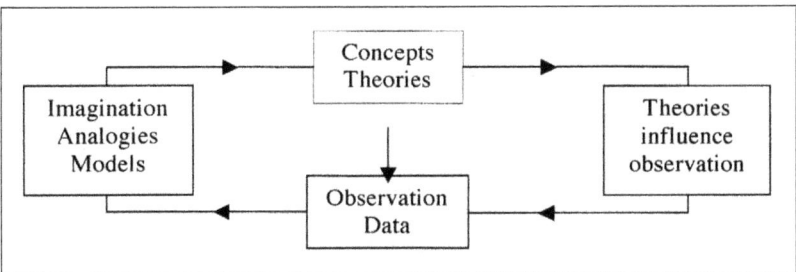

Figure 2. *The Structure of Science*[84]

Thus, while observational data provide some stimulus to the formation of models and theories, it must be said, on the other hand, that theories impinge on the researcher's observation of these data as well. This double movement is captured by Barbour in the above diagram (Fig. 2). The vertical arrow (if abstraction is made of the two side loops) represents the hypothetico-deductive view of science, explains Barbour. The most popular theory in the 1950s and early 1960s, it assumes that data can be described in a way that is theory-free. In particular, Karl Popper maintained that, while agreement with data is not sufficient to verify a theory, disagreement with data most certainly falsifies them. Popper represented the school of logical positivism, which taught that there must be one universal scientific method.[85] Its certain results are derived from its empirical method (sure knowledge comes entirely from observing data) and its insistence that all science—including the social sciences—must be value-free (metaphysics are a meaningless category).[86]

What happens in reality, however, is that all theories demonstrate a great tolerance for discordant data. These are classified either as "ad hoc auxiliary hypotheses," or as "unexplained anomalies." This is so, because a theory cannot be tested on its own, but necessarily as a part of a larger network of theories. All theories are bolstered by a cluster of background

[84] Barbour, *Religion and Science*, 107.
[85] As mentioned above, Bertrand Russell and the early Wittgenstein pioneered the application of logical positivism to the field of linguistics and philosophy of language in the 1920s. Wittgenstein veered from this path shortly thereafter. He rejected bottom-up reductionism in favor of a more holistic approach—meaning stems from usage in particular contexts.
[86] Harold Kincaid, "Positivism in the Social Sciences," in *Routledge Encyclopedia of Philosophy*, vol. 7:559. In a quite consistent manner, Popper argued that both Marxism and Freudian psychology were pseudoscientific because they were non-falsifiable. Positivists have raised similar objections to economics, because theories are maintained in spite of contradicting data.

theories, and when one encounters a discordant datum, there is usually a way to adjust other members of the cluster in order to improve the data-theory fit. When competing theories clash, both sides can usually agree on a method of experimental testing, because it is consonant with the background theories that are common to both. Yet in some cases, when the two theories involve a wide range of data, much of which does not overlap, and differ in their interpretation of the data, then adjudication between the two becomes more problematic.

Furthermore, as the above figure shows by its right loop, "all data are theory-laden."[87] Among the ways theories influence observation, according to Barbour, (1) theory decides the data selection and the choice of variables deemed relevant to the study; (2) the questions we ask shape the answers we get; (3) a theory brings its own assumptions to the study, both about the instruments to be used and about the language in which the result will be couched; (4) finally, the data themselves can be significantly affected by the observation process—a truism, for instance, in the study of ecosystems. Hence, "We are not detached observers separate from observed objects; we are participant observers who are part of an interactive system."[88]

This was one of Thomas Kuhn's main contributions. Kuhn (d. 1996) devoted his career to the study of the history of science and concluded early on in his influential work, *The Structure of Scientific Revolutions*,[89] that, contrary to traditional imagination, science does not grow by continually adding new bricks of knowledge to its edifice. Again, the metaphor of a building with foundations is misleading. Both the creativity of the individual and the dynamics of social interaction are more relevant factors. Typically, scientific disciplines mature through the competition between rival schools. These competing research orientations are called "paradigms," and "normal science" eventually confronts a sufficient body of discordant data so that the routine anomalies become "significant anomalies" and the consensus of the practitioners in a particular scientific field begins to unravel. As a group of researchers focus their attention on these anomalies, they engage in "extraordinary science," or "crisis science."[90] A scientific revolution occurs if the new theory that emerges attracts a sizeable number of adherents and presents enough possibilities

[87] Barbour, *Religion and Science*, 108.
[88] Ibid.
[89] Thomas Kuhn, *The Structure of Scientific Revolutions* (Chicago: University of Chicago Press, 1970 [1962]).
[90] Paul Hoyningen-Huene, "Thomas Samuel Kuhn (1922–1996)," in *Routledge Encyclopedia of Philosophy*, vol. 4:316.

for further research. A paradigm shift then takes place—as occurred with Newtonian physics, Einstein's theory of relativity or quantum physics.

What is to be noticed here is that the said transformations did not proceed because one theory was falsified (as Popper argued). Rather, a particular research tradition—including its preference for a particular cluster of theories and assumptions and its "proven" methods and instruments of research—is gradually superseded by a competing paradigm. This is far from the empiricism of the logical positivists. As in Barbour's right loop, there are no "raw" data—just observations that give rise to analogies, which, in the working of creative minds, become new models that can then lead to new concepts and theories. Kuhn was in essence dispelling the notion of a "scientific method." Instead, in its normal phase, science progresses with the help of exemplary problem-solving strategies. Agreeing with Polanyi, he emphasized the hidden agendas of individuals and the politics within and between research institutions, as well as the implicit analogies that are made when a research tradition puts forward new problems to solve or sorts out which solutions are deemed adequate.[91] In a quite postmodern fashion, Kuhn saw paradigms evolving in the context of specific historical communities.[92]

At this point Barbour presents an account of theory assessment that I find extremely useful for this present project. In common scientific practice, four criteria serve to evaluate theories:

1. *Agreement with data:* despite the fact that "theories are always underdetermined by data" (there may always be other theories that fit the data better or at least as well; theories tolerate an inordinate amount of conflicting data), agreement with data, which usually is coupled with an ability to predict novel phenomena with a measure of success, constitute an important element of support for any theory.

2. *Coherence:* (a) external coherence: a theory connects well with other accepted theories; (b) internal coherence: its constituent parts easily connect with one another and the theory is simple ("simplicity of formal structure, smallest number of independent or ad hoc assumptions, aesthetic elegance, transformational symmetry, and so forth").

3. *Scope:* theories should be comprehensive both in their ability to unify "previously disparate domains" and to apply to a wide spectrum of relevant variables; a theory is valued too if it can be supported by several kinds of evidence.

[91] Ibid., 317.
[92] Barbour, *Religion and Science*, 109.

4. *Fertility:* a theory is judged by its ability to spin off new research both now and in the foreseeable future: can it generate new hypotheses, lead to a wider theoretical understanding and suggest new experiments? In other words, it will be valued to the extent that it promotes the fruitful collaboration of a scientific community.[93]

This approach raises the question of truth, contends Barbour, as it has been pursued in the Western tradition since the Enlightenment. In turn, the idea of truth raises the issue of realism in metaphysics. The first criterion, agreement with data, connects with the correspondence view of truth—the idea that a proposition is true to the extent that it corresponds to reality. This is both the common sense view and the view of classical realism (or what Barbour calls "naïve realism"). The sentence "there is a book on the chair" is true if in fact a book is on the chair in question. But the problem arises from the fact that all data are theory-laden. And theories often "postulate unobservable entities only indirectly related to observable data. We have no direct access to reality to compare it with our theories."[94] In addition, the coherence theory of truth can be linked to the second criterion: a proposition is true if it is internally coherent. This is the position adopted by the rationalists who decried the empiricism of the logical positivists. By emphasizing the theoretical side of science as opposed to its observational component, the rationalists rightly valued theories that connect with a larger group of theories and hence add scope to their coherence. But in practice, coherence and scope are not always compatible—several competing theories could be equally coherent and, while coherence deals with logic, scope deals with data agreement issues. Reality is decidedly more complex.

The third view of truth that emerges from this list of criteria is that of pragmatism: a proposition is true if it solves a problem that needs solving. As with the criterion of fertility, a theory is judged on the basis of its potential consequences. Here truth is a function of usefulness. This is akin to the position adopted by the later Wittgenstein and J. L. Austin in reaction to logical positivism. Both views reject metaphysical reasoning, but while for the positivists it represents an *a priori* facet of their position, for the linguistic analysts who focus on language games, questions of truth are simply irrelevant. Yet that is the point: if truth is only what works in practice (or how language happens to function in a specific cultural setting), then it is plainly inadequate. Other criteria are needed.

[93] Ibid.
[94] Ibid.

This is what leads Barbour to link the questions of truth and meta-physics—as is required by the first criterion, agreement with data. He concludes that "the *meaning* of truth is correspondence to reality. But because reality is inaccessible to us, the *criteria* of truth must include all four of the criteria mentioned above."[95] Objective truth exists, in that as human beings we perceive a world independent of our own existence. But unlike the classical, or naïve version of realism, critical realism insists that other criteria besides correspondence to reality must be used to generate truth. Reality is knowable, but remains partially opaque to the human beholder, whether at the level of biochemistry or physics, or at the level of stars millions of light years away; whether in the complexity of human behavior itself or in the macrodynamics of social movements; or in the metaphysical questions of life beyond death, God, creation and eschatology. People of faith of all stripes will want to assert that all of these elements combine to form one reality. They may differ on the degree of certainty with which one can come to know aspects of this reality, and in particular, whether the truth of "my" system cancels out the incompatibilities of "your" system (cf. Appendix D).

This is precisely the move Barbour makes. Despite the obvious dif-ferences, he chooses to highlight the similarities between scientific knowledge and religious knowledge. As models are constructed in science to explain observed patterns in the world, so religious models enhance and aid beliefs "that correlate patterns in human experience."[96] In the sciences, models display three characteristics: (1) they are "ana-logical," as the Bohr model of the atom, which borrows planetary move-ments in space to describe the orbiting of electrons around a nucleus; (2) they are open-ended and suggestive, enabling scientists to extend exist-ing theories: when new data cast doubts on a model, it is modified and science moves forward;[97] finally (3) "models are intelligible as units". Barbour's explanation here is worth quoting at length, as it applies beyond the physical sciences, and indeed, beyond science:

> Models provide a mental picture whose unity can be more readily unders-tood than that of a set of abstract equations. A model can be grasped as a whole, giving a vivid summary of complex relationships, which is useful in extending and applying the theory as well as in teaching it. Images are

---

[95] Ibid., 110, emphasis original.

[96] Ibid., 119.

[97] As when the billiard ball analogy suggested how the diffusion and viscosity of gases might be calculated, but later proved inadequate when high pressure was applied; the theory had to be modified, yet this led to a revised model ("elastic spheres with finite volume and attractive forces) (ibid., 116).

creative expressions of imagination in the sciences as well as in the humanities. The intuitive intelligibility of a model is, of course, no guarantee of its validity. Deductions from the theory to which the model leads must be tested carefully against the data, and more often than not the proposed model must be amended or discarded. Models are used to generate promising theories to test by the diverse criteria outlined earlier.[98]

Classical realism holds that models and theories describe reality as it is—sidestepping Kant's distinction between noumenal vs. phenomenal reality. George Berkeley (d. 1753) in the preceding generation of Enlightenment philosophers had introduced the concept that matter is in fact mind-dependent—the position of idealism. Thus the physical world is nothing more than a collection of ideas that come to us through sense perception.[99] Physical objects appear to our senses with qualities that one cannot imagine could be divorced from them. In other words, their very existence depends on the mind of an observer that perceives them. For Berkeley this way of thinking was a strong demonstration of the existence of God: "it showed that there must be a God who is responsible for those ideas which (after acting in a certain way) we have no choice but to experience and who keeps the whole system of ideas available to each individual spirit in conformity with a universal system of laws determining the appearances available to each."[100] Naturally, many idealists have subsequently dropped the divine factor, concentrating either on ontological idealism (common sense misleads us, because it is absolutely true that the physical world is constituted by the ideas of perceiving minds) or on epistemological idealism (the only truth we can obtain about the world is truth-for-us). Kant, then, went down the second path, remaining skeptical about the claims of Berkeley's ontological idealism. Absolute truth about things-in-themselves is not knowledge we humans are equipped to gather. Rather, we are confined to obtaining truth-for-us. Yet, as we learned in the previous chapter, Kant does posit a certain leap of faith in the noumenal realm—the obligatory character of morality's "categorical imperatives."[101]

The German idealists who came after Kant were more sanguine about the abilities of human rationality. With Hegel as the apex of what came to be known as absolute idealism (of the ontological kind), human

---

[98] Ibid.

[99] T. L. S. Sprigge, "Idealism," in *Routledge Encyclopedia of Philosophy*, vol. 4:662-69. Sprigge explains that Berkeley's position is based on Locke's conviction that ideas is all that we humans really perceive, and the distinction between primary and secondary qualities (p. 663).

[100] Ibid., 664.

[101] Ibid., 665.

history becomes both the battlefield for creative ideas that through a dialectical process move toward perfection and the focal point of revelation of Absolute Spirit (or Mind). Though there remains some controversy about whether Hegel actually believed that the physical world was entirely a product of ideas, absolute idealism did become popular in the US and UK in the early twentieth century. Then, as we know, idealism was largely displaced by those who believed metaphysics of all kinds were a doomed enterprise, whether of the logical positivist mindset, or of the linguistic analytical one.

I bring up idealism in order to illustrate how widespread the antirealist position has been for a long time in the Western world. Ironically, one might say, from Berkeley to Hegel and beyond, strident philosophical voices denying the reality of the physical world have been raised in a society exceedingly proud and reliant upon its technological prowess. What is more, among philosophers who deal specifically with the meaning of science, there has been a turn away from the naïve realism still noticeable in the positivism of Karl Popper. Though nobody wants to be caught doing metaphysics, all bring metaphysical assumptions to their task. Thus the positivists are classical realists who contend that their models and theories correspond to the world as it is; the instrumentalists are pragmatists who see theories merely as "useful intellectual instruments for organizing research and for controlling the world";[102] constructivists rely on the Marxian idea of knowledge production to assert that scientific knowledge is constituted by people in their social contexts; finally, the conventionalists similarly hold that "the truths of science ultimately rest on man-made conventions."[103]

Figure 3 attempts to summarize the above discussion. Two separate axes depart from the same point, the Enlightenment episteme of representation and its metaphysical corollary, the disembodied subject observing the world outside itself. The first axis, then, represents the modern philosophical move toward the strong version of postmodernism, as Benhabib has put it. In terms of academic disciplines, we are dealing here mostly with the humanities. Yet, as it has often been observed, the social sciences are somewhat in between the humanities and the physical sciences. The work of Habermas and Benhabib, for instance, involves a conscious linking of philosophy with the social sciences. In addition, when it comes to the issue of realism, all disciplines across the spectrum take positions on the issue. For the sake of clarity, however, I am here

---

[102] Barbour, *Religion and Science*, 117.
[103] Arthur Fine, "Scientific Realism and Antirealism," *Routledge Encyclopedia of Philosophy*, vol. 8:581.

marking off the second axis as that of "science" on the specific issue of realism/antirealism. My contention is that the issue of metaphysics is unavoidable. The theory of discourse ethics that Habermas advances in support of a "postmetaphysical" democratic politics is in fact an attempt to bracket the truth issue and its relationship to the world-as-it-is.

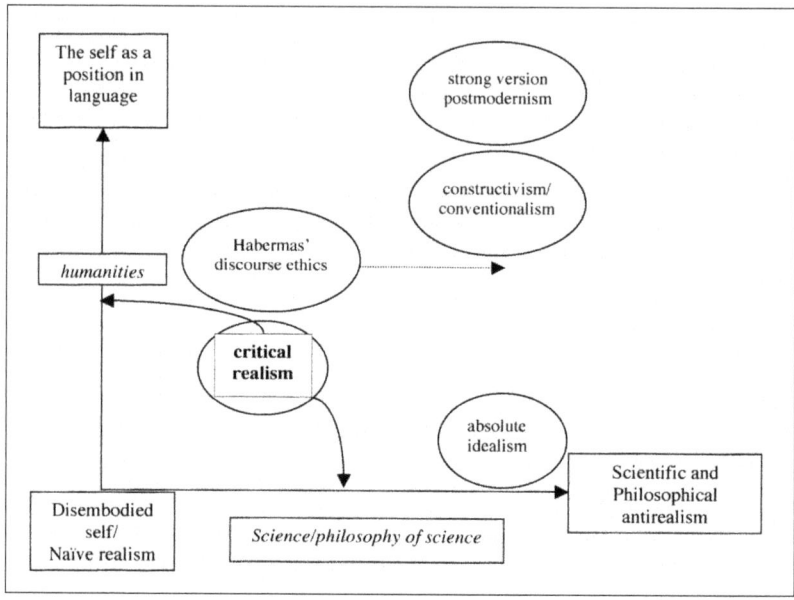

Figure 3. *Critical Realism and Postmodernism*

I will come back to the implications of language and hermeneutics in the next chapter, but already with this figure I hope to suggest that coherence and fertility/pragmatism versions of truth are not satisfactory by themselves. So, depending on how one reads Habermas, one might see his project either as a middle-of-the-road critical realism (as charted above) or as an instrumentalist/conventionalist on the issue of realism (moving him horizontally to the right).[104] Graphic attempts like this one

[104] My contention is that Habermas is a critical realist in Barbour's terminology. James Bohman ("Two Versions of the Linguistic Turn: Habermas and Poststruc-turalism," in *Habermas and the Unfinished Project of Modernity: Critical Essays on The Philosophical Discourse of Modernity*, ed. Maurizio Passerin d'Entrèves and Seyla Benhabib [Cambridge: Polity Press, 1996], 209) examines Habermas's use of hermeneutics, and in particular his pointing to the world-disclosure function of language, in his critique of poststructuralism. Social critics such as Martin Luther King Jr provide fresh approaches to old problems often using new vocabulary and powerful rhetorical devices in order "attempt both to disclose things about the initial

are only heuristic devices intended to further an ongoing conversation. Yet, hopefully, they also serve to approximate reality more closely. This is precisely how Barbour characterizes the critical realist perspective.

On my view, it is significant that the scientific community has expressed in recent years a growing interest in realism. Popper's realism was originally embedded in arguments against the instrumentalists. Though his extreme realism fell out of favor in the 1960s, others picked up the realist torch, but now had to frame their arguments in the light of what some perceived to be the constructivism or conventionalism of Thomas Kuhn and Paul Feyerabend—Barbour's modified realism, or critical realism. Though scientists in general tend to be "incurably realist" (especially when it comes to observable phenomena, assuming that "existence is prior to theorizing"), when models attempt to describe the hidden dynamics of postulated entities, they are likely to proceed with more humility.[105] Ernan McMullin, for instance, argued that "a good model gives us insight into real structures, and that the long-term success of a theory, in most cases, gives reason to believe that something like the theoretical entities of that theory actually exist."[106] Though geologists cannot be direct witnesses to prehistoric dinosaurs or conclusively prove the existence of tectonic plates, they usually believe that these "really" did or do exist. Barbour notes that "[a]s we move further from familiar objects, instruments greatly extend our powers of direct or indirect observation." Thus, for instance, when Mendel postulated in 1866 the existence of " 'units of hereditary transmission,' which were later identified as genes in chromosomes and more recently as long segments of DNA," we witness a scientific model and theory being extended and modified in significant ways over time. Our knowledge of the truth (at least partially defined as correspondence to reality) is never exhaustive, but it can progressively be expanded.

---

interpretations of the common world and to recontextualize them in new patterns of relevance." He advises Habermas to stop restricting this world-disclosure function to art. It would be entirely consistent for him to enlarge this role of language to the whole gamut of thoughts and actions, in which a social agent decides to engage: "Suitably expanded, Habermas' appropriation of the concept of disclosure helps him to account for the plurality of cultural worlds and for the possibility of transformative agency within them" (ibid., 215-16). The critical realist position, as I see it, is a necessary assumption for any kind of sociopolitical activism in the multicultural world of today.

[105] Barbour, *Religion and Science*, 118.

[106] Ernan McMullin, "A Case for Scientific Realism," in *Scientific Realism*, ed. Jarret Leplin (Berkeley/Los Angeles: University of California Press, 1984), 39, cited in Barbour, *Religion and Science*, 118.

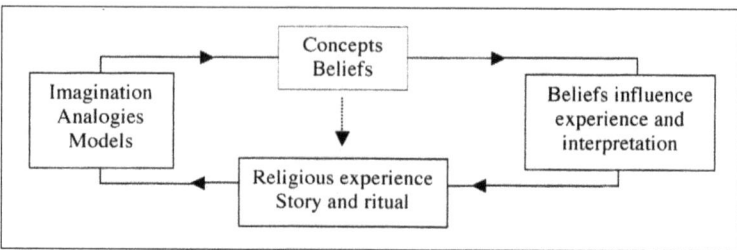

Figure 4. *The Structure of Religion*[107]

What this means is that our present state of knowledge in any area remains tentative and subject to further modification. In this sense it is cumulative, but following a much more circumlocutory route than classical realism had imagined. This is also the case for religion, Barbour argues—admittedly, a more controversial claim. Yet I offer his view in our ongoing Muslim–Christian conversation as food for thought. Since I will be advocating the application of critical realism in our approach to religion in the next chapter, I start with his view as an introduction. Figure 4 is Barbour's model, now refitted to describe religious knowledge.

The assumption here is (1) that the data from which a religious community draws its basic concepts and beliefs are primarily religious experience (particularly that of a founding individual), story (as recounted in sacred texts) and ritual; (2) that these provide models and analogies, which then enable the community to shape its particular beliefs; and (3) that in reexamining their religious data, communities are influenced by the beliefs they hold in the first place. The downward arrow in the figure above is a dashed line showing that the testing of religious beliefs (which are even more theory-laden than in science, he says) is more problematic—though not impossible. The four criteria used to validate the truth of scientific theories can be marshaled for theories and models in religion as well: agreement with data, coherence, scope, and fertility.[108]

Barbour's diagram is also predicated on a taxonomy of religious experiences that has broad support among scholars working in the field of phenomenology of religion:[109]

[107] Ibid., 111.

[108] Cf. Ibid., 113.

[109] He cites two books in particular: Frederick J. Streng, *Understanding Religious Life* (Belmont, CA: Dickenson, 1976); and Ninian Smart, *Worldviews* (New York: Charles Scribner's Sons, 1983).

1. *Numinous Experience of the Holy:* religions that emphasize the transcendence of God and the finitude of people as creatures; thus a sense of awe and wonder before the holiness and mystery of the divinity is combined with a human person's sense of dependence, limitation and weakness; the three monotheistic faiths fall in this category, as well as the *bhakti* tradition of Hinduism and Mahayana Buddhism.
2. *Mystical Experience of Unity:* occurring both with impersonal and personal models, mysticism represents the universal quest for unity of the soul with the Divine.
3. *Transformative Experience of Reorientation:* whether following the model of a spectrum from guilt to forgiveness, or that of estrangement to reconciliation, or brokenness to healing, this represents the life-changing power of religious experience in many traditions.
4. *Courage in Facing Suffering and Death:* sometimes the experience of ultimate reality provides serenity in the place of anxiety, and sometimes (especially in Buddhism) attachment to the world gives way to detachment.
5. *Moral Experience of Obligation:* monotheistic religions tend to emphasize the good as God's will, along with a prophetic denunciation of social injustice; others may emphasize the duty to pursue justice and love as a way to obtain harmony with the cosmic order.
6. *Awe in Response to Order and Creativity in the World:* in the numinous mode, this leads to dependence on the Creator; in the mystical mode, "it is more often articulated as dependence on a creative force immanent within nature."[110]

If by this taxonomy Barbour means that these are types of religious experience, then it serves as a useful catalog, and, as such, is uncontroversial. All six of these features can be observed as part of the faith and practice of Muslims, Jews and Christians. Each community of faith includes various branches and a diversity of expressions across time and cultures, all of which encompass this kind of spectrum of religious experiences. If, on the other hand, however, Barbour means that the "essence" of each religion can be captured by one of these experiences, then the claim would be a typically modern one, an exercise in "essentializing" a reality that is much broader and subject to change. The rest of his discussion shows that this is not what he is implying. Parallels

---

[110] The titles are Barbour's, but the content is my own summarizing (ibid., 111-12).

between science and religion, he maintains, can be drawn most clearly in two areas: models and paradigms.

First, as in science, models build on analogies; they allow an extension of thought and application to new situations; they are "unitary," that is, "they are grasped as a whole with vividness and immediacy."[111] In particular, they provide images that help to interpret the flow of narrative in the sacred texts. God as creator is one key model in both Bible and Qur'an. God as father appears in the Hebrew Bible, but is further developed in the New Testament. Besides its key function in the doctrine of the Trinity (a model taken out of, but not directly from, the text), the image of God as father serves as a model to express and draw out the implications of his tender love for his children. Most importantly for our purposes, the model of humankind as God's "caliphs" on earth is found in the Qur'an as applying to Adam and David, and in other passages more generally that seem to emphasize the "succession" meaning of the verb *khalafa* and its derivatives. As we shall see in later chapters, the caliphate of humanity is an interpretive model that was coined (mostly) in the modern period to make sense of the divine empowerment of the human race in an age when nature had been mastered to an unprecedented degree. Some classical commentators gave the foundational verse (Q. 2:30) a universal interpretation that already could be seen to move in this direction. But it was not until the twentieth century that this model became accepted and amplified in the way it is today.

Second, in both scientific and religious communities paradigms encompass "the broad set of metaphysical and methodological assumptions" that stand behind each community's tradition. "Here too, new members enter a tradition by being initiated into the assumptions and practices of the community, and they normally work within its accepted framework of thought, which we can call 'normal religion,' corresponding to 'normal science.'"[112] As in science, religious paradigms display the three following features:

1. *Religious experience is paradigm-dependent*, yet even between rival paradigms one can find some common features of experience that allow discussions, arguments in rational terms and evaluation.

2. *Religious paradigms are highly resistant to falsification*, meaning that when discordant data appear, rather than invalidating the reigning paradigm, either the leading figures resort to ad hoc modifications, or they bracket those data as "anomalies."

---

[111] Ibid., 119.
[112] Ibid., 128.

3.   *There are no rules for paradigm choice in religion*, but the criteria presented above are often used to evaluate belief and practices within paradigms and sometimes between paradigms. As in science, historical, sociopolitical and personal factors can also determine or speed up the advent of paradigm shifts in religious traditions.[113]

Barbour cites the work of Catholic theologian Hans Küng, who posits five major historical paradigms for Christianity, each supplying the conditions for normal theological work and growth of the community, and each extending the scope of its paradigm while resisting change: "Greek Alexandrian, Latin Augustinian, Medieval Thomistic, Reformation, and Modern-Critical."[114] Küng shows how each paradigm "arose in a period of crisis and uncertainty," and how "conversion to a new paradigm involved subjective factors and personal decisions as well as rational arguments." Naturally, not everything was changed from one paradigm to the next—only some of the models and methodologies were modified.

As I have attempted to argue throughout this chapter, historical events have a way of bringing certain issues to the fore, issues which had previously been seen as tangential, so that amid the various movements that vie for pre-eminence, shifts occur in the very fabric of a generation's intellectual landscape. The metaphysics of the isolated self that contemplates the world from a distance is now behind us; yet this does not mean that a consensus has gelled around the postmodern self "as a position in language." Critical realism in some form, one that incorporates some of Murphy's top-down holism and Barbour's composite theory of truth, is closer to what we might call a consensus in various fields at the opening of the twenty-first century. This is the sense in which I use "postmodern world" as the context in which Muslim–Christian discussion and action must develop today. As the final section of this chapter will argue, this term also corresponds to an awakening of consciousness about meaning in history as primarily arising from the struggle of oppressed peoples, including women, impoverished classes and marginalized ethnic and racial groups. It is an awareness that grew out of the ideals of equality, dignity, freedom and democracy, embedded in the Enlightenment writings and revolutions. Yet now, in the aftermath of the Cold War era, the

---

[113] Ibid. Again, only the headings are his.
[114] Ibid., 129, referring to Küng, "Paradigm Change in Theology," in *Paradigm Change in Theology*, ed. Hans Küng and David Tracy (Edinburgh: T. & T. Clark, 1989).

emancipation and liberation of these various constituencies is in most cases tied to issues of capitalism and globalization—the straightjacket of postmodernity, as I have suggested.

Postmodern Activism: From Self to Global Identity

Malcolm X neatly illustrates how one person can pass through three paradigms, and because he was such a powerful leader, his successive shifts impacted many others. His conversion to the ideology of the Nation of Islam was the first major shift; then came Islam as an historic faith practiced by a worldwide community (recall the illumination that came from the Hajj experience); then, on the level of activism, came the transition from a community-based advocacy (civil rights for the African-American community) to a global focus on human rights. Arguably, the second and third shifts are only different sides to the same coin: theology directly impacting Malcolm's activism. Islam's doctrine of creation (reinforced by what he experienced in his world tour) led him to widen the scope of his advocacy. And beyond theology, as with Martin Luther King Jr after a couple of years, the international scene provided some direct impetus for the direction he would take. For Malcolm it was the ascendancy of "third-worldism" (the non-aligned movement and the Organization of African Unity he was exposed to); for King, it was his personal wrestling with the Vietnam War and how it should relate to the civil rights movement. Both men struggled with the theological implications of their changing contexts and made some bold decisions that probably cost them their lives.

The social unrest of the 1960s was a more general phenomenon, as we have seen. I argued that the upheavals of 1968 in Europe, coming as they did at the apex of the welfare state, revealed a popular malaise among the youth about, on the one hand, a consumer society determined by a market economy, and on the other, the continuing rigidity of the old hierarchies of wealth and privilege in an increasingly bureaucratic society—the fabric of modernity was beginning to unravel. With the post-Fordist transition occasioned by the 1973 oil crisis and the American take-over of international finance as a result, the neoliberal vision became the new paradigm of an increasingly global economy. By the time the "Iron Curtain" fell in 1989, it was reigning supreme. In parallel fashion, and in the wake of momentous shifts in the philosophy of science (Polanyi 1958 and Kuhn 1962), Derrida, Foucault and others were focusing on the dimension of power in all discourse and claimed to have found a link between the epistemology of the detached subject and the atrocities committed by a colonial and imperialist West. We also saw how work in the literary field (Said 1978) contributed to a more generalized focus on the role of language as the key to unlocking the oppression of people at the margins—thus the mushrooming growth of such disciplines as feminism and postcolonialism.

I am not suggesting causal links here in a linear fashion. All these phenomena and many more should not be construed as material conditions determining intellectual life, but rather a complex interdependence of factors interacting throughout the period—rather like the web image Murphy takes from philosopher W. V. O. Quine.[115] As in the following statement by Burns on King's view of human rights (quoted above), there has been an observable two-way crossover between postmodernity and postmodernism: "King came to have hardly more affinity for such individualistic rights than he had for unbounded freedom or democracy, coins of the same realm." On the one hand, King realized that, for all the progress they measured, civil rights inscribed in US law reflected no more than "the classic liberal ideology of unbounded rights, owned by isolated, unencumbered selves devoid of community ties."[116] On the other hand, he expressed his disgust with the hollow freedom of choosing between myriads of consumer products in a capitalist system that pressured one's soul into the worship of Mammon. He would have agreed with Harvard theologian Harvey Cox, who recently wrote about the sovereign power of the neoliberal ideology on a global scale as the religion of "the new dispensation." The divine omnipotence of this new god, The Market, he muses, "means the capacity to define what is real. It is the power to make something out of nothing and nothing out of something. The willed-but-not-yet-achieved omnipotence of The Market means that there is no conceivable limit to its inexorable ability to convert creation into commodities."[117] The Catholic doctrine of transubstantiation states that ordinary bread and wine are transformed into the very presence of the crucified Christ. The Market, with its own set of priests and soothsayers, reverses the process, suggests Cox. Things deemed holy or held sacred by people for millennia—take land, for instance—"transmute into interchangeable items for sale." Thus capitalism comes, invades and distorts the values long held dear by traditional cultures:

---

[115] See her Chapter 6, "Philosophical Resources for a Postmodern Conservative Theology," in *Anglo-American Postmodernity*, 113-29.

[116] Cf. n. 11.

[117] Harvey Cox, "The Market as God: Living in the New Dispensation," *The Atlantic Online* (March 1999): http://www.theatlantic.com/issues/99mar/marketgod.htm.

It has been Mother Earth, ancestral resting place, holy mountain, enchanted forest, tribal homeland, aesthetic inspiration, sacred turf, and much more. But when The Market's Sanctus bell rings and the elements are elevated,[118] all these complex meanings of land melt into one: real estate. At the right price no land is not for sale, and this includes everything from burial grounds to the cove of the local fertility sprite. This radical desacralization dramatically alters the human relationship to land; the same happens with water, air, space, and soon (it is predicted) the heavenly bodies.[119]

In this light, it is significant that the 1992 Earth Summit in Rio marked the beginning of a global awakening of civil society on these issues. The development theorist David Korten writes that since Rio "millions of citizens are now learning, sharing information, developing common strategies, and strengthening the human bonds that are forging the foundation of a new global civilization."[120] Arthur Mitzman looks back on the social effervescence from Rio to the Seattle protests in 1999 and in Washington in 2000 and tries to determine the significance of these new social movements.

For Mitzman, it was the preindustrial classes of artisans and peasants who represented the backbone of revolutionary opposition to the autocratic regimes of France, Russia, England and China. Their "millennialist aspirations" grew out of a worldview shaped by the cyclical nature of seasons and traditional religious calendars. Before 1850, those dreams were channeled into action by the new elites, imbued "with a linear mentality of progress through work and scientific rationality"—whether in the new business, professional or bureaucratic classes, with the objective of establishing "democratic nation-states."[121] Yet the early nineteenth century ideals of the Romantic movement ("the aesthetic equivalent of the French Revolution")[122] and its attendant ideals of human reconciliation with nature, beauty, peace and creativity, gave birth to two new

---

[118] An allusion to the rites of a Catholic Mass: the priest, having said, "holy, holy, holy," lifts up the cup of wine.

[119] Ibid. Recall Barber's coining of the term "McWorld" for what I am calling postmodernity. Note its distorted view of freedom: "This politics of commodity offers a superficial expansion of options within a determined frame in return for surrendering the right to determine the frame. It offers the feel of freedom while diminishing the range of options and the power to affect the larger world. Is it really liberty?" In effect, McWorld "severs the 'freedom' to buy and sell from the right of women and men to choose in common their common goods or the social character of their shared world" (*Jihad Vs. McWorld*, 220-21).

[120] Korten, *The Post-Corporate World*, 279.

[121] Mitzman, *Prometheus Revisited*, 189.

[122] Ibid., 27.

ideologies: nationalism and socialism. Thus it was after 1850 that the guiding principles of revolutionary fervor turned more often to social justice and a socialist vision of the nation-state. In fact, argues Mitzman, this movement in western Europe "ended with the Keynesian welfare state—a compromise, compatible with the Fordist stage of capitalist production, that was trashed in the post-Fordist, post-cold war era."[123]

Since the rise of a dissident and dissatisfied youth in the 1960s, however, "the preindustrial strata with their millenialist aspirations...return to the fore in the recent upsurge of resistance to the commercialization of the world."[124] Significantly, the vanguard of opposition to neoliberal capitalism in the 1990s was located in the disparate branches of the Green movement—people who made common cause with the peasant and indigenous rebellions that gained momentum after the 1992 Rio Summit. In a rather striking way, José Bové, the French sheep farmer and union leader, began his ascent as one of the most articulate symbols of the antiglobalization movement by vandalizing in August 1999 a half-built McDonald's restaurant in Millau, France. Since then, he has become a ubiquitous emblem in the protests staged in Seattle and Genoa against corporate capitalism and in the World Social Forum gatherings in Porto Allegre (2001, 2002, 2003, 2005) and in Mumbai (2004).[125] In 2003, he was sentenced to ten months in a French jail for destroying two fields of genetically modified maize and rice. He explains why food is so politically charged today:

> We reject the global [trade] model dictated by the multinationals. Let's go back to agriculture; less than 5% of agricultural production goes on to the world market. Yet those responsible for that 5% of international trade dominate the other 95% of the production that is destined for national consumption (or neighbouring countries) and force this sector to submit to their logic. It's a totalitarian exercise. Agriculture should not be reduced to mere trade. People have the right to be able to feed themselves and take precautionary measures on food as they see fit.[126]

Since the Seattle protests, the WTO is still intact, he admits. Yet this global opposition movement seeks to hold it accountable. "Why should

---

[123] Ibid., 189-90.

[124] Ibid., 191.

[125] Its gathering in 2006 was polycentric: in Bamako, Mali; Caracas, Venezuela; Karachi, Pakistan. For information on the hundreds of organizations that make up and support the World Social Forum, visit: http://www.forumsocialmundial.org.

[126] Extract from Jose Bové and Francois Dufour's book, *The World Is Not for Sale: Farmers Against Junk Food* (London/New York: Verso, 2001), in *The Guardian* (June 13, 2001): http://society.guardian.co.uk/societyguardian/storyo,,505636,00 .html.

the global market escape the rule of international law or human rights conventions passed by the UN?," he asks. He then connects this struggle with that of the French Revolution: "The WTO has arrogated the functions of legislature, executive and judiciary solely for itself. In the 18th century such an anti-democratic concentration of power provoked the French revolution."[127] The WTO, therefore, should be pressured by the people to conform its policies to the UN's charter on human rights.

Moreover, the 1990s witnessed a growing convergence of opposition forces to global corporatism in the "core" (Western nations) and in the "periphery" (other nations), from Bové's Confédération Paysanne (the farmers' union he founded in 1987), to the environmentalist activists, to the Students Against Sweatshops (USA), to the various indigenous movements (most prominently in Mexico). I quoted Roger Burbach in the Introduction, who labeled this grassroots revolt "postmodern," partly because of its incredibly diverse nature; partly because of the realization that the traditional political tools of party and state were now meaningless; and partly because it came from "below": "It functions from below as an almost permanent rebellion, placing continuous demands on all the powers that be."[128] I would add the notion of global solidarity as a crucial paradigm shift for this post-cold war movement. Here, I believe, Bové's points are indicative of this new ethos. Among the ten points he offers as "a guide to good farming," these two indicate his global perspective: "(2) worldwide solidarity with all farmers; ... (10) always respectful of the long-term and global context."[129]

For Gustavo Esteva and Madhu Suri Prakash, the revolution sparked by the Zapatista uprising—"the first revolution of the twenty-first century"—was the needed impulse for all opponents of the neoliberal regime to understand the necessary dialectical movement from local to global action, and back to local action.[130] We do not live in one world—a *universe*, they contend, but rather in a *pluriverse*. We must put a stop to the steamrolling of the Western market economy over the multitude of local economies and cultures. Traditional development theory dictated that Western-trained elites would bestow the secrets of Western know-how and technology on the "less fortunate" Third-World peoples. Belatedly perhaps, noted theorist Robert Chambers came to see that development practitioners ought to proceed very humbly: putting the last first—that is, the poor, vulnerable and powerless—should be their first

---

[127] Ibid.
[128] Burbach, *Globalization and Postmodern Politics*, 11.
[129] Society Guardian: extracts from Jose Bové.
[130] Esteva and Prakash, *Grassroots Postmodernism*, 36.

consideration. From the "uppers" that they were taught to be, they need to learn to become the "lowers," in order to learn from local cultures, and once this is done, then maybe they could find ways to complement the age-old wisdom of local practices.[131] Thus in a pluriverse, the word "development"—an arrogant modern invention, if there ever was one—becomes obsolete.

In light of this, Esteva and Prakash advocate a "radical democracy" which takes stock of the current reality. In modern centralized states, the "social majorities" are not represented by their governments. These are "supported by middle and upper classes, elected through manipulations by parties and media."[132] The Marxist mirage consisted in fighting the state. But this only leads to consolidating its legitimacy, or at least its structures, whether capitalist or socialist. Korten would say civil society should get busy filling up the space that the state has not reserved for itself. Esteva and Prakash would retort that the "social majorities" must reverse that process: "attributing only those functions to central political bodies that cannot be absorbed by the commons or communities where the people exert their power."[133] In ways that can only be defined in each locale, communities must find creative avenues for bypassing the "global project."

Significantly, both they and Rasmussen[134] designate the Chiapas "rebellion" of 1993 as a watershed experience in the shifting of the development paradigm. The Zapatistas, gathered solemnly in their tiny village in the middle of their rainforest, "hosted the inauguration ceremony of the Intercontinental Encounter for Humanity and Against Neoliberalism."[135] The nearly three thousand who answered their invitation from all points of the globe experienced first-hand "a good illustration of the new political styles being created at the grassroots."[136]

Struggling social movements around the world instinctively felt a kinship with the Zapastitas. Not that it was not an authentic cultural renewal of a particular native people. Indeed it was that, first and foremost. Nor was it the universal rhetoric and the plastic sounding generalities of the "Global Project." But through the buzzing of the Internet a new web of global solidarity was forming.

---

[131] Robert Chambers, *Whose Reality Counts? Putting the Last First* (London: Intermediate Technology Publications, 1997).
[132] Esteva and Prakash, *Grassroots Postmodernism*, 158.
[133] Ibid., 164.
[134] Rasmussen, *Earth Community, Earth Ethics*, 127-37.
[135] Esteva and Prakash, *Grassroots Postmodernism*, 173.
[136] Ibid., 177.

Another perspective on the mushrooming of these new social movements comes from sociologist Jackie Smith, who defines grassroots
organizations that seek to educate and organize people locally to solve
global problems as "transnational social movement organizations"
(TSMOs). When it comes to making courageous decisions in favor of
protecting the environment and easing the plight of the most indebted
countries, governments are notoriously resistant. Leaders know it will
cost them votes in the next elections. Only citizens who band together
to pressure their elected officials to act for the greater good can effect
change.

For Smith as well, the Rio Earth Summit represented the ascending
wave of such TSMOs. In Rio, 1400 NGOs gathered to discuss urgent
global issues under the umbrella of the UNCED. They held over sixty
daily meetings and forums. After Rio, the forward momentum experienced by these numerous groups only got stronger.

EarthAction is one of the more visible of these TSMOs. From late
1992 to the beginning of 1995 they alone initiated more than twenty
global campaigns highlighting issues of ecology, development, peace
and human rights—over half of which explicitly called for the implementation of goals set out by the Agenda 21. To get an idea of the current
scope of these organizations I offer Smith's description of EarthAction—
a nice illustration of grassroots democracy configured in such a way that
it can shape decisions on environmental issues on the international level:

> Founded in 1991, EarthAction is a TSMO that mobilizes local, national,
> and international nongovernmental organizations (NGOs) around issues
> of peace, human rights, and sustainable development. Its more than 1,500
> partner organizations in over 40 countries are linked around common
> campaigns through routine mailings and telephone contacts from
> EarthAction's regional international offices...
> ...EarthAction's environmental campaigns have aimed to generate
> pressure on governments to strengthen their commitments to more equita-
> ble and sustainable development. These campaigns supported multilateral
> negotiations on the two major conventions that resulted from UNCED—
> the biological diversity and the climate change conventions—and
> EarthAction has also supported the development of the Desertification
> Convention, which was another goal of the Agenda 21. EarthAction's
> treaty-oriented campaigns are complemented by its efforts to support
> institutional reform and treaty implementation.[137]

[137] Jackie Smith, "Building Political Will after UNCED: EarthAction International,"
in *The Globalization Reader*, ed. Frank J. Lechner and John Boli (Malden, MA:
Blackwell, 2000).

Much more could be said, naturally, about the width and depth of such movements of solidarity between TSMOs, local peasant and indigenous initiatives, and the growing network of opposition to the neoliberal globalization agenda represented in the gatherings of the World Social Forum, for instance. My only point in this work is to signal the existence of this global movement and its growing impact, despite its often divided and inchoate state. The following quote from Esteva and Prakash illustrates well why they and others call this faction "postmodern." It is a distinctly post-cold war, extremely diversified grouping of people and organizations, committed to a more holistic and local vision of community:

> These grassroot initiatives sustain the hope that after modernity we will not be oppressed by universal, unique truths nor by the global certainties of the globalists: whether the conventional or alternative managers of the development discourse, including those offering the salvation of the Universal Declaration of Human Rights to all peoples, regardless of culture, caste, color or creed. After modernity, once again, we may have the flourishing of diversity: with a "good life" defined in local, rooted terms; incommensurable truths or perceptions regarding the nature of Nature, of Reality. "The people" are revealing a multiplicity of different cosmic visions conceived at the local level, emerging from the ruins left by modernity. After "the end of history," we can have the continuation and regeneration of thousands of histories.[138]

On the other hand, in light of what I have arguing in this chapter, I do not believe that the alternative to the UDHR is the abandonment of the notion of universal human rights, but rather a continuous conversation (recall Habermas' communicative action), in which both secular and religious people can carve out a common discourse of human solidarity. Nor do I wholeheartedly echo Esteva and Prakash's utter contempt for the modern notion of classical realism and truth. I fear that in a glib celebration of diversity and the incommensurability of worldview paradigms we might end up in the kind of fog Benhabib warned might obfuscate any rationale for human liberation. Cultural relativism, as we have seen, might prove helpless to mobilize people against some of the more heinous crimes perpetrated against humanity. So, while we may wish to jettison the notion of an arrogant and subjugating "Reality," we should do well to seek together what are some of the necessary contours of a "reality" that allows civil rights leaders to advocate human rights in a way that has resonance with people from among the variegated colors of the human tapestry. Thus while Christians and Muslims might opt (as

---

[138] Esteva and Prakash, *Grassroots Postmodernism*, 193.

I advocate) for a weak version of postmodernism, they can at the same time—and more effectively—shape a theological argument for jointly confronting today's corporate globalization and strengthening democratic bonds of solidarity between all levels of global civil society.

In the next chapter I sketch the outlines of a hermeneutic of sacred text that corresponds to what I see as a "weak version of postmodernism." It will also involve a summary of how critical realism may fruitfully contribute to the issue of interfaith dialogue.

Chapter 4

QUR'AN, BIBLE, HERMENEUTICS AND THEOLOGY

The last two chapters began with Martin Luther King Jr and Malcolm X as examples of Muslims and Christians who used the theological notion of inherent human dignity in order to broaden their appeal from civil to human rights. In some ways, King was the pioneer of a genre that came to be known as "liberation theology." By way of reminder, this theological movement grew out of the mostly Marxist-inspired post-colonial ferment in Latin America, in which ethics and development theory met in creative ways. Though some Catholic priests, including Camilo Torres, ended up opting for violence, most Latin American theologians did not. But for all these thinkers, the key concept was not economics but power—based on the tragic realization that a handful of elites monopolized power while oppressing society's majorities.[1]

The discourse of liberation was initiated in the Catholic Church, yet by the early 1970s it had been co-opted by Latin American social scientists, planners, and even some political leaders. It was a call to set aside the theories of the elite "and replace them with a deliberate stress on self-development as opposed to aid, foreign investment, and technical assistance."[2] This perspective of sociopolitical liberation in theological circles spread to Africa and Asia as well. Particularly interesting were the theological storms swirling around the issue of apartheid in South Africa,

---

[1] For Gustavo Gutierrez, for instance, the foremost advocate and articulator of liberation theology, the very word "development" is suspect and should be replaced by "liberation." "Development," he writes, "obfuscates the asymmetrical distribution of power in the world and the inability of the evolutionary model of social change to reach the necessary goals of the oppressed populations" (no reference provided; quoted in Denis Goulet, *The Cruel Choice: An New Concept in the Theory of Development* [Lanham, MD: University Press of America, 1985], xv). The correct word, says Gutierrez, is liberation, "a term which directly suggests domination, vulnerability in the face of world market forces, weak bargaining positions, the need for basic social changes domestically and for freer foreign policies" (ibid.).
[2] Ibid.

where some Muslims began to join the mostly black Christians in liberation circles. I begin this chapter, therefore, with this intriguing case study of theological cross-pollenization and then continue with some broader remarks about theological currents among Muslims.

Hermeneutical Rumblings in Muslim Circles

Farid Esack, a South African Muslim who has lectured in universities in the US and Europe, writes about the impact of modernity on educated Muslims in a Western context and, in particular, on South African Muslims who learned to do theology in the anti-apartheid movement. They saw the Qur'an as impelling them in the way of justice and yet were horrified to discover other Muslims appealing to the Qur'an to defend apartheid—just as many Afrikaner Christians were appealing to the Bible. Esack expressed the dilemma this way: "The South African engagement with the Qur'an in recent years has suggested that it is possible to have perfectly orthodox understandings of what the Qur'an is about and yet use these texts in rather perverse ways, e.g. justifying racism."[3] He argues for the use of "reception hermeneutics" when reading the Qur'an,[4] trying to avoid "textualism—focussing literally on texts" but on the other hand, to espouse "contextualism—focussing on patterns in the texts and contexts."[5]

---

[3] Farid Esack, "Qur'anic Hermeneutics: Problems and Prospects," *The Muslim World* 83.2 (1993): 122.

[4] Esack explains, "Reception hermeneutics focus on the process of interpretation and the appropriateness of interpretation rather than on the fixed literal text... Reception hermeneutics would thus transform the analysis of the reception of the text 'into a task of the study of the meaning of the text'. It challenges historical positivism in that it requires that diverse receptions of the texts, 'including present popular understanding of the text as concretizations of its meaning, be included in the problem of interpretation'" (ibid., 123; the quotes are from Francis Schüssler Fiorenza, "The Crisis of Scriptural Authority," *Interpretation* 44 [1990]: 23).

[5] Ibid., 122. Another Muslim writer who is aware of contemporary hermeneutics is French Muslim, Roger Garaudy. In a passage that exhorts fellow Muslims to take what is good from the West, Garaudy gives an example. Emmanuel Kant can teach us, he writes, that all that we might say about God, humanity, the world and history is "susceptible to criticism and revision, something that must be relative and conditional. This truth is the other side of another truth which Muslims have contributed to humanity, namely, that although it is God Himself who dictated the Qur'an, it is nonetheless humans who read, understand, and comment upon it. Their word can never be of the same status as His word. Humans are the products of history, of their problems and needs, of their time and environment. It is therefore always a difficult task to distinguish what is divine and eternal from what is human and relative" ("The

Esack relates his own experience, offering us details about his own context:

> Along with three friends, I spearheaded the founding of the Call of Islam in 1984. This affiliate of the United Democratic Front (UDF, established in 1983), the major internal liberation movement, soon became the most active Muslim movement, mobilizing nationally against apartheid, gender inequality, threats to the environment and to interfaith work. In the UDF itself, the Call was one of many religiously based organizations engaged in "the struggle." For these organizations, religion had always been a contested terrain and the struggle was as much about regaining ideological territory from religious conservatism and obscurantism as it was about political freedom.[6]

While the powerful white minority claimed the Bible as their guide in support of apartheid, many Muslim clerics quoted the Qur'an as exhorting the faithful to political quietism. Nelson Mandela, as is generally known, was outspoken about how his Methodist roots helped to shape his liberationist perspective and activism. Yet he was also adamant about his support for the interfaith character of the struggle against the official accommodationist versions of religion. Within this context, Esack provides a helpful definition of his theology and how it was applied to the political arena:

> In contrast to accommodation theology, liberation theology is the process of praxis for comprehensive justice, the theological reflection that emerges from it and the reshaping of praxis based on that reflection.[7] In South

Balance Sheet of Western Philosophy in this Century," *American Journal of Islamic Social Sciences* 2.2 [1985]: 175, 176). By contrast, Isma'il R. al-Faruqi is a good example of a Muslim scholar who worked from a modern—not postmodern—perspective. In an article on Muslim–Christian dialogue, al-Faruqi deplores the skepticism that has come upon the Western mindset of late. Both Islam and Christianity "make exclusivist claims to truth, and, therefore, assume that theoretical and axiological reality is knowable... The discovery by man of God Himself would not have taken place without reason. In the Holy Qur'an, God has presented the case of religion itself not as a myth, not as a stumbling block, but as a rational, critical, apodeictically certain truth" ("On the Nature of Islamic *Da'wah*," *International Review of Mission* 65.2 [1976]: 21).

[6] Esack, *Qur'an, Liberation & Pluralism*, 6.

[7] My quarrel with the "liberation hermeneutic" is not political (unless it advocates violence—which it does not), but rather that it focuses so much attention on the horizon of the present sociopolitical landscape of oppression that all of Scripture is read from that perspective. The horizon of the text and the horizon of the present context are not fused, but collapsed into one. At best it may distort the message of the Bible as a whole (e.g. human sin can be seen almost exclusively in the social

Africa, liberation theology was manifested in the growing numbers of religious figures and organizations who confessed the sin of silence in the face of oppression, acquiesced in the face of exploitation and power in the face of want. They sought a God who is active in history, who desires freedom for all people and the simultaneous conversion of hearts and social structures, a God whose own unity was reflected in the oneness of people.[8]

Esack and his colleagues in the Islamic movement for social justice in South Africa found many clues for their action in the Qur'an: from the Prophet's early siding with the poor and oppressed in Mecca, to his injunctions to protect their rights in Medina, and from the Exodus theme to a truly contextual understanding of the *wilāya* of the "other."[9] Above all, what is distinctive in liberation theology (whether in Islam, Judaism or Christianity) is the hermeneutic of the marginalized and the oppressed who stand in a privileged position to understand and experience the truth of God's word. It is a hermeneutic of action, or "praxis." The constant dialectic between praxis-reading-praxis is what continues to fuel the

structures). At worst, it can lead to a non-realist hermeneutic in which only the meaning "created" by the reader counts. In this study I come to similar conclusions, but following a hermeneutic that is at the same time more holistic and more faithful to what I see as the overall intention the Bible's teaching.

[8] Ibid., 8. This was the point of departure for his qur'anic hermeneutics: "My present search for a South African qur'anic hermeneutic of pluralism for liberation was rooted in the fusion of our nation's crucible and in my own commitment to comprehensive justice" (ibid., 9). Esack easily adopts the Christian liberation discourse: "To engage in qur'anic hermeneutics in a situation of injustice is to do theology and to experience faith as solidarity with the oppressed and marginalized in a struggle for liberation." On the next page he quotes from theologian Rebecca S. Chopp, who writes that "theological reasoning is uttered upon truth that is a way, upon a Word who has pitched...tent in the midst of history" (*The Praxis of Suffering* [Maryknoll, NY: Orbis, 1989], 61). This vision, he asserts, has not ceased to guide the "progressive Islamists" (like himself) in their stand against the traditional clerics: "It was and remains inevitable that this word of God that has pitched its tent in the midst of history would be affected by the storms, rain, wind and, yes, the sunshine, surrounding it" (ibid., 111). This is "incarnational" theology without Christ, yet with much the same force—after all, in both cases, it is God who comes down in the squalid trenches of the poor, gets himself "dirty," and fights for them through those who would obey his call.

[9] Drawing from Jane McAuliffe and others, Esack combs through all the verses in the Qur'an that warn Muslims about their *wilāya* ("association") with unbelievers, and sometimes Christians and Jews. He concludes that these passages, especially when read along with the Exodus passages, actively support "solidarity with the other...but also embraced those among the poor and downtrodden who actively rejected the religious beliefs of Islam" (Esack, *Qur'an, Liberation & Pluralism*, 203).

"base communities" in Latin America and Asia—and apparently also among many Muslims in South Africa.[10]

Building on the foregoing analysis of postmodernism and postmodernity, my argument throughout this study is that both qur'anic and biblical interpretation directly impinge upon, and are immediately affected by a host of factors, including socioeconomic and political ones. To these, of course, must be added personal factors from the life-world of the interpreter, such as his or her personality, his or her upbringing and culture. Hermeneutics is an interdisciplinary project that seeks to explore these dimensions. My objective in this chapter is that with this added perspective of theoretical hermeneutics, readings of Bible and Qur'an will gain in clarity and open new horizons of thought, while at the same time being chastened by a critical realist perspective.[11]

It needs to be said that the kind of sociopolitical readings of the Qur'an Esack is promoting is rare in Muslim circles, and should it become more widely known, it would likely meet with some resistance. Yet in the long run, since conservatives constitute the majority on both sides, it would behoove us to find ways to approach the question of hermeneutics from a fresh perspective. One advocate for this position is Mohammed Arkoun, an Algerian Kabyle (a dominant Berber group in that country) and an islamicist with a brilliant career behind him at the Sorbonne. Arkoun has been writing on epistemic and hermeneutic issues for over four decades. In a landmark article published in the first volume of the *Encyclopaedia of the Qur'an*, he has recently pleaded for a narrowing of the gap between critical studies of the Bible and similar studies of the Qur'an. One would have hoped, he writes, that the kind of "modern historical criticism" that flourished in North Africa and the Middle East "during the so-called Renaissance (*Nahḍa*, 1830–1940), would have grown to incorporate subjects as taboo as qur'anic studies, including the sacralised areas of law appropriated by the *sharīʿa* and its legal statutes and rulings" in the post-independence period that followed.[12] This was not to be the case, however, because of the powerful spread of "so-called fundamentalist movements," sparked in part by the

---

[10] "The Qur'an bears testimony to the idea that it is a book of understanding through praxis, rather than one of doctrine and dogma... The South African experience taught the progressive Islamist that liberative praxis in solidarity with the oppressed is the initial act of understanding the Qur'an" (ibid., 257).

[11] Here I mean producing a theology better adapted to the context, and therefore better able to guide people in their day-to-day lives, and to strengthen their faith accordingly.

[12] "Contemporary Critical Practices and the Qur'an," in *Encyclopaedia of the Qur'an*, vol. 1, ed. Jane Dammen McAuliffe (Leiden: Brill, 2001), 414-15.

1979 Islamic revolution in Iran. To make things worse, with regard to "the already very complex and inadequately explored area of qurʿānic studies," this new wave of revivalism (reviving "the rather archaic combination of the violent and the sacred") has been radicalized in its confrontation with a fundamentalism of another sort, particularly in France: "the global civilization of disenchantment, desacralization and the supremacy of sciences over all dimensions of human reality."[13] Arkoun does not in fact label this specifically modern credo "fundamentalism," but strongly implies it. Its "assumptions of scientific socialism and militant secularism" have led it on a mission "to eradicate [the religious imagination] through teaching official atheism or through eliminating the concept of the religious event (*fait religieux*) from an educational system run by a state that self-proclaimed its neutrality."[14]

Not unlike Christians in the nineteenth century, most Muslims continue to oppose "the earlier works marked by historicist-philologist positivism" and even the contemporary research that has tried to be more sensitive to the phenomenology of religion, and specifically to the wide variety of Islamic beliefs and practices. This last category, however, is also found wanting by Arkoun. The field of Islamic studies is monopolized by a narrow perspective of "pure research," a perspective which refuses "to integrate theological reasoning…into a methodological program for an epistemology of historical research (*épistémologie historique*) which would include all aspects and dimensions of reason and its products."[15] What is needed, he argues, is an honest quest for truth (without catering to or cowering before the authoritarian guardians of orthodoxy), which will not shy away from the thorny questions of history, but use a multidisciplinary approach. Here is the program Arkoun advocates, in his own words:

> This is not a question of establishing the true meaning of texts as lived by the faithful…nor is it a matter of articulating the certitudes recorded in a long process of sacralization…and systematized in the great products of theological, philosophical, legal or historiographical thought inherited from the Middle Ages. Rather, the task of the contemporary researcher is to problematize all systems which claim to produce meaning, all the

[13] Ibid.

[14] Ibid.

[15] Ibid. Arkoun adds by way of explanation: "A necessary correction to this narrow perspective would mean moving toward the use of historical psychology, historical sociology and historical anthropology for vast territories of the past, long ignored by the historian interested in narration, description and taxonomy." He believes that the German scholar Joseph van Ess (*Theologie und Gesellschaft*) is pointing in the right direction.

forms, still existent or not, which offer meaning and assumptions of meaning. This is an essential distinction that encompasses many problems yet to be raised or, if they have been, only poorly or without full recognition.[16]

If this sounds like a negative project, it is—at least in this initial phase. Arkoun uses the word "deconstruction," while quoting Derrida and Lyotard. But it is abundantly clear from his other writings that, in the end, he aims for a "humanizing" of theology. Hence, in the same paragraph he insists that in qur'anic studies, "the scholar approaches the activity of the human spirit that most closely expresses its own utopian vision, its hopes…its struggle to push back the limits of servitude and to attain the full exercise of its 'will to know', combined with its critical and creative freedom."[17] Clearly, the Enlightenment ideals of freedom and emancipation from oppressive structures—be they religious or political—live on in his project.

While it is doubtful that Arkoun can convince the religious establishment in Muslim countries (and 'ulama, in particular)[18] to apply modern critical tools to their foundational texts, as Muhammad Qasim Zaman has shown, there is considerable debate and change taking place even in these most traditional circles.[19] Though virtually none of these traditional scholars (the 'ulama) would consider any new approach to the very tradition that justifies their own social and professional standing, Zaman notes two important phenomena. The first is that incremental changes are taking place continually in the most conservative circles, if only as responses to the changes of society around them. As it is with every tradition, and particularly in the long history of the 'ulama in Muslim societies, the Islamic tradition is one that has continuously been "updated and variously reconfigured."[20] Yet, the movement for change is hampered in many cases by state structures (he singles out Pakistan in this) that seek to wrest religious authority from their hands and complicated by islamists who often compete with them for the allegiance of the masses. The second phenomenon is a small but growing number of 'ulama who raise critical voices from within the religious establishment. He notes the

---

[16] Ibid., 414.

[17] For more details on his perspective, see also ibid.

[18] The 'ulama are the legal experts in traditional Muslim lands. Already in the third century of Islam, they emerged as a power to counterbalance that of the state, particularly after the caliphs' political power was gradually transferred to the sultans. It was from the ranks of the 'ulama that judges and muftis were named.

[19] Muhammad Qasim Zaman, *The Ulama in Contemporary Islam: Custodians of Change* (Princeton, NJ/Oxford: Princeton University Press, 2002).

[20] Ibid., 180.

minority yet influential voices of the Indian scholar Mawlana Wahid al-din Khan, the Iranian Hujjat al-Islam Sayyid Muhsin Sa'idzadeh, and the politically progressive Indonesian movement Nahdlatul Ulama. My own research has pointed to the seeds of hermeneutical change planted in the heart of Islamic legal theory by Muhammad Abduh (d. 1905), seeds which are now beginning to blossom again and, in particular, the emphasis on the "purposes of shari'a" (*maqāṣid al-sharī'a*).[21] Hermeneutics will always be at the heart of theology, either subverting traditional pathways, or affirming them while at the same time indicating new directions.

All these movements that seek to renegotiate traditions of jurisprudence and theology do so under the push and pull of sociopolitical and economic conditions—postmodernity as laid out in this work. As we have seen from several angles, economic issues press hard on the political and social ones in many areas of the world, even as they overshadow—or, alternatively, exacerbate—the religious tensions. A Dutch writer talks about his experiences in Indonesia and Holland, which have taught him that attention to both theology and issues of economic justice are necessary for fruitful Muslim-Christian dialogue. The conclusion of his article sums up a good part of my own project:

> The first part of this article focuses on a radical shift in the theology of religions and the interreligious debate: international economic issues and peace problems have become more urgent than metaphysical and dogmatic problems in the *theologia religionum*. Interreligious relations are no longer considered part of a project that can be postponed, since what is urgent has a definite priority. Peace and physical well-being are too closely related to interreligious harmony.[22]

In his Introduction to Jacob Neusner's edited book, *Sacred Texts and Authority*, William Scott Green rightly states that all the major religions have both written and "living" texts. By "living" he means that "religious intellectuals and virtuosi read, understand, interpret, mediate, and exemplify the texts in the life of a religious community."[23] The written

---

[21] David L. Johnston, "An Epistemological and Hermeneutical Turn in Twentieth-Century Uṣūl al-Fiqh," *Islamic Law and Society* 11.2 (2004): 233-82, and "*Maqāṣid al-Sharī'a*: Epistemology and Hermeneutics of Muslim Theologies of Human Rights," *Die Welt des Islams* 47.2 (2007): 149-87.

[22] Karel Steenbrink, "Quranic Guidelines for Economy as a Basis for Interreligious Solidarity in Favor of the Poor? Some Reflections on the Indonesian and Dutch Contexts," *Mission Studies* 15.2 (1998): 117.

[23] William Scott Green, "Introduction," in *Sacred Texts and Authority*, ed. Jacob Neusner, The Pilgrim Library of World Religions 3 (Cleveland, OH: Pilgrim Press, 1998), xv.

texts of Judaism, Islam, Buddhism, Hinduism, and Christianity are made to connect with the adherents of their respective religions through the work of interpretation carried out by their scholars over the centuries. As a result, "All the religions...have developed methods, criteria, and communities of interpretation of sacred texts."[24]

Muslims and Christians traditionally have approached the Qur'an and the Bible assuming that there is a one-to-one correspondence between the words in front of them on the page and the meaning of these words. All that is needed when ambiguities arise is to look up the meaning of a word in the dictionary, or take a closer look at the grammatical structure of a sentence. Certainly this is easier for Christians, who readily find translations of the Bible in the current idioms of their tongue.[25] For Muslims, the task is more difficult in that only the Arabic Qur'an is considered God's dictated Word (eternal, in Ash'arite theology) to the Prophet, so that any translation is at best a useful paraphrase. Yet even for a contemporary native Arabic speaker, the literary form of the Holy Book is difficult. Regardless of ethnic backgrounds, however, all serious readers rely on classic works of commentary (*tafsīr*) as a companion to the Qur'an.

For both traditional Muslims and Christians, the assumption is that the barrier to meaning retrieval is only a semantic one. Words mean certain things, and those "things" just need to be looked up in a dictionary or commentary in order to grasp the meaning, which by definition is unchangeable. Words put together form sentences that yield meaning in the same mechanical way. Thus if syntax is a hindrance, then one consults a grammar—Islamic commentaries specialize in these kinds of explanations. Traditional hermeneutics, then, was always the work of exegetes and philologists.

In his Introduction to *Islam and Modernity*, the late Fazlur Rahman presented his own work as dynamic, "genuine, original, and adequate Islamic thought," which can serve as a useful criterion to judge and improve an Islamic educational system.[26] He then explained that the question of method and hermeneutics are central, first because the

---

[24] Ibid., xvi.

[25] See Yale scholar Lamin Sanneh's classic work on the sociopolitical effects of vernacular translations of the Bible in colonial and postcolonial Africa, *Translating the Message: The Missionary Impact on Culture* (Maryknoll, NY: Orbis, 1989); also, more recently: *Whose Religion is Christianity? The Gospel beyond the West* (Grand Rapids, MI: Eerdmans, 2003).

[26] Fazlur Rahman, *Islam and Modernity: Transformation of an Intellectual Tradition* (Chicago: University of Chicago Press, 1982).

Qur'an is the very basis of Islamic thought, and secondly because these issues have not been adequately addressed by Muslim scholars. His method is twofold: go to the text and seek out its meaning in its original historical context, and then systematize those answers God gives to questions arising out of concrete situations through the mind of the Prophet. This allows theologians to apply these general principles to their own contemporary situation, yet not in a slavish way. It might be that in this *jihād* of mental endeavor they find that a particular principle cannot be honestly applied. This method serves as a dynamic corrective for the first part of the process. Believing scholars are engaged in a dynamic conversation with the text between the two horizons of the qur'anic situation and their own. While this certainly qualifies as "doing theology in context," it offers no significant interaction with postmodern herme-neutics, in spite of his attempt to disqualify Hans Georg Gadamer's subjectivism.[27]

It may be worthwhile to quote another Muslim scholar, Vincent J. Cornell, who deplores the contemporary islamist's belief in the inerrancy of scripture and yet "is more likely to find his inspiration in the cut-and-dried lessons of *ḥadīth*," thus positioning himself closer to his Jewish orthodox counterpart (rather than the Christian fundamentalist), "in that the scripture which engages him most is nomological [tied to law] rather than revelatory." The main problem with this excessive reliance on a body of literature crafted a millennium ago is that "the Muslim fun-damentalist suffers from an epistemology that denies the legitimacy of modern hermeneutical methods."[28] He then goes on to propose his solution:

---

[27] Ibid., 8-11. In his Introduction to Rahman's posthumously published work, *Revival and Reform in Islam*, ed. and with an Introduction by Ebrahim Moosa (Oxford: Oneworld, 2000), Moosa emphasizes the metacritical direction of both Gadamer's and Rahman's work: "Fazlur Rahman, like Gadamer, also believed in the metaphysi-cal distinction between objective meaning and ceaselessly changing expression. At the end, the truth for him was singular. Even though the truth may have multiple or overlapping expressions, it did not necessarily mean a multiplicity of truths. In that sense he was every bit a child of the Enlightenment" (ibid., 21). My contention throughout will be that one can leave behind the Enlightenment without letting go of objective truth, through the perspective of critical realism.

[28] Vincent J. Cornell, "Where is Scriptural Truth in Islam?," in *Holy Scriptures in Judaism, Christianity and Islam: Hermeneutics, Values and Society*, ed. Hendrik M. Vroom and Jerald D. Gort, Currents of Encounter: Studies on the Contact between Christianity and Other Religions, Beliefs, and Cultures 12 (Amsterdam: Rodopi, 1997), 74.

A new generation of savants is needed who are as much a product of their times as their predecessors were in the innovative days of early Hanafism and Malikism. This is not to say that Muslims have to abandon all of their traditions. Only that the traditions which count are not to be trivialized by a traditionalism that makes tradition itself appear ridiculous.[29]

Indeed, there are a few rumblings of change in contemporary Muslim hermeneutics, but they have not yet affected the masses of conservative Muslims, whether in the west and or Muslim-majority nations. To what extent the current interest in Islamic jusrisprudence on the "purposes of the shari'a" will open new avenues of reasoning deserves to be monitored. Broadly speaking, the difficulty with the traditional view is that interdisciplinary research in the philosophy of language, meaning, texts and their interpretation (all of which is included in "hermeneutics") in the last two centuries has shown that this process is much more complex and that, as a result, the view most people hold is naïve—in exactly the same way classical realism is "naïve realism." Much of the meaning intended by the author can be lost, distorted, or simply changed, when those original ideas are committed to writing, on the one hand, and, on the other, when a reader picks up the text and reads it.[30] There is a big difference, of course, between a note written by one's friend about a

[29] Ibid., 75. The problem is well illustrated by reading through the various essays on qur'anic interpretation published in cooperation with the University of London School of Oriental and African Studies (*Approaches to the Qur'an*, ed. G. R. Hawting and Abdul-Kader A. Shareef [London: Routledge, 1993]). Though a variety of methods are presented, nowhere is the theoretical issue of hermeneutics addressed. More recently, however, there has been some movement. In the same issue of *Islamic Studies* (41.4 [2002]), two conservative writers raise hermeneutical issues that lead them both to state that, in some cases, clear legal rulings in the texts can be laid aside today (more on them in the next chapter): Mohammad Hashim Kamali, "Issues in the Understanding of Jihad and Ijtihad," 617-34; Soualhi Younes, "Islamic Legal Hermeneutics: The Context and Adequacy of Interpretation in Modern Islamic Discourse," 585-615.

[30] Esack in the above mentioned monograph, *Qur'an, Liberation & Pluralism*, castigates reformist Muslim scholars whose solution to the current crisis in qur'anic hermeneutics is to bypass tradition and go back to the Qur'an: "This argument does not take into the account the fact that exegesis is not entirely independent from the text but actually belongs to its historical productivity... No scripture, least of all a text simultaneously abounding with symbolism and an all-pervasive contextuality as the Qur'an evidently is, emerges from a vacuum and comes to us unencumbered by 'the plural and ambiguous history of the effects of its own production and all its former receptions' [quoting from David Tracy, *Plurality and Ambiguity: Hermeneutics, Religion, Hope*, San Francisco: Harper & Row, 1987, 69]" (*Qur'an, Liberation & Pluralism*, 77).

familiar topic and picking up a religious book written centuries ago in a foreign culture. Distance in relation to context and culture is indeed an important consideration. And yet, as the following discussion aims to demonstrate, several other crucial factors are at work, each rendering the task of meaning retrieval more difficult and complex.

I begin my presentation of hermeneutics with a brief summary of its history in modern times. This in turn will lead to a summary statement on the semiotic nature of the human person, and then to my own proposals for the interpretation of Bible and Qur'an in my quest to articulate a common theology of creation.

The Modern Trajectory of Hermeneutics

A cursory survey of the development of hermeneutical studies in the last two centuries shows that the focus of study began with the author of a text, continued with the text itself, and then at a more recent stage has emphasized the role played by the reader. Hence, the hermeneutical continuum:

Figure 5. *The Hermeneutical Continuum*

Focus on the Author

As we have seen, one of the central tenets of Enlightenment philosophy was the conscious dismissal of the symbiotic relationship between human reason and divine revelation that had prevailed from medieval times and throughout the Renaissance. It represented the enthronement, in a sense, of the human person as an autonomous being, capable of surveying the world and marshaling its knowledge and rational powers in order to shape it according to its designs and perceived needs—whether in fact there was a God or not. This turn toward the human subject as the agent of human freedom and progress implied that the modern self is not only the creator of its own norms and values, but also is the author of texts, in which meaning is found in the subjectivity of unique individuality and subjectivity of its author. As theologian Kevin Vanhoozer has it, "[t]hanks to the light of reason, the knowing subject, like Adam before the fall, sees the world as it is and names it truly."[31]

---

[31] Kevin Vanhoozer, *Is There Meaning in This Text? The Bible, the Reader, and the Morality of Literary Knowledge* (Grand Rapids, MI: Zondervan, 1997), 45.

In this regard, one must point to German theologian Friedrich Schleier-macher as the father of modern hermeneutics.[32] By examining his con-tribution several key elements stand out, ones which were picked up and reworked by subsequent thinkers. First, he isolated hermeneutics as a subject of inquiry in itself. Second, by asking the question "How can the human mind understand at all?" Schleiermacher placed hermeneutics in the context of epistemology. Third, he emphasized the creative role played by the human mind in the process of interpretation. In that sense, all interpreters begin afresh (though also building on some of the previous historical and cultural research done before) in their task to understand the text, which means that they must put themselves in the position of the author, both in the objective (historical, cultural and intellectual) and subjective sense (the psychology and thoughts of the author). Fourth, for the first time Schleiermacher raises the issue of language itself. If the hermeneutical task involves both critical reflection and "divining" (gaining an intuitive connection with the author) then the language of the text is not the only criterion for understanding.[33]

Finally, Schleiermacher was the first to offer the idea (though not the term) of a hermeneutical circle. This circular process always takes place in the context of trying to assess the author's thought "as a whole":

> Complete knowledge always involves an apparent circle, that each part can be understood only out of the whole to which it belongs, and vice-versa... To put oneself in the position of an author means to follow through with this relationship between the whole and the parts... The more we learn about an author, the better equipped we are for interpreta-tion. A text can never be understood right away. On the contrary, every reading puts us in a better position to understand.[34]

[32] A helpful definition of hermeneutics is provided by D. E. Klemm: "The term 'hermeneutics' derives from the Greek *hermeneuein*, which carries the senses of expression (uttering a thought or intention), explication (interpreting an utterance), and translation (mediating meanings from one language to another). Hermeneutics denotes the theoretical and methodological process of understanding meanings in signs and symbols, whether written or spoken" ("Hermeneutics," in *Methods of Biblical Interpretation*, Foreword by Douglas A. Knight, excerpted from the *Dictionary of Biblical Interpretation*, ed. John H. Hayes (Nashville, TN: Abingdon Press, 2004), 147.

[33] Though he is not yet formulating the by now classical quandary ("what is lan-guage?"), his wrestling with the interaction between both poles anticipates Wittgen-stein's emphasis on language as "word-games." Again, the aspect of meaning as arising at least in part from the function of language in human society becomes a central concern in the contemporary debate.

[34] F. D. E. Schleiermacher, *Hermeneutics: The Handwritten Manuscripts*, ed. H. Kimmerle, trans. J. Duke and J. Forstman, A.A.R. Text and Translation Series 1

The other thinker who is best known for his focus on the author side of interpretation is Wilhelm Dilthey (1833–1911), who chose to concentrate on the psychological dimension of Schleiermacher's work. Reading therefore leads to new levels of self-discovery, with the reader coming into a psychological/historical union with the author. While one could object that the text itself is obscured in the process, Dilthey was the first to coin and develop the idea of worldview—the universal human attempt to "solve 'the riddle of life,'" yet always within a particular historical and cultural context.[35] One can point to several universal structures of the human psyche, including mind, emotion and will, or the impulses toward religion, poetry and metaphysics. His overall objective was to find a way to ground the human sciences—with history at the top—in a way that parallels Kant's effort to ground the physical sciences. Dilthey believed that despite a multiplicity of human worldviews across cultures, enough commonality in human experience and understanding guaranteed a true—though not exhaustive—accessing of knowledge between readers and authors in different historical settings. This is so, because the "[h]istorical understanding involves emphatically transposing oneself into the inner experiences of other human beings by means of and within their proper worlds of cultural and linguistic meanings."[36]

Focus on the Text
Early in the twentieth century, Gottlob Frege initiated a shift in hermeneutics away from the author. He rejected any psychological content, which had ruled the concept of textual interpretation since Locke's famous insistence that language is the mind of those both sending and receiving messages, be they oral or written. What is crucial here is that Frege's work banishes the usefulness of digging into what an author might have intended to mean, while at the same time leaving the door open to the possibility that reference is not all there is to meaning (recall Benhabib's discussion of the episteme of representation).

In a similar vein, Nancey Murphy has argued that modern thought is characterized by three strategies: (1) foundationalism (there is an indubitable foundation all knowledge can be built upon—knowledge

(Missoula, MT: Scholars Press, 1977), 112-13, quoted in Anthony C. Thiselton, *New Horizons in Hermeneutics: The Theory and Practice of Transforming Biblical Readings* (Grand Rapids: Zondervan, 1992), 221.

[35] Naugle, *Worldview*, 86. Naugle quotes Michael Ermath, a specialist on Dilthey, who wrote that the concept of worldview had "come into wide usage through Dilthey's work" (ibid., 82; Ermath, *Wilhelm Dilthey: The Critique of Historical Reason* [Chicago: University of Chicago Press, 1978], 15).

[36] Klemm, "Hermeneutics," 149.

pictured as a multi-storied building); (2) referentialism (meaning in language is derived from its smallest units, which refer to simple ideas in the minds of persons); (3) reductionism (the end of Aristotelian metaphysics—the vision of things as composed of both form and matter, and the idea that everything ultimately boils down to physics, or "atomism"). One might have thought, contends Murphy, that with Frege's distinction between *Sinn* (sense) and *Bedeutung* (reference) and his assertion that meaning is to be found in the former, that philosophers of language would have left "reference" behind them. Yet Frege was not consistent, in that he defined "sense" as that which points to truth in a sentence and that truth comes only by way of reference. Of course, it is not a one-to-one correspondence with things in the world, but it is "reference" nonetheless, in his particular view of representation:

> The logical atomists—Bertrand Russell, Ludwig Wittgenstein in his early work, and others—followed Frege in supposing that philosophy of language was to be done by devising formal artificial languages rather than by analyzing natural languages. They also followed Frege in recognizing the sentence or proposition as the smallest unity of meaningful discourse. Thus, in place of a referential theory of the meaning of words, we have a representative or "picture" theory of the meaning of sentences or propositions.[37]

According to Murphy, this rather wooden approach to meaning betrayed a hidden metaphysical agenda, one very much in line with Kant's distinction between what we perceive in our minds and "reality out there as it is" (the noumenal *Ding an sich*). The language of science was tied to sense experience phenomena and therefore had fairly reliable and predictable meaning, whereas the language of ethics and religion had to be considered a "second-class" language. Thus for logical positivists such as A. J. Ayer, religious, aesthetic or ethical judgments carry no factual content but only serve to express the attitudes, intentions, or feelings of the speaker/writer. While scientific discourse is referential, then, religious discourse is only expressive of the speaker's inner state.[38]

Meanwhile, in the 1920s, the debates on the issue of hermeneutics were spilling over from one discipline to the next. Schleiermacher was a theologian concerned with biblical exegesis, and at the same time a

---

[37] Murphy, *Anglo-American Postmodernity*, 11.

[38] Murphy shows how the modern "picture" of foundationalism could only provide the theologian with two options with regard to language: "language that is descriptive, representational, and propositional [fundamentalist or evangelical position] or language that is merely expressive of the inner awareness, attitudes, and existential orientation of the speaker [liberal position]" (ibid., 110).

philosopher concerned with formulating an adequate theory of textual interpretation. Frege was a pioneering philosopher of language, as were Ayer and Wittgenstein. In the 1930s the field that came to dominate the hermeneutical debate was literature, and the reigning school for nearly three decades was the Anglo-American New Criticism.[39]

Another influential current of thought comes from the discipline of linguistics. As with the literary critics, these theorists kept the focus chiefly on the text. With Charles S. Pierce (1839–1914) in the United States and the Swiss language theoretician Ferdinand de Saussure (1857–1913), not only did semiotics ("study of signs") get underway as a modern discipline, but a foundation was laid for more radical interpretations: the distinctive American hermeneutical tradition,[40] and, in Europe, structuralism followed by poststructuralism (or as it was called, the "poststructuralist school of semiotics"); and finally, Jacques Derrida and the later Roland Barthes's deconstructionism.

Saussure laid the foundation for the modern discipline of linguistics. He showed how words function as signs pointing beyond themselves, and which are, by definition, arbitrary. Here Saussure contributed an important distinction between diachronic linguistics (how language changes over time) and synchronic linguistics (how signs relate to one another at a given time). His unique contribution concerned the latter category: meaning is generated when one sign is contrasted with another, which might at first sight seem similar. Language is a collection of many such interdependent systems and sub-systems. "For example, in a sub-system of colour-words, 'orange' derives its meaning from its difference from its next-door neighbours in the continuum, 'red' and 'yellow,' rather than from pointing to oranges on trees."[41] This led him to differentiate between *langue* (the existing language one may draw upon) and *parole* (a concrete act of speech). *Langue*, of course, does not "exist" as such, except in the conventions of a society, and as such it is constantly changing, evolving.

---

[39] Osborne, *The Hermeneutical Spiral,* 369.

[40] Thiselton lists three aspects of his work on meaning that were built upon by others in the American context: "First, Pierce stressed the fallible character of all human knowledge, beliefs and statements. Beliefs amount largely to 'habits of behaviour.' Second, thinking or thought has to do with the use of signs. Yet signs point beyond themselves to other signs and sign-relations. Finally, meaning is to be seen primarily in terms of meaning-*effect*. It is here that Peirce's pragmatism has its most far-reaching effects. What is important and 'cashable' about meaning is its bearing on the conduct of life" (*New Horizons*, 84, emphasis original).

[41] Ibid., 832.

This leads me to a short comment on the only school of hermeneutics that could truly be classified as "text-only": the structuralists. Structuralism developed out of Saussure's view of meaning, but now generated exclusively within a synchronic sign-system, and thus with a metaphysical twist. Claude Levi Strauss, for instance, widens these insights to the field of anthropology.[42] For these structuralists, the surface of a text yields no meaning. One has to dig beneath it to uncover universal structures common to humankind. These universal patterns unconsciously determine a writer's worldview and impinge on his or her writing unwittingly—a position remarkably similar to that of Dilthey.

Structuralism's main contribution came with its preoccupation with semiotics, or the theory of signs. All language, after all, presupposes a code, one which is both verbal and non-verbal. As Thiselton puts it, "The code is the sign-system, lattice, or network, in which the linguistic choices which convey the message are expressed."[43] A music sheet, for instance, is a good example of a code which is uniquely structured in such a way that when it is deciphered, its meaning also includes its performance.

Further, texts may comprise several levels of code. Texts often refer to other texts that come to enrich their meaning. This phenomenon in semiotics is called "intertextuality." Thiselton offers several helpful illustrations:

> For example, the Apocalypse of John at one level presupposes the range of possible lexical and grammatical choices available in hellenistic Greek (albeit the Apocalyptist's Greek presses the code to its limits!). But it also operates on the basis of a system of conventions used by earlier apocalyptic. Some allusions to earlier texts such as Ezechiel, Zechariah, and Daniel are not merely reminders about earlier traditions. Sometimes they perform not a stylistic but a *semiotic* function, providing yet another level of encoding in terms of which a message is to be read.[44]

---

[42] "Structuralism was finally applied to other disciplines and brought to the centre of the stage in Lévi-Strauss' book *Structural Anthropology* (1958)... Linguistics and social phenomena are 'the same' in the sense of being a language which structures or codes... Lévi-Strauss' widest interest, however, was in the structure and significance of myth. Here the fundamental oppositions include life vs. death; man vs. God; good vs. bad... Myth is a deep structure, a universal narrative model freed from temporal and cultural conditioning" (Thiselton, *New Horizons*, 90).

[43] Ibid., 80.

[44] Ibid., emphasis original.

Structuralism as a movement, however, represents the dead-end of the "text-only" approach to interpretation.[45] First, it arbitrarily posits the existence of universal traits in humankind, which, by implication, must be trans-cultural. But, even more indefensibly, contends Murphy, it denies any relation between the signs and "reality." Literary critics who have used the structuralist method deny that there is some kind of "given" in the text per se. For them, to focus on the author's intent is to commit the "intentional fallacy," or to focus on how the text affects the reader is to commit the "affective fallacy." "The key to a text is form: structures of sound, rhythm, rhyme, sequences of images, even ratios of passive to active verbs."[46]

There is yet another dimension to the structuralist strategy, contends Vanhoozer, and yet another discipline which must be brought into the debate: theology. The humanists of the fifteenth century, and indeed all literary work since the Greeks, sought to plumb the depths of classical literature by recovering the mind of the author. And this has certainly been the aim of traditional biblical research as well. This conviction continued to drive the moderns in their literary pursuits. For historical critics such as James Barr and Benjamin Jowett, for instance, true exegesis seeks to uncover the original sense of the biblical text, thus preserving it from the corrupting influence of dogmatism. "We may conclude, therefore, that the concept of the author as the 'home of meaning' lies at the center of premodern and modern interpretation alike."[47]

I agree with Vanhoozer's argument, one that people of faith—Muslims and Christians in the context of this work—will find attractive: deconstruction is the last battle in a long war against "the author as the determiner of textual meaning" and a logical consequence of Nietzsche's pronouncement of God's death. I believe that an exclusive hermeneutical focus on the text (or as is now more fashionable, on the reader) represents a metaphysical choice away from realism. In fact, Nietzsche, long

---

[45] It was easily undone by Derrida. Vanhoozer offers a helpful explanation: "For Derrida, structuralism is just another attempt, like that of Descartes, to stave off the threat of relativism by finding some stable ground for meaning. In the case of structuralism, however, it is not the sovereign subject but the language system itself that accounts for the stability of meaning by positing distinctions (e.g., male/female; white/black; rational/irrational). Structures too—whether linguistic, familial, social, or philosophical—are ultimately arbitrary and artificial constructions. The point is that differences—even the distinction between truth and falsity—are not natural but man-made" (*Is There a Meaning in This Text?*, 32-33).

[46] Murphy, *Anglo-American Postmodernity*, 136.

[47] Vanhoozer, *Is There a Meaning in This Text?*, 48.

before Derrida and Rorty, was a non-realist, in that he believed in a real world, but the human mind and the language that it creates cannot apprehend it in any meaningful way. In that vein, Judaism, Christianity and Islam are dismissed in one sweep by Nietzsche, who asserts that "Christianity is Platonism for the people." Deconstruction rejects all metaphysics. Derrida called this "overthrowing idols," and as a consequence, there are no facts, only interpretations; no gods, only persons using "God" to oppress other persons; no authors, only texts whose meaning are created by readers.[48]

Again, this is the "strong version" of postmodernism about which Benhabib was warning her feminist colleagues in a secular setting. For the religious with monotheistic convictions, without a Creator-God to guarantee human communication, speech and knowledge, logically there could be no real author in any text. There certainly could be no responsibility, no human caliphate, no accountability for how we relate to the world and to one another. While Benhabib argued on the basis of ethical principles and a progressive politics, people of faith would opt for the weaker version by adding theological reasons. But we still have to consider the reader side of the hermeneutical continuum. Clearly the reader, especially if we come back to the agenda developed by Schleiermacher and Dilthey, is actively involved in the meaning-extraction process.

Focus on the Reader
Martin Heidegger's *Time and Being* (1927) reframed much of the previous work on hermeneutics. Instead of a focus on the epistemic nature of textual interpretation, he proposed to connect hermeneutics to ontology. Understanding should be seen as one of the basic modes of human being, *Dasein*, the human impulse to care about its own existence, which acts as a kind of preunderstanding in the reading of texts— whether written or the texts of life as people experience it. Thus, for Heidegger, "the goal of hermeneutics is philosophical self-understanding, conceived as a way of combating a pervasive forgetfulness of what it means to be human."[49] One of the fundamental ways *Dasein* expresses itself is through human interpretation. Hence, hermeneutics can be understood as the projecting of one's existential concerns onto a text that one has already apprehended in a kind of preunderstanding, and as that process of interpretation which gradually reveals what was projected in the first place. Every text (or worldly thing in general) comes to its interpreter already mixed with his or her own presuppositions and concerns.

[48] Ibid., 57.
[49] Klemm, "Hermeneutics," p. 149.

In essence, there can be no object we can mention that does not owe its existence to our human interpretation or apprehension.[50]

Even more influential than Heidegger's writings in the subsequent flurry of interest in hermeneutics was the publishing of Hans-Georg Gadamer's *Truth and Method* in 1960.[51] Without discounting Heidegger's involvement with being (though for Gadamer it did turn out to be a less than useful exercise), Gadamer (1900–2002) came back to Schleiermacher and Dilthey's agenda: how one might legitimate the methods of the human sciences. In so doing, he questioned the quintessential modern dogma of proper method leading to truth. Heidegger was right about preunderstanding, concedes Gadamer, but his later preoccupation with language rather than being was more to the point. People seek truth in three separate but related spheres: art (we experience its truth as we allow ourselves to be drawn into its meaning-games); historical consciousness (truth comes via the self-critical awareness of the historian who inevitably recreates a new historical moment when writing history); and language as the "universal medium of the self's dialogical openness to the other."[52] Thus, in coming to a text, a reader brings into each act of interpreting a set of preconceived notions, which in turn color the understanding process.

Thiselton sees Gadamer's main contribution in what he calls a "horizon of expectation," emphasizing both the provisional nature of one's network of assumptions and the fact that many of these are unconscious. In that sense, like the reading of a musical score, a certain knowledge is required if understanding is to take place at all. Yet each new piece of music invites discovery and surprise. Hence, understanding texts is a dual movement between the strange and the familiar:

> My horizons must contain a space within which the text can be intelligently "slotted" in terms of provisional linkages with the familiar that allow patterns of recognition. On the other hand the realization that what seems familiar is not quite what I had expected or assumed it to be, necessitates an expansion of my horizons to make room for what is new.[53]

---

[50] Ibid., 150. The New Testament scholar and theologian Rudolf Bultmann made direct use of Heidegger's hermeneutics. For him, faith in Jesus Christ "is a decision the self as *Dasein* makes to entrust itself to the grace of God that breaks into the world through the *kerygma* of the New Testament." Faith, then, entails an authentic throwing of the self onto the liberating power emanating from the gospel.

[51] See his *Truth and Method*, 2d rev. ed., translation rev. Joel Weinsheimer and Donald G. Marshall (New York: Continuum, 1993).

[52] Ibid., 151.

[53] Thiselton, *New Horizons*, 45.

One of the directions to which Gadamer's thought has pointed is the importance of a reader's response to a given text. According to the more radical postmodern view, texts are actualized only in the reading of persons in specific cultural and historical contexts. Whatever truths emerge, consequently, remain context-relative. This is what the "Reader-Response Criticism" school of literary criticism asserts, with Richard Rorty and Stanley Fish remaining its best-known representatives. Subjectivity is no longer a liability but rather an asset—which comes, however, with a heavy metaphysical price: in its extreme form, the text disappears and the reader creates meaning. Yet, as Murphy shows, reader-response criticism (as deconstructionism) is a reaction against the self-assured proclamations of structuralism. Any structure that structuralists claim to be there, they would counter, is "in the eye of the beholder."[54]

The historical sketch of hermeneutics as it developed in the twentieth century would not be complete without one more major figure, the French philosopher Paul Ricœur (b. 1913). I will summarize his contributions to the understanding of hermeneutics in three points. The first is his insistence that, contra Gadamer, "explanation" (method) and "understanding" (truth) must be kept together. Gadamer

> has as a matter of principle left no metacritical or even critical procedure for testing the validity of traditions in effective-historical consciousness. For Ricœur, hermeneutics properly remains a "metacritical discipline" which sets into motion both the unmasking role of explanation and the creative role of understanding.[55]

In his own words, the fundamental task of hermeneutics is "to destroy the idols, to listen to the symbols."[56] Keeping these two axes of interpretation in creative tension (a "dialectical relationship") enables the interpreter not only to critique a text at face-value, either from a socio-critical perspective or from a metacritical perspective but also to function at a "post-critical" level (all three are functions of "explanation"). Explanation involves the will to unmask and abolish the idols that the fallible human person is prone to create and worship. Then comes understanding, the active listening to symbols and all forms of indirect speech.

Like Heidegger and Gadamer, Ricœur is interested in the ontology and phenomenology of human being. And, as such, he "has been a key figure in mediating German philosophical hermeneutics to the much broader domains of Anglo-American analytic philosophy and cross-disciplinary

---

[54] Murphy, *Anglo-American Postmodernity*, 137.

[55] Thiselton, *New Horizons*, 344.

[56] Cited in ibid., 348.

methodological discussions."[57] But while Heidegger put ontology before epistemology, Ricœur did the opposite, going even further in that direction than had Gadamer.[58] At the same time, even as suspicion is deployed to excavate levels of arrogance and oppressive power in texts (à la Foucault)—indeed as a Christian he often emphasizes the fallenness of humanity—Ricœur aims at a "post-critical retrieval embodying openness towards a new 'possibility' which may entail renewal and change."[59]

Ricœur's second contribution is in the area of metaphor and narrative. Leaving behind the traditional (positivistic) view of metaphor as a distortion of reality (or, like Aristotle, that is simply "giving the thing a name that belongs to something else"), he argues that metaphor is a "dual linking of 'fiction' and 're-description.'"[60] Thiselton quotes what he sees as the heart of Ricœur's theory of metaphor, warning that one must not entirely disassociate "fiction" from "saying something about reality": "Metaphor presents itself as a strategy of discourse that, while preserving and developing the creative powers of language, preserves and develops the *heuristic* power wielded by *Fiction*."[61] His greatest contribution in this field, however, may be his linking of metaphor with narrative. Time, for Ricœur, is an all-defining category with respect to human existence. Though much indebted to Aristotle, he nevertheless takes a postmodern turn in describing narrative as the very possibility of human action as played out in holistic fashion. In a story, the narrator selects events, actions, and elements of speech and "reconstructs" them to form a "plot" which unites the whole. The text projects a "re-figured" world, and when it is grasped by the reader in its wholeness, its effect are "revelatory and transformative."[62]

But, one might object, in what sense is Ricœurian narrative "true"? Fiction and history are both true, he would answer, but in a different

---

[57] Klemm, "Hermeneutics," 152. We learn from the article on Paul Ricœur in the *Internet Encyclopedia of Philosophy* (ed. James Fieser, 2004, http://www.iep.utm .edu) that as a Christian philosopher he remained active in the French academy (he was even dean of the Faculté de Lettres at Nanterre for a while); he also taught in Belgium and at the University of Chicago.

[58] As Klemm puts it, Ricœur wants to recover Schleiermacher's methodological interest in articulating the principles, concepts, and rules of interpreting written texts truthfully. Ricœur does not want to eschew method in favor of truth but to place method in service of truth" (ibid.). This is exactly the kind of critical realism for which Barbour is arguing.

[59] Thiselton, *New Horizons*, 344.

[60] Ibid., 351.

[61] Ibid., 352, emphasis original.

[62] Ibid., 355.

sense. The referential claim of a narrative via its dynamic flow of emplotment is particularly appropriate to poetic discourse. Its new re-configuration of the world nevertheless reveals truth in that it projects possibilities in the future and connects them to the reader. On the other hand, Thiselton worries that the idea of fiction re-making reality could mean the expulsion of reference in historical narrative.[63]

James Fodor picks up this concern in his writing on Paul Ricœur. He argues in parallel fashion to Murphy that part of the problem of reference is that the debate has more often than not been couched in "modern" or Cartesian terms: the positing of an autonomous self on one side, and the external world on the other, with language as the only possible link between the two. The very real anxiety here, writes Fodor, is that "there is no way of determining whether or not the linguistic bridge will reliably and safely span the epistemological chasm between the mind and the external world."[64] The problem is not to find an ingenious way to strengthen the bridge but rather to question the epistemological assumptions behind such a view in the first place. In fact, self, language, and "reality out there" are distinguishable but at the same time very much intertwined. The Cartesian model is actually a prevalent Western paradigm, but precisely because it is so, it must be called into question. Notice too Fodor's treatment of the triangle of subject, language and the world:

> It is simply not the case that linguistic assertions first of all occur in a mental domain which then require some sort of ontological confirmation to be admitted as true. Rather, the self, language, and world coexist in relations of mutual implication. They can only be conceived together, the reality of one being contingent upon the reality of the other two. Of course, language, self, and world are, for reasons of analysis and clarification, distinguishable but they are not separable. It makes no sense to speak of one without invoking, if only implicitly, the others.[65]

Fodor shows that Ricœur "produces a sophisticated and highly nuanced account of reference that has much to offer for theological reflection."[66] Together with such Christian writers as Thiselton, Vanhoozer, Osborne

---

[63] This is certainly Vanhoozer's argument in his *Biblical Narrative in the Philosophy of Paul Ricœur: A Study in Hermeneutics and Theology* (Cambridge: Cambridge University Press, 1990). Thiselton concurs: Ricœur has only the human self left as "object of reference" (*New Horizons*, 358).

[64] James Fodor, *Christian Hermeneutics: Paul Ricœur and the Refiguring of Theology* (Oxford: Clarendon Press, 1995), 11.

[65] Ibid.

[66] Ibid., 12.

and Murphy, Fodor recognizes that none of the above-mentioned movements in recent hermeneutical theory can be ignored; instead, he contends that their insights have to be put alongside those proponents of the
Anglo-American philosophy of language, from the latter Wittgenstein, to
Austin, and John Searle. This is also a point strongly made by Murphy to
which I return below. No doubt, the issue of "reference" is an important
aspect to this ongoing debate. And thus the central issue in this section
on the text is epistemology: if no "real" (though not complete in any
way) knowledge of the world can be apprehended by the human mind,
then there can be no stable meaning in a text.

Following Gadamer and yet going beyond him, Ricœur's third contribution is in insisting that the text expands the horizon of the reader.
Having taken into account many of the concerns of the structuralists,
poststructuralists and deconstructionists, he nevertheless forges ahead
with the objective modes of explanation along with the more subjective
elements of understanding. The text, then, through the creative and
courageous decisions of the reader, can yield unimagined possibilities,
which point to the notion of eschatological hope in his biblical hermeneutics. This emphasis surfaces particularly in his writing on the Parables
of Jesus and other biblical themes.[67]

Hope, however, is not the monopoly of any of the authors we have
surveyed. As noted previously, Jacques Derrida never lost hope for a
more progressive politics, from the time of the May 1968 riots till the
end of his life.[68] Hope for a more just world is what propels the writing of

---

[67] See especially *Fallible Man*, trans. Charles Kelbley (Chicago: Regnery, 1965); and
Symbolism of Evil, trans. Emerson Buchanan (New York: Harper & Row, 1967); and
more recently, see his co-authored book with André Lacocque, *Thinking Biblically:
Exegetical and Hermeneutical Studies*, trans. David Pellauer (Chicago: University of
Chicago Press, 1998).

[68] Derrida is a rich and multifaceted author. See John D. Caputo's essay comparing
his rendering of Augustine's Confessions in ("Circum., Circumfession: Fifty-Periods
and periphrases," in *Jacques Derrida*, Geoffrey Bennington and Jacques Derrida,
trans. Geoffrey Bennington [Chicago: University of Chicago Press, 1993]), and
Martin Heidegger's version in a 1921 lecture ("Toward a Postmodern Theology of
the Cross," in *Postmodern Philosophy and Christian Thought*, ed. Merold Westphal
[Bloomington, IN: Indiana University Press, 1999], 202-25). What stands out of this
last essay is Derrida's constant concern and compassion for the "other"—here, the
death of his mother, who, like Augustine's mother Monica, died in exile (Nice,
France, far from her native Algiers). As he is often prone to do, he returns to the
theme of circumcision. Too young to remember the pain, it represents the scar of
displaced guilt, like that of original sin for Augustine, but which haunted him at
every turn (first, as the Jewish "other," and then as the Jew who broke with his religion). In the end, circumcision for him is the pain of knowing that there is no Truth,

other secular writers considered here, including Mitzman, Harvey, Habermas and Benhabib. Hence, in this work I am arguing that people of faith—and here specifically Muslims and Christians—can add substance and vigor to this rising chorus of yearning and hope amidst the global suffering and devastation wreaked by postmodernity. The present discussion of hermeneutics is urgently needed: it will enable us to rethink the caliphate of humanity in a way that both connects to tradition and extends its insights by applying them to today's context. But first, the next three sections outline what I see as an adequate hermeneutic in our post-Enlightenment situation.

An Approach to the Hermeneutics of Sacred Texts

The above historical survey leads me to make some observations about the human person that recapitulate the content of the last two chapters: How do human subject, world and language interact? I will argue that as a species our entire existence is experienced through signs that enable us to communicate with one another and make sense of our existence. This metaphysical and theological observation provides the framework for a more detailed presentation of how I conceive hermeneutics in our post-Enlightenment environment.

Humans as Semiotic Beings

Both Gadamer and Ricœur illustrate well the movement of late in the social sciences and humanities toward a kind of critical realism. Both believed it was necessary to pick up again the effort deployed by Schleiermacher to ground the human sciences, which had gone adrift with the impact of empiricism. Thus Gadamer turned to historiography, and while he maintained that historians only fool themselves if they think they can rise above the flow of historical phenomena in any kind of "objective" way, he believed that the application of self-critical strategies could enable them to uncover many of their own preunderstandings and apprehend a measure of historical truth.[69] On the other hand, he was less sanguine about the role of language, which he connected in a Heideggerian manner to our being-in-the-world. Language mediates all human endeavors to make sense of the world and their own existence. Thus

---

as he has been cut off from the book and the promise, and in spite of which, he continues to cry and to pray—Derrida's own theology of the cross, argues Caputo (ibid., 216).

[69] Klemm, "Hermeneutics," 151.

human ontology is wrapped up in language and, hence, steeped in hermeneutics: "it is the universal medium of the self's dialogical openness to the other."[70]

We saw that Ricœur, by placing ontology at the end rather than at the beginning of his project, places a greater emphasis on epistemology, and that though he started with many of the same presuppositions as Gadamer, he believes that by paying attention to time, metaphor and symbol, research in the humanities and human sciences could uncover more truth than Gadamer thought possible. Moreover, Ricœur willingly raises metaphysical questions that bear at least some connection to his Christian faith. In his earlier works, for instance, he dealt with the fallibility of human perception as a justification for the use of phenomenology in understanding the moral issues that relate to our sense of guilt and the reality of evil in the world.[71]

My point here is to return to the beginning of Chapter 2, when I quoted Ashworth's contention that philosophy "aims at intellectually responsible accounts of the most basic and general aspects of reality."[72] Chapters 2, 3 and 4 have leaned heavily on philosophy as a guide to chart a course out of the conundrums of modernity—and especially away from the picture of the autonomous and disembodied self, setting out to control the world. This is because theology necessarily relies on what Catholic theologian Francis Schüssler Fiorenza calls "background theories"—"implicit assumptions, philosophical or scientific, about the world and science."[73] Thus I noted that a different relation must be found between the triangular elements of self (or mind), world and language. In this quest, I found much help from Habermas's discursive rationality and communicative ethics and in Benhabib's use of these to free the modern self from its self-imposed and artificial isolation from history, culture, society and state. To the contrary, as human beings we are profoundly shaped by the conversations taking place all around us in our particular contexts. At the same time, as members of the human race, we share similar aspirations for communication through common discourses that involve "norms of universal moral respect and egalitarian reciprocity."[74]

---

[70] Ibid. Klemm says it well: "For Gadamer, we always understand more than we can say."

[71] Ricœur, *Symbolism of Evil*; *Fallible Man*.

[72] Cf. Chapter 2, n. 11.

[73] Francis Schüssler Fiorenza, "Systematic Theology: Task and Methods," in *Systematic Theology: Roman Catholic Perspectives*, vol. 1, ed. Francis Schüssler Fiorenza and John P. Galvin (Minneapolis: Fortress Press, 1991), 74.

[74] Benhabib, *Situating the Self*, 45.

It follows that all who cry out for human rights and social justice choose (consciously or not) to disassociate themselves from the strong version of postmodernism that logically calls for "a retreat from utopia." These people affirm a human vision in the name of a reality jointly perceived: this world reels under the blows of many injustices visited on the weak by the more powerful, whether through racism, nationalism, economic greed fueling new wars, the oppressive structures of global neoliberalism, or the consumerist fever promoted by the latter, which heats everything up from the international arms trade to the addiction of urban teenagers to expensive labels in clothing. Amidst the din of emerging social movements around the globe we hear that "a better world is possible."

I then turned to Ian Barbour to explore why the radical changes in the philosophy of science of the last century so profoundly affect the way we think about religion. His understanding of the use of models and paradigms in science directly impacts how we think about theology. Thus critical realism asserts the "real" connection between models and the phenomena they seek to describe and explain, but also warns us that models, theories and paradigms are only tentative approximations of reality. Hence, since theology, or "religious knowledge," grounds its reflection in texts from the past, the issue of hermeneutics looms very large indeed.

This is where I would like to insert the helpful comments made by David Naugle about semiotics and culture, communication between people and how we map reality through signs. Naugle, an evangelical philosopher and theologian, takes his cue from the prolific and creative Italian philosopher, Umberto Eco (b. 1932).[75] Citing from his seminal work, *A Theory of Semiotics*, Naugle supports his two founding propositions: "(i) the whole of culture *must* be studied as a semiotic phenomenon; (ii) all aspects of culture *can* be studied as the contents of semiotic activity."[76] Culture, then, should be seen as the manifestation of communicative acts posited by individuals in society using discreet "signification systems." As the mechanisms of those systems are unveiled, we catch a

---

[75] Eco started off as a medievalist philosopher in the 1950s, but moved on to journalism, a position from which he became an astute observer of modern culture and the media. He reentered the academic world in the 1960s and gradually became one of the founding fathers of semiotics, with several books on this theme published in the 1970s. While keeping up with many academic appointments internationally, he launched an enormously successful career as a novelist, starting with the publication of *The Name of the Rose* (1980).

[76] Umberto Eco, *A Theory of Semiotics* (Bloomington, IN: Indiana University Press, 1976), 22, emphasis original, cited in Naugle, *Worldview*, 292.

better glimpse of how worldviews are constructed in a social context. This observation leads Naugle to "connect semiotics as the science of signs with human subjects who use them so profusely." The creation and management of signs in such a natural way in human society points to "the essential semiotic nature of human persons."[77]

Indeed, signs and symbols permeate human reality and experience. They underpin our efforts to make sense of the world around us, both in explaining how it functions (science) and what it means for humankind (philosophy and religion); they provide the necessary medium for human knowledge and communication, and, at a certain level of culture, a particular network of central and guiding symbols come to be what is commonly known as "worldview":

> As an individual's or culture's foundation and system of denotative signs, they are promulgated through countless communicative avenues and mysteriously find their way to the innermost regions of the heart[78]... They are the putative object of faith and the basis of hope, however it may be conceived. They are embraced as true and offer a way of life. They are the essential source of individual and sociocultural security. They are personal and cultural structures that define human existence. Thus, when they are in crisis or are challenged, people respond anxiously, and even with hostility.[79]

[77] Naugle, *Worldview*, 292.

[78] In a preceding chapter on "Theological Reflections on 'Worldview'," Naugle designates the biblical "heart" (a pervasive Semitic notion in the Hebrew Bible and New Testament as humanity's locus of will, thought and emotions) as the seat of worldview. A fruitful comparative study has yet to be made between Bible and Qur'an on this issue. The closest word to the Hebrew is the Arabic *lubb*, which in the Qur'an appears in the plural (*ulū-l-albāb*, "those that are wise") sixteen times, and the very common word *qalb* (132 times) is used in a strikingly similar way to the Hebrew meaning of "heart." Tellingly, neither Toshihiko Izutsu (*Ethico-Religious Concepts in the Qur'an* [Montreal: McGill University Press, 1966]) nor Fazlur Rahman make any mention of this (*Major Themes in the Qur'an*, 2d ed. [Minneapolis: Bibliotheca Islamica, 1994 (1980)]).

[79] Ibid., 296. Recall my mention of influential scholar of religion Ninian Smart in the Introduction. In a way that parallels Naugle's thesis here, Smart considers the study of religion as "worldview analysis." In the second chapter of his *Worldviews: Cross-cultural Explorations of Human Beliefs*, he produces an inventory of the world's major worldviews: the modern West, Islamic societies, South and Southeast Asia, the Latin South, Black Africa and the Caribbean, East Asia, the Pacific, the Marxisms. Added to those "great tectonic plates of human civilization," one should also mention the impact of quasi-religious ideologies on these blocks, like nationalism and secular humanism (pp. 33-54).

The primary mode of fixing and transmitting the way a particular social group solves the mystery of human life is through stories. At the heart of a culture's worldview we almost invariably find stories and myths about the origin of the world and/or of the people or nation. Plato and Aristotle, however bent they were on inculcating the virtue of philosophy on their hearers, also advocated narrative as the means to educating the young.[80] Rollo May argues that myth is the very framework that upholds society's sense of identity and belonging:

> Whether the meaning of existence is only what we put into life by our own individual fortitude…or whether there is meaning we need to discover…, the result is the same: myths are our way of finding this meaning and significance. Myths are like the beams in a house: not exposed to outside view, they are the structure which holds the house together so people can live in it.[81]

Yet, with the promises of the scientific method newly ringing in their ears, "the architects of the modern project did their best to rid *homo narrator* of their troublesome tales and banish them from cultural significance."[82] Partly because competing metanarratives had led Europe into a series of protracted and bloody struggles (recall the importance of the Peace of Westphalia in 1648), and partly because modern man felt empowered to sever his ties to primitive religious narratives and launch out in a bid for emancipation, myth was evacuated. It was Nietzsche who realized the hidden cost of such a project: "…without myth every culture loses the healthy natural power of its creativity: only a horizon defined by myths completes and unifies a whole cultural movement."[83] Indeed, the Cartesian idea that modern man could leverage his use of reason in order to refashion the world to his own advantage was itself a myth that has now been called into question.

---

[80] Naugle cites Bruno Bettelheim (*The Uses of Enchantment: The Meaning and Importance of Fairy Tales* [New York: Random House/Vintage Books, 1977]) on the essential role of narrative: "fairy tales and myths are the basic means by which children fashion and refashion their worlds" (*Worldview*, 297). Recall Benhabib's recounting of the narrative self, as she learned it from the communitarians (cf. Chapter 2).

[81] Rollo May, *The Cry for Myth* (New York: Bantam Doubleday Dell, Delta, 1991), 15, cited in Naugle, *Worldview*, 298.

[82] Ibid., 299.

[83] Friedrich Nietzsche, *The Birth of Tragedy and the Case of Wagner*, trans. and commentary by Walter Kaufmann (New York: Random House/Vintage Books, 1967), 135, cited in Naugle, *Worldview*, 300. Yet, for Nietzsche, it was a complete step away from realism—not a move I am recommending here.

The contemporary thinker who has best helped us to gauge the importance of narrative for human ethics and life, asserts Naugle, is Alasdair MacIntyre. In his seminal work, *After Virtue*, MacIntyre argued "that a virtuous life is possible only to the extent that it is conceived, unified, and evaluated as a whole."[84] Narrative is the necessary moral glue that provides integrity to the self's personal history—from birth and every new turn in life to death—and connects it to the wider story of the community. Each culture possesses a stock of stories that weave together the mythology that lies at the heart of its worldview. This is the reason, on MacIntyre's view, why Aristotelian ethics must be revived today:

> A central thesis then begins to emerge: man is in his actions and practice, as well as in his fictions, essentially a story-telling animal. He is not essentially, but becomes through his history, a teller of stories that aspire to truth. But the key question for men is not about their own authorship; I can only answer the question "What am I to do" if I can answer the prior question "Of what story or stories do I find myself a part?"[85]

MacIntyre cropped up in Chapter 2 as a communitarian for whom the idea of virtue as embodied in a community's narrative militated against the possibility of articulating any sort of deontological ethics. Interestingly, both Murphy and Naugle resort to MacIntyre in their bid to construct a theory of truth and rationality that takes into account both the situated character of the self (the shared narratives of traditions) and the requirement of critical realism (necessary to their evangelical commitments).[86] Take the notion of rationality, for instance, the paragon principle of the Enlightenment, defined by its leading lights as a building consisting "of epistemic planks that are securely nailed together and established upon an unshakable foundation of solid cognitive concrete."[87] Why

---

[84] Ibid., 301.

[85] Alasdair C. MacIntyre, *After Virtue: A Study in Moral Theory*, 2d ed. (Notre Dame, IN: University of Notre Dame Press, 1984), 204-5, cited in Naugle, *Worldview*, 301.

[86] Cf. Murphy, *Anglo-American Postmodernism*, 123-29, for his discussion of truth. Murphy advocates critical realism, not in its modern form, but rather in its postmodern form as the best ally for new directions in conservative theology. She contends that MacIntyre's project helps to bring together postliberals such as George Lindbeck (*The Nature of Doctrine: Religion and Theology in a Postliberal Age* [Philadelphia: Westminster Press, 1984]) and postevangelicals like herself. It was the foundationalism of modern philosophy that created the rift between conservative and liberal theologians. Embracing a holistic postmodern theory of truth and epistemology provides the means to heal much of that rift.

[87] Naugle, *Worldview*, 305. Fiorenza writes that theology must be articulated in a context in which rationality is facing a crisis. Christians (as well as Muslims, I would add) have always held on to the complementarity of faith and reason. Yet

would that notion of rationality be more believable a priori than a score of premodern beliefs drawn from the worldviews of traditional cultures that still survive in many parts of the world? For one, its secular bias leads to its rejection by the great majority of religious adherents, whether Christians, Buddhists or Muslims. Thus Freud and Marx easily came to the conclusion that religion was a projection of human desires that functions as a kind of pain-numbing drug because their starting assumptions were shaped by their humanist worldview—there is nothing above or beyond the human person.[88] I would have to agree with Naugle: "Thus the Enlightenment's very prejudice against prejudice as well as its antitraditionalism has become (at least until recently) the predominate modern prejudice and the new cultural and intellectual tradition!"[89] By contrast, rationality is a context-dependent enterprise, always under construction, continuously negotiated by people in community who have tacitly (and unconsciously for the most part) committed themselves to its core principles. As such, rationality is part of the network of signs and stories that constitute a human society and enable it to make sense of life's most pressing questions.

Drawing on philosopher R. G. Collingwood, Naugle differentiates between propositions (which are falsifiable or verifiable) and presuppositions (which are not); and further, between relative presuppositions (which answer previous questions but lead to new questions) and absolute presuppositions (which are not subject to proof or disproof).[90] Thus, "people argue *from* but not *to* presuppositions," because the rationality of any system of thought is worked out by reference to the starting point provided by its absolute presuppositions.[91] Since the latter cannot be proven, it follows that worldviews (as the core myths, symbols and narratives of a culture) are adhered to, not on the basis of logic, but rather by an act of commitment.

This is what MacIntyre asserts relative to competing conceptions of rationality in a later book, *Whose Justice? Which Rationality?*[92] Each

today "we face a crisis of rationality. This crisis of rationality was articulated forcefully in the critical theory of the 1940s, and the crisis is at the center of the current postmodern critique of the technocratic as well as scientific rationality" ("Systematic Theology: Task and Methods," 69).

[88] Smart, *Worldviews*, 66-67.

[89] Naugle, *Worldview*, 305.

[90] This is from Collingwood's early work, *Essay on Metaphysics* (Oxford: Clarendon Press, 1940).

[91] Naugle, *Worldview*, 307.

[92] Alasdair C. MacIntyre, *Whose Justice? Which Rationality?* (Notre Dame, IN: University of Notre Dame Press, 1988).

notion of justice is tied to a particular view of rationality. The more one investigates the issue, the greater the number of versions one uncovers. Hence, "disputes about the nature of rationality in general and about practical rationality in particular are apparently as manifold and as intractable as disputes about justice."[93] The reason Enlightenment rationality turned out to be a dead end is that the epistemological foundationalism it espoused completely ignored "the inescapably historically and socially context-bound character which any substantive set of principles of rationality, whether theoretical or practical, is bound to have."[94] The modern paradigm of a tradition-free system of rational justification was bound to fail, concludes MacIntyre, because it is a contradiction in terms. To the contrary, all rational inquiry takes place within a specific tradition—at the deep level of worldview assumptions, adds Naugle.[95]

Here we find ourselves back to the subject's knowledge of the world. MacIntyre obviously rejects the classical or naïve view of realism, in which nothing comes between the perceiver and the physical (or social, or economic, etc.) object perceived. It is like the joke about the three baseball umpires, suggests Naugle. The first umpire says, "'There's balls and there's strikes, and I call 'em the way they are.' Another says, 'There's balls and there's strikes, and I call 'em the way I see 'em.' The third says, 'There's balls and there's strikes, and they ain't *nothin'* until I call 'em.'"[96] The first umpire represents the commonsense (naïve) realist position: "I call 'em the way they are."[97] The second umpire expresses the critical realist perspective: "I call 'em as I see 'em." An objective reality exists outside the mind of the perceiver and humans can truly know aspects of that reality, yet because of the nature of human cognition that knowledge will always be partial and tentative, and always to

---

[93] MacIntyre, *Whose Justice?*, 2, cited in Naugle, *Worldview*, 308.

[94] MacIntyre, *Whose Justice?*, 4.

[95] Naugle, *Worldview*, 309. For Naugle, this seems to be what MacIntyre is saying. For instance, he quotes MacIntyre toward the end of *Whose Justice?*: "it has become evident that conceptions of justice and of practical rationality generally and characteristically confront us as closely related aspects of some larger, more or less well-articulated, overall view of human life and of its place in nature. Such overall views, insofar as they make claims upon our rational allegiance, give expression to traditions of enquiry which are at one and the same time traditions embodied in particular types of social relationships" (p. 389).

[96] Ibid., 322.

[97] Naugle notes that some philosophers have of late taken up this position once again, because of difficulties with the other positions. He cites Thomas Reid in the eighteenth century, and more recently, D. M. Armstrong, John Searle and William Alston (ibid., 323).

some extent relative to the knower.[98] Finally, the third umpire's boast about creating the ball or the strike by his call reveals a "creative antirealism," in Naugle's words. Here reality disappears entirely from human sight. Worldviews in this mode are "belief systems that are reified."[99] There may be an external world, but all we can make of it comes to us in patterns of thought mediated through human language and shaped by culture and context. What we are left with is in fact very little: "[t]he speciousness of the so-called 'given,' the creative power of the mind, the variety and formative function of sign systems, and the multiplicity of symbolic worlds."[100]

So we arrive once again at the strong version of postmodernism Benhabib deplored for its inability to provide any crosscultural ethical guidance in today's globalized world. At the same time, we realize that we can no longer return to the premodern epistemic view of unmediated knowledge. We will have to admit, once for all, that there is "no view from nowhere!"[101] As we stand necessarily somewhere, that location will in fact determine much of what we claim to know from there. On the other hand, as people of faith, observes Naugle, we must wager on the possibility of knowledge—a stance particularly crucial to the present project:

> If God exists and is the maker of heaven and earth; if he has created all things by his word and designed all things by his wisdom and law; if he is the architect of the human mind and its cognitive powers; and if he has so made people that their lives and perspectives consist of the belief content of the human heart (the system of semiosis or narrative framework embraced in faith that dominates it), then it is reasonable to assume that knowledge of the cosmos is possible, though it is always conditioned by human finitude, sinfulness, and the experience of redemption.[102]

Here is, on my view, a compelling statement on the human person, which can both be read as a metaphysical/epistemological position and a

---

[98] Naugle later enlists the insight of Russian literary theorist Mikhail Bakhtin, who posited the "dialogical imagination" as critical to the knowing process. It is only in our situatedness that we can apprehend the reality of others in their own context— what he calls "creative understanding." As people begin to converse from their various and disparate locations and truly listen to one another, they can teach one another about the blind spots in their own systems to which they are themselves blind. A gradual process of enlightenment can then take place, though it will always remain tentative and partial in the final analysis (ibid., 326-27).

[99] Ibid., 323.

[100] Ibid.

[101] Ibid., 325.

[102] Ibid., 325-26.

theological one. This is unsurprising, since theology cannot avoid recruiting philosophy in issuing statements about reality. What I find particularly rewarding, as I come to the end of Naugle's section on philosophy and worldview, is that the three criteria he offers for "testing" worldviews are virtually identical to the criteria that Barbour proposes in order to judge the relative value of scientific or religious models:

1. *The rational, or coherence test*; rational coherence is important (though not sufficient in itself), as Naugle explains: "While statements that agree with each other do not necessarily demonstrate the truthfulness of a worldview perspective, propositions that are patently contrary to one another would falsify the worldview, or at least certain claims within it."[103]

2. *The empirical, or correspondence test*; here Naugle telescopes two of Barbour's points, *correspondence* and *scope*: the worldview must explain in a cogent way what can be known of reality, both in its particular aspects and in its entire sweep as it is presented to human consciousness (through the physical and human/social sciences, and the totality of human experience).

3. *The existential, or pragmatic test*; this is Barbour's "fertility" criterion, with the added dimension of "cash value," or the answer to the question, "does it actually lead to success (however defined) in one's life if it is applied?" It must respond to basic human aspirations, "provide a sense of peace and well-being," and actually turn out to be "useful and existentially satisfying."[104]

I underline the convergence of the two authors on this issue in order to signal a welcome phenomenon, at least from my own perspective. Barbour is a theologian, scientist, and a philosopher of science, while Naugle is primarily a philosopher who in this book proceeds intentionally as a Christian. Further, they illustrate what Murphy has consistently argued—that the postmodern context of thought offers new opportunities for evangelicals (or conservatives) and liberals to come together in a way that was impossible under conditions of modernity. Naugle, an evangelical Christian, and Barbour, a liberal Protestant theologian (his process theology would still be anathema to most evangelicals, and certainly to fundamentalists), rethink the central metaphysical and epistemological issues at the heart of theology today and, though coming from different fields of study, arrive at amazingly similar conclusions.

---

[103] Ibid., 327.
[104] Ibid., 328.

But this section is not merely about intra-Christian debates on the human person; it is also about their wider applicability to Muslim–Christian dialogue. Clearly, one observes a growing number of conservative Christians scholars pondering the implications of a postmodern episteme and hermeneutic for their theology.[105] Little movement in this direction can be seen within the ranks of conservative Islam, as mentioned earlier. In the Western academies, however, liberal Muslim scholars willing to publish in such volumes as *Progressive Muslims* are aplenty. In fact, this chapter opened with the experiences and views of Farid Esack (who also contributed to that volume). My own research here will initially appeal more to the latter category. Yet I continue to entertain the hope that with the excellent work of scholar/activists such as Tariq Ramadan, a growing number of Muslim conservatives will take advantage of the new opportunities for reinvisioning the world alongside people of other faiths. Ramadan, whose work I discuss in the next chapter, is a good example of a scholar in the West (Switzerland) who can also galvanize the youth to rethink their Muslim faith in their own context.[106] Whether for Muslims, Christians or others, however, this task is closely linked to hermeneutics.

Language, Distortion, and the Hermeneutical Process
If it is true that we humans are fundamentally semiotic beings, then our use of language is not only for the purpose of communicating with one another, but even prior to that, it is the medium through which we ascribe meaning to the world around us, including our own self and the society of which we are a part. Much of this construction of reality is inherited through culture—we grow up in a particular culture, which functions within a particular worldview. And especially now in our pluralistic global societies, this worldview overlaps with a good deal of other worldviews, as people move from childhood to a state-sponsored education (with all of its identity-forming myths, symbols and rituals) to

---

[105] Cf. n. 68 on Merold Westphal's 1999 edited book, *Postmodern Philosophy and Christian Thought*. It was the fruit of a six-week seminar hosted by Calvin College and funded by the Pew Charitable Trusts. Westphal's mandate was to bring together leading evangelical philosophers and to discuss the issue of what he calls, "the appropriation of postmodernism." For Westphal and most of the participants, appropriation means both acceptance and rejection; a sorting out of what can be of use, and what cannot (ibid., 2). For a conservative attempt to refute all the claims of postmodernism (with a readable, systematic presentation of all aspects of philosophy), see J. P. Moreland and William Lane Craig, *Philosophical Foundations for a Christian Worldview* (Downers Grove, IL: InterVarsity Press, 2003).
[106] On Ramadan, see also my article, "*Maqāṣid al-Sharīʿa*.

various other adult settings—professional, neighborhood, church or mosque, or various civil society organizations. We are often forced to rethink and rearticulate our basic understanding of life's meaning for us through the symbolic import of other systems of thought, both secular and religious. Hence, the deliberate Muslim–Christian theological project proposed here.

In this section, I focus on the dimensions of language as a vehicle for interpersonal communication. Having pointed out the complexities of the hermeneutical process—we can no longer entertain a naïve view of the various roles played by author, text and reader—I nevertheless choose to look at the reading of a sacred text (at least from a monotheistic perspective) as comparable in many ways to the reading of any text passed down through generations.

I begin, however, by prefacing this discussion with a remark on the role of models in religion and the limits of language in apprehending the numinous. Ninian Smart's typology of religious experience (Appendix C) highlights not only the rather obvious differences by which the divine is apprehended in Eastern religions (in Buddhism and most strains of Hindu thought) by contrast to the three monotheistic religions, but also the more subtle differences within the later faiths. On the numinous/mystical spectrum (Figure in Appendix C), for instance, we find Muslims, Jews and Christians (Catholics, Orthodox, mainline Protestants) worshiping God through their respective rituals in such a way that God's otherness, loftiness and awesomeness is emphasized (the numinous model). Yet we also find Sufis in their *dhikr* ceremonies ardently seeking to experience God within themselves in very much the same fashion as do Pentecostals and Charismatic Christians in their own contexts, where music, prayer and the laying on of hands can contribute in producing transe-like states and effect physical and emotional healings.

Most Christians, however, would deny that distinction. The central doctrine of the Holy Spirit—God indwelling each and every believer—would seem to mitigate any great chasm between liturgical worship and more free-flowing forms of worship. Nevertheless, I would argue that the language used by each side about God (and the choice of biblical passages is seen as central) and about their experience of him is distinctive. Both are using different models—the numinous vs. the mystical. Though clearly compatible for countless Christians and Muslims who operate on both ends of the spectrum, the difficulty we have (in theory or in practice) in delineating the numinous from the mystical is itself indicative of the paradox of theism. How can finite human beings come to know anything about—let alone enter into a relationship with—the Infinite Being who set the universe into motion? The answer has to do with revelation, and thus we come back to the language of the revealed texts. If God is Wholly Other, then how can we say anything meaningful about him? The language of revelation, consequently, is laden with symbolism. Think of the debates between Muslim Mu'tazilites who emphasized the symbolic nature of the Qur'an's depiction of the "hand of God," or "throne of God" in the heavens. For them also, this meant that the Qur'an was created by God for the purposes of revelation. The Ash'arites, by contrast, refused to make any pronouncement on the status of anthropomorphisms in the Qur'an, and their insistence that the Qur'an was co-eternal with God became the orthodox position still held officially today.[107]

However, beyond the symbol-laden quality of the language of revelation, there are other dimensions of language in general in which it also shares. Any reader or interpreter coming to a text does so through what I will call five "lenses." In a sense, I follow Ricœur's procedure of starting with "explanation," or promoting a "hermeneutic of suspicion," before moving on to the more positive task of "interpretation". With each lens comes an additional layer of potential distortion. In addition, the five lenses point to five dimensions of meaning production—what Schüssler Fiorenza calls "reality domains."[108] They are: language (the particular language used, Saussure's *langue*), external reality, social reality, academic discipline, internal reality.

Fiorenza begins by asking a question: How is one to proceed with the interpretation of a doctrine, a scriptural passage, a praxis or a tradition? One conclusion drawn from current research is that the hermeneutical process must include both a search for meaning and the specifying of truth conditions—a definite critical realist project. A consensus seems to have emerged (from Wittgenstein to Donald Davidson and Quine), that "in order to understand the meaning of utterances, it is necessary to know the truth conditions of this utterance. For instance, if the utterance is a command, in order to understand it one would have to know what constitutes obedience and disobedience to that command."[109] Yet meaning extraction is also more than a mental process in the reader's mind. It goes beyond a correspondence between sentences and a possible state of

---

[107] The great Egyptian reformer Muhammad 'Abduh (d. 1905) had included a sentence in his *Risālat al-Tawḥīd* on the created nature of the Qur'an, but omitted it for all subsequent editions, so controversial was such a statement deemed to be (Albert Hourani, *Arabic Thought in the Liberal Age, 1798–1939*, 2d ed. [Cambridge: Cambridge University Press, 1983], 142).

[108] Francis Schüssler Fiorenza, *Foundational Theology: Jesus and the Church* (New York: Crossroad, 1984).

[109] Ibid., 292.

affairs. It is connected in a broader sense to the process of communication. Fiorenza explains:

> In order to communicate individuals must not only produce grammatically correct sentences, they must also perform successful speech-acts. Communication involves not only propositional content, but the force of performative actions. A hermeneutical theory that abstracts from the performative aspects of communication fails to grasp precisely what makes an act of communication successful or how a speech-act successfully communicates.[110]

Successful communication, therefore, includes not only propositional content (on the level of objects) but also a dynamic of personal interaction (on the level of intersubjectivity). "This intersubjective relation is established because in every act of communication a speaker explicitly or implicitly raises certain validity claims and these validity claims are reciprocally acknowledged, sometimes explicitly, sometimes implicitly."[111] Further, following Fiorenza's model, this dynamic takes place within the interaction of two distinct axes: validity claims and the aforementioned reality domains. Both axes have five components (cf. Fig. 6 below).

The five validity claims are the following:

1. The speaker/writer claims that the utterance is intelligible. By this Fiorenza means the language, about which rules of grammar and syntax are applicable. This is the work of traditional exegesis.

2. The propositional content refers to objects or events, which make the utterance verifiable (at least potentially) as true or false. Beside this obvious narrative-type of text, John Goldingay would include three more types: prescriptive, commissive and expressive. For Goldingay, "prescriptive statements attempt to lay obligations on others; in commissive statements I commit myself to some act; expressive statements put into words my convictions, feelings, or attitudes."[112] Though not referential in the same way, speech in the last three dimensions points to common human experiences with which the reader can identify. It is analogous to Ricœur's "second order reference."[113] However,

---

[110] Ibid., 293.

[111] Ibid., 185.

[112] John Goldingay, *Models for Interpretation of Scripture* (Grand Rapids: Eerdmans; Carlisle, PA: Paternoster, 1995), 5.

[113] Fodor sees this point as perhaps Ricœur's greatest asset to the theologian's task: "He chooses to call the Bible a 'poem', but he intends by this term considerably

such statements cannot be proven true or false. They would have to be judged "reliable," or "trustworthy," considering the source of the text and its immediate context. Bishop Lesslie Newbigin would say they are an authentic part of the tradition that I choose to "indwell."[114] Therefore, with regard to the Qur'an or the Bible, room must be made here for the dynamic, personal, and faith dimensions of interpretation. I use Fiorenza's model, therefore, with this caveat in mind.

3. The utterance is a speech act performed according to the norms and values of the surrounding society. Its truth claim is judged, in Fiorenza's wording, according to its cultural "rightness"—a mixture of culturally accepted conventions of speech, social customs and norms of ethical behavior. A written text, just as much as a public speech or a conversation between two people, is a speech act. It is a message encoded in particular cultural forms, according to the values and norms of the author's world-view. Yet this is still only half of the hermeneutical process. While interpreting an ancient text, for instance, I the interpreter am made aware of the fact that language is "a symbolic structuring of the experienced world."[115] Through this text I am necessarily plunged into the symbolic scaffolding of a particular language that is organically connected to its particular social environment. Fiorenza puts it this way:

> Interpreters, therefore, retrieve the meaning of past texts insofar as they learn to differentiate between the life-world of the author and their own life-world. The text to be interpreted must be located within the life-world of the past in order that its diverse truth claims with their reference to distinct reality domains be acknowledged...

more than the customary notion of a particular literary genre. He uses the notion in a broader, technical sense to designate the kind of discourse of which the Bible is an instance. In this technical sense, poetics includes the totality of literary genres, and its distinctiveness resides in the manner in which these genres 'exercise a referential function that differs from the descriptive referential function of ordinary language and above all of scientific discourse' [his 'first order reference']" (*Christian Hermeneutics*, 150).

[114] Lesslie Newbigin, *The Gospel in a Pluralist Society* (Grand Rapids: Eerdmans; Geneva: WCC, 1989).

[115] Fiorenza, *Foundational Theology*, 295. Fiorenza refers his readers to Habermas's seminal work, *Theorie des Kommunikativen Handelns*, 2 vols. (Frankfurt: Suhrkamp, 1981), in which he discusses the notion of life-world and its relation to the thinking of Durkheim and Mead.

> The knowledge of the life-world that has become symbolically structured in the language of the text is also a requirement in that precisely what is affirmed in the text can be acknowledged insofar as the claims are either accepted or rejected.[116]

The modern worldview (roughly Fiorenza's use of "life-world," borrowed from Habermas), and especially in its more positivistic versions, posits the world of sense experience (the only "reality" amenable to scientific research) as the only realm about which meaningful statements can be made, because science alone can redeem truth claims. Thus, as Murphy noted, all religious language is dubbed "expressivist," in that it can only be the expression of feelings and subjective values. Indeed, if one assumes a secular stance as a starting assumption, not only God, but the other creatures of which Bible and Qur'an are replete—angels, demons, jinn—and the regular occurrence of miracles (whatever the definition) will have to be relegated to the realm of imagination. Yet research in anthropology has shown us that in most traditional societies this other realm, which draws simultaneously from the world of sense perception and the transcendental world of the spiritual realities (generally labeled as "Myth" by anthropologists since Levi-Strauss), is part and parcel of people's worldview. Religious practices at the grass-roots level, furthermore, assume two categories of reality, the personal and non-personal: beings who are this-worldly (spirits, jinn, ghosts, ancestors, earthly gods and goddesses who live in various places around us) and powers that interact in everyday life—some of which are good (helpful magic and astrology, power transmitted through charms and amulets), and some of which is dangerous (evil eye, curses, bad omens). On the higher level, there is God, or the high gods of Hinduism, for example; but very often people assume that their lives are being affected

---

[116] Catholic theologian Fiorenza, like Osborne (an evangelical), is more optimistic than I am regarding the possibility of closing the gap between writer and reader. Though I think poststructuralists such as Barthes and Derrida have overstated their case, the more moderate position taken by Ricœur on the nature of texts must be seriously heeded: "The reader is absent from the act of writing; the writer is absent from the act of reading. The text thus produces a double eclipse of the reader and the writer. It thereby replaces the relation of dialogue, which directly connects the voice of one to the hearing of the other" (*Hermeneutics and the Social Sciences* [Cambridge/New York: Cambridge University Press, 1981], 147, cited in Thiselton, *New Horizons*, 56).

by transcendental forces such as fate, *qadar* (in folk Muslim practices), *karma* and *kismet*.[117]

It is highly probable that because of this worldview discrepancy, a Western interpretation of the Qur'an will clash rather sharply with one by either a Western educated Muslim intellectual or a traditional *shaykh* in a village of Upper Egypt. Further, the traditional *shaykh* would consider his own interpretation—based as it is on centuries of Muslim commentaries—as the only possible one, and hence discount any relationship between worldview and interpretation. A critical realist view would enable interpreters to take into account their own culture and the "distortions" or "lens" through which previous interpreters examined the text.

4. For this reason, I will interject another validity claim, not mentioned by Fiorenza. It has to do with the discipline or combination of academic disciplines used to produce the interpretation of the text. The results will be different whether one uses only philology, like al-Tabari, or whether one relies heavily on philosophy, as in the case of Ibn Sina (cf. next chapters).[118] A particular lens, or a configuration of lenses must be employed. Any protest to the contrary is an exercise in self-delusion. One needs to make

---

[117] Paul G. Hiebert, *Anthropological Reflections on Missiological Issues* (Grand Rapids: Baker, 1994), 193-98. This concept of the "middle zone" which Westerners have "missing" (the "modern" worldview—as opposed to "premodern" and "postmodern"), is what Hiebert has coined the "excluded middle." The original article entitled "The Flaw of the Excluded Middle," *Missiology* 10.1 (1982): 35-47 was reprinted with permission in *Anthropological Reflections*, 189-201.

[118] Peter Heath's article, "Creative Hermeneutics: A Comparative Analysis of Three Islamic Approaches," can be seen as "metahermeneutics," in which an inter-disciplinary approach is advocated: "Obviously, adumbrating the fourteen century history of individual Islamic hermeneutic traditions is a task which will require decades of sustained scholarly inquiry. But while engaging in such long-term investigation, it is useful to pursue simultaneously analytic approaches which attempt trans-disciplinary comparison" (*Arabica* 36 [1989]: 173-210). This has two advantages, he writes. First, people working in different fields are able to bring together a wider perspective enabling new comparisons and contrasts between various hermeneutical traditions over the centuries. And second, particular diachronic studies provide in time rich material for a broader study of Islamic hermeneutics. "This provides a critical basis for inter-cultural comparison which, in turn, opens the possibility of ultimately establishing a transcultural metahistory of hermeneutics" (ibid., 174). If the principle is more accepted by Muslim scholars, then one might see the acknowledgment and development of local Islamic theologies—parallel to the current trend of theological contextualization among Christians.

clear what the disciplinary approach is actually favored by the interpreter. Here the truth criterion, or truth claim, is that of appropriateness.

5. An utterance expresses the speaker/writer's own internal world, his or her intentions and subconscious feelings and attitudes. The interpreter's own life-history, or the sum of his or her experiences, including mental and emotional responses, must be taken into account. This kind of psychologizing may seem an unjustifiable throwback to the Romanticism of Schleiermacher, yet in some instances it may well be justified. In any case, I agree with Vanhoozer that deconstruction has gone too far when it denies the author any determination of the text's meaning. Both Muslims and Christians believe that in some sense God is the author of scripture, seeking to communicate a message that human persons, as semiotic beings, can understand. Otherwise, they would not be responsible for neglecting the teachings and commands contained therein. At the same time, Muslims are necessarily more reticent when it comes to discerning the intent or feelings of the Qur'an's author, since God, its direct author, is also totally transcendent.[119] Yet for Christians with regard to the Bible, this may be an appropriate tack with the Pauline epistles, for example—particularly those in which Paul defends his apostleship in the crossfire of opposition. We have enough data on his biography and several writings from which to launch a tentative appraisal of his mental state.

To sum up, each of these components raises distinct truth claims. In order they are: intelligibility, truth, rightness, appropriateness and truthfulness. Another way of putting this is that truth is affected by all of

---

[119] Having said that, the texts of Islamic law, and especially the theoretical manuals (*uṣūl al-fiqh*), constantly refer to "rationale" (*ʿilla*) for various commands found in the Qur'an and Sunna. This then makes possible the establishment of new rulings on the basis of analogy: a parallel is drawn between the putative reason for a ruling found in the texts and this serves as the basis for a new ruling in a situation that has recently arisen. From the prohibition of drinking wine (the rationale being its potential to disable the mind), for instance, new rulings can be enacted to prohibit all intoxicants and mind-impairing drugs. The use of analogical reasoning is only one part of the accepted methodology for the sake of adapting Islamic law to new situations and changing times. This effort can be carried further in the direction of discerning the "purposes of shariʿa"—a hermeneutical turn I have argued elsewhere, which will deemphasize specific rulings of the law and favor a more elastic and ethically sensitive Islamic legal theory.

these reality domains, but that each area has its own criteria for validation. On the surface at least, these are related one to one, each validity claim corresponding to one particular reality domain. On the other hand, language, for example, takes on new meaning as it is passed through the grid of the four other domains, the external, the social, the academic and the internal (or individual) reality of the person communicating. The other domains as well will need to be examined in terms of all five of the validity claims, depending on the nature of the utterance.

Figure 6 attempts to capture this process in graphic form. Though certainly oversimplified, it does call attention, at the very least, to some of the crucial factors that enter the hermeneutical process. Each field is in reality a new lens, or level of meaning, through which interpreters access the text. Each reading too unfolds within the dynamic process of interpretation, and in particular as it shapes all subsequent readings, just as the interpretation of others will most certainly influence one's own understanding of the text. We are moving away from the autonomous modern subject and moving toward a more situated and community-oriented one.

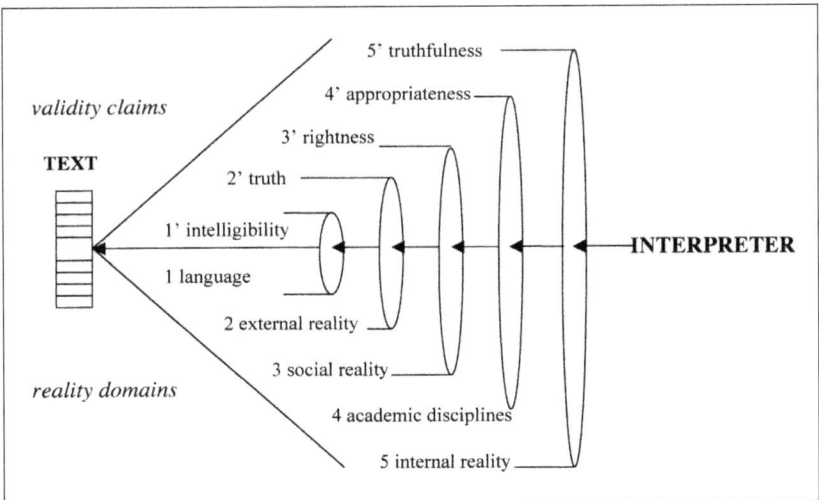

Figure 6. *The Speech-Act Complex and the Hermeneutical Process*

A sacred text has been interpreted for centuries and the collective memory of those interpretations will inevitably—as it ought to—color one's personal encounters with the text. This is why I turn again to Mohammed Arkoun, offering his diagram (cf. Fig. 7), which is admittedly controversial, to show that some Muslims, while maintaining a high regard for their sacred text, are also willing to deal with the human/community dimensions of the *mushaf* (the Qur'an in the form we have it today).[120]

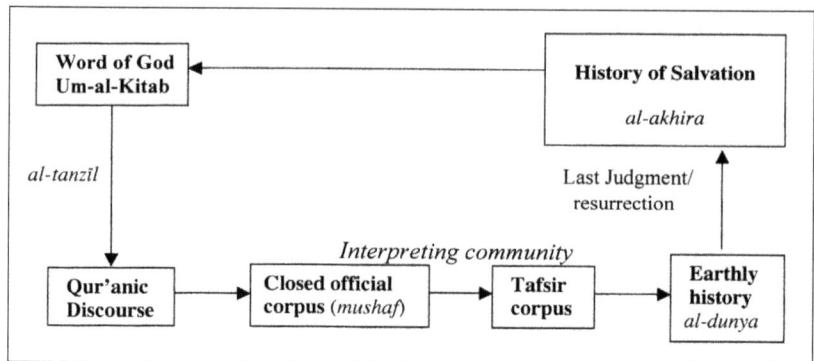

Figure 7. *Arkoun's Islamic Revelational Movement*

---

[120] This is also a hermeneutical model that Christians can easily identify as parallel-ing their understanding of canon formation with regard to the New Testament. Much as was the case among Christians over the centuries, hermeneutics have often bitterly divided believers. Some of the tensions would be lessened if people had realized the dynamics of the process of meaning production in the first place. Fiorenza shows that even religious experience—like all human experience (immedi-ate sense perception followed by practical judgment, and experience in the sense of a person being able to draw on the memory of multiple previous practical judg-ments)—is primarily an act of interpretation. It is this experience, both personal and collective, that allows one to come to a text with a 'pre-understanding' (Fiorenza borrowing from Heidegger). "This projective pre-understanding may be revised as a result of the encounter with a text or event, and what may emerge is a further understanding. Nevertheless, without this pre-understanding, further understanding is not possible. Moreover, the pre-understanding influences how understanding will take place... This Heideggerian insight has been expanded by Gadamer to show that the Enlightenment belief in the freedom of human subjectivity from all prejudice is illusory. Persons stand within the history of a tradition of ideas and values" (Thiselton, *New Horizons*, 299). But this also means that the interpretation of the sacred texts is vulnerable to many winds and currents. I used the idea of a lens, with the resulting idea of distortion; Arkoun adds the idea of manipulation (as Habermas), a judgment of people's motives: "As a result of confusions characteristic of the activity of the religious imaginary ['imaginary,' for Arkoun, is synonymous with 'worldview'] (and to a great extent, political imaginary: they are inseparable, contrary to the assertions of the secularizing ideology), the proper values and the irreducible functions of the Archetypal Book, the Qur'anic Discourse, the closed official corpus and the interpretational corpus have been projected into the *Mushaf*. It

Arkoun's model of revelation deserves more explanation. The emphasis, plainly enough, is on the eternal cycle of revelation from the "preserved tablet," or "the Mother of the Book" (*umm al-kitāb*) to the repeated act of *tanzīl* (lit. "sending down" of revelation) in the Prophet's lifetime, to the actual history of life and interpretation of the believing community; and finally, to the resurrected state of the believers after Judgment Day. Presumably at that point God's Word has fully accomplished its eternal purpose. What is relevant here is the distinction between the three facets of revelation: the heavenly (which in his scheme includes all the books that were sent down—*Tawrāt, Zabūr, Injīl* and *Qurʾān*),[121] the "Qur'anic discourse" (the writing down of what the Prophet heard and then repeated himself), and the "closed official corpus," or the *mushaf.* This distinction represents what most Muslim scholars would regard as an unacceptable bending to the conclusions of Western textual criticism of the Qur'an.[122] But Arkoun is only concerned with the integrity of the hermeneutical process. Here is what he has to say about the next phase, the body of many diverse commentaries:

> These commentators sought to elaborate on revealed truths to illuminate the conduct of human beings through the course of earthly history in the world down here (*al-dunyā*). This earthly history is entirely lived as a passage toward the Other World (al-ākhira) after the test of Resurrection... All these meanderings, all these mental and cultural exercises, and all these images depend on the *mushaf* for their concrete references and their field of projection.[123]

becomes the object of unlimited manipulations, something within the reach of all 'believers'; the ideological constructions are then trumpeted as theology and 'orthodox' truths" (Arkoun, *Penser L'Islam Aujourd'hui* [Algiers: Éditions Laphonic ENAL, 1993], 98, my translation).

[121] In order: the Torah (five books of Moses), the Psalms (written by David), and the gospel (always in the singular form, a point that is sometimes made in Muslim apologetics to impugn the fact that Christians have four gospels).

[122] Arkoun argues that for Muslim scholars to incorporate the conclusions of the German school of qur'anic criticism would actually enhance the scientific credibility of the text itself. But for political and psychological reasons, they generally do not do so (ibid., 35). For a recent presentation of the traditional view, see Yaqub Zaki ("The Qur'an and Revelation," in *Islam in a World of Diverse Faiths*, ed. Dan Cohn-Sherbok, Library of Philosophy and Religion [New York: St Martin's Press, 1997 [1991]). The role of Arabic as "the sacred language" is crucial from this perspective (ibid., 48), whereas for Arkoun, the Heavenly Archetype contains the source of all the previously revealed books as well as the Qur'an. Nevertheless, Zaki follows al-Ghazali (d. 1111) in distinguishing between the eternal *kalām Allāh*, the *mushaf* (letters and writing), and the articulated sounds (language and utterance) (ibid., 53).

[123] Arkoun, *Penser L'Islam Aujourd'hui*, 37.

Speaking both as a believer and as a poststructuralist philosopher and anthropologist, Arkoun advocates the same kind of project advanced by Peter Heath (cf. n. 118):

> There does not yet exist any exhaustive history of Qur'anic exegesis (*tafsīr*) that could speak to two concerns:
>
> 1. Defining the genesis, kinship, and historic diversification of a rather large literature, with special attention to the beginnings…
>
> 2. Studying the conditions for the exercise of Islamic reason in each ancient or contemporary commentary. It is indeed essential to show how the theological, historical, and linguistic postulates of this reasoning have led to confusion about levels of signification in the Qur'an.[124]

I now offer Farid Esack's interpretation of Arkoun (Fig. 8), in which he privileges the double movement of revelation (coming down) and the interpreting community (moving upward toward salvation). The difference in the diagrams also illustrates the difficulty of interpretation, especially when it comes to the theology of others, and then putting it in the form of a diagram. Particularly interesting is his version of Arkoun's view of revelation—a view that I believe brings him closer to Cragg's functional comparison of Jesus in Christianity with the Qur'an in Islam.[125] Yet he, like Arkoun, gives it a definite pluralist twist.[126]

---

[124] Heath, *Rethinking Islam*, 38.

[125] See, for instance, Cragg's *The Event of the Qur'an Islam in Its Scripture* (London: George Allen & Unwin, 1971).

[126] In his words, "The first is the word of God as transcendent, infinite and unknown to humankind as a whole with only fragments of it having been revealed through the prophets. Second are the historical manifestations of the word of God through the Israelite prophets (in Hebrew), Jesus of Nazareth (in Aramaic) and Muhammad (in Arabic). (It was memorized and transmitted orally during a long period of time before it was written down… Third, textual objectification of the word of God takes place (the Qur'an became a *mushaf*, i.e., written text) and the scripture is available to the believers only through the written version of the book preserved in the officially closed canons" (Esack, *Qur'an, Liberation and Pluralism*, 70). Clinton Bennett gives Esack a prominent place in the conclusion of his book on Muhammad. He notes that his pluralistic reading of the Qur'an brings him close to both Knitter and Hans Küng (*In Search of Muhammad* [London/New York: Cassell, 1998], 239). His closing remarks rejoin my own emphasis here. Instead of trying to forge a common theology that forces Muslims to abandon the infallibility of the Qur'an and Christians to jettison the divinity of Christ and the Trinity (as Knitter and Hick seem to be asking), Bennett urges Christians and Muslims to accept the fact that none of their theological formulations exhaust the divine mystery, and to concentrate instead on "liberating the oppressed, seeking the welfare of the poor, and meeting our neighbours' needs" (ibid., 242).

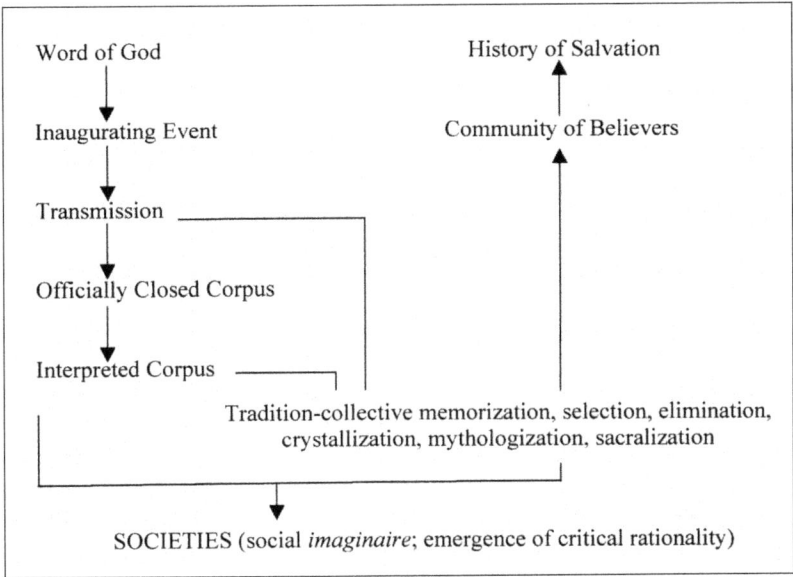

Figure 8. *Esack's Version of Arkoun's Revelational Moment*[127]

The present discussion on the interpretation of scriptures continues in the next chapter, along with some remarks about the issue of religious dialogue and pluralism.

[127] Esack, *Qur'an, Liberation and Pluralism*, 71.

Chapter 5

READING HOLY SCRIPTURES

This chapter closes the first part of this project. In the preceding pages I have described the socioeconomic and political landscape of postmodernity, pointing simultaneously to the urgent nature of Muslim–Christian cooperation in fostering a more inclusive, just and peaceful alternative to the Western-led, neoliberal McWorld, and to the philosophical issues within the postmodern paradigm that Muslim and Christian theologians cannot afford to sidestep. Foregoing the scientism of the modern mind-set, I have urged both parties to take into account how knowledge is produced in science, through theories, models, paradigms and research projects—a process that owes a great deal to the dynamics of individuals working within the framework of received traditions in particular communities. I hope to have shown that Barbour's comparison between the production of knowledge in science and the working out of theology in religion is not so far-fetched as it might have first seemed. Despite the obvious differences in subject matter, one does encounter a great deal of similarity in the methodologies employed. In both cases, the core beliefs draw from and feed into the data through the use of imagination, symbols, analogies and models. Though I am here highlighting the reading of sacred texts, Barbour's characterization of the data of religion as being "religious experience, story and ritual" is a good reminder that all four elements are in dynamic interaction and mutually influence one another in surprising ways.[1]

---

[1] Recall Ninian Smart's dimensions of religious experience and my own Fig. 6, through which I attempted to chart religious practices along the numinous–mystical and personal–impersonal axes. Along with the often-conflicting strands within religious traditions, these elements of experience, together with the regular practice of community rituals, impinge on the understanding of foundational texts from one context to the next.

In the following section, I draw out the conclusions of the last chapter on the hermeneutics of sacred scriptures and apply them to the task at hand. I end by making explicit my own approach to the Qur'an in the following chapter.

The Dynamic Nature of Sacred Texts

One of the consequences of the foregoing analysis is that only a community of faith—be it the Church or the Islamic *umma*—can read its Holy Book authentically. This is so because the text "projects a world" into which readers must not only enter, but commit themselves to embody in the practices of their tradition and in the application of its ethical imperatives. By contrast, outsiders can read and construct their own meaning. They can even study the interpretation of that community's experts over the centuries, yet they (my case in this study) cannot "indwell" that tradition,[2] understand it from the inside or hope to carry any weight among the interpreters of that community.

A corollary to that point is that textual interpretation within the community of faith will evolve as it grapples with the changing dynamics of the cultural and sociopolitical dimensions of its own context. As we have seen, adherents of both Islam and Christianity are facing a new situation (which I have called postmodernity) that will require fresh interpretations —readings that will open up cooperative courses of action to manage the earth's resources, both human and natural, for the collective survival of our planet. Having looked at the distortion factor and the hermeneutical process in general, I now seek to construct a hermeneutics that could potentially enhance a common Muslim–Christian theology of humankind in creation. This task was begun with Fiorenza's dynamic concept of

---

[2] This is an ideal that Bishop Newbigin borrows from Polanyi. Referring to the Christian and the Bible, he writes: "the important thing in the use of the Bible is not to understand the text but to understand the world through the text... What is required...is that one lives in the text and from that position tries to understand what is happening in the world now" (*The Gospel in a Pluralist Society*, 98). Just as we indwell our own body, or the language we speak, and therefore use them quite unconsciously to obtain knowledge or communicate with other people, so we seek to indwell the Bible "and from within it seek to understand and cope with what is out there. In other words, the Bible furnishes us with our plausibility structure" (ibid., 99). He goes on to explain that the structure is in the form of a narrative which began in the first pages of the Bible but which continues as we today allow God to work in and through us, and follow his lead as a community in a specific context, while keeping before us the horizon of our Lord Jesus' second coming.

speech acts, and now, with the help of Murphy and Vanhoozer, I wish to explain this theory of interpretation in more detail.

A Complex Communicative Act
Nancey Murphy finds a philosophical ally in J. L. Austin,[3] who turned away from the Fregian view of meaning as exclusively reference (which led directly to the dead-end of deconstruction) to a view of meaning as use. The philosopher must ask the broader question, "How does language actually function?" and its corollary, "What counts as a successful use of it?" She then turns to McClendon and Smith, who summarize Austin's criteria for a "happy" speech act in four points:

> (1) Preconditions—speaker and hearer must share a common language and be free from relevant impediments to communication. (2) Primary conditions—the speaker must issue a sentence in the common language that is a conventional way of performing that kind of speech act. (3) Representative or descriptive conditions—the sentence must bear a relation to a state of affairs that is appropriate to that sort of speech act. (4) Affective or psychological conditions—the speaker must intend to perform the speech act by means of the sentence and have the relevant attitudes or affects; the hearer must take the speaker to have the requisite intentions and affects (uptake).[4]

I will now use the key verse in this study as an example: "Behold, thy Lord said to the angels: 'I will create a vicegerent on earth'" (Q. 2:30).[5] Without trying to answer the question as to who the implied speaker of this sentence is (more easily, the privileged addressee is the Prophet), I will start at the level of the obvious speech act denoted by the sentence as a whole: God says to the angels, "I will create a vicegerent on earth." This is a dramatic declaration that could also be taken as a promise. Notice too, that the words *refer* to specific beings. Reference cannot be evacuated from the role and function of human language without the latter becoming nonsensical. Finally, in the following paragraphs, as an outsider I only make certain remarks that I believe Muslim colleagues and friends might find helpful.

---

[3] J. L. Austin, *How to Do Things with Words* (Cambridge, MA: Harvard University Press, 1962).

[4] Murphy, *Anglo-American Postmodernity*, 132; a summary of James W. McClendon and James M. Smith, *Understanding Religious Convictions* (Notre Dame, IN: University of Notre Dame Press, 1975).

[5] See the details given in the next chapter. Unless otherwise specified, I will be using throughout Yusuf Ali's translation (*The Holy Qur'an: Text, Translation and Commentary by Abdullah Yusuf Ali* [Elmhurst, NY: Tahrike Tarsile Qur'an, 1987 (1934)]).

What are the conditions for this speech act to be successful? First, the angels and God seem to be sharing a common language, as the next verse indicates. But the angels are at least apprehensive at the thought of this declaration: "Wilt Thou place therein one who will make mischief therein and shed blood? Whilst we do celebrate Thy praises and glorify Thy holy (name)?" To which God answers, "I know what ye know not." The second condition is conventional speech appropriate to the social context. God Almighty is here taking the angels into his confidence and revealing to them a coming event of momentous proportions. The context may be similar to the "Let us" clauses of Genesis 1, in which God declares that he will create a being superior to all the rest: "Let us make man in our image" (v. 26). Many commentators have felt that this was an allusion to the ancient Near Eastern concept of the Heavenly Court (see Chapter 8). God as Ruler calls his angels together as a council and informs them of his plans.

The third condition is that a state of affairs recognized by both speaker and hearer is actually the case. Here, we find a suitable state of affairs involving the earth (that the angels knew about) and the creature "Adam," presumably standing before them, since in the next verse we read, "And He taught Adam the nature of all things." When a person says, "I will do such and such," the hearer understands his or her intentions, and if he or she is trustworthy and the said action is within the realm of possibility, then it becomes a promise—whether or not it entails some benefit for the hearer. In this case the angels acknowledge being in the presence of the Creator God who said, "Let there be light!" and there was light. The qur'anic "*kun!*" ("let there be…") of God (e.g., Q. 3:47) is automatically followed by the coming into being of whatever the command might entail. So, there is no doubt in the receptors' mind that this creature will become God's trustee on earth. Then as the speech act opens up a whole new world in their imagination, the angels are not sure in what direction it will take them. At first glance, God's project is a risky one. Somehow, in their limited knowledge, they have experienced this creature to be one who creates trouble and violence. Their surprise is indeed appropriate.

Finally, for a speech act to be successful, Austin stipulates that, (1) the speaker holds to his or her word and intends to perform whatever is specified in the speech act, and (2) the hearer will adopt an attitude required by the content and context of the speech act. Here, God, by the sheer fact of uttering this promise, has already performed it. The response he expects from his hearers is confession of faith and worship. Interestingly, its actual fulfillment waits until God has asked them, "Tell me the nature of these if ye are right." Then the confession of faith spontaneously follows, "Glory to Thee: of knowledge we have none, save what Thou hast taught us: in truth it is Thou who art perfect in knowledge and wisdom."[6]

Murphy's point is that expressivist and referential theories of meaning are only partial truths. The factual content is necessary but so is the intended direction and force of the speech act. Here is a good example of discourse that aims at bringing into being a state of affairs and eliciting the proper response in the hearers' attitudes and actions (an attitude of worshipful awe and a confession of praise). With this introduction in mind, I present three characteristics of Kevin Vanhoozer's concept of discourse as "communicative act," inspired by Austin, but more directly, taken from John R. Searle.[7]

A Non-Reductionist View. Vanhoozer relies on philosopher Paul Ricœur to move beyond the premises of deconstruction that would deny the existence of anything outside textuality. He urges his readers to wager that beyond the surface signs of language they may find meaningfulness, some form of transcendence exists. Ricœur helpfully adopts an Augustinian stance: "You must understand in order to believe, but you must believe in order to understand."[8] On my view, then, Vanhoozer rightly exposes the theological nature of the deconstructionist project and thereby declares that "unfaith" must in the end be countered by faith of another kind.

Recall Saussure's distinction between *langue* and *parole. Langue,* as the given structures of language, includes the cultural, political and economic context of our human environment as well. The code is totally arbitrary throughout, and it has no connection to the world—only to other parts of itself. A consistent postmodern view would conclude that *langue* controls every human endeavor, and that under such a deterministic existence no act is reprehensible. People are no better than animals; they are here by chance, and only survive, hopelessly pushed around by fate.

---

[6] It is important to note for what follows that the meaning of this passage flows from a position of faith. Though it makes little sense to an agnostic, a Christian shares enough of the related beliefs broadly to enter into the world projected by this text.

[7] Vanhoozer quotes from two of his books in particular (besides several articles): *Expression and Meaning: Studies in the Theory of Speech Acts* (Cambridge: Cambridge University Press, 1979); *Intentionality: An Essay in the Philosophy of Mind* (Cambridge: Cambridge University Press, 1983).

[8] Cited in Vanhoozer, *Is There a Meaning in this Text?*, 31. Thus, he argues, hermeneutics begin with faith.

Few pursue the deconstructionist scheme to its logical conclusion.[9] Yet the postmodern ethos has heavily borrowed from Nietzsche, who asserted that it is the strong people, the poets who rewrite the world and invent new systems of differentiation. Here the temptation is not sloth but pride, not responsibility but pleasure. Clearly, one's view of language reflects one's view of the human person. I follow Vanhoozer here in his depiction of the "strong version" of postmodern deconstructionism:

> The modern view pictures the subject in conscious control; the post-modern picture shows the subject in a textually bound induced coma. The alternate picture defended in the present work portrays the subject as communicative agent, neither sovereign nor slave, but rather citizen of language. To repeat: A picture of language as an arbitrary system of differential signs holds deconstruction captive. Postmodern views of language are concerned more with semiotics (the science of signs) than with semantics (the science of sentences).[10]

By contrast to Derrida, Ricœur points out that a sentence is of a higher order than a word, or even a set of words. In fact, it is entirely a new

---

[9] I would not want to sound dismissive of the potential contributions of postmodern approaches to biblical studies—as I feel Vanhoozer sometimes is. A. K. M. Adams characterizes the more recent methodologies (such as narrative criticism, rhetorical and literary criticism, reader-response criticism, feminism and the use of structuralism and poststructuralism) as partaking of a general postmodern ethos: a suspicion of totalizing methods and worldviews (including objectivity, historicity and universality); a questioning of authority in established academic fields and thus a willingness to explore new lines of inquiry (imagination, popular readings, intertextuality, allegory and the like) and new media for biblical interpretation (such as poetry, film, sculpture, or graphics of various kinds). Though postmodernism is not likely to dominate biblical studies as of yet, Adams writes, "the field has already registered the impact of postmodern biblical criticism as claims of determinacy, universality, univocity, and legitimacy sound increasingly muted and defensive" ("Postmodern Biblical Interpretation," *Methods of Biblical Interpretation*, Foreword by Douglas A. Knight, Excerpted from the *Dictionary of Biblical Interpretation* [Nashville, TN: Abingdon Press, 2004], 176). This view is confirmed in another article in the same volume, which emphasizes the impact of literary criticism on biblical studies. The above-mentioned approaches all participate in this shared perspective. The three authors of this essay, "Literary Theory, Literary Criticism, and the Bible," themselves widely published in the field, conclude with the hope that the engagement of literary theory and criticism will enable biblical studies to find its way out of what has often been its own self-made ghetto of extreme specialization and into a larger academic and intellectual conversations, which have too often both given religious texts a central place and have caricatured them and made simplistic assumptions regarding them" (T. K. Beal, K. A. Keefer and T. Linafelt, in ibid., 165).

[10] Ibid., 203-4.

thing, much more than the sum of its parts. For him this basic distinction between semiotics and semantics unlocks some of the intractable philosophical problems related to language. Vanhoozer carries this concept over to persons: "The human person is a 'basic particular'—a primitive concept that cannot be explained by something more basic... The sentence is a 'basic particular' in much the same way."[11] As we saw in the last chapter, the physical/mental composition of humans can be related to the semiotic/semantic descriptions of sentences.[12] In the last two chapters I have been careful to jettison the modern disembodied self, while at the same time refraining from embracing the postmodern "situated self," imprisoned within the incommensurability of language and culture. Recall too Seyla Benhabib's disagreement with postmodernist feminist Judith Butler: the self as subject cannot be reduced to "yet another position in language." As social actors, we may be to a large extent "constituted by language," but unless "the subject retains a certain autonomy and ability to rearrange the significations of language," no communication is possible, and worse yet, there would be no discussion of ethical principles that might provide a basis for social actors to engage cross-culturally with others on issues relating to justice and peacebuilding.[13] But whereas Habermas and Benhabib construct a philosophy of human agency within a secular framework in order to justify the politics of communicative action, people of faith can do so more naturally, starting with theology.

Back to Augustine: faith seeking understanding—hence theology. This is precisely where Adam comes in: God creates a human being who can not only enter into conversation with God and his angels, but who is given the capacity to name things.[14] His language, in other words,

---

[11] Ibid., 204.

[12] Recall Murphy's contention that modernity is characterized by "atomism," and the postmodern paradigm in the physical sciences, correspondingly seeks holism relying on at least three strategies: supervenience, decoupling and top-down or whole-part causation (*Anglo-American Postmodernity*, 18-28). This is the kind of anti-reductionism Vanhoozer is advocating when he says that a sentence cannot be reduced to the amalgamation of its sounds, and not even to its words or grouping of words.

[13] Benhabib, *Situating the Self*, 216.

[14] In the Bible (Gen. 2) Adam gives names to the animals. Besides the reference to the animals and not "things" in general, the main difference is that the biblical account has Adam inventing names out of his own God-given creativity, whereas in the Qur'an he presumably enunciates what God has now planted in his memory. Yet the difference should not be overly emphasized. Both instances function theologically in the text as support for the human caliphate. David W. Shenk, a veteran of Muslim–Christian dialogue in Africa and in the USA, seems more pessimistic about

literally connects with the world as God created it to be. This is both a metaphysical and an epistemological assertion, true both for the qur'anic and biblical versions of humanity's creation. It has profound ethical implications as well:

> Given the centrality in Scripture not only of naming, but of other uses of language (e.g., praising, prophesying, promising, preaching, etc.), and given the many biblical passages on the correct use of the tongue (e.g., James 3), it is clear that God holds speakers accountable for what they say. More importantly, the Bible represents God as the pre-eminent speaker. Much of what he does takes the form of speech: promising, forgiving, commanding, and so on. The God of the Christian Scriptures is a God who relates to human beings largely through verbal communication.[15]

the rapprochement I am seeking. He sees in the "soul" of Islam a more fatalistic spirit. Yes, "Humans are created to be God's caliph, caretakers of the earth. Islam in its truest sense joins hands with all humanity in modern ecological commitments." But there is also a fundamental difference, he asserts: "In the Qur'an, God teaches Adam the names of the animals. In the Bible, Adam names the animals. These divergences in the primal accounts of creation may seem insignificant. But they are not, for these differences nurture a full-orbed worldview. In Islam God provides guidance which instructs humankind in every aspect of life. In biblical faith Adam names the animals, while God looks on, perhaps with some amusement, seeing what these people will do" (*Global Gods: Exploring the Role of Religions in Modern Societies* [Scottdale, PA: Herald Press, 1995], 313). I agree to some extent, but the dynamics of the global postmodern context is such that increasingly this "worldview" of Islam is evolving (it never was monolithic to begin with). Chapters 7–9 amply witness to that fact.

[15] Vanhoozer, *Is There a Meaning in this Text?*, 205. A useful discussion of the qur'anic discourse is found in Neal Robinson, *Discovering the Qur'an: A Contemporary Approach to a Veiled Text* (London: SCM Press, 1996). His chapter entitled "The Dynamics of the Qur'anic Discourse" discusses the baffling changes of pronouns (certainly for the non-initiate) denoting the "implied speaker" of the Qur'an. He uses Roman Jacobson's typology of communication as expressive (the speaker designates himself by oath or by "We" or "I"); conative (the speaker addresses his audience using the vocative "O" or the pronouns "thou" or "you"); and the cognitive—those passages in the Qur'an "when the speaker refers to Himself as 'He', or 'Allah', or mentions one or more of His names… This function is vital in a Scripture which is intended as a message for humankind. For if God had restricted Himself to expressive or conative communication, there would have been no universal message, no statements about Him which human beings could reiterate" (ibid., 229). While this is helpful in sorting out the complexities of qur'anic speech, this typology obscures the fuller and more dynamic nature of text as speech acts. Moreover, every sentence has content (his "cognitive" category). Whether it is addressed to someone else, or is simply a story, is more a function of the literary type of the passage (recall Goldingay's descriptive, prescriptive, commissive and expressive categories).

So far, any Muslim theologian could have written the same words. The last two sentences in this paragraph are strictly Christian, however: "Of course, God can embody his Word with a completeness that humans cannot: God's Word was made flesh. *God's Word is thus something that God says, something that God does, and something that God is.*"[16] I attempt to show in Chapters 10 and 11 that as Christians and Muslims we can go further than is normally assumed in this vein. Just the same, the common ground is staggering—great enough to build a common herme-neutic of human agency for the building of God's purposes of peace and human dignity in the world.

Hence, in terms of our common Muslim–Christian theology of crea-tion, because God speaks, we can speak. What then is a speech act—that deliberate attempt to communicate with another person? In Vanhoozer's words, "A text is a complex communicative act with *matter* (proposi-tional content), *energy* (illocutionary force), and *purpose* (perlocutionary effect)."[17] It is the author, therefore who holds one key element to the meaning of the text: "The author is not only the cause of the text, but also the agent who determines what the text counts as. In other words, the author is responsible both for the existence of the text (*that* it is) and for its specific nature (*what* it is)."[18] So, a text is the momentum of one person's willful act. It carries not just the propositional matter, but also the force and intended effect of its author. In a sense, it "is an extension of one's self into the world, through communicative action." The text is an embodiment of the author.[19]

What connects the words of a sentence to the world is the author's intended illocutionary force. If I read, "Adam named the things," I understand an assertion that is intended to describe the world as it is—a word-to-world fit (the words are meant to match the world). If, one the other hand, I read, "Adam, name the things!," the intended force of the sentence is different. It aims at getting the hearer to do something, to

---

[16] Vanhoozer, *Is There a Meaning in this Text?*, 205, emphasis original.

[17] Ibid. Murphy and McClendon have also built on Searle's speech-act theory, which stipulates three components to each speech event: locutionary (the sound uttered by the person); illocutionary (the social context in which it is uttered); and perlocution-ary (the force that utterance has according to the social rules applied to that context). In the end I prefer theirs to Vanhoozer's version, because they add "uptake" as being the reception of the intended meaning by the reader/hearer—the text producing its intended effect on them. Their use of "illocutionary" has the sociocultural setting in mind. Vanhoozer's illocution, then, is their perlocution. But his analysis of the author's intention is especially useful.

[18] Ibid., 228, emphasis original.

[19] Ibid., 229.

conform the world to the word, or a world-to-word fit. "Interpretation…
is largely a matter of following directions: the direction of the author's
attention (e.g., to a proposition), the direction of fit between words and
world (e.g., the kind of illocution)."[20]

That this is a statement about reality is clear: just as the self and the
sentence are irreducible entities, so too is human action. As humans we
all find ourselves acting as agents in our everyday lives—agents with
intentions. This says something important about language, thanks to the
concept of "supervenience," which Vanhoozer borrows from Murphy:

> The author's intention is the originating and unifying power that puts a
> linguistic system (the infrastructure) into motion in order to do something
> with words that the system alone cannot do…
> …To use rather technical language of the mind-body debate, we can say
> that meaning, like the mind, is an "emergent property." An emergent
> property is one that characterizes a higher order phenomenon (e.g., the
> brain) that has attained such a level of organizational complexity that it
> displays new properties (e.g., mental rather than physical) and requires
> new categories (e.g., the mind) to describe them…
> …Mental states, that is, "supervene" on physical states. This way of con-
> ceiving of mind has the advantage of being the most adequate scientific
> explanation of consciousness and of conforming to a Christian under-
> standing of human beings as created in the divine image.[21]

The concept of supervenience is a paradigm that neatly makes room,
both for the possibility of real knowledge through language, and for the
moral responsibility that comes with such a God-given capacity—and
thus taking a radical turn away from modern reductionism. This way of
thinking also has implications for the process of extracting meaning from
a text, as the next sub-section aims to show.

The Author, Intention and Meaning. Whether in speaking or writing (and
only often with "body language"), an agent intends to "do" something
specific. That action is in turn discernable for others, because (1) they
also are human beings, and (2) the systems are culturally encoded. Add
to that the complications of cultural and time distance, and the task

---

[20] Ibid., 247.

[21] Ibid., 249. Notice the adjective "adequate." This is a critical realist position. In this
explanation the categories of science, philosophy and theology are necessarily
blurred. In what sense can it be "true"? As Barbour would note, it is a useful paradigm, mainly because of its explanatory power. But it does approximate "truth" in
two ways as well. First, it conforms to our understanding of ourselves as persons
with agency. And second, it reflects the revealed "Truth" of the Christian (or
Muslim, in this case) holy text.

becomes more difficult, but not impossible. Vanhoozer argues that "the meaning of a text emerges only against the backdrop of the author's intended action and the background of the author's context":

1. "Every text is the result of an enacted intention."[22] The text before my eyes is more than symbolic marks of ink on a page. A person with a purpose created this piece of writing at a particular time in history in a specific cultural context. It is a meaningful act, an intention in the sense of "an emergent property that is required to explain what illocutionary act has been performed in a text."

2. "Every text is an embodied intention." The intention is not some mental state of the author preceding the actual writing of the text. Rather, "writing fixes the author's enacted intention in a stable verbal structure." It is not an action plan concocted in the author's mind, nor even the result that the author hopes to achieve by the writing, "but what the author was doing *in* writing, *in* tending to his words in such and such a fashion."[23] A reader will never fully know the intention of the author, but he or she can propose "relatively adequate descriptions of communicative acts." Of course, this is where the previous section on "distortion" must be brought in to bear. But in principle, it is possible to communicate meaningfully with an author through his or her text. Perhaps Vanhoozer is too sanguine about the stability of texts, yet in light of the previous discussions of critical realism, we are simply at two places on the same spectrum, with myself closer to Murphy on this issue.

From Illocution to Perlocution. Getting the meaning of a text, then, is to grasp the illocutionary directedness of the text—"*the meaning of a text is what the author attended to in tending to his words.*"[24] The speech act was successful if the reader understands whether it was meant as a command, a statement or a promise, and acts accordingly. This was good communication, one could also say—according to the function of the proposition's energy, its illocutionary effect. In the case the author wrote a series of arguments to persuade his or her readers that they must change their opinion on an issue, or, better yet, to adopt a costly course

---

[22] Ibid., 252.

[23] Ibid., 253, emphasis original.

[24] Ibid., 262, emphasis original.

of action, this purpose behind the speech act is called the perlocutionary effect. The intended effect might in fact fail, but the perlocutionary force (I use Murphy's sense here, cf. n. 17) is still part and parcel of the meaning. A text that seeks to persuade (most books do in one way or another) cannot be understood without taking that dimension into account.[25] In fact, it is this dynamic character of the speech act committed to writing (first brought up in connection with Fiorenza) which becomes crucial with regard to sacred texts. They are a communication from the Creator who wishes his creatures to live in a certain way. Figure 9 seeks to portray this process in graphic form.

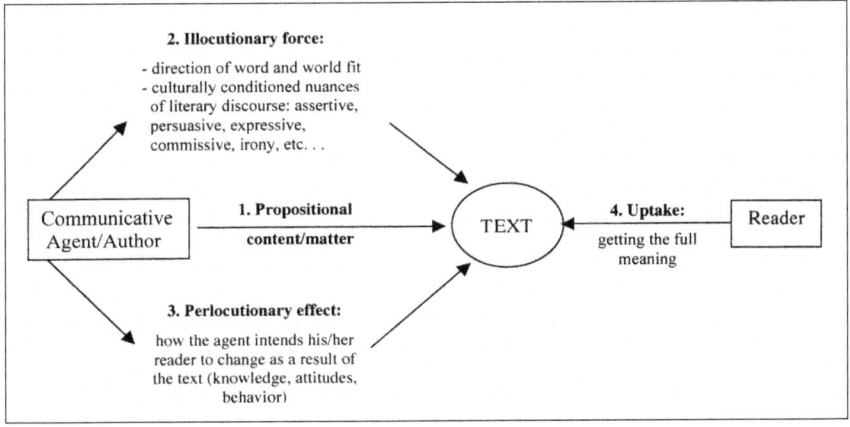

Figure 9. *The Dynamic of a Communicative Act*

[25] Here I disagree with Vanhoozer's word usage: "Perlocutionary intent fails regularly, *but this does not threaten the possibility of communication, for perlocutionary intents pertain not to the act but to the effects of meaning*" (ibid., 261, emphasis original). I prefer McClendon and Smith's version of the "happy" speech act, according to which the fourth condition is the affective or psychological conditions. Whether one obeys the scriptural injunction or not, understanding it as a command to be obeyed is necessarily central to its meaning. Yet Vanhoozer is also right when he distinguishes meaning from significance. In a sense, perlocution is intended significance. How-ever, texts can be applied quite differently, according to the historical and cultural context. There may well be unintended consequences to a writer's text. Hermeneutic (critical) realism constrains us to differentiate between the author's intended meaning and the manner in which the text's significance might be articulated way down the road. It is not only possible, but essential that one aim at recovering the original meaning (though always imperfectly).

Reading a Sacred Text

How might a "sacred text" be different? Though I cannot discuss theories of revelation and doctrines of inspiration in the scope of this study, I will briefly make four points. First, however one defines the relation between the human authors and the divine source of inspiration, in all cases Muslims, Christians and Jews affirm the divine origin of the Qur'an, the Bible and the Torah.[26] Further, because of their doctrine of creation, they all believe in the meaningfulness of what was recorded. Both the teaching and the commands of scripture would make no sense if God had not created human beings capable of understanding what he had revealed to them.[27] How could they be held accountable for obeying something they could not comprehend?

But positing God as the implied Author does not solve all our problems, Khaled Abou El Fadl reminds us. After a brief consideration of the recent movement of hermeneutics between author, text and reader, he declares that "[a]t a symbolic level, the interpretive inquiry in Islam should begin and end with authorial intent because the Divine intent is determinative of all meaning."[28] Yet in light of the historically contextualized discourse of the Qur'an and the nature of language itself, it cannot be said that the qur'anic text contains "an authorial intent," but rather that it "contains an *attempt* at authorial intent or a *partial* view of the authorial intent."[29] What the text reveals is the author's response to specific circumstances and chains of events. Thus authorial intent "is bounded by its audience, historical context, and language."[30] Another way of putting it is to differentiate between the qur'anic text embodying fully the Divine Will (something impossible), and the text embodying indicators to the Divine Will and authorial intent. In the end, it will never be possible to reduce the complexity of the reader/text relationship:

> From the Islamic perspective, the integrity of the legal text is derived from the possibility that it contains the indicators of the Divine. The integrity of the reader is derived from the fact that the reader is the Divine's agent.

---

[26] Traditionally, Muslims see the Prophet as simply the mouthpiece, the agent who receives and passes on the words recited to him by the angel Gabriel. In both cases one could use Robinson's cautious and respectful phrase, the "implied author" of the Qur'an.

[27] As many Muslim writers point out, the first revelation of the Qur'an to Muhammad in the cave of Hira was the divine command: "Recite [or 'read'—from the same root as 'Qur'an'] in the name of thy Lord who creates!" (Q. 96:1).

[28] Khaled Abou El Fadl, *Speaking in God's Name: Islamic Law, Authority and Women* (Oxford: Oneworld, 2001), 125.

[29] Ibid., 127, emphasis original.

[30] Ibid.

> The text and the reader are in a constant state of negotiation and construction. This negotiative process is ultimately what determines the meaning.[31]

The second point about sacred texts is that the meaning of a particular sentence or paragraph cannot be fully understood apart from the whole of the scripture. Christians call this the canon, but the same principle applies to Muslims and Jews as well. Therefore, the *sensus plenior* (the "fuller meaning") emerges only at the level of the entire canon: "If we are reading the Bible as the Word of God, therefore, I suggest that the context that yields this maximal sense is the canon, taken as a unified communicative act. The books of Scripture, taken individually, may anticipate the whole, but the canon alone is its instantiation."[32] Naturally, this causes the biggest tension between Jews and Christians. Sharing as they do the Tanakh, Christians see Christ anticipated throughout. But in all three faiths, the implied divine authorship constrains the reader to seek integration of all aspects of theology that remain in tension, like the equal affirmations of God's sovereignty and the freedom of human choices, or the mercy of God and his holiness and punishment of evil doers. Each text must be read in its context and in the light of the whole.

Allow me to interject here a brief parenthesis about canon and religious authority. The word "canon" has acquired a wider and more technical meaning recently in the discipline of religious studies. This idea is astutely developed by Brannon M. Wheeler in his study of the

---

[31] Ibid., 131.

[32] Vanhoozer, *Is There a Meaning in this Text?*, 265. The sources of legal authority in Islam were first set down by Muhammad b. Idris al-Shafi'i (d. 820). First the Qur'an, then the Sunna, then the consensus of the religious leaders, and then human reason (early on called "opinion," or *ra'y*) or the personal interpretation of the judge in a new case (*ijtihād*). Variant paths were devised over time and according to the various schools of law or Islamic sects. But the Qur'an has always remained on top of the list. Its content was fixed in writing in the decades following the death of the Prophet, and it holds a universal place of pre-eminence in the spiritual lives of Muslims worldwide. As Jonathan Brockopp puts it, "The primacy of the Qur'an over the sunna as a sacred text can be seen in the fact that sunna is never utilized in a ritual fashion. According to al-Ghazali, chanting even a single letter of the Qur'an is equivalent to ten good deeds to be added to all one's deeds at the end of time. Chanting 100 verses of the Qur'an in the morning is like all the good deeds in the world" ("Islam," in *Sacred Texts and Authority*, ed. Jacob Neusner, The Pilgrim Library of World Religions [Cleveland, OH: Pilgrim Press, 1998], 53). There is no merit attached to the memorization of the *hadith*, except that, depending on the soundness of each one, they are more or less useful for the making of legal pronouncements. For a thorough epistemological distinction between Qur'an and Sunna, see Khaled Abou El Fadl, *Speaking in God's Name*, 98-115.

Hanafi legal tradition.[33] He borrows from Jonathan Z. Smith the insight that from an anthropological perspective, "the radical and arbitrary reduction represented by the notion of canon and the ingenuity represented by the rule-governed enterprise of applying the canon to every dimension of human life is that most characteristic, persistent, and obsessive religious activity."[34] Thus research on divination in the African Ndembu tribe reveals that it is the community of diviners that establishes both a fixed scope of objects used in divining and the meaning given to these objects in the process of divination.[35] Just as the Ndembu diviner can only activate his divination in a particular case by referring to the "canon" of common practice of former generations of diviners, so the Hanafi jurist "must be able to demonstrate, using inductive reasoning, a link between his definition of practice and the Qur^cān through the medium of certain precedents considered to have canonical authority by the Ḥanafī school."[36]

Wheeler's study of the hermeneutics that guide the gradual accumulation of Hanafi tradition over the first centuries demonstrates that the Qur'an is interpreted through a double grid: first the Sunna, and then the corpus of Hanafi writings that have interpreted the Sunna's own interpretation of the Qur'an. In this perspective, the authority of the canon is tied both to the power of those mandated by society to be its guardians and to the fact that this canon continues to be applied to the daily lives of the larger community. This would be true of not only all schools of law in Islam, but also of all religious communities—whether the canon is a body of texts, rituals handed down, or the practices of religious officials in determined settings or a combination of the above. Hence, "canonical authority is less about the canon itself than it is about the process surrounding the institutionalization of the canon's authority and the method allowing the use of the canon to perpetuate the present aims of those responsible for institutionalizing its authority."[37]

This leads me to the third aspect of sacred texts: as authoritative texts they come to us already wrapped in particular interpretations, yet they

---

[33] Brannon M. Wheeler, *Applying the Canon in Islam: The Authorization and Maintenance of Interpretive Reasoning in Ḥanafī Scholarship*, SUNY Series, Toward a Comparative Philosophy of Religions (Albany, NY: State University of New York Press, 1996).

[34] Jonathan Z. Smith, *Imagining Religion* (Chicago: University of Chicago Press, 1982), 43, cited in Wheeler, *Applying the Canon in Islam*, 1.

[35] Wheeler's discussion is based on Victor Turner, *Revelation and Divination in Ndembu Ritual* (Albany, NY: Cornell University Press, 1975).

[36] Wheeler, *Applying the Canon in Islam*, 10.

[37] Ibid., 239.

are open ended at the same time. As Wheeler's examination of Hanafi scholarship over time seeks to illustrate, a religious community's canon derives its authority from its inherent applicability to changing circumstances. Thus both Christian and Muslim scholars plumb the depths of revelation to "divine" how believers should act out their faith in the midst of the pluralistic cities of the West or how they should respond to the ethical challenges posed by recent applications of genetic engineering. Again, this is similar to Wheeler's conclusions about Hanafi jurisprudence:

> Although the canon epitomizes how to argue it does not stipulate what to conclude. Hanafi scholarship fixes the epistemology and methodology to be used but leaves the practical significance of the revelation open to its continuing application to novel circumstances. This means that each generation of scholarship must be prepared not to imitate the past but to demonstrate that the application of the revelation, although its own, has the authority of previous generations going back to the prophet.[38]

Put differently, the open-endedness of sacred texts also implies that they are inherently dynamic. On the level of perlocution, sacred texts seek not only to provide knowledge about God, about his world, and about the human self and its need for repentance and faith. The scriptures not only aim to teach, but also to stir up human emotions (awe in worship, gratitude, sorrow for sin and the like) and provide commands to be followed. The author of the Epistle to the Hebrews wrote that "the word of God is alive and powerful."[39] This is especially true from the Christian's perspective, for two reasons. First, the written word is made alive by the power of the Holy Spirit: "He will teach you everything and will remind you of everything I have told you," Jesus says reassuringly to his disciples the night on which he was handed over to the Jewish authorities.[40] And second, the Bible can only be the word with a lower case "w." I agree with Vanhoozer: "Jesus Christ is the preeminent interpreter of God's self-communication, the unique and definitive embodiment of God's self-communicative act or 'Word.'"[41]

Where I see the most common ground, however, is in Jesus' Parable of the Sower, which finds an echo in the Qur'an. It is particularly significant, since Jesus provides himself the interpretation: the seed is the word of God; it will be received by a variety of people; some are like the hard path where it has no chance even to take root; others are like the

---

[38] Ibid.
[39] 4:12, New Living Translation (NLT).
[40] John 14:26, NLT.
[41] Vanhoozer, *Is There a Meaning in this Text?*, 440.

rocky soil, with only shallow roots (life's difficulties cause them to wilt); others are like the thorny ground (worries, riches and greed crowd out the word); finally, those who accept the word, live it out and persevere in it are like those who produce fruit thirty, sixty, or even a hundred times as much as was planted.[42]

The qur'anic parallel (Q. 48:29) is different in form but similar in meaning:

> Thou (O Muhammad) seest them bowing and falling prostrate (in worship), seeking bounty from Allah and (His) acceptance. The mark of them is on their foreheads from the traces of prostration. Such is their likeness in the Torah and their likeness in the Gospel—like as sown corn that sendeth forth its shoot and strengtheneth it and riseth firm upon its stalk, delighting the sowers—that He may enrage the disbelievers with (the sight of) them. Allah hath promised, unto such of them as believe and do good works, forgiveness and immense reward.[43]

The word of God in the Qur'an is compared to his word as revealed in the Torah and the Gospel. In all cases it is like a seed of corn that grows up vigorously. Presumably, the delight of the sower comes from the fruitfulness of the stalks. In this verse (as in the Qur'an throughout) faith is tied to the practices of the faithful (here prayer is singled out) and good works. Perhaps the most telling picture of the perlocutionary effect of the sacred text is in Psalm 1. The righteous who delight in doing everything the Lord wants are "like trees planted along the riverbank, bearing fruit each season" (v. 3, NLT). The Apostle James closely echoed his master, when he exhorted his readers to "be doers of the word, and not hearers only."[44]

---

[42] N. T. Wright is surely correct in seeing in this parable a cryptic (his enemies are looking for ways to put him to death) but clear double reference to the story of Israel and to his own: "the parable *tells the story of Israel, particularly the return from exile, with a paradoxical conclusion, and it tells the story of Jesus' ministry, as the fufilment of that larger story, with a paradoxical outcome*" (*Jesus and the Victory of God*, Christian Origins and the Question of God 2 [London: SPCK; Minneapolis: Fortress Press, 1996], 230, emphasis original). On the other hand, the parable as the story of Jesus' ministry encapsulates the story of Israel (and not merely marking its climax; see the parallel Parable of the Wicked Tenants, Mt. 21:33-46). Jesus is slated to die, as he brings God's word to a recalcitrant people. More than that, he is the word being sown, the seed, which, as it dies in the ground, bears much fruit (as John says in 12:24). He is the rejected stone that becomes the cornerstone, says Matthew (Mt. 21:42), quoting Ps. 118:22.

[43] *The Glorious Koran: A Bilingual Edition with English Translation, Introduction and Notes by M. Pickthall* (London: George Allen & Unwin, 1976), 682-83.

[44] James 1:22, Revised Standard Version (RSV).

I started the previous section with McClendon's four conditions for a "happy" speech act, to which I applied God's announcement to the angels that he was about to place Adam, his deputy, on earth (Q. 2:30). I now conclude with Murphy's rendering of his "Baptist Vision," in order to make my fourth and final point. Reading is not an individual exercise, as "modern" Westerners are prone to assume. It is primarily a community experience—a corollary to Wheeler's definition of "canon." To explain what I mean, allow me to first re-establish the felicity of a communicative act committed to writing centuries ago, as is the case with the monotheistic scriptures. Murphy says that to presuppose the felicity of the text's performance of these speech acts is to posit that an author successfully used the linguistic conventions of his or her time and that his or her hearers and readers at the time would have understood his basic meaning. After all, it is reasonable to believe that these conventions were "somewhat widespread and invariate over time," since these texts were collected, circulated and preserved by several communities. And then also, "if we know what an author *did* through or by means of a text, and if we have no reason to suppose that the author was dishonest or incompetent in using the language, then we have public access to what the author intended."[45] Getting the meaning or "uptake," is, then, like having an intention, which is really a communal affair:

> But on the account of language here presented, getting the meaning is "operationalized" as a communal (shared, intersubjective) response to a text—a living of its import rather than a mere hearing of it. The intrinsic relation between language and action, emphasized by both Austin and Wittgenstein, is crucially important here. So conceived, it would have been a public matter whether the texts "meant" the same to everyone. And in some cases, as shown in Paul's letters, the author could and did object when his writings had been taken wrongly.[46]

So far, according to Murphy, we have only solved half of the hermeneutical problem. We have only established that the texts had a reasonably determinate meaning in their original contexts. The other half is to

---

[45] Murphy, *Anglo-American Postmodernity*, 143, emphasis original. She makes clear, as does Vanhoozer, that "there is no mental component—the intention or the meaning—to be sought beyond the speech act itself (a Wittgensteinian point). So it is true, *in a sense* that we can never recover the mind of the author—it is true in the trivial sense that we never have immediate access to anyone's thoughts but our own. But it is obviously false in another sense since in the normal, 'happy' state of affairs, people's speech acts are public enactments of their intentions" (ibid., 143, emphasis original).

[46] Ibid., 144.

show how that meaning can be accessed in today's context. This is more than the issue of fusing the present horizon with the text's horizon, she argues. The act of understanding must be seen as "uptake." First, today's believing community understands itself to be addressed by those texts, and second, in reading them, they act as the same "interpretive community" as that of the author.[47]

This is precisely McClendon's "Baptist Vision"—"Baptist" in the vein of the denominations that rather loosely trace their roots to the Anabaptist tradition of the sixteenth-century Reformation. This vision was the "guiding stimulus," the "touchstone" which points to authentic Baptist theology and practice. For this vision to be authentic, the present community must see itself in a double dimension. It is both aligned with the primitive community though its belief and practices, yet it is also the eschatological community to be judged by the standards of the coming kingdom of its Lord Jesus Christ. Just as the Apostle Peter proclaimed to the crowd on Pentecost, "This is that which has been spoken through the prophet Joel" (Acts 2:16), so the present church believes itself to be that same community. The "then" is also "now." In Austinian terms, "we might describe it as the present church's *determination* to take the texts to be addressed to itself, despite awareness of historical distance, so that the illocutionary force *then* is to be the illocutionary force *now*."[48]

Murphy then applies the Ricœurian view of the text as "projecting a world" into which the reader is invited.[49] This means that to grasp a text's

---

[47] Murphy quotes at length from the "later" Stanley Fish's *Is There a Text in This Class? The Authority of Interpretive Communities* (Cambridge, MA: Harvard University Press, 1980): "Interpretive communities are made of those who share interpretive strategies… The notion of interpretive communities thus stands between an impossible ideal and the fear which leads to so many to maintain it. The ideal is of perfect agreement and it would require texts to have a status independent of interpretation. The fear is of interpretive anarchy, but it would only be realized if interpretation (text making) were completely random" (ibid., 182, cited in Murphy, *Anglo-American Postmodernity*, 146).

[48] Ibid., 148, emphasis original.

[49] Though he is weak on McClendon's third condition ("referential") in his *Texts Under Negotiation*, Walter Brueggemann's "postmodern imagination" is an edifying exhortation to the contemporary Bible believer: "The purpose of preaching and of worship is transformation" (*Texts Under Negotiation: The Bible and Postmodern Imagination* [Minneapolis, MN: Fortress Press, 1993], 24). This will only be accomplished, not through "doctrinal argument" nor through "moral appeal" but by letting the dynamic models, pictures and images of the sacred text (carried by their narrative form) capture the hearers' imagination, offering them a "counterstory" so subversive to the consumerist perspective they ingest on a daily basis that they will be powerfully drawn to change. But this does not happen in one sitting, but only

meaning is to enter into its worldview. "To understand biblical language is to enter sympathetically into that world, just as one becomes *absorbed* in a good novel. This reading strategy is the opposite of modern interpreters, for whom the goal was to bring the Bible into the modern world, to understand it by the lights of modern thought."[50] This fact makes it crucial, then, for the readers to take up the practices of the first community. This is precisely the Baptist version of the hermeneutical circle. As she notes: "Historical criticism helps to recover the practices of the early church in their own setting; attempting to live out those practices in the contemporary setting sensitizes readers to new meanings in the texts."[51] In a real sense, what is required of the believer, individually and in community, is truly to "perform" the sacred text.

Some readers, Christians in particular, will ask how the foregoing might be compatible with a conservative view of scriptural inspiration. A very brief answer will have to suffice here. Following Murphy, I claim that the modern picture of knowledge as a building resting on indubitable foundations has captured Christian theology since the nineteenth century.[52] In its liberal version—starting with Schleiermacher's definition of religion (all across the board) as "awareness of absolute dependence," or "God-consciousness"[53]—the foundation was related to religious experience. Naturally, this view went hand in hand with an expressivist view of language and an immanentist perspective on God's action in the world (miracles make no sense). For the conservatives, the foundation had to be more objective than that. In particular, the Princeton theologians Charles Hodge and A. A. Hodge (father and son) relied upon the work of American philosopher and founder of commonsense realism, Thomas Reid, as well as on John Locke's confidence in the truth of deductive reasoning, both in science and in the demonstration of God's existence. Along with Benjamin B. Warfield, they laid the groundwork for American fundamentalism: God revealed himself most fully in the Bible, such that each and every word is inerrant in the original text (plenary verbal inspiration). Hence, "if Scripture is to provide an *indubitable* foundation for

---

"one text at a time, one miracle at a time, one poem, one healing, one pronouncement, one promise, one commandment" (ibid., 25). This will enable the hearers or readers to comment, to discuss, to apply this new "collage" in community.

[50] Murphy, *Anglo-American Postmodernity*, 148, emphasis original.

[51] Ibid., 149.

[52] For a more extensive development of this theme, see her *Beyond Liberalism and Fundamentalism: How Modern and Postmodern Philosophy Set the Philosophical Agenda* (Valley Forge, PA: Trinity Press International, 1996).

[53] Ibid., 95.

theological construction, then all of its teachings must be free from error, lest the theologian make erroneous judgments in distinguishing true teachings from false ones or essential teachings from incidental cultural assumptions."[54] In contradistinction to liberal foundationalism, then, conservatives have propounded a representational (or propositional) theory of language and an interventionist theory of divine action in the world (miracles happen).

British theologian and New Testament scholar N. T. Wright agrees with this assessment. With him, I would argue that the words most bandied about in conservative circles—infallible (scripture will not deceive us) and inerrant (one step further: it makes no errors)—appeared in a specific cultural context in which the Bible in North America became a tool against both Roman Catholics and liberal modernism. Yet in fighting those two alleged enemies, conservative Protestants adopted the assumption of their foe, that is, Enlightenment rationalism.[55]

In American Protestantism today, the fundamentalists are few, while the majority of conservatives call themselves "evangelical"[56]—a label that applies to a spectrum of views. Clearly, as can be seen from this and previous chapters, I have tried to distance myself from the modern foundationalist perspective and find much affinity with Ian Barbour's critical realism and his four criteria for assessing competing truth claims both in science and religion (agreement with data/correspondence; coherence; scope; fertility). I remain conservative in the sense that I believe the Bible is the primary source and highest authority for my self-understanding as a follower of Jesus in the Christian tradition. These are the texts that shaped the first Christian community. The biblical canon, then, whether through its narrative or its more didactic passages, projects a reality that goes beyond our human experience and therefore gives meaning to it. Hence I believe it is wrongheaded to measure its reliability by its ability to predict and harmonize with the findings of contemporary science (a modern view); rather, I believe that God inspired individuals over many centuries and in several historical and cultural contexts to testify to his acts in history, and supremely to the birth, crucifixion and

---

[54] Ibid., 91, emphasis original.
[55] N. T. Wright, *Simply Christian: Why Christianity Makes Sense* (New York: HarperSanFrancisco, 2006), 183. Wright then offers this insightful comment: "It is no accident that this Protestant insistence on biblical infallibility arose at the same time that Rome was insisting on papal infallibility, or that the rationalism of the Enlightenment infected even those who were battling against it."
[56] For the distinction, see George Marsden, *Understanding Fundamentalism and Evangelicalism* (Grand Rapids, MI: Eerdmans, 1991).

resurrection of Jesus, Israel's awaited Messiah.[57] Again, the emphasis is on the *Word* (Jesus, the incarnate Son) and on the *word* (scripture) as a witness to him. And more importantly, the Bible is more than just a reliable record of the words and mission of Jesus of Nazareth. It is a story that begins with creation, continues with God's focus on a particular people through Abraham and Moses, and climaxes with the life, death and bodily resurrection of Jesus, Israel's Messiah and humanity's redeemer, who ushers in God's plan to redress the horrendous damage done to his world and heal the wounds of a broken humanity, his agents in the world. It is rather like "an unfinished novel beckoning us to become, in our own right, characters in its closing episodes."[58]

Thus I subscribe to the most widespread formulation (arguably) of evangelical theology today, the "Lausanne Covenant," when it confesses the biblical canon as "the only written Word of God, without error in all that it affirms, and the only infallible rule of faith and practice."[59] The emphasis here is on perlocution—the action effected by the God-breathed text. The effect of the Holy Spirit touching the heart, mind, will and emotions of the readers through the text he originally led the authors to write (and at times the work of collectors and editors) is mostly to provide the necessary energy to carry out the mission to which God has called them. As the Apostle Paul put it, "All scripture is inspired by God and is useful for teaching, for reproof, for correction, and for training in righteousness, so that everyone who belongs to God may be proficient, equipped for every good work" (1 Tim. 3:16-17).[60] As was emphasized in the last chapter, the reader's response (never infallible) must be included into the framework of meaning, which is upheld by three factors: the Holy Spirit's inspiration of the authors' work, his guiding the consensus of the church (as the hermeneutical community which gives recognition to this inspiration by delimiting the canon) and finally, the Spirit's inspiring the reader who, in faith, seeks guidance from God in order to grow into the maturity of Christ, who is at once the pioneer of our faith in God as the New Adam and God himself.

---

[57] The Apostle Paul writes in Greek that "All Scripture is inspired by God [lit. 'God-breathed'] and profitable for teaching, for reproof, for correction, and for training in righteousness, that the man of God may be complete, equipped for every good work" (1 Tim. 3:16, RSV). Here is a strong emphasis on the perlocutionary power of the text.

[58] Wright, *Simply Christian*, 183.

[59] John Stott, *The Lausanne Covenant: An Exposition and Commentary* (Minneapolis, MN: World Wide Publications, 1975), 10. Note that "inerrancy" covers only the broad strokes of doctrine and historical events ("all that it affirms"—no theory on *how* creation took place, for instance).

[60] Quoting Wright's own translation, *Simply Christian*, 182.

I therefore conclude this section with the observation that in going beyond the modern vision of the solitary knower who first had to rid his or her mind of all tradition (as a form of prejudice), we have come back to Polanyi's original insight that knowledge and, all the more, interpretation, is like an apprenticeship. Just as Bishop Newbigin was fond of repeating, the church is the "hermeneutic of the Gospel."[61] It is in this setting that believers stake their faith in the God who communicated to them in scripture and bestowed on them the gift and responsibility of communication. In a true sense, the meaning of the formative texts can be accessed, but in the end, it can only be done through a community that is ready to live for—and die for—the truth of the world it has chosen to enter. Understanding for the Christian is inevitably linked to discipleship.[62] For the Muslim and the Jew as well, true exegesis of the sacred text will entail authentic and embodied commitment to its Straight Path.

The distinction made above between the Word of God (Jesus the Son) and the word of God (the Bible) is also made by Palestinian theologian and Anglican priest Naim Ateek. He notes how the Bible has long been abused by Christian Zionists (the land of Palestine was promised to the Jews; therefore Palestinians must either emigrate or submit to Israeli rule)—so much so that among Palestinian Christians "the Old Testament has generally fallen into disuse among both clergy and laity."[63] Various alternative readings have been proposed, but the fundamental question for a people daily absorbed by their own burning experience of injustice remains: Is God a partial God or is he a God of justice and righteousness for all his creatures? As a pastor, Ateek adds the pressing question: How can the biblical message be adequately presented to a Palestinian audience? He continues:

> The answer lies largely in the doing of theology. The only bridge between the Bible and people is theology. It must be a theology that is biblically sound; a theology that liberates; a theology that will contextualize and interpret while remaining faithful to the heart of the biblical message...
>
> For Palestinian Christians, the core question that takes priority over all others is whether what is being read in the Bible is the Word of *God* to them and whether it reflects the nature, will and purpose of *God* for them...

---

[61] Murphy follows Alasdair MacIntyre's concept of "tradition-constituted enquiry"—the insight that "all traditions, religious and secular alike, are shaped by their interpretation and application of a formative text (ibid., 151).

[62] Ibid., 440-41.

[63] Naim S. Ateek, *Justice and Only Justice: A Palestinian Theology of Liberation* (Maryknoll, NY: Orbis, 1989), 77.

> Palestinian Christians are looking for a hermeneutic that will help them
> to identify the authentic World of *God* in the Bible and to discern the true
> meaning of those biblical texts that Jewish Zionists and Christian funda-
> mentalists cite to substantiate their subjective claims and prejudices.[64]

That central hermeneutic, claims Ateek, is actually as old as the church:
it is the person of Jesus Christ himself, who fulfilled in his birth, life,
ministry, passion and resurrection the fullness of God's redemptive purposes for humankind and revealed the true nature of God's character. It
is in this sense that "the *Word* of God incarnate in Jesus the Christ interprets for us the *word* of God in the Bible." Due to human sin, distortions
of all kinds are still possible, but such a strategy allows the Palestinian
believer to retain the full canon of the Bible, "while its contents would be
judged by this hermeneutic and scrutinized by the mind of Christ.[65] As
past situations brought up by the sacred text are studied in their historical
context, today's believer reads through the life and teaching of Jesus. In
some cases, for example when Joshua at the fall of Jericho is commanded
by God to destroy completely everything and everyone,[66] we will conclude that the writers reflected "an early stage of human understanding of
God's revelation that conflicts with the Christian's understanding of God
as revealed in Jesus Christ."[67] But even negative examples can reinforce
the overall picture of who God is and how he expects us to act as his
representatives on earth. "The Bible, therefore," concludes Ateek,
"remains the Word of God, 'profitable for teaching, for reproof, for
correction, and for training in righteousness';[68] but it is continuously
submitted to an authoritative concept, that is, the revelation of God in
Christ."[69]

Ateek's presentation of a Palestinian liberation theology displays
many parallels with the theological approach espoused by Esack for the
Muslim community in Apartheid South Africa. Both in essence are
extensions of McClendon's "Baptist vision" or Murphy's "uptake."
Three consequences follow this emphasis on liberation. First, the character of God unfolds in such a way as to illumine ethical norms that in turn
enable us to integrate a strong ethical component to our theology of
human trusteeship of the earth—a point I pursue in some detail in Chapter 11. Thus, in the context of our increasingly pluralistic global socie-

---

[64] Ibid., 78-79, emphasis original.
[65] Ibid., 80, emphasis original.
[66] Josh. 6:17, 21.
[67] Ibid., 82.
[68] Quoting from 2 Tim. 3:16.
[69] Ibid., 83.

ties, this tenacious faith in the dignity of the human person represents a vital energy that can be directed to peacebuilding. Second, it reinforces the observation that sacred texts are read in communities facing particular concerns. Divine guidance comes in particular through the questions posed by concrete historical situations. Finally, such an emphasis on deliverance and release from oppression can also highlight differences in perspectives between various communities. As the apostle Paul noted, on this side of eternity "we see but a poor reflection as in a mirror."[70] Hopefully, such differences would increase our desire to learn humbly from others, strengthening our resolve to settle disputes by deepening our dialogue with the "other," both inside our community of faith and between communities of faith. This is the spirit in which I set out to explore how the issues of truth and salvation are connected to inter-religious dialogue (cf. Appendix D).

My Own Approach to the Qur'an

Here is the crux of the matter: What place is there for a non-Muslim to venture into a reflection on the meaning of the Muslims' sacred scripture? Or, as Cragg has it, "Can it be properly read and pondered without commitment to Islam as its institutional consequence and meaning?"[71] Since dialogue is my purpose, I will be using Cragg as my guide in presenting three elements of response to this very legitimate question.[72]

First, my purpose is to take the text as I see it. Most of the Qur'an has God speaking through Muhammad to the people. I accept the inner logic that this entails and attempt to let the Qur'an interpret the Qur'an, much as I do for the Bible. I look at a text and first seek to understand its plain meaning, as much as that is possible; I examine it in the context of the surrounding passage and then consider the historical context; finally, I consider parallel passages and try to integrate the particular within the theological whole.[73] As a Christian, I do not ascribe to the qur'anic text

---

[70] 1 Cor. 13:12. On must remember that mirrors were a good deal more opaque in those days.

[71] Cragg, *The Event of the Qur'an*, 20.

[72] I am reminded of Ebrahim Moosa's comment on Rahman's call to renew religion in the light of the developing socioeconomic and political realities: "In rethinking Islam Fazlur Rahman has argued that people outside the tradition can also play a meaningful role" ("Introduction," in Fazlur Rahman, *Revival and Reform*, 28).

[73] This is similar to what John Goldingay calls the "formalist" approach to hermeneutics: "Formalist criticism seeks to understand works as wholes, even long works such as the Gospels or Genesis–Kings, and to see the place of individual sections within such wholes—the way people normally read books!" He decries what he con-

the same revelatory value as I do to the Bible, but that does not hinder me from seeking to report what the text is saying. Thus, for the sake of this project my approach to the Qur'an and the Bible will be very similar.

Having said this, it could be rightly objected that in stringing together passages dealing with similar themes I am assuming that an original text lies behind the various expressions of those themes or events evoked in the Qur'an. In many cases this is simply not possible to substantiate, especially because the Qur'an points to a revelatory process over twenty-two years, with some modifications along the road from Mecca to Medina. Muslims themselves have readily accepted this in their doctrine of abrogation (*naskh*). What I am attempting in this chapter is a "naïve" reading, or, better, a "narrative" approach, in the spirit of Muhammad Mahmoud's article, "The Creation Story in *Sūrat Al-Baqara* [Sura 2]." Without following his literary scheme in detail (my purpose is different), I will keep his distinction between "the primary text level" (the overall diachronic sequencing of events, or "external time'), and the "sublevel text"—each passage and its "internal time."[74] Mahmoud's thesis is that the primary text, or the "Qur'anic Text" (QT in his article) assumes a narrative voice. This is particularly helpful when dealing with the seven qur'anic passages describing creation.[75]

Though this kind of holistic reading is coming back to biblical studies as well, it is not new. In his discussion of the two qur'anic terms related to Muhammad's Jewish counterparts, *rabbaniyyūn* and *aḥbār*, Gordon D.

---

siders to be an overemphasis on historical criticism applied to the historical books, which has produced very little consensus and has in fact obscured the very meaning of the text itself (*Models for Interpretation of Scripture*, 21-23).

[74] Muhammad Mahmoud, "The Creation Story in *Sūrāt Al-Baqara*, with Special Reference to Al-Tabari's Material: An Analysis," *Journal of Arabic Literature* 26 (1995): 201.

[75] Following Mahmoud on one of those passages—the one selected for this study, Q. 2:30—the narrative voice functions on two levels: independently, and as the voice of God: *Wa idh qāla rabbuka...* (And when your Lord said...). "On announcing that He will set on earth a viceroy, God posits an opposition between earth (*al-ard*)/the Garden and earth/heavens. This opposition carries within it seeds of tension and conflict that will soon burst forth. However, God's statement proclaims that man will be his *khalīfa* on earth, implying thus a continuousness of relationship that dispels tension and affirms harmony" (ibid., 203). Though different from the Judeo-Christian notion of canon, Mahmoud sees this literary device as seeking to bring harmony (and tension at times) between the different subtexts within the Qur'an on the one hand, and between the Qur'an and the commentaries in the Hadith literature on the other. As I said above, this is also a somewhat naïve view since it does not take into account the *asbāb al-nuzūl* (the causes for divine revelation in specific circumstances), thus the profoundly historical and contextual nature of the qur'anic text.

Newby rightly points out the importance of focusing on the shared worldview perspectives and hermeneutical strategies of the Jewish and early Muslim communities. A pivotal verse in this regard is *Sūrat al Māʾida* (Sura 5), v. 47: "It was We who revealed the law (to Moses): therein was guidance and light. By its standard have been judged the Jews, by the Prophets who bowed (as in Islam) to God's Will, by the Rabbis and the Doctors of the Law."

No one disagrees with the translation of *rabbaniyyūn* as "rabbis." This is not the case for the term *aḥbār*, however. Al-Tabari (d. 923), for instance, along with the traditionists[76] who followed him, seems to be confused as to whether it refers to the same group of people as the *rabbaniyyūn* or not. He does note, however, that it must refer to those who possess knowledge (*ʿilm*). Though it is not possible to distinguish firmly the two terms today, argues Newby, we do have strong evidence that Arabian Rabbinic (he prefers "Rabbinite") Judaism helped shape many of the debates, already in the Qur'an,[77] but also in the *tafsīr* (commentary) literature in later centuries. Significantly, many of the key early exegetes of the Qur'an were converts from Judaism. Their extensive use of both Christian and Jewish scriptures, as well as their knowledge of the Jewish Arabian dialect (*al-yahūdiyya*) was officially sanctioned by Islam's prophet:

---

[76] This is the common term for those experts in the Sunna—the reports (*aḥadīth*) that were circulated about what the prophet Muhammad said and did. This work, which often involved a good deal of travel in order to substantiate certain reports, involved an important critical methodology: attempting to discern the authentic from the spurious reports, which in the end turned out to be far more numerous than the former. As a group, the traditionists (*ahl al-ḥadīth*) became the major group behind the movement that was to form the theological consensus undergirding Sunni thought.

[77] Newby correlates the qur'anic verse 5:64 ("The Jews said that the Hand of Allah is fettered; their hands are fettered and cursed be what they say…!") with two passages in extant rabbinic literature which mention this very concept. One of the more significant examples he offers is that of the extra son of Noah who is said in Q. 11:42-43 to have drowned in the flood. Centuries later 'Umar al-Zamakhshari quotes Qatada in saying that "the People of the Book do not disagree in the fact that he is his son" (quoted in Newby, *A History of the Jews of Arabia* [Columbia, SC: University of South Carolina Press, 1988], 65). Newby argues that this might parallel certain rabbinic texts which reflect on Noah's grandson Canaan's curse as if he (and not his father Ham) had actually committed the impropriety mentioned in Gen. 9:20-27. As scholars have shown with the Ethiopic text and other Jewish legends from more isolated regions, many midrashic texts never found their way into the Talmud. The Arabian Jewish community would certainly be a good example of this isolation, and, interestingly, the Qur'an remains virtually the only reflection of that community's thinking.

The companion, Abū Hurayrah, although illiterate, had extensive knowledge of the Torah, as did ʿAlī, Salmān al-Farīsī and, of course, the
"Ocean of *Tafsīr*," Ibn ʿAbbās, who is often called the "*ḥibr* [singular of
aḥbār]*al-ʾumma*," or "Rabbi of the [Muslim] community," on account of
his extensive knowledge of Judeo-Christian as well as Muslim Scripture
and commentary acquired in Arabia. Muḥammad, Abū Bakr, and ʿUmar
are reported to have made several trips to the Bet Midrash in Medina, and
Muḥammad's amanuensis, Zayd b. Thābit, who was so central in matters
Qurʾānic, is reported to have gone so far as to learn *al-yahūdiyya* in a Bet
Midrash at Muḥammad's behest in order to read Jewish material. More
to the point, converted Jews like ʿUbay b. Kaʿab [21/642] and Kaʿab
alʾAḥbār [or sometimes, *ḥibr*, the probable paradigm for Ibn ʿAbbas's
appellation], who converted to Islam under Abu Bakr, transmitted much
information originally derived from rabbinic tradition. Some of this
material can be found in the Talmud and the Midrashīm, but some of it is
preserved only in Islamic versions.[78]

My point here is not to harken back to the orientalist desire to show
how much material in the Qur'an is borrowed from outside sources, but
rather to point out that in contrast to the later centuries when polemics
between various "People of the Book" had led to opposing positions
cemented into place, the first few generations of Muslims felt freer to
consult and discuss theological issues with one another. No doubt there
are strongly polemical passages in the Qur'an itself, but one also finds
amazingly irenic ones such as the following: "And we did not send
before you any but men to whom We sent revelation, so ask the followers of the reminder [*ahl al-dhikr*] if you do not know" (Q. 21:7; see also
16:43 and 43:45). So, my first reading strategy in this work is to follow a
narrative hermeneutic, using the kind of midrashic approach common to
Bible, Torah and Qur'an interpretation that was assumed in the days of
Muhammad.[79]

---

[78] Ibid., 66.
[79] I owe this point to Newby (also the outside reader for my dissertation), who in a
personal email note (March 4, 2001), suggested that this was what I was in fact
doing. He states that all three faiths shared at the time a "midrashic" worldview, "in
which the text is regarded as an expression of 'truth,' and each chapter, verse, word,
or letter is reflective of that whole 'truth'... [O]ne could do what you have done here
and produce a narrative composed of all the verses you could adduce, but that would
not be understood as the only narrative you could or should construct, because there
would be a great number of parallel readings possible, some of them contradictory.
Indeed, multiple contradictory readings is one of the goals of the method, because it
is in the tension between the contradictions that more commentary is produced and
greater understanding comes."

The second response comes from the acknowledgment that the Qur'an's meaning cannot be apprehended apart from the understanding of the Muslim community. The bulk of this study will consist of reports of what qur'anic commentators have written over the centuries. In the next chapter I seek only to say what the text is saying—presenting qur'anic data in the barest possible form.[80] The attitude I bring to this project is best expressed in Cragg's own words:

> Our hope, in these pages, is more specifically "religious", using the term in no restrictive sense. It is to seek the event of the Qur'an with a steady awareness of centuries of devotion, of personal existence and collective possession, gathering around it through the long generations of Islam. For the Qur'an as event in the seventh century has become an unbroken history of patient, jealous, tenacious recollection and religious existence. As such it is a document of faith looking for a faith to receive it. The scholar must be alive to it as such.[81]

My third response has to do with the desired outcome of this research, its very *raison d'être*. It is to provide incentive, meaning and substance to Muslim–Christian cooperation in projects of social justice. Yet, contrary to common wisdom, cooperation can be strengthened by common theologizing—hence the focus on the common ground between Bible and Qur'an, which, in fact, is much greater than the disagreements.[82] And accordingly, the atmosphere of this study is not polemic in nature, but rather irenic and sympathetic—much more conducive, hopefully, to mutual understanding and concerted action in relieving the suffering and upholding the dignity of our fellow human beings.

---

[80] Nobody can claim objectivity, especially on the heels of the last chapter. Nevertheless, I do consult a variety of scholars already in this chapter, most of whom are Western islamicists. Yet I do so mainly to present the data and the main questions that have been raised. The next three chapters bring the rich intra-Islamic conversation to the fore.

[81] Cragg, *The Event of the Qur'an*, 20.

[82] I am saying that the verses in the Qur'an that we as Christians would disagree with are actually very few in number, but by that I am not stating that the disagreements are unimportant. For a more recent discussion of these Muslim–Christian debates, see my Review Essay, "Advancing Muslim–Christian Dialogue Today," *American Journal of Islamic Social Sciences* 25.4 (2008): 108-15. One of the three books reviewed in that essay is Mahmoud Ayoub, *A Muslim View of Christianity: Essays on Dialogue*, ed. Irfan A. Omar (Maryknoll, NY: Orbis, 2007). Ayoub's work is especially significant as he is a Lebanese Shi'ite scholar of Islam with decades of dialogue experience and an intimate knowledge of Christianity.

Part II

HUMAN TRUSTEESHIP:
MUSLIM AND CHRISTIAN INTERPRETATIONS

Chapter 6

ADAM AS GOD'S *KHALĪFA*: QUR'ANIC DATA

This chapter explores the use of *khalīfa* as applied to Adam in the Qur'an, and, by extension, how it may apply to all his ancestors, the human race. Leaving aside for a time some of the hermeneutical insights of Chapters 4 and 5, I simply look at the qur'anic data and list the passages in which this word appears. I mostly leave the work of classical qur'anic commentators (*mufassirūn*) for the next two chapters. At first glance, this seems to be quite easy: there is only one verse in which *khalīfa* in the singular is applied to Adam (sura 2, v. 30—from now on, Q. 2:30), and another in which it refers to David. My focus is on Adam, and therefore I will look at the contexts that contain material about creation, Adam's disobedience, and his subsequent calling to make his home on this planet. This will also call for a brief look at some parallel passages.

The next step is to examine each of the seven instances in which the term is used in the plural. Once this is done, it will be helpful to examine more carefully the possible meanings of the root *kh-l-f*. In turn, this will lead to a brief overview of ten other passages in which derivatives of *kh-l-f* are employed, in order to understand and better evaluate the "unitary" argument put forward by German scholar, Rudi Paret.

Finally, one more passage will have to be examined (Q. 33:72), in which God offers the "Trust" (*amāna*) to the heavens and the earth and they refuse to bear it; yet humanity accepts the challenge. This verse must be included in the discussion for two main reasons: first, it is potentially the only direct parallel to Q. 2:30, in its application to the doctrine of humanity; second, it is invariably quoted in this context in contemporary literature, as will be seen later.

The Foundational Passage: Q. 2:30

Q. 2:30 reads, *Wa-idh qāla rabbuka li-l-malāʾikati ʾinnī jāʿilun fī l-ʾarḍi khalīfatan*, "When your Lord said to the angels, 'I am setting on earth a vice-regent.'"[1] Other translations are similar, but already the nuances are indicative of the difficulty in rendering the word *khalīfa*: "Behold, thy Lord said to the angels: 'I will create a vicegerent on earth'" (Yusuf Ali). Rashad Khalifa's translation is almost a paraphrase: "I am placing a representative (*a temporary god*) on earth."[2] Then one of the oldest translations in English, by Sale: "I am going to place a substitute on earth."[3] A. J. Arberry, the only other non-Muslim I am quoting, renders it like this: "I am setting in the earth a viceroy."[4] Marmaduke Pickthall too translates *khalīfa* as "viceroy."[5] Finally, here are four more translations by Muslims: "I am placing an overlord on earth" (T. B. Irving, or Al Hajj Ta'lim 'Ali);[6] "I am going to place a representative on earth" (Sheikh Fadhlallah Haeri);[7] Zafrulla Khan (as Yusuf Ali) has "vicegerent" for *khalīfa*,[8] and Ahmed Ali has "trustee."[9]

The context clearly presents Adam as the recipient of this title of *khalīfa*. Despite the differences, what all these translations have in common is the loftiness of the position.[10] God is addressing the angels and, four verses later, he asks them to bow down before Adam. The weakest rendering is, perhaps, "representative," yet Rashad Khalifa in his translation inserts an explanatory parenthesis, "a temporary god." Clearly, even this word "representative" is very strong, since it entails that Adam is God's ambassador on earth, a god (with a small "g").

---

[1] *The Bounteous Koran: A Translation of Meaning and Commentary by Dr. M. M. Khatib* (London: Macmillan, 1986), 7.

[2] Rashad Khalifa, trans., *Quran: The Final Testament* (Tucson, AZ: Islamic Productions, 1989), 6.

[3] E. M. Wherry, *A Comprehensive Commentary of the Quran, Comprising Sale's Translation and Preliminary Discourse, with Additional Notes and Emendations* (London: Kegan Paul, Trench, Trubner & Co., 1896), 300.

[4] A. J. Arberry, trans., *The Koran Interpreted* (London: George Allen & Unwin, 1955), 33.

[5] *The Glorious Koran*, 6.

[6] T. B. Irving, trans., *The Qur'an (Al-Hajj Ta'lim Ali)* (Brattleboro, VT: Amana Books, 1991), 4.

[7] Shaykh Fadhlallah Haeri, *The Cow: A Commentary on Chapter 2: Surat Al-Baqarah* (Reading: Garnet, 1993), 29.

[8] Zafrulla Khan, trans, *The Quran* (London: Curzon Press, 1981), 9.

[9] *Al-Qur'an: A Contemporary Translation by Ahmed Ali* (Princeton, NJ: Princeton University Press, 1984), 14.

[10] It is to be noted that none of these contemporaries translates *khalīfa* as "successor," as I point out below.

The Context

Almost all the translations set apart this verse as the beginning of a new paragraph. It clearly begins a new section on the theme of the creation of Adam. Here is the full text (vv. 30-37), according to Yusuf Ali's popular translation:[11]

> 30. Behold, thy Lord said to the angels: "I will create a vicegerent on earth." They said: "Wilt Thou place therein one who will make mischief therein and shed blood? Whilst we do celebrate Thy praises and glorify Thy holy (name)?" He said: "I know what ye know not."
> 31. And He taught Adam the nature of all things; then He placed them before the angels, and said: "Tell me the nature of these if ye are right."
> 32. They said: "Glory to Thee: of knowledge we have none, save what Thou hast taught us: in truth it is Thou who art perfect in knowledge and wisdom."
> 33. He said: "O Adam! tell them their natures." When he had told them, God said: "Did I not tell you that I know the secrets of heaven and earth, and I know what ye reveal and what ye conceal?"
> 34. And behold, We said to the angels: "Bow down to Adam:" and they bowed down: not so Iblis: he refused and was haughty: he was of those who reject Faith.
> 35. We said: "O Adam! dwell thou and thy wife in the Garden; And eat of the bountiful things therein as (where and when) ye will; but approach not this tree, or ye run into harm and transgression."
> 36. Then did Satan make them slip from the (Garden), and get them out of the state (of felicity) in which they had been. We said: "Get ye down, all (ye people), with enmity between yourselves. On earth will be your dwelling place and your means of livelihood for a time."
> 37. Then learnt Adam from his Lord words of inspiration, and his Lord turned toward him; for He is Oft-Returning, Most Merciful.

I will briefly make seven points on this passage, as they help to illuminate the meaning of Adam's call as God's *khalīfa*. The first three, interestingly, have no parallel in the rest of the Qur'an.

God's Announcement to the Angels. The angels seem to object to God's announcing humanity's creation and their placement on the earth.[12]

---

[11] As mentioned in the previous chapter, I will be using Yusuf Ali's translation, unless indicated otherwise. He closely follows the Cairo versification of the Qur'an (though not exactly).

[12] Ali has a long footnote on this verse, which is entirely devoted to the contrast between the angels and "man" (*The Holy Qur'an*, 24). They are "holy and pure" and without passion or emotion": "We may suppose the angels had no independent wills of their own: Their perfection in other ways reflected God's perfection but could not raise them to the dignity of vicegerency."

However, since the angels in the Qur'an do not seem to have free will (at least, according to the prevailing view since the third Islamic century),[13] it is unlikely that they are expressing even a twinge of rebellion in this statement.[14] It is more probable that they are genuinely surprised and spontaneously asking a sincere question.

God Answers the Angels' Question. God's answer to the angels' question is instructive. He does not deny the truth that it implies, namely, that human beings will bring "mischief and bloodshed" to earth. Rather, he knows more about the total implication of the decision he has made. Yusuf Ali addresses this issue in a plausible way:

> The angels in their one-sidedness saw only the mischief consequent on the misuse of the emotional nature by man; perhaps they also, being without emotions, did not understand the whole of God's nature, which gives and asks for love. In humility and true devotion to God, they remonstrate: we must not imagine the least tinge of jealousy, as they are without emotion. This mystery of love being above them, they are told that they do not know, and they acknowledge (in 2:32 below) not their fault (for there is no question of fault) but their imperfection of knowledge. At the same time, the matter is brought home to them when the actual capacities of man are shown to them (2:31, 33).[15]

[13] I am indebted here to Newby who, in the same correspondence mentioned above (March 4, 2001), informs me that the question of free will and determinism only starts with the controversies between the Muʿtazilites and the Ashʿarites theological positions. Before that time, the *muḥaddithūn* and the *mufassirūn* shared the "mid-rashic" worldview of Arabian Jews and many of the Christian interpretive strategies in relation to their scriptures. One of those common assumptions revolved around the identity of Satan and the origin of evil in general. This is usually called the "War in Heaven" tradition, according to which a faction of angels follows Satan in his rebellion against God and thus create the distinction between heaven and hell. In this perspective all angels have free will, and not only the leader of the rebellion.

[14] As Ali notes, Iblīs seems to be one of the angels in this passage. Yet he obviously is endowed with freedom of choice, as this episode illustrates. Elsewhere in the Qur'an, Iblīs is one of the *jinn* (e.g. Q. 18:50), and thus his appellation, *al-shayṭān* (e.g. Q. 36:60). By associating him more with the angels, Ali represents the general Muslim consensus today on this issue. Iblīs, as Arthur Jeffery pointed out, comes from the Greek diabolos (*Foreign Vocabulary in the Qur'an* [Baroda: Oriental Institute, 1938]). Angelika Neuwirth states that this "double affiliation" is parallel to the earlier Gnostic writings in which "both groups, demons and angels, are closely related" ("Cosmology," in the *Encyclopaedia of the Qur'an* [*EQ* from now on], ed. Jane Dammen McAuliffe, vol. 1:447). At the same time, these two species are sharply differentiated in the Qur'an when it comes to revelation. Contrary to what his opponents allege, Muhammad was not inspired by a *jinn* (and thus *majnūn*), as poets were said to be in contemporary Arabian culture (e.g. Q. 69:41).

[15] *The Holy Qur'an*, 24.

Following Ali, it would seem to me that part of Adam's calling as trustee is his ability freely to respond to God and shoulder the responsibility of caring for the created order. Quite obviously, there was a great risk involved and, in light of v. 36, where we see Adam and Eve expelled from the heavenly Garden, it is easy to see why the angels asked this question.[16] Logically, the announcement to the angels would have followed Adam and Eve's disobedience. Presumably, if they had not eaten from the forbidden tree they would never have been banished from Paradise and so would not have come to earth. So the trusteeship of man would be God's way of redeeming a less than perfect situation. This may also help to explain the angels' perplexity.[17]

Adam is Taught All the Names. God teaches Adam—literally—"all the names" (*al-asmā'a kullahā*). Yusuf Ali renders it "the nature of all things"; Zafrulla Khan, "the names of all his attributes"; Rashid Khalifa (*Quran: The Final Testament*) and M. Khatib (*The Bounteous Koran*) translate the phrase literally,[18] but Khalifa has an explanatory note: "It is said that these 'names' were the knowledge of sciences, the nature of things and names of different creatures."[19] Ahmed Ali offers a longer paraphrase: "Then He gave Adam knowledge of the nature of reality of all things and every thing."[20]

What is clear, amid the variety of interpretations, is that this knowledge or ability, once it was demonstrated before the angels, left them speechless with admiration and wonder. When Adam had told them the

---

[16] Cragg makes this point rather emphatically in *The Mind of the Qur'an*: "The very theme of history is thus the question-mark of human worth, understood as a vital question-mark of divine wisdom and power. The wisdom of God is staked on the credibility of man as its supreme test and venture" (p. 142).

[17] Shaykh Haeri represents the majority opinion today when he comments, "In the case of man, Allah's representative, the angels wondered at Allah's decree to place Adam on earth. Although angels are sanctified, and at the same time limited, man is not sanctified, but has limitless potential, for he reflects the boundless might and power of Allah. It was the knowledge of man's potential for creativity and destructiveness, obedience and disobedience, belief and denial, which caused the angels surprise" (*The Cow*, 29). But I have not yet found anyone to comment on the (apparent?) paradox of Adam being rewarded for having sinned. Certainly, it is God's mercy that, having sinned and caused his own banishment from the Garden, he is given a mission. Yet this in contrast to the biblical narrative, in which humanity is empowered before the fall. There, his mission is more a part of his intrinsic purpose and destiny.

[18] *The Bounteous Koran*, 7.

[19] Khalifa, trans., *Quran: The Final Testament*, 8.

[20] *Al-Qur'an*, 15.

names, they fell to their knees and exclaimed: "Glory to Thee: of knowledge we have none, save what Thou hast taught us: in truth it is Thou
who art perfect in knowledge and wisdom." It is highly probable that
Adam's gift is directly related to his being called *khalīfa*. The exact
connection is variously debated in the commentaries, but a connection, at
least for myself, there must be.

Angels in Prostration before Adam. The fourth point brings me to the
issues that are taken up elsewhere in the Qur'an. The first is that God
commands the angels to bow down to Adam (see also Q. 7:11; 15:29;
18:50; 20:116; 38:72). The root of the word *sajada* is used in some form
every time, implying that this is not a mere salute or polite bow. This
root is always used for worshiping God in the Qur'an. One derivative,
for example, serves to denote a mosque—a place where Muslims prostrate themselves before God (*masjid*). As will be clear from the commentaries, this is a difficult passage to interpret for the adherents of a
religion that is so strenuously focused on *tawḥīd*, the unicity of God.
This has no direct parallel in the Bible, yet I will argue in Chapter 8 that
it could well be related to humanity's being created in God's image.[21] In
the immediate context, it is connected to Adam's ability to inform the
angels about "the names." Indeed, humanity is the pinnacle of God's
creation, in the words of Pakistani writer Muhammad Zubair Farooqi:

> Man is the apex of Divine creation. For his physical survival as well as
> for enabling him to fulfil the assigned role of Allah's representative on
> earth he has been endowed with a unique blend of physical, mental and
> intuitive faculties which Allah has not given to any of His other creatures.
> In their life on earth, human beings apply these faculties to acquire
> knowledge about themselves and their environment. In this way they
> endeavour to achieve self-awareness and know the meaning and purpose
> of human life on earth.[22]

---

[21] Yusuf Ali hints at this possibility when he comments on humankind's gift of
moral freedom: "The power of will or choosing would have to go with them, in
order that man might steer his own bark. This power of will (when used aright) gave
him to some extent a mastery over his own fortunes and over nature, thus bringing
him nearer to the God-like nature, which has supreme mastery and will" (*The Holy
Qur'an*, 24). There is an indirect parallel in the Bible, where we read in the Epistle to
the Hebrews: "And again, when God brings his firstborn into the world, he says, 'Let
all God's angels worship him' " (1:6). Also, the Apostle Paul writes to the believers
of Corinth: "Do you not know that we will judge angels?" (1 Cor. 6:3).

[22] Muhammad Zubair Farooqi, *Islam, The Muslim World-Community and Challenges of the Modern Age* (Islamabad: Isti'ara, 1997), 122.

Again, the angels bow to Adam in obedience to God's command—a command which in the narrative flows from his introduction of the new creature as his *khalīfa*, and then from Adam's esoteric knowledge of "all the names."[23] Gisela Webb rightly points out the centrality of "man's vicegerency" in this narrative, but as we will see in the next chapter, commentators did not all agree on humanity's superiority over the angels. Also, when she affirms that "[a] traditional reading of the narrative" attributes Adam's superiority to his calling to make moral choices (whereas angels exhibit perfect obedience), she is likely projecting a modern reading onto the classical period.[24] As we shall see, the early commentators were uncomfortable with Adam being named God's *khalīfa* and as a result tended to avoid a theological reading.[25] Another cause for caution on their part is that Adam's deputyship disappears from the other six parallel creation passages. The closest parallel, however, is possibly a significant one. In Q. 7:10 we read: "It is We who have placed you with *authority* on earth, and have provided you therein with means for the fulfillment of your life: small are the thanks that ye give!" The emphasis on Ali's use of "authority" is mine, pointing out the obvious correspondence with Q. 2:30. This is a warranted addition, according to the dictionaries. Literally, one might translate "We have established you in the earth with authority and power" (*wa laqad makannākum fi l-ard*).[26] Is Yusuf Ali reading into the text? I think not, especially in light of the

---

[23] Mark Hillmer believes that the great qur'anic emphasis on this theme (six different passages) is more related to the fall of Satan motif in the Qur'an. The same story "is found in the Christian legend of The Life of Adam and Eve, datable between A.D. 60 and 300" ("The Book of Genesis in the Qur'an," *Word and the World* 14 [1994]: 201).

[24] "Angel," *EQ*, vol. 1:86.

[25] See Cornelia Schöck's article, "Adam and Eve," *EQ*, vol. 1:22-26). The Qur'an is replete with moral injunctions. Two passages in fact (early Medinan) connect creation and God's testing of humankind, in order to hold them accountable for their works (Q. 11:7; 67:2; cf. Thomas J. O'Shaughnessy, S.J., *Creation and the Teaching of the Qur'an* [Rome: Biblical Institute Press, 1985], 61-62). But this was not explicitly linked to humankind's calling as God's on earth until the modern period.

[26] In Abdallah al-Bustani's dictionary (*Al-Bustān: Muʿajam Lughawī Muṭawwal* [Beirut: Maktabat al-Lubnān, 1992]), the expression *makannahu min shayʾ* should be translated, "to put someone in charge of something with authority and power" (p. 1046). According to Hans Wehr, the root means "to be or to become strong; to become influential, gain influence, have influence, have power" (J. Milton Cowan, ed., *The Hans Wehr Dictionary of Modern Written Arabic* [Ithaca, NY: Spoken Language Services, 1994], 1076). Here we have the second form with a causative sense: "to make strong…to firmly establish." If anything, Arberry's translation is a little weak: "We have established you in the earth and there appointed for you livelihood."

overall theological framework of the Qur'an. But I will add: this is a thoroughly modern view.

Iblīs Refuses to Bow before Adam. Iblīs, or Satan, refuses to bow down. Elsewhere God asks him why he is so obstinate and insolent. His answer has two parts. First, he will not bow down to someone made of clay (Q. 7:12; 15:33; 17:61; 38:76). And then, in two passages he answers, "I am better than he: Thou didst create me from fire and him from clay" (Q. 7:12; 38:76).[27] Iblīs's arrogance is only matched by his willingness to disobey God. Perhaps this is why he is elsewhere identified as one of the *jinn*, those made of fire, and not of light as the angels. Yet, however one might try to explicate his "double affiliation" (Neuwirth, cf. n. 14), he plays a central role in all of the creation passages.

Adam and Eve's Fall from the Garden. For both Qur'an and Bible, the Garden of Eden is a place of innocence and bliss. But whereas the Genesis account locates it on earth, the qur'anic garden is a heavenly one. Jean Louis Déclais, in a fascinating article, "La Tenue d'Adam," contrasts the biblical description of Adam and Eve's disobedience in which they discover that they are naked and that of the Qur'an:

> In the Qur'an, the Garden is equally a place of happiness. But in thinking about a paradise-like happiness, one does not imagine a nudity of innocence; to the contrary, one sees oneself in the shade, well dressed and well satisfied (20,118-119). *After* having given in to Satan's temptation and tasted of the tree allegedly promising immortality (20,120; 7,20), the human couple find themselves stripped of their clothes (20,121; 7,22); and from 7,20 on, the Qur'an informs its listeners that this was indeed Satan's objective (Satan tempted them so that…), but of this Adam and his wife were ignorant; they had only heard Satan offer them a path to immortality and assure them that he was a wise counselor. As in the Bible, they have only one recourse left—to protect their offended modesty: to sew on themselves leaves from the garden.[28]

---

[27] O'Shaughnessy points out that the long sura 38 passage on creation (71-85) is from the late second Meccan period (cf. Gerhard Böwering, "Chronology and the Qur'an," *EQ*, vol. 1:316-35). The use of the three terms, *bashar* (humankind), *ṭīn* (clay), and "I created (*bashar*) with My own hands," all three point directly to a Christian Syriac origin. However one might interpret the relationship between the Qur'an and similar material in prior religious sources in the same region (according to one's theological *a prioris*), *bashar* and *ṭīn* are close cognates with Syriac words and the Syrian theologian Aphraates wrote in the fourth century about God creating man "with His own hands," in contradistinction to his other creatures (*Creation and the Teaching of the Qur'an*, 17.

[28] J. L. Déclais, "La tenue d'Adam," *Arabica* 46 (1998): 111-12, emphasis original.

He then quotes Ibn Kathir (d. 774/1373), who finds this difference puzzling, particularly when Wahb b. Munabbih[29] had written, "Their clothes were made of the light that covered their sexual parts." Déclais goes on to show that the theme of Adam's original robe of light (sometimes called "robe of glory") was a common rabbinic theme, appearing early in the Palestinian Targum.[30] In fact, a consensus emerged on the material from which God wove Adam and Eve's original wardrobe: human nail. This clothing began to rip apart after the couple sinned, revealing their nakedness for the first time. Yet, after they repented, God restored their initial glorious state. According to the Palestinian Targum, "Yahweh Elohim made for Adam and his wife clothing of glory from the skin of the serpent that had been taken from him and covered their skin with it, replacing the splendid garments of which they had been stripped."[31] Adam's superb, God-crafted costume is then passed on from generation to generation and appears, among other places, as Esau's outfit, which Rebecca gave to Jacob in order to fool his father, and as Joseph's coat of many colors. Tabari apparently knows this tradition, as he reports that Adam's first clothes were made of nail and that they were torn when the couple disobeyed God's orders. The difference in his narrative is that they sewed a new outfit for themselves from the leaves of the garden, as the Qur'an indicates (Q. 20:121).[32]

As to the nature of Adam and Eve's[33] disobedience, the Qur'an is clear: "approach not this tree."[34] Satan led them to disobey the order, with the

---

[29] He was from the generation following Muhammad (the Followers) and was known as one of the most knowledgeable sources on the Talmud and Jewish oral traditions (*Midrashīm*). Many of the links (*isnād*) traced back from various hadiths end with him (Newby, *A History of the Jews of Arabia*, 66-67).

[30] Déclais, "La tenue d'Adam," 114. Déclais believes that the Rabbinic interpretation of Gen. 3 (in line with the more optimistic vision of Ps. 8), in which the first couple saw their original clothing of light restored, was formulated in the matrix of Jewish-Christian polemics (ibid., 117).

[31] Ibid. The reference given by Déclais is: R. Le Déaut, *Targum du Pentateuque, tome 1 Genèse*, Sources Chrétiennes 245 (Paris, 1978), 97.

[32] Déclais quotes from Tabari's *History of the Nations and Kings* (*Tārīkh al-uman wa-l-mulūk* [Beirut: Dār al-Kutub al-ʿIlmiyya, 1987], vol. 1:71). Later Tabari quotes a tradition traced back to Ibn 'Abbas: "The tree that God forbade to Adam and his wife was wheat. When they had eaten of it, their sexual organs appeared to them. What had previously hidden them from their sight was their nails" (p. 83).

[33] Though not mentioned directly, she is very present in the text. In v. 35 we read "thou and thy wife" and in the next verse both verbs have a dual pronominal suffix ("he caused the two to slip...and got the two out...").

[34] No more precision as to the nature of the tree is given anywhere else in the Qur'an. Yusuf Ali comments: "The forbidden tree was not the tree of knowledge, for man

consequence that they were cast down to earth. Yusuf Ali calls this the
"Fall," but only in the sense that by sinning they had disqualified themselves from living in a place with no sin. And thus, they were sent down
to earth.

The Qur'an has no concept of "original sin"[35]—every human being
comes into the world free, innocent and pure. Hence, the original pattern
has not been altered by this one slip: "So set thou thy face steadily and
truly to the Faith: (Establish) God's handiwork according to the pattern
[*fiṭra*] on which He has made mankind: no change (let there be) in the
work (wrought) by God: that is the standard religion: but most among
mankind understand not" (Q. 30:30). There is, nevertheless, the beginning of enmity between humans (literally, "get down as an enemy one to
the other") and, as in other passages, Satan vows to stand in ambush and
attack Adam's descendants in order to make them swerve from the
straight path (Q. 7:16-17; 15:39; 17:62; 38:82).[36] Nor is the earth cursed
in any way, as it is in Genesis. The earth is to be their "dwelling place"
and their "means of livelihood." This is an important point, which will
come up again, with regard to the ubiquitous "signs" of God to be read in
nature and history by the faithful.

But here the question must be asked as to the relationship between v.
30 ("I am placing a *khalīfa* on the earth") and v. 36 ("Satan, however,
brought about their fall, causing them to be ousted from their state of life.
For We said: 'Go out from here in mutual enmity. Out in the earth is a
habitation and needful provision for a season'").[37] God's announcement

was given in that perfect state fuller knowledge than he has now (2:31): it was the
tree of Evil, which he was forbidden not only to eat of, but even to approach" (*The
Holy Qur'an*, 25).

[35] It must be admitted, however, that because Adam and Eve sinned, all their descendants have been refused the chance to live in *janna*, or the heavenly Garden. On one
level, Christians do not see any more than that in the Fall: through Adam sin came
into the world and, as matter of fact, all have sinned. But we are not held accountable for Adam's sin. Of course, on a deeper level, this makes all the difference. In
the Bible, the fact that Adam and Eve are chased from the Garden means that their
nature is changed: they have lost the capacity not to sin—precisely that generalized
state of condemnation which in the Bible calls for God's drastic remedy, redemption
through Christ's sacrifice.

[36] The parallel passages in sura 7, 15 and 38 all have a dialogue between God and
Iblīs in which the latter pleads with God to give him a delay for his final condemnation until the Day of Judgment. God grants him this respite. Satan then vows to
assault and corrupt humans until that day. God says that he may, but that he will
have no authority over the faithful. The Qur'an makes it clear that those who follow
Iblīs do so freely and thus justly bear their guilt (e.g. 17:63).

[37] Cragg, *Readings from the Qur'an* (Brighton: Sussex Academic Press, 1999), 96.

to the angels seems more glorious and lofty—at least as far as human-kind is concerned. Six verses later Adam and Eve come to earth because they have forfeited their right to stay in the Garden of bliss.[38] This point is not missed by the commentators, but I think it is worth pointing out the fact that even if in the end both Bible and Qur'an see humans as entrusted by God with a mission of trusteeship on earth, in the Qur'an this endowment follows the Fall whereas in the Bible it precedes it (Gen. 1:26-30).[39]

God Forgives Adam. The last point of this passage is God's response to the Fall: he teaches Adam "words" (Ali adds "of inspiration"), which certainly corresponds to the "guidance" Adam is offered in the many parallel passages (e.g. Q. 20:122).[40] The text says: "and his Lord turned toward him; for He is Oft-Returning, Most Merciful." The word translated "turned toward" is the same verb that means, on the human side, "to repent." When applied to God, it means "to restore to His grace, forgive."[41] That this forgiveness is God's answer to Adam and Eve's repentant prayer is clear from Q. 7:23: "They said: 'Our Lord! We have wronged our own souls: if Thou forgive us not and bestow not upon us Thy mercy, we shall certainly be lost.'"

---

[38] I find it interesting to compare Ali and Cragg on this expression. Both, in essence, are paraphrasing, because the Arabic literally says "and he caused both of them to go out of where they were in." Ali even has a footnote explaining "(of felicity)": "bliss, delight." He is obviously supplying this meaning from other passages. Cragg, on the other hand, by translating "the state of life," might have tipped his hand more than he should have. True, Adam and Eve would have lived forever in the Garden whereas now they will know death on earth. Yet only in the Bible are there two trees in the middle of the Garden, the tree of life and the tree of the knowledge of good and evil. Also the penalty for eating the fruit, as Eve recounts to the serpent, is death. This is brought up by Satan himself in the Qur'an: "Your Lord only forbade you this tree, lest ye should become angels or such beings as live forever" (7:20). However, this is not the main emphasis of the qur'anic account.

[39] To my knowledge, no qur'anic commentator would agree with me on this point. They see this passage as basically chronological. Yet God says he will "place" a *khalīfa* on earth, not "create" him. Thus if Adam had not disobeyed he would have stayed forever in the Garden. Further, this explains why the angels question the wisdom of a disobedient *khalīfa*.

[40] The very next verse (38) makes this idea explicit: "We said: 'Go down from here, all of you. Guidance will come to you from Me. There will be no fear for any who follow My guidance, nor any reason to grieve.'" This translation is from Cragg (*Readings from the Qur'an*, 96) and he rightly includes this verse in the above passage.

[41] Cowan, ed., *The Hans Wehr Dictionary*, 119.

The Other Singular Instance: Q. 38:26

Q. 38:26 reads: "O David! We did indeed make thee a vicegerent on earth: so judge thou between men in truth (and justice): nor follow thou the lusts (of thy heart), for they will mislead thee from the Path of God: for those who wander astray from the Path of God, is a Penalty Grievous, for that they forget the Day of Account." The context speaks of David, the psalmist (vv. 18, 19) and king (v. 20). In v. 17 his qualities are summarized as "the man of strength: for he ever turned (to God)," or, as Arberry has it, "the man of might; he was a penitent."[42] And then in v. 20 we read how God gifted him as a king: "We strengthened his kingdom, and gave him wisdom and sound judgment in speech and decision."

The one feature that this passage adds to our foundational verse on Adam and *khalīfa* is that, apparently, it clarifies the content of what is meant by *khalīfa*: "judging between men in truth and justice" and following God's path scrupulously. This does seem to imply that as God's vicar on earth David was to use his kingly authority wisely and reverently.[43] To what extent does the word *khalīfa* imply authority? This is the question I will now begin to examine, starting with the lexicons and dictionaries.

The Arabic Root KH-L-F

The first form of the verb *khalafa* means "to be the successor, to succeed; to follow, come after; to take the place of, substitute; to replace; to lag behind; to stay behind."[44] The following is a common expression in the first centuries of Islamic literature, if one had just lost a father or mother or someone who could not be replaced: *khalafa allāhu ʿalayka khayran*, "May God supply to thee well the place of him whom thou hast lost."[45] John Penrice, who is specifically looking at the qur'anic meaning, writes: "to be behind, come after; to succeed; to do a thing behind one's back; to act as a deputy."[46] The verbal noun (*maṣdar*) *khalfun*, accordingly, means

---

[42] A. J. Arberry, trans., *The Koran Interpreted*, 159.

[43] Cragg's translation is characteristically thoughtful and well crafted: "David, We have appointed you as a viceroy in the earth. Therefore, judge rightly between men. Do not give way to passion, lest it divert you from the path of God. For those who stray from God's path, forgetting the Day of reckoning, incur a stern retribution" (*Readings from the Qur'an*, 157).

[44] Cowan, ed., *The Hans Wehr Dictionary*, 297.

[45] Edward William Lane, *Arabic–English Lexicon* (New York: Frederick Ungar, 1955), vol. 3:792.

[46] John Penrice, *Dictionary and Glossary of the Kor-an* (Beirut: Lebanon Book Shop, n.d.), 44.

"back, rear; successors," and the preposition *khalfa* means "behind, after, in the rear of."[47]

The second verb form *khallafa*, as might be expected in common Arabic usage, means "to appoint as successor; to leave behind; to have descendants, have offspring."[48] In the same way, the tenth form (*istakhlafa*) means "to make a successor, cause to succeed."[49]

The original form of *khalīfatun* is *khalīfun*. The *tā marbouṭa* "is to denote intensiveness of signification."[50] E. W. Lane's definition is worth quoting: "a successor; a vice-agent, vicegerent, lieutenant, substitute, proxy, or deputy; one who has been made, or appointed to take the place of him who has been before him."[51] Penrice has a similar definition: "a successor, lieutenant, vicar." In the Classical Arabic dictionary *Lisān al-ʿArab*, surprisingly little is said about the word *khalīfa*,[52] and the definition is succinct: "one who is appointed in the place of someone before him."[53] Lane had just defined the use of the tenth form (*istaklafa*) thus: "to put someone in someone else's place." As we shall see, the idea of succession, when it comes to political settings, carries with it the idea of rulership and authority.[54] More surprising still, though other verses are cited, the Q. 2:30 passage is not.

Lane is again very helpful in bringing to light the idiomatic uses of *khalīfa* in the writings of the early Muslims. For instance, *khalafahu fī*

[47] Cowan, ed., *The Hans Wehr Dictionary*, 298.

[48] Ibid., 297.

[49] Ibid.

[50] Lane, *Arabic–English Lexicon*, 798.

[51] Ibid.

[52] The section on the root *kh-l-f* is ten pages long (pp. 1234-43) and the discussion pertaining to *khalīfa/khilāfa* fills only half a page.

[53] *Alladhī yustakhlafu mimman qablahu*, 6 vols. (Muḥammad b. Mukarram Ibn Manẓūr, Lisān al-ʿArab, Cairo: Dār al-Maʿārif, 1981), vol. 2:1235.

[54] Ibn Manẓūr quotes the lexicographer and philologist Ibn Sida (d. 455/1066) regarding the David verse (Q. 38:16): "the greatest authority" (al-sulṭān al-aʿẓam)—no doubt the historical caliphate. Then, as he quotes the second-century Kufan grammarian al-Farra', the discussion becomes even more political. Commenting on the qur'anic expression, "God made you his deputies (*khalāʾif*) on earth" (Q. 6:165; 10:14; 35:39), al-Farra' writes, "He made Muhammad's umma the successors of all the nations." Further, according to philologist al-Sikkit (d. 244/858), there are only three successors/deputies (*khalāʾif*)—yet no explanation is given. Already in the middle of the second century the debates between the partisans of Ali (the Shi'ites) and the followers of traditionist Abu Hanifa (a movement that was later to be called "Sunnism") were heating up. The three caliphs here would have to be the first three of the "rightly guided caliphs," and thus excluding the fourth, 'Uthman. The word *khalīfa* was indeed a theological and political minefield.

ahlihi would mean "He was, or became, his *khalīfa* (ie, his successor, or vice-agent, etc.) among, or in respect of his family." The same can be said for the verbal infinitive *khilāfa* (succession or deputyship). Thus, according to Lane, *awᶜā lahu bil-khilāfati ᶜalā ahlihi wa-mālihi* would best be translated, "he charged him by his will with the being his successor, or vice-agent over his family and property."[55]

The noun *khalīfa* has two plurals in the Qur'an: *khalāʾif* and *khulafāʾ*. The former, in Lane's view, is "generally applied to any people that have succeeded others, and supplied their places," whereas the latter generally applies to "the successors of the Prophet."[56] Lane does note, however, that, "[s]ome say that application of the title *khalīfat-Allāh* (The Vice-gerent of God) is not allowable, except to Adam and David because there is express authority in these instances."[57] What he means is that these are the only two instances in the Qur'an. This question did arise, however, with the first dynasty of caliphs (*khulafāʾ*), the Umayyads in Damascus, who took for themselves the title *khalīfat-Allāh*.[58] This was contested by the opponents of the regime, the Shi'i and the Kharijites.[59] But also during this period the term *khalīfa* came to mean any lieutenant of a general or governor.[60]

W. Montgomery Watt notes that even before Islam *khalīfa* was commonly used to signify a "deputy" or "viceregent." He points to an inscription in South Arabia dating back to about 543 CE in which the

---

[55] Lane, *Arabic–English Lexicon*, 792.

[56] Ibid., 798.

[57] Ibid.

[58] D. Sourdel, "Khalīfa: (1) The History of the Institution of the Caliphate," in *The Encyclopaedia of Islam, new ed.* (*EI¹*), ed. E. van Donzel et al. (Leiden: Brill, 1978), vol. 4:938.

[59] T. W. Arnold notes that "Muslim historians commonly assert that it was first so used by Abu Bakr (Muhammad's first successor); it is doubtful, however, whether he ever assumed it as a title" ("Khalīfa," in *E. J. Brill's First Encyclopaedia of Islam* [*EI²*], ed. M. T. Houtsma et al. [Leiden: Brill, 1987], 881). There is good evidence that the first four caliphs ("the Rightly Guided Caliphs") did use the title *khalīfat rasūl Allāh* ("successor of the apostle of God") but that the title *khalīfat Allāh* excited Abu Bakr's indignation. But as I argue in the next chapter, much Muslim historiography of this period is colored by the intense political struggles between the partisans of election (*bāyᶜa*), the Sunnis, the partisans of Ali (the Shi'ites) and the partisans of *sharīᶜa* over political power (the Kharajites). One of the consequences is that the doctrine of humanity as God's *khulafāʾ* on earth has not been clearly developed by Muslims until the last one hundred and fifty years.

[60] R. Dozy, *Supplément aux Dictionnaires Arabes*, vol. 1 (Paris: Maisonneuve Frères, 1927), 397.

corresponding dialect word for *khalīfa* is used to mean "viceroy."[61] Watt believes that this may well have influenced the later Arabic meaning as well. Yet he doubts that this was the qur'anic meaning regarding both Adam and David:

> Since Abu-Bakr was not appointed by Muhammad except to deputize for him in leading the public prayers, the phrases "*khalīfa* of the Messenger of God" cannot have meant "deputy." The primary meaning must have been merely "successor," except that there would be a suggestion of "one succeeding in the exercise of authority." Thus the word *khalīfa* as applied to Abu-Bakr was vague; but this vagueness was an asset, since the meaning of the word was able to develop as the office itself grew in importance and changed its character.[62]

When applied to the successors of Muhammad in general the term *khalīfa* undergoes some transformation with time. And with the advent of the second dynasty, the Abbasids, based in Baghdad, the title "Caliph of God" (*khalīfat Allāh*) was taken for granted by the rulers, yet strongly resisted "by many of the *ʿulamāʾ*, who rejected the idea that the *khalīfa* was the representative of God and the implication of autocratic power contained in the title."[63]

The title also came to be used in the Sufi religious orders over the centuries. For the Qadiriyya, for example, "the *khalīfa* is the delegate of the *Shaikh* of the order and is invested with a certain amount of his powers and represents him in countries remote from the parent *zawiya*."[64] Among the Tidjaniya the title implies even more. The *khalīfa* is the *shaykh* of the order because he has inherited his spiritual power (*baraka*).

How does all this information help with a better definition of *khalīfa* in the verse, "Behold, thy Lord said to the angels: 'I will create a vicegerent on earth'"? Plainly, the first meaning of the root is "to come after, to succeed, to take someone else's place." But right behind this meaning is the idea of being commissioned to take authority in someone's place. Both Lane and Penrice agree on this. Terms like "viceroy," "vicar," "lieutenant," certainly imply a delegated authority. Yet this is contested by a respected qur'anic scholar, the late German islamicist Rudi Paret. His argument is carefully crafted and therefore deserves some scrutiny.

[61] W. Montgomery Watt, *Islamic Political Thought: The Basic Concepts* (Edinburgh: Edinburgh University Press, 1968), 33.
[62] Ibid.
[63] A. K. S. Lambton, "Khalīfa: (2) In Political Theory," *EI*[1] 4:948.
[64] Arnold, "Khalīfa," 881.

Rudi Paret's Unitary Interpretation

In 1970, Paret published an article in which he contended that *khalīfa*, its two plural forms and its derivatives, are imprecisely translated in the Qur'an (17 instances in all). He begins by pointing out the wide divergence in various translations and then offers a long quote from Watt's book, *Islamic Political Thought*. The gist of the quote is: "The meaning of *khalīfa* has so many facets that it is hard to know which is dominant in certain contexts... In the singular however, *khalīfa*, besides meaning 'successor,' may also have the suggestion of one who exercises authority, though possibly in a subordinate position."[65]

Watt argues that the basic meaning of *khalīfa* is "successor," but that in certain plural instances it is virtually synonymous with "settlers" or "inhabitants," and that in the two cases where it is in the singular it means one who exercises authority (though probably in a subordinate position)—a position argued more recently by University of Chicago scholar Wadad al-Qadi.[66] Paret's contention is that Watt is making the situation more complicated than it really is. "In the following pages," he writes, "I would like to undertake the attempt of an unitary interpretation of the respective texts and submit it to discussion.[67]

To the 17 verses where one form or another of the root *kh-l-f* appears, Paret adds five more where "analogous expressions such as *istabdala* are used."[68] *Istabdala* means "to replace one thing/person/people with another." This is precisely the meaning of *kh-l-f*, he argues, and these five other verses only reinforce the common thread between all 22 verses. I propose to look at the three categories he proposes—all of which connote the same basic meaning of "succession."

---

[65] Rudi Paret, "Signification coranique de khalīfa et d'autres derives de la racine khalafa," *Studia Islamica* 31 (1970): 212.

[66] Wadad al-Qadi, "The Term 'Khalīfa' in the Early Exegetical Literature," *Die Welt des Islams* 28 (1988): 392-411. I come back to her argument in the next chapter.

[67] Paret, "Signification coranique," 212. Paret's use of "unitary" here expresses his conviction that all instances of *kh-l-f* in the Qur'an are best understood under the "successor" umbrella. Wadad al-Qadi differentiated between Paret and Watt's approaches in terms of the aims they had in mind. Paret, she notes, was a Qur'an translator who sought a kind of "homogenous" and "neutral" definition of the root *kh-l-f* as used in Qur'an without indicating his "supporting authorities" among the classical *mufassirūn*. Watt, on the other hand, was an historian, who was keen to investigate the connection between the Umayyad caliphs' claims, the exegetical claims of a broad sweep of commentators, and the qur'anic text itself ("The Term 'Khalīfa'," 392-93).

[68] Paret, "Signification Coranique," 213.

The Succeeding Peoples. As I see it, the category for which Paret makes his strongest case is that of the "succeeding peoples." These passages all have in common the qur'anic doctrine of God replacing unbelieving peoples by other peoples, who then are given the same chance to undergo the divine testing. The foundational passage is sura 10, vv. 13 and 14:

> Generations before you We destroyed when they did wrong: their Apostles came to them with Clear Signs, but they would not believe! Thus do We requite those who sin!
>
> Then we made you heirs in the land after them, to see how ye would behave!

"Heirs in the land" translates *khalāʾif fī-l-arḍ*. M. Khatib translates the same expression "we caused you to succeed them on the earth" with a footnote offering the alternative, "We appointed you vice-regents in the earth after them." Paret simply has "successors."

In Paret's favor, admittedly, there are a great number of verses that express this idea. Speaking of Noah, the Qur'an says: "They rejected him, but We delivered him, and those with him, in the Ark, and We made them inherit (the earth), while We overwhelmed in the Flood those who rejected our Signs" (Q. 10:73). As might be expected, what Ali translates "We made them inherit (the earth)," Paret translates: "We made them the successors." That indeed is the idea. He then cites three similar passages concerning the 'Ad and Thamud peoples (Q. 7:69, 74; 11:57), and three passages about Moses (Q. 7:128-29, 169; 19:59). The first of these is instructive (Q. 7:129). Moses addresses the people in these words: "It may be that your Lord will destroy your enemy and make you inheritors of the earth; that so He may try you by your deeds." For "inheritors of the earth" the original has *wayastakhlifakum fī-l-arḍ*, which is the tenth verbal form of *kh-l-f*. Following Penrice, I would translate "and will cause you to succeed on the earth."

Warnings to Muhammad's Contemporaries. Paret singles out eight passages in which the divine message is addressed to Muhammad's direct audience, either in Mecca or in Medina. All except one repeat in different ways the same theme: believe and do good works, and you will not be replaced by another nation. Perhaps Q. 24:54 is most representative of these: "God has promised, to those among you who believe and work righteous deeds, that He will, of a surety, grant them in the land, inheritance (of power), as He granted it to those before them." "Grant them inheritance" is a translation of the tenth form of the verb *khalafa* and is used again in the last phrase as well. The verb translated by "promise" is the word for covenant and all the verbal forms have the *lām* prefix indicating that an oath is being made. Lane translates this as Paret does: "He will assuredly make them to be successors in the earth, like as He made to be successors those who were before them."[69]

Several of these verses emphasize the fact that this replacing of a people by another people is a sign of God's power, as, for instance, in Q. 4:133: "If He wills, he can cause you to pass away, O People, and bring others, for God is Omnipotent over that."[70] This time the verb *khalafa* is not used, but the idea is certainly the same—literally, "He will take you away and bring in others."

One of these verses, however, does not support his argument. Ali's translation of sura 57, v. 7 reads: "Believe in God and his Apostle, and spend (in charity) out of the (substance) whereof he has made you heirs." Paret, however, translates "whereof he has made you successors." The Arabic uses the present participle of the tenth form of *kh-l-f*: literally, "spend from that which He has made you to inherit of." The idea of succession here seems forced. Who are the people "you" are succeeding? The context makes no mention of this. Who would have passed on their inheritance to you? Clearly God is the one who is rewarding his faithful and, as often the case in Bible, they inherit their blessings from him (e.g.

---

[69] Lane, *Arabic–English Lexicon*, 793. This verse has been the object of another German scholar's study more recently, Wolfdietrich Fisher ("Das Geshichliche Selbstvertändnis Muhammads und seiner Gemeinde zur Interpretation von Vers 55 der 24. Sure des Koran," *Oriens* 36 [Festschrift for Franz Rosenthal] [2001]: 145-59). Fisher argues that the Qur'an has taken over the biblical theology of the Hebrew Bible, in which "successorship" and "the inheritance of the land and the scriptures" are two sides of the same coin—with one main difference: what was applied to Israel is now applied to humankind in general (p. 152). He shows that the "successor" verses all belong to the Medinan period after the victory of Badr (624), during which progress has been made to establish the nascent Medinan Muslim community, and yet not without serious dangers looming on the horizon. The last of those was the threat of attack in 627 on the part of a large Meccan-led coalition (leading to the so-called Battle of the Ditch, or Battle of the Confederates), and that is precisely the context of the revelation of sura 24. Muhammad comforts his people with a reminder of the pattern of God's dealings in history with his people, from the Israelites to his own followers now ("holy history," or *Heilsgeschichte*). He has promised his chosen people, specifically those "who believe and work righteous deeds," that he would "deputize them on the earth as he had deputized those before them" (using "deputize" for *istakhlafa*, tenth form of *kh-l-f*). Hence, Fisher's unpacking of that term "deputize" as "making them successors by inheriting the land and the scriptures." While I agree with his *Heilsgeschichte* reading, I disagree with his assertion throughout that "successorship" is the only meaning in all instances of *kh-l-f* and its derivatives in the Qur'an. Though he nowhere cites Paret, he is clearly of the same mind.
[70] *The Bounteous Qur'an*, 125.

Ps. 37:9, 11, 18, 22, 29, 34). Significantly, Jesus picks up this theme of "inheriting the earth" and widens its application (Mt. 5:5). Cragg, not surprisingly, translates Q. 57:7 in this way: "what He has brought into your possession expend for His sake."[71] Surely that is closer to the original meaning.

The Other Passages. First, Paret cites several passages that obviously point to the primary meaning of the root *kh-l-f*. It is clear that a succession is taking place because the preposition "after" is added. One example is Q. 6:133: "If He pleases he shall do away with you and cause whom He pleases to succeed you, as We raised you from the progeny of other people." Another one is Q. 7:150, where Moses returns from his time on the mountain with God and explodes with anger at the sight of the golden calf. He exclaims, "Evil it is what you have done in my place in my absence!" Yet in this case, Ali's translation is poor: "In my absence" translates "after me." And "In my place" is ambiguous: did they do evil in Moses' stead? Here I find Paret's translation very much à propos: "How badly you have behaved after I had gone away and left you behind!"[72] He is right, in that here *kh-l-f* expresses both temporal and spatial succession.

Paret's last examples, however, seem to work against him. I mentioned one instance already in which "succession" did not fit. There are two more categories of verses in which the root *kh-l-f* must mean more than temporal and/or spatial succession. The first group includes the two singular instances of *khalīfa* to which I have already given much attention. With regard to Q. 2:30, Paret reflects a tradition which is given serious consideration in the next chapter: "Here, when God says to the angels, concerning Adam, that he will establish on earth a successor, one must understand that, on the face of it, Adam (and with him the human race) will from now on replace the angels (or the spirits in general) as inhabitants of the earth."[73] This is simply the reverse of what is said in Q. 43:60, where he points out: "And if it were our will, We could make angels from amongst you, succeeding each other on the earth."

Though Paret's is not an implausible rendering of this key verse, it does not do justice to the context, which makes the title *khalīfa* loftier than simply "taking over from somebody else." Speaking of the Umayyad caliphs' interpretation of this passage, Watt noted the fact that in

---

[71] Cragg, *Readings in the Qur'an*, 254.
[72] Paret, "Signification coranique," 216.
[73] Ibid., 215.

addition to succession, *khalīfa* carries with it the idea of authority, albeit a subordinate one:

> They allowed the use of the title "caliph of God" in the sense of ruler or viceroy appointed by God. They further justified their claims to be divinely appointed by quoting the Qur'anic verse in which God addresses David saying, "O David, we have made thee a *khalīfa* in the earth." From the other verse in which the word *khalīfa* occurs (2:30/28), where God says to the angels, "I am placing (or making) in the earth a *khalīfa*, namely Adam, it was inferred that the office of *khalīfa* was higher than that of angels and prophets."[74]

The last category comprises the verses in which one of the plurals of *khalīfa* is used with "on earth" or "in the earth." To start with, it is good to remember that this combination is also made in 2:30: "a *khalīfa* in the earth," as well as Q. 38:26 with regard to David. I see four passages of this type (Q. 6:165; 10:14; 27:62; 35:39). Only one of them obviously reflects the kind of "holy history" Paret is alluding to, that is, God replacing corrupt peoples or generations by new ones who in turn will be tested. This is Q. 10:13 and 14, which I already quoted: "And We annihilated generations before you when they became iniquitous... Then We caused you to succeed them on the earth that We might behold how you would do."[75] Yet even here, Yusuf Ali translates: "We made you heirs in the land after them." There seems to be more than just succession at stake here. Responsibility often implies some measure of authority. Their test may include more than obedience and right belief. They likely have been given the ability to manage their affairs and will be judged accordingly. After all, as mentioned above, Khatib has a footnote on "succeeded them" explaining the alternative translation, "We appointed you vice-regents in the earth after them."[76]

Furthermore, the other three references do not require any idea of succession to be fully understood. The next to last verse in sura 6 speaks of everyone needing to bear his own burden. Then follows the exhortation, which ends the sura:

> Your goal in the end is toward God: He will tell you the truth of the things wherein ye disputed. It is He who hath made you (His) agents, inheritors of the earth: He hath raised you in ranks, some above others: that he may try you in the gifts he has given you: for thy Lord is quick in punishment: yet is He indeed Oft-Forgiving, Most Merciful (6:164, 165).

---

[74] Watt, *Islamic Political Thought*, 33.
[75] *The Bounteous Koran*, 267.
[76] Ibid.

Ali makes explicit his interpretation of *khalāʾif al-arḍ*. Whereas Khatib has "who has made you successors on the earth," Ali emphasizes the notion of authority and responsibility ("agents, inheritors of the earth").

In the next passage, Khatib renders *khulafāʾ al-arḍ* as "vice-regents of the earth" (Q. 27:62). The context here is the anguished call of the brokenhearted, which God hears with compassion: "Or, is it not He who responds to the harmed when he calls to Him, and does away with evil (and iniquity), and makes you vice-regents of the earth?"[77] The last passage, too, has no hint of succession (Q. 35:38-39). To the contrary, it is another meditation on God's knowledge of the *ghayb* (secrets/ hidden things) of the heavens and the earth and the *ghayb* of the human heart. Then v. 39: "He it is that has made you inheritors of the earth: if, then, any do reject (God), their rejection (works) against themselves." Taking the passage at face value, it would seem to me that *khalīfa* here is applied to humankind and connotes a moral and spiritual gifting in them—a responsiveness to God and a responsibility to the world, for which they are accountable. This calling of humanity, as it is described in all these passages as "vice-regents of the earth," reaches far beyond Paret's "unitary interpretation." I agree with him that there is a theology of replacement in time and space in the Qur'an. Evil generations and nations are replaced with new ones who are then given a chance to live out their responsibility as *khulafā* on the earth. To the extent that they faithfully live out their calling, they are either rewarded or punished, but in every case they are held accountable.

The Amāna *Verse: Q. 33:72*

Q. 33:72 reads: *innā ʿaraḍnā l-amānata ʿalā al-samawāti wa-l-arḍi wa-l-jibāli wa-abayna an yaḥmilnahā wa-ashfaqna minhā wa-ḥamalahā al-insānu innahu kāna ẓalūman jahūlan,* "We did indeed offer the Trust to the Heavens and the Earth and the Mountains: but they refused to undertake it, being afraid thereof: but man undertook it—he was indeed unjust and foolish." The word *amāna* is rendered "trust" by most translators. Khatib has "We presented the trust to..." and Cragg has "We offered the trust to..."[78] Zafrulla Khan proposes the same translation as Cragg's. Yet Rashad Khalifa prefers to expand on the meaning: "We have offered the responsibility (freedom of choice) to..."[79]

---

[77] Ibid., 502.
[78] Cragg, *Readings in the Qur'an,* 109.
[79] Khalifa, trans., *Quran: The Final Testament,* 10.

I chose to include this verse within the discussion because of its close parallel to Adam's calling as *khalīfa*. The immediate context, first of all, is that of exhortation to obey God, the community's leaders, and God's messenger in particular, and the penalty for not doing so (v. 68).[80] Then an example is made of Moses, whom his people vexed and insulted repeatedly; yet God restored his honor. The preceding verse explains how the believer might follow whole and sound conduct and be forgiven of his or her sins: "He that obeys God and His Apostle has already attained the highest Achievement." The context would indicate, then, that v. 72 is a commentary on how humankind can be expected to live up to this "highest Achievement," namely, that man dared to undertake the challenge God offered him in the beginning—the challenge put to the material world and which it refused out of fear.[81]

The definition Lane gives for *amāna* is the following: "Trustiness; trustworthiness; trustfulness; faithfulness; fidelity."[82] From there, it can also mean "a thing committed to the trust and care of a person; a trust; a deposit." Lane quotes this verse and proceeds to various interpretations put forward by the classical commentators. I will simply mention the more certain parameters of this verse: some kind of responsibility was offered to humankind at the start; this was connected to their ability to obey God and follow his directives; it included the possibility that they would fail to do so, and, as a matter of course, people have often failed ("man is surely a creature of sinful folly").[83]

---

[80] The sura itself ("The Confederates") is considered Medinan in the Cairo system of ordering, and is therefore considered such in all Muslim versions. Yet Régis Blachère puts it in the first Meccan period, an opposite view. This would put a different twist on its comparison with the Q. 2:30 verse, which is unanimously Medinan. Is the "trust" an earlier form of the idea of humanity's *khalīfa*? Or is it simply a complementary way of looking at the issue? In either case, however, the comparison strengthens the wider application of the *khalīfa* calling from Adam to all of humanity.

[81] This verse shocks the first-time Arabic reader: Arabic grammar requires the use of the feminine singular when a verb or adjective qualifies the plural of inanimate objects. Here the three verbs that describe the action of the heavens, the earth and the mountains are in the feminine plural, very likely a convenient way of personalizing them. Though this is not uncommon in the Qur'an, Ali seems to lean in this direction as he capitalizes the three nouns, Earth, Heavens, and Mountains.

[82] Lane, *Arabic–English Lexicon*, 102.

[83] Cragg, *Reading in the Qur'an*, 109.

Kenneth Cragg on Khalīfa[84]

Bishop Cragg, in his various books over the years, has been my mentor of sorts and his reading of the Qur'an—especially in its doctrine of humanity—led to this project. In this last section I summarize some of the main themes of his qur'anic anthropology. This serves as foundation, in turn, both for the subsequent chapters on the career of Q. 2:30 in Muslim commentaries and for my own theologizing in the last three chapters.

Commenting on Iblīs's refusal to bow before man, Cragg states that to reject humanity's status is a direct attack on God's authority. In this poignant drama played out in the heavenly council, the esteem of the creature is somehow tied to the honor of the Creator:

> There could hardly be, in terms of myth and symbol, a clearer affirmation of the lordship of man as that which may not be spurned without impugning the sovereignty of God. The celestial insurrection, if we may put it this way, turns on a terrestrial issue: God is flouted where man is despised. Or, in positive terms, Adam is, so to speak, the test and crux of the Divine will and wisdom.[85]

However eloquent and inspiring this may be, how does this lofty view of humanity compare with the qur'anic data? Questions have been raised of late as to how much of Cragg's doctrine of humanity in his understanding of Islam is really qur'anic. Some of these doubts have been expressed in his own circles.

Christopher Lamb's Critique
In his recent book, *A Call to Retrieval: Kenneth Cragg's Vocation to Islam*, Christopher Lamb, a long-time colleague and student of Cragg, expresses much admiration for his mentor's tireless, sensitive and

---

[84] Bishop Cragg read through the whole manuscript, leaving notes here are there, which I gratefully took into account. I finally had the privilege of meeting him in his home in Oxford in June 2008.

[85] Kenneth Cragg, *The Privilege of Man: A Theme in Judaism, Islam and Christianity* (London: Athlone Press, 1968), 28. In his *Event of the Qur'an*, Cragg writes in passing that this verse illuminates Islam's view of the meaning of civilization, which is found in "the entrustment of the world to man." Here he has a footnote: "This passage about man accepting 'the trust of the heavens and the earth' is at the centre of Islamic thought about man. Surah 2:30 likewise" (p. 177). He says this on the basis of the contemporary Muslim consensus on this. In the end, the crux of this project revolves around current Muslim interpretation of the Qur'an. And Khalifa's translation is very representative in this regard.

penetrating study and exposition of Islam over the last five decades. He praises him for the consistency and cogency of his portrayal of Islam to Christians and his bridging of Christianity to Muslims. Nevertheless, he does present a critique of Cragg's perspective, echoing what many Christians have felt for some time. In being so intent on dialogue, may there not be the danger of diluting each side's doctrines in order to make them more palatable to each other?[86] At least with regard to his interpretation of Islam, is he not raising the status of humanity above the level that is taught by the Qur'an itself?

Lamb looks at the passage quoted above and compares it with Zamakhshari, Baidawi and other classical commentators and finds that the focus of the heavenly drama for them is far removed from Adam's status and entirely on God's honor, which is mocked by Iblīs yet upheld by the angels, who, out of sheer loyalty to the Creator, obey and bow before a creature obviously inferior to them. Lamb then surveys the mystical commentators, Nisaburi, Ibn al-Arabi and even al-Ghazali, whose more moderate Sufi stance still held a place for humankind fit to be God's vicegerents because of their connection of spirit to him.[87]

Coming to the modern period, Lamb recognizes that there is a shift, and this he attributes to Muhammad Iqbal, the Indian philosopher-poet. Citing authors as diverse as Vahiduddin, Tabataba'i, Maududi and Sayyid Qutb, he sees a consensus forming around a new emphasis on humanity's role as God's *khalīfa* on earth, but not based on their inherent connection to God's nature. Humanity is God's representative because it alone can bear the responsibility of free moral choice. Yet his conclusion is less than enthusiastic: "Modern, as compared with classical Muslim exegesis of the Qur'an would then appear to give some support to Cragg's theological programme of promoting the dignity of human status by emphasizing its divine support."[88]

My feeling is that Cragg, though at times going a bit beyond, is nevertheless very much in touch with contemporary Muslim thought on humanity and *khalīfa*. It may be worth reiterating that the purpose of this study is not to identify what the "correct" doctrine of *khalīfa* is according to the Qur'an; rather, it is to understand how this doctrine has in fact

---

[86] Admittedly, as mentioned in the second chapter, there have been students of Islam who have successfully described Islam from such an insider's perspective that they have broken down barriers and allowed people on both sides to understand one another in a new light. Such was Louis Massignon, whose lifelong work on the life of Hallaj sparked and continued to nourish his own Catholic faith.

[87] Lamb, *The Call to Retrieval*, 42.

[88] Ibid., 44.

evolved over time and, with that as a backdrop, to focus on the current consensus of opinion on the issue. The overall objective is to construct a viable theological platform solid enough to carry the weight of concerted Muslim and Christian engagement in issues of peace and development. I have found Cragg to be very helpful in this regard. This next section, then, is a summary of his many-faceted exposition of humanity and *khalīfa* in the Qur'an.

Cragg's Exegesis of Q. 2:30
As previously mentioned, this theme is liberally sprinkled throughout Cragg's voluminous writings. *Sūrat al-Baqara*, v. 30, plainly, is the "definitive" passage, he writes.[89] The significance he sees in the angels' protest has already been underlined. What is also important is the teaching of the "names": "always a Semitic image for sovereignty and anticipating the vital role of nomenclature in the processes of science."[90] This ability, once proven by Adam, is reason enough for the angels to acknowledge their own limits and fall prostrate before the Creator's new representative on earth.[91] In actual fact, they are recognizing "that the Divine lordship itself is in some sense staked in the human role, that the due recognition of humanness is inseparable from the proper, angelic acknowledgement of God."[92] Then, in passing, he mentions the other singular instance of *khalīfa* this time applied to David (Q. 38:26), with a definite overtone of "power and trusteeship," the same two elements that applied to the previous verse.

Cragg has returned to this theme in the aftermath of the September 11 attacks on American soil. With a title echoing the rhetorical question of the Creator at the primordial covenant (Q. 7:172, "Am I Not Your Lord?"),[93] he devotes the entire book to the theological implications of the divine drama unfolding in earthly space and time, as God launches

---

[89] Cragg, *The Privilege of Man*, 27.
[90] Ibid., 28.
[91] Rahman's description of this scene is very similar. After the angels "protest," "God did not deny these allegations against man but replied, 'I know what you do not know.' He then brought about a competition in knowledge between angels and Adam, asking the former to 'name things' (to describe their natures). When the angels could not do so, Adam could (2:30ff.). This demonstrated that Adam possessed the capacity for creative knowledge that angels lacked, whereupon God asked all angels to prostrate themselves before him to honor him" (*Major Themes in the Qur'an*, 18).
[92] Cragg, *The Privilege of Man*, 28.
[93] Kenneth Cragg, *Am I Not Your Lord? Human Meaning in Divine Question* (London: Melisende, 2002).

humankind on earth as his "lieutenants"[94] and Satan vows to derail this risk-fraught enterprise. The angels had, from the beginning, recognized the divine wisdom of entrusting humans with the "knowledge of the names"—the "[n]aming, classifying, identifying" that give rise to empirical investigation and science in all its forms.[95] They understood the locus of human dignity as resting in "the ethical, social, political and spiritual tenancy of God's world."[96] So did Iblīs—but his conclusion was that God was making a mistake. He decided then and there to stake his whole existence on bringing to ruin God's good plan—hence the qur'anic epiteth of *al-shayṭān al-rajīm*, "the accursed devil," or "the devil who deserves stoning."[97] It follows also that messengership and prophethood are costly callings that entail suffering and hardship at the hands of people who have made common cause with the devil by falling into his trap of *shirk*—worshiping the creature instead of the Creator and latching onto position and power, the better to oppress the weak and the poor.[98]

The Plural Instances of khalīfa. The dominion of humanity over the created order, or their *"imperium,"* as Cragg likes to put it, is something received by God and exercised on his behalf. Accordingly, "the Quranic 'caliphate' is not the political institution later developed to serve the continuity of Muhammad's achievement after his death, but the general dominion of Adam in the world."[99]

---

[94] Cragg deliberately uses this word once, while commenting in a footnote: "Not an improper word—its meaning prior to the military sense, namely 'one who has the place of'" (ibid., 36). He also notes how in two of the earliest suras (113 and 114) Satan is abjured as "the evil whisperer who whispers in the bosoms of men" (ibid., 48).

[95] Ibid., 26.

[96] Ibid., 42.

[97] Hence the stoning of the devil at the *Jamrat al-'Aqaba*, which from the beginning became a central rite and symbol of the Hajj (ibid., 34).

[98] "Shirk, as the Qur'an accuses it, the diversion of worship away from Allah as the One to whom alone worship belongs, is the most perverse of all the evils into which humans are beguiled and entrapped" (ibid., 37). Muhammad's Meccan opponents unwittingly stand with Satan in disparaging humanity's high vocation. All their evil strategies reveal how they "are distorting that human autonomy, how their habits in trade and idol-owning involve them in assertive pride and social wrongs which they hate to hear denounced. Hence the vehemence of their reaction as they aim to give the lie to his words and to him bringing them. They have too avidly imbibed the very mood of Satan" (ibid., 48).

[99] Cragg, *The Privilege of Man*, 29.

As was pointed out by Paret, the plural *khalāʾif* is always used "with the general sense of temporal successiveness and thus of replacement."[100] Generation succeeds generation and a nation is set in place to replace another one, which, because of its *fasād* ("corruption"), had been disqualified by God to exercise its dominion any longer. Here is, of course, where Cragg parts ways with Paret:

> "Taking the room of" is throughout the initial sense of *khalafa*, but always with the accompanying ideas of either "following after" and/or "occupying in lieu of." The second meaning is necessarily detached from "succession" in all that has to do with God. The Eternal is never superseded but He may entrust His will to deputies on His behalf.[101]

One of the examples given is in Q. 10:73 where Noah and his family emerge from the ark and are "appointed viceroys" in the renewed earth. "It is not a continuity of rulers that is in mind but the moral sequences of requital and renewal."[102] Thus, "to replace" always carries with it the idea of responsibility in the same breath.

Humanity and the Earth. At the beginning of sura 7, v. 10 lays out three aspects of humanity's vocation on earth: "It is we who have placed you with authority on earth, and provided you therein with means for the fulfillment of your life: small are the thanks ye give!" We humans have been deliberately placed here, or, as Q. 2:36 puts it, "on earth will be your dwelling place" (literally, "your settlement"). Second, we have been made to rule over the physical world with our intellect, our skills and our wise use of moral choice. Third, we are to accomplish this task with the appropriate attitude of worship. Above all, the qur'anic idea of human *khalīfa* must be linked to their calling as God's servants, or "worshipers" (*ʿabīd*):

> The role of man as *khalīfa* both validates his empire and expects his empire and expects his hallowing, and both in essential unity. For if he wielded no mastery he could bring no submission. He would have nothing to offer or to consecrate. His very culture and all his works are the substance of his Godward obligation"[103]

---

[100] Ibid., 31.

[101] Cragg is assuming with most modern Muslim commentators that the primary meaning of *khalīfa* as successor even when used in these passages still carries with it the notion of authority invested by God. This is why a particular group is called to succeed another one. The preceding one did not live up to its calling as God's viceroys on earth.

[102] Ibid.

[103] Kenneth Cragg, *The Mind of the Qur'an: Chapters in Reflection* (London: George Allen & Unwin, 1973), 141.

Tenants on the Earth. In Q. 6:165 we saw the expression "He has made you the inheritors of the earth," or "*khulafā* of the earth." Akin to the idea of "inheriting the earth" is that of settling it and exploiting its riches in a responsible way. Cragg points out that when the prophet Salih is sent to the people of Thamud, his message is: "O my people! Worship God: ye have no other god but Him. It is He Who hath produced you from the earth and settled you therein: then ask forgiveness of Him, and turn to Him (in repentance): for my Lord is always near, ready to answer" (Q. 11:61).[104] In Arabic, "settled you" is *istaᶜmarakum*, or, literally, "He has made you to colonize it" (this is the root for the word "colonialism"), or, as Cragg has it, "He has planted you as tenants..." This prolific Arabic root, writes Cragg, "combines the same twin ideas of time-occupancy and place tenancy and yields terms for a span of years and for an abode, a dwelling, an establishment... Men in this sense are all empire builders, exploiting the occasions of the years and of the lands, and all by the Divine design and leave."[105]

The Sacramental Earth. The "sacramental" nature of the earth is another re-occurring theme in Cragg's works and it is tied to an oft-repeated emphasis of the Qur'an on God's signs in nature (*ayāt*). Humanity's calling to exploit the earth is only one side of the coin. The other side is our calling to recognize God's hand in it at all times. As we look around at the beauty, order and perfection of the world, we must deliberately enter into its meaning and significance. The particle *laᶜalla* ("perhaps") occurs in this connection 72 times in the Qur'an and is invariably used with verbs expressing intelligence, recognition, understanding, thankfulness and worship: "Perhaps you may come to your senses," "peradventure you may realize...," "if perchance you may know that mercy is being done to you..." And it is always used with the *ayāt* of God.[106]

Another way of expressing this double movement is this: on the one hand, people witness natural events on a daily basis, but on the other

---

[104] The asking of forgiveness is significant for Cragg in that, for him, this is consonant with "the constant re-iteration in the Qur'an of the obligation on the part of men to a recognition of the hallowedness of the material world" (*The Privilege of Man*, 32). This remark naturally opens up his discussion of the all-pervasive role of *ayāt* ("signs," or "miracles") in the qur'anic narrative, to which Rahman pays a good deal of attention as well.

[105] Cragg, *The Privilege of Man*, 32. The other word for "settle," which figures so prominently in the qur'anic Fall narratives is the verb istaqarra, which means "to settle down, establish oneself...to come to rest" (Cowan, ed., *The Hans Wehr Dictionary*, 880). It does not carry with it the tenancy idea that *istaᶜmara* does.

[106] Cragg, *The Mind of the Qur'an*, 147.

hand, the perceptive soul is exhorted to reach out to God and respond in gratitude and wonder. "Perhaps you may give thanks," is the oft-repeated qur'anic phrase. Here Cragg rightly connects "thanklessness" and "unbelief" in the Qur'an:

> That last dimension of gratitude is central to the Quranic perception of what has to obtain from humankind Godward. To be "thankless" is very close to being "unbelieving." For it is part of that wily miscreant's strategy to delude us into supposing there is no debt we owe in being here, no bearings outside a selfish interpretation of ourselves.[107]

Ayāt is sometimes translated as "miracles," but if we mean by this extraordinariness, then we have missed the point: "It is rather in the very ordinariness of natural phenomena that the sign quality resides."[108] But since *ayāt* also refers to the verses of the Qur'an, Cragg sees in this confluence of terms "the conviction that the external world is a kind of 'scripture,' intimating in its own realm and within its own order that divine knowledge which, in history and prophecy, in word and action, speaks Quranically to mankind, so that the one is not unworthy to be denominated in the same term with the other."[109]

Responsible Earthkeeping. Islam denies any curse on the earth as a result of Adam and Eve's disobedience and thus humankind is not to treat it as a foe, but rather as a blessing and a bounty from God. In the chapter he contributed to the volume *Three Faiths—One God: A Jewish, Christian, Muslim Encounter*, Jamal Badawi wrote, "As a bounty, it must be used wisely so as to prevent the quality of life from deteriorating, lest future generations be deprived from partaking their shares of such blessing."[110]

---

[107] Cragg, *Am I Not Your Lord?*, 36.

[108] Cragg, *The Mind of the Qur'an*, 147.

[109] Ibid., 148. Rahman supports this view in part: "The parallel (or even identity) between the revelation of the Qur'an and the creation of the universe has been pointed out by several medieval Muslim authors who have noted the numerous passages in which the revelation of the Qur'an and the creation of nature are coupled" (*Major Themes in the Qur'an*, 71). On the other hand, he is quick to point out that when the term *ayāt* refers to the verses of the Quran, "these *ayāt* are usually said to be 'recited' (*natluha*, or *tutla*, etc.); they are often said to be 'clear' *ayāt* [*ayāt bayyina*]. This latter expression is applied to signs other than the Qur'anic verses only three times... [I]t seems never to be applied to nature, presumably because natural signs lie buried beneath natural causation until the Qur'anic verses resurrect them and clarify them as signs of God" (ibid., 72). Cragg may be stepping out on an Islamically fragile limb here.

[110] Jamal Badawi, "The Earth and Humanity: A Muslim View," in *Three Faiths—One God: A Jewish, Christian, Muslim Encounter*, ed. John Hick and Edmund S. Meltzer (Albany, NY: State University of New York Press, 1989), 98.

Cragg also sees humanity as God's *khulafāʾ*, as having profound implications for community development and ecology, in particular:

> Nature offers both delight and duty but only in unison. Economy and ecology, wealth and habitation, are as it were a constant interrogation of his environment by the mind of man. The questioner is himself questioned. The answers to man have to be matched and sanctified by the answer from man. It is these together which are the essence of the sacramental. The good earth is the earthly good: they require each other.[111]

As God's *khulafāʾ* on earth we are not only to go about our work in wonderment and gratitude, discerning behind the natural the supernatural, but also fulfill our calling responsibly, knowing that we must answer to its owner who is also our Creator. In this light, we are to use our God-given gifts to explore and exploit responsibly the resources of our planet. Perhaps this is the sense of Q. 2:36, "on earth will be your dwelling place and your livelihood for a time." We are temporary residents, which is certainly a biblical theme as well. The implication is that we are to take care of the earth, because it has been entrusted to us for a time, and of our use and management of it we will be held accountable.

Cragg's qur'anic exegesis and theologizing here is in basic harmony with current Muslim thought on this topic. Badru D. Kateregga, an African Muslim scholar committed to dialogue with Christians, writes in the book he co-authors with David W. Shenk: "As a *khalīfa*, humans are chosen to cultivate the land and enrich life with knowledge and meaning. Nature is subject to humans. The superior position man holds in the eyes of God makes man an authority over all God's creation."[112] In his conclusion he emphasizes accountability: "Nature is the provision of the merciful Allah for the sustenance of humans; they are therefore commanded to make the best use of God's resources on earth. Responsible stewardship in obedience to God's Divine commands is the key to the Muslim approach to development."[113]

The next three chapters summarize some of the historical meanderings of Islamic exegesis and theologizing on these verses. I trust that with the wider hermeneutical framework in the background exegetes and theologians, both Muslim and non-Muslim, will be able to build on this knowledge and help to chart our way more effectively as we seek to tackle together some of the pressing challenges facing humanity in the twenty-first century.

---

[111] Cragg, *The Mind of the Qur'an*, 153.
[112] David W. Shenk, *A Muslim and Christian in Dialogue* (Scottdale, PA: Herald Press, 1997), 39.
[113] Ibid., 40.

Chapter 7

TAFSĪR OF Q. 2:30: CLASSICAL PERIOD

As I mentioned in Chapter 4, much effort has been invested in both Muslim and non-Muslim scholarly circles to approach the study of the Qur'an—and especially hermeneutics—in a new light, even in mainstream conservative Muslim circles. Muhammad Ibrahim Surty of the London Islamic Foundation offers an interesting survey of *tafsīr* (commentary) in this respect. On the one hand, he recognizes that *tafsīr* in each generation highlights contemporary issues, and that because the variety of disciplines through which the Qur'an was approached, the *tafsīr* literature "illustrates the diversity of Muslim thought, the academic freedom and intellectual tolerance of Muslims, secure in fundamental unity."[1]

On the other hand, however, that intellectual tolerance would not encompass the liberation perspective of Farid Esack, or the radical revision of the theory of abrogation put forward by the late Sudanese cleric Mahmud Muhammad Taha,[2] or the views of many other con-

---

[1] Muhammad Ibrahim Surty, "The Qur'an in Islamic Scholarship: A Survey of *Tafsīr* Exegesis Literature in Arabic," *Muslim World Book Review* 7.4 (1987): 51.

[2] Taha reversed the traditional view that later verses abrogated earlier verse in the Qur'an when contradictions arose. This becomes crucial in legal matters. For instance, Muslim consensus since the classical period held that the more violent verses concerning *jihad* in the Medinan period abrogated the more universal and peaceful ones revealed in the Meccan period. Taha argued that it was precisely the Meccan verses that laid out the universal norms of human dignity and reconciliation that were meant to serve as the ethical foundation of the Islamic faith. The Medinan verses were revealed for a particular setting, and therefore are not applicable today. Taha was executed for "apostasy" by Sudanese president Numeiry in 1985. See his work, *The Second Message of Islam* (Syracuse, NY: Syracuse University Press, 1987), translated and introduced by his disciple, Abdullahi Ahmed An-Na'im, who wrote an updated version of this hermeneutic, *Toward an Islamic Reformation: Civil Liberties, Human Rights, and International Law* (Syracuse, NY: Syracuse University Press, 1990).

temporary reformers. Surty writes in a devotional vein, "It is required of all Muslims equally that they should strive to comprehend the meaning of the Qur'an... The Qur'an, over and over again, invites *tadabbur*, contemplation/reflection, *tafaqquh*, comprehension and *tafakkur*, observation and study of measured divinely ordained and sustained natural phenomena."[3] While Surty's exhortation to look at the world through the lens of the Qur'an is certainly edifying, he nowhere comments on the theoretical aspects of the hermeneutical process that takes place between the reader and the Qur'anic text. Despite great diversity of interpretation over the centuries, one gains the impression that for him, the sacred text is fixed and static—we simply extract what is there, as it is.

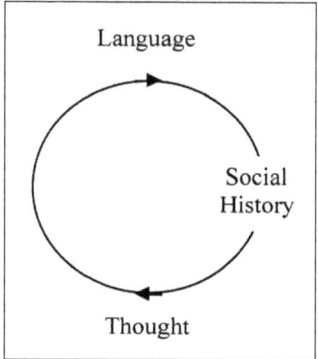

Figure 10. *Arkoun's Form and Content of Social-Historical Existence*[4]

Contrast this age-old confidence (a mixture of pre-modern and modern) in hermeneutical simplicity with the following diagram proposed by Mohammed Arkoun (Fig. 10). In a long chapter on Islam and society, Arkoun warns that the Western "islamological viewpoint" has perpetuated some of the assumptions of traditional Islam. The most crucial of these might well be the postulate "that society and its past are a product of the religion taught by God and not the reverse." Hermeneutics, Arkoun contends, cannot be divorced from sociocultural and political influences played out in various historical contexts—again, a matter of importance, when the question of Shari'a is brought up:

> The recent updating of the religious law (Shari'a) in Pakistan, Egypt, Iran, Algeria, Morocco...confirms the penetration of additional schemes of perception, comprehension and action. The return to 'Muslim' law is

[3] Surty, "The Qur'an in Islamic Scholarship," 51.
[4] Ibid., 143.

necessitated by the difficulty in establishing the legitimacy of powers arising from violent confrontations and facilitated by the absence of social sciences which present a critical vision of social-historical phenomenon.[5]

Peter Heath proposes what I see as a mediating position between Surty and Arkoun, and thereby offers a useful introduction to this chapter through his own comparison of Tabari, Ibn Sina and Ibn al-'Arabi with regard to their styles of *tafsīr*. In agreement with Surty, he avers that part of the differences in approach relates to the variety of Islamic sciences through which these scholars approached the Qur'an.[6] But also, much of the richness—and therefore diversity—of this interpretation arose "in response to the needs of particular intellectual environments, the demands of different historical conditions."[7] In particular, his use of these three authors is mainly to highlight three principal approaches to qur'anic interpretation in the classical period: a traditional, a philosophical, and a mystical one.[8] Heath cogently argues that these styles have profound implications for the way the divine text is approached. In this chapter, beginning with the early stage of qur'anic commentary and then following up with Heath's three divisions, I try to show how various disciplines and historical contexts affected various Muslim understandings of

[5] Mohammed Arkoun, "Religion and Society: The Example of Islam," in *Islam in a World of Diverse Faiths*, ed. Dan Cohn-Sherbok, Library of Philosophy and Religion (New York: St. Martin, 1999 [1991]), 140. Arkoun is arguing here (as he was in the *EQ* article quoted in Chapter 4) that what is often passed off as "Islamic"— whether by regimes seeking to enhance their legitimacy, and hence their hold on power, or by opposition islamist groups—turns out to be either political agendas clothed in religious garb, or, depending on one's emphasis, religion as political ideology. What the social sciences can provide is a sifting out of the various levels of discourse production—from language and culture to the specifics of a people's history of ideas and politics. This is the critical dimension that is needed, contends Arkoun, in order to bring into focus the hermeneutical dynamics at play.

[6] Heath lists these disciplines: "Qur'anic commentary, prophetic tradition, jurisprudence, dialectical theology, historiography, the study of Islamic sectarianism, grammar, rhetoric, mysticism, and philosophy" ("Creative Hermeneutics," 173).

[7] Ibid.

[8] I will follow his typology for the classical commentators singled out in this chapter. This is not dissimilar to the nomenclature adopted by Claude Gilliot in his article, "Exegesis of the Qur'an: Classical and Medieval," *EQ* 2 (2002): 99-124, except that he offers two earlier periods ("the beginnings of qur'anic exegesis," and "the formative period") and breaks down the "classical" period between (1) "an intermediary and decisive stage: the introduction of grammar and the linguistic sciences," and (2) "constitutive Sunni corpora based upon traditions and later developments." Further, he adds a short section on "special legal exegesis," and finally, he offers a substantial section on Kharijite and Shi'ite exegesis.

Q. 2:30. I must add, however, that I will merely be scratching the surface, seeking only to illustrate in broad strokes some of the hermeneutical material of the previous chapters.

Early Tafsīr *Literature*

Here I briefly summarize the arguments presented by Wadad al-Qadi on the term *khalīfa* in the early exegetical materials on the Qur'an. She specifically sought out commentators who wrote during the Umayyad caliphate (661–750 CE). In so doing, she examined one of three exegetical works from that period in manuscript form (*Tafsīr* of Muqatil b. Sulayman, d. 150/767) and three works published in the 1980s: *Tafsīr Mujāhid* (d. 103/721); *Tafsīr Sufyān al-Thawrī* (d. 161/777), and the first volume of Muqatil's *Tafsīr*. Her findings are particularly relevant to this study, first because of her focus on the Umayyad period (hence the connection between the qur'anic *khalīfa* and the actual claims of Umayyad caliphs), and second because of her hypotheses about why specific exegetes chose particular meanings of the root *kh-l-f*. She distinguishes five possible meanings—all closely related:

1. "To follow another, to succeed, as people naturally do, generation after generation."[9]

2. "To substitute for, or replace another; to deputize for temporarily" (like Moses asking Aaron to stand in for him as he goes up the mountain, Q. 7:142).

3. "To replace someone after he or she has gone" (the *Heilsgeschichte*, or salvation history theme noted in the last chapter, so prevalent in the Qur'an).

4. "To inhabit, to cultivate" (*sakana*, *ʿamara*): she notes that Tabari (cf. next section) quotes the famous biographer of Muhammad, Ibn Ishaq (d. 151/768) as giving this meaning for *khalafa* in

---

[9] Al-Qadi quotes two early exegetes who take *khalīfa* in the "Adam verse" (her term for Q. 2:30, as opposed to the "David verse," the only other instance of *khalīfa* in the singular, Q. 38:72) to mean "the sons of Adam," or humanity. Qatada (d. 118/736), a scholar from Basra, commented thus on Q. 35:39 ("It is he who has made you *khulafāʾ* on the earth"): "A nation (*umma*) after a nation, and a generation (*qarn*) after a generation." Then the other famous exegete/theologian also from Basra, al-Hasan al-Basri (d. 110/728), as quoted by another contemporary scholar, Ibn Sabit (d. 118/736), declared that *khalīfa* in the Adam verse referred to the "children of Adam," or "people" (*al-nās*), because they follow one another in time ("The Term *Khalīfa* in Early Exegetical Literature," 398).

Q. 2:30[10]—an interpretation given a generation earlier by the
Meccan exegete, Mujahid b. Jabr.[11]

5.   "To rule or govern," as the exegete from Kufa, al-Suddi (d.
128/745), translated in the David verse. More surprising however, is Sufyan al-Thawri's commentary on Q. 24:55 ("God has
promised those of you who believe and do righteous deeds that he
will surely *la-yastakhlifannahum* on earth, as he *istakhlafa* those
who were before them, and that he would surely establish for them
their religion which he has approved for them..."): the people to
be "deputized" here are the governors (*al-wulāt*).[12]

It is clear from the above that Paret's homogenous scheme for the
qur'anic root *kh-l-f*, once again, was overly simplistic. Among some of
the most relevant additional findings in al-Qadi's article, I will single out
the following:

1.   None of the early exegetes denied that Adam was the first
intended referent for the term *khalīfa* in Q. 2:30.[13] Yet because the
angels immediately raise the objection of mischief and spilling of
blood on the part of this person (or group of people), several
posited that Adam was merely a symbol of the human race.
Nonetheless, this may be seen as a positive as well: "Qatada
narrated in his commentary a report on the authority of 'Umar b.
al-Khattab [the third caliph], in which 'Umar said, 'Our Lord
spoke the truth; He did not make us "*khulafā*" except to see how
[we carry on with] our deeds [commenting on Q. 10:14]. Thus,

---

[10] Ibid., 396.

[11] This is not directly in relation to the Adam verse, but to Q. 57:7 ("And spend from
that which He had made you *mustakhlafina* [past participle of tenth form of *khlalafa,
istakhlafa*) in it..."). From two sources, Mujahid translates the preceding as "cultivating it by means of the fortunes [God has granted you]" (*Tafsīr Mujāhid*, ed. Abd
al-Rahman al-Tahir b. Muhammad al-Surati, 2 vols. [Beirut, n.d.], vol. 2:656; and al-
Tabari, *Jāmiʿ al-bayān ʿan taʾwīl āy al-Qurʾān*, ed. Mahmud Muhammad Shakir, 30
vols. [Cairo, 1954–1968], vol. 27:218).

[12] *Tafsīr Sufyān al-Thawrī* (Rampur, 1385/1965), 185, cited by al-Qadi in "The Term
Khalīfa," 404. This is precisely the verse around which Fisher built his article on
qur'anic *Heilsgeschichte* (cf. Chapter 6, n. 69)—such an early opinion should have
added some nuance to his "unitary" translation of *khalīfa* (encompassing al-Qadi's
first three meanings at the most).

[13] Al-Qadi states that "almost all of the exegetes accepted this identification... Consequently: all men are created as *khalīfas*" (ibid., 408). Ironically, this is true of the
first few generations of Muslims as well as for those of the modern period. There
was no such consensus from the third/ninth centuries onward, as we shall see.

show (pl.) God the goodness of your deeds by night and day, in secret and in public.' "[14]

2.   The two dominant meanings chosen were "to succeed" and "to replace," mainly because the close semantic connection between the verb *khalafa* and the preposition *khalfa* ("behind"), and the fact that most qur'anic instances naturally lend themselves to this meaning. The noun *khalīfa*, however, is never formed with a construct, that is, "the *khalīfa of somebody*." That difficulty, in combination with the angels' obvious reticence, caused several commentators to gloss over this verse completely. For others, relying on the readily available stock of Jewish and Christian stories (with Ibn 'Abbas as the chief expert in these *isrāʾiliyyāt*), it was possible to propose several theories as to the beings humanity was replacing on earth when it was created. Yet, the crucial fact to notice here, according to al-Qadi, is that nobody advances the possibility that humanity was God's *khalīfa* on earth.[15]

3.   When the question of what it means to be *khalīfa* is raised, the early exegetes turn to other meanings of the root *kh-l-f* (besides al-Qadi's meanings 1-3).[16] Thus, humankind's "distinguishing function" on earth is "to cultivate" and "to rule."[17] Chronologically, these answers date respectively to the middle and late Umayyad period "with the ever increasing urbanization and complexity of Islamic societies." Here al-Qadi ventures into the kind of hermeneutic I have been advocating in this work: is there some connection, she asks, between the choice of meanings and the socio-cultural, political or economic realities of the time? On her view, this seems to be so in light of the fact that, from a philological viewpoint, the interpretation of *khalīfa* as someone who cultivates or rules is unsound, and can only be justified by appealing to the context.[18] In addition, those last two meanings were always combined with the idea of succession and replacement in time and space.

---

[14] Ibid., 408-9. Al-Qadi shows that Tabari, who records this tradition, immediately follows up with a "forged" tradition about the same caliph and companion of the prophet, mainly because he was seeking an early authority to link the qur'anic *khalīfa* and the caliph—more on this in the next section.

[15] Ibid., 407.

[16] Not surprisingly, al-Qadi authored the *EQ* article on "Caliph," summarizing again her article (vol. 1:276-78). The first three meanings she gave before are now telescoped into one—the meaning of "successor, substitute, replacement, deputy" (277).

[17] Ibid.

[18] Ibid., 407-8.

4. According to the present sources available to us, none of the
    commentators from this period linked the qur'anic *khalīfa* with the
    political caliphate of the time. This fact, however, may be due to
    the huge gaps in our current knowledge of early exegetical litera-
    ture. According to the backward chains of transmitters (*isnād*), our
    sources all stem from hotbeds of political opposition to the
    Umayyad caliphs: Iraq, the Hijaz and Khurasan. What if we were
    to uncover the work of Syrian exegetes, who by definition were
    pro-Umayyad (its capital being Damascus)? This question takes
    even greater importance in the following sections.

The Traditional School of Tafsīr

In this section I aim to present a suggestive sampling in diachronic
fashion of Muslim interpretations of the phrase, *wa-idh qāla rabbuka li-
l-malā'ika innī jā'ilun fi-l-arḍi khalīfatan*: "Behold, thy Lord said to the
angels: 'I will create a deputy on earth'" (Q. 2:30). Since all interpreters
assume that *khalīfa* refers in the text to Adam, the question becomes,
"Does this concern only Adam, or are others implicated as well?" Also,
"How does the interpretation affect the interpreter's theology of humanity?" Conversely, "How is the commentary affected by the interpreters'
theology of humanity?" And then, "Are these interpretations affected by
political and cultural factors as well?" Finally, how is the qur'anic text
treated? Are there appreciable differences in approach within the larger
movements to be scanned in this chapter?

More specifically, I will be testing a working hypothesis: a commentator's view of humankind and *khalīfa* is directly affected by his own
belief in humankind's relationship to God (what connection is there
between Creator and human creature, if any?), to other spiritual beings
(are humans higher or lower than the angels?), and to the physical world
(is it a positive one; or is there any sense of human control over it?).
Finally, in this section I will highlight two men as representatives of this
movement, one in the third/ninth century (Tabari), and one in the
eighth/fourteenth century, (Ibn Kathir).

Tabari: Who is Adam Succeeding?
Standing at the beginning of the *tafsīr* period (end of the Islamic third
century), Abu Ja'far Muhammad al-Tabari (d. 310/923) defines the key
word *khalīfa* as "successor," but, as an historian of Islamic dynasties, he
connects the term to its political usage—something none of the early
exegetes had done:

Khalīfa is the form *faʿīla* derived from the verb *khalafa*, meaning to take someone's place after him in some matter, as in His words: 'Then We appointed you as *khalīfa*-s on earth after them, that We might behold how you would do' (10:14), meaning that He replaced them with you on earth, and appointed you as *khalīfa*-s after them. Because of this, the supreme ruler (*al-sulṭān al-aʿẓam*) is called the *khalīfa* (= caliph), because he replaces the one who was before him, and takes his place in the affair, and is his successor (*khalaf*).[19]

By identifying Adam's title of *khalīfa* as "the supreme ruler" (*al-sulṭān al-aʿẓam*), Tabari tips his hand right from the start: from now on (recall that Sufyan al-Tahwri was the first to point in this direction, though without saying it directly) the qur'anic *khalīfa* in the Adam verse connects to the actual religio-political leader in Baghdad—yet without any sense of rulership. [20] In Tabari's mind, then, it was clear that for the early commentators *khalīfa* meant a "successor," someone who would stand in someone else's place. Almost immediately thereafter,[21] Tabari asks this question: "What resided on earth before mankind, so that mankind could have replaced them and been successors to them?"[22] One might infer from the above that Tabari also rejoined his predecessors in seeing Adam as representing all of humanity as *khalīfa*. Yet this deduction proves to be misleading in the light of the rest of his lengthy discussion.

The first opinion cited is that the *jinn* first populated the earth but began to spread wickedness and bloodshed. As a result, God sent angels to fight them and expel them from the earth. Adam was to take their place. One of the Companions consulted is al-Rabiʿ b. Anas: "God created the angels on Wednesday, and He created the *jinn* on Thursday, and He created Adam on Friday. And a group of the *jinn* disbelieved, so the angels came down to them on earth and fought with them. There was blood, and there was corruption on earth."[23]

---

[19] Abu Ja'far Muhammad b. Jarir al-Tabari, *The Commentary on the Qur'an* (An Abridged Translation of *Jāmiʿ al-Bayān ʿan Taʾwīl al-Qurʾān*), with Introduction and Notes by J. Cooper, vol. 1 (New York: Oxford University Press, 1987), 208.

[20] Cooper, *The Commentary on the Qur'an*, 208. This passage is also quoted by al-Qadi in her *EQ* article, "Caliph" (p. 278).

[21] Al-Qadi notes that at this point Tabari has already quoted Ibn Ishaq's opinion that *khalīfa* in the Adam verse means "to inhabit and to cultivate." Yet he dismisses it out of hand, she notes, and moves on, dogmatically assuming that succession is the only meaning worth retaining ("The Term *Khalīfa*," 396).

[22] Ibid., 209. This for al-Qadi illustrates well what she calls Tabari's "dogmatism": politico-historical dogmatism (the *khalīfa* in Q. 2:30 is the historical caliph); and philological dogmatism (*kh-l-f* means only "succession") (ibid., 395-96).

[23] Ibid.

A second opinion is that the *khalīfa* is Adam and his children succeed him as *khulafāʾ* after him. So each generation succeeds each preceding generation. The third opinion does not really add anything, except to raise the question of who is the one who works corruption and bloodshed on earth.[24] It seems that, uncharacteristically, Tabari inserts his own opinion, thereby rendering the discussion a little confusing. Even the translator has a note pointing out that the text is unclear and that, as it stands, it "is plainly contradictory."[25] Apparently, the various commentators find it difficult to explain why the *khalīfa* chosen by God would act in such an evil way. Most of them conclude, as Tabari does in the end, that the sin could not refer to Adam but to his descendants.

The next sixteen pages are devoted to a discussion mostly relative to the angels. How could they have known beforehand what was going to happen? Were they given hidden knowledge (*ghayb*) or did they simply guess? What was their motive for asking the question in the first place? Was Iblīs an angel, and if not—the Qur'an states elsewhere that he was one of the *jinn*—how did he come to be among them? The *aḥādīth* (pl. of *ḥadīth*) that are cited address these issues, but several also give more details about why God had the angels fall prostrate before Adam. Two main reasons are given. The first is that at the moment of creation God breathes his spirit into Adam: "When the time came when God wished to breathe the spirit into him, He said to the angels: 'When I have breathed of my spirit into him, prostrate before him'" (p. 210). The second reason is that Adam is able to recall "all the names" while they are not capable of this (pp. 220-21). But nowhere does he state that the bowing of the angels is connected to Adam's *khilāfa* (vicegerency/trusteeship). Should it be assumed? Among the various positions proffered about the meaning of "the names" and of the prostration before Adam, Tabari declines to state his own opinion, as he often does. Options are left side by side, many of them contradictory, often leaving the readers to make up their own minds.

The rest of the next forty or so pages devoted to this passage deal with a variety of questions: the creation of Adam, the subsequent creation of Eve, the Fall and the circumstances around it. Many of the traditions are quite different one from another, and often even contradictory. Yet nothing is said about who Adam is and the purpose for his being sent to earth

---

[24] The second part of Q. 2:30 seems to justify this: "[The angels] said: 'Wilt Thou place therein on who will make mischief therein and shed blood? Whilst we do celebrate Thy praises and glorify Thy holy (name)?' He said: 'I know what ye know not.'"

[25] Ibid., 210.

(except as a punishment for disobedience). I would have to agree with Mahmoud Ayoub when he writes: "It may be argued that the purpose of the entire drama of creation was for God to manifest His knowledge and power and to expose the pride of Iblīs."[26] Manifestly, the focus is on God and his conflict with Iblīs, not on any theological statement about the worth of the human person.

Allow me a quick aside on Tabari's hermeneutics. Heath examines the creation story as interpreted by Tabari both in his commentary and in his universal history and shows that in the former his exegetical method is essentially philological, whereas in the latter it is a diachronic narrative method (recall my "academic discipline lens" in Fig. 7). The task of interpretation will arrive at different results depending on the chosen discipline and the methodology used. In his commentary, Tabari examines the qur'anic date phrase by phrase, and when one set of options are given for each phrase, he moves on to the next. The result is a method of hermeneutic pluralism. Of course, he has certainly discounted several other versions on theological grounds, but among those he puts forward he rarely discounts any of them—though he usually, if only implicitly, tips the reader off as to the one he espouses. On Heath's account, by not censoring opposing views he demonstrates a "humanistic" attitude.[27]

Then, with regard to his historiographical method, Heath compares Tabari's *History of the Kings* to al-Kisa'i's *The Tales of the Prophets*. In both cases, details are added to fill in the blanks in the scriptural text and the flow is strictly diachronic. Yet, "the width of the gap separating the two approaches should not be overlooked. At-Tabari is writing serious history, whereas al-Kisa'i is writing popular fiction. At-Tabari maintains sophisticated control over his materials, holding firmly in check the narrative forces he employs."[28] Nevertheless the result is quite different from his commentary. He has been much more selective in his history, while he has also filled in much of the gaps. The finished product is a narrative in which time sequencing is respected throughout, and at the same time, a work of *tafsīr*—only now it includes both the Qur'an and the Sunna. So, in the end, from the same author, two very different types of *tafsīr* come down to us—the difference being rooted in the choice of disciplines and methodologies.

---

[26] Mahmoud M. Ayoub, *The Qur'an and Its Interpreters* (Albany, NY: State University of New York Press, 1984), 75.

[27] Heath, "Creative Hermeneutics," 181-90.

[28] Ibid., 189.

Back to Tabari's *Tafsīr*. Heath calls Tabari's theological choice to highlight Iblīs and the angels the hermeneutical "Principle of Privilege." He places side by side the creation narrative of *Sūrat al-Ḥijr* (Q. 15:26-35) and that of *Sūrat al-Baqara* (Q. 2:30-34), while highlighting the three themes that stand out: "God creates [Adam] physically molding him from clay and breathing His spirit into him; He endows him with knowledge, instructing him rather than the angels the names of things; and He gives him dominion, appointing him His earthly viceroy."[29] Then he points to the fact that in the Genesis creation account there are only two narratives, ones which approximate the themes of the two Qur'anic passages mentioned above: "In the first Genesis story, God forms him in His image and bestows on him dominion over the earth (Genesis 1:26-27). In the second, He shapes him from clay, blows in him the breath of life, and brings him animals to receive names (2:7, 18-20)."

The point here is not to compare the content in detail, but rather to show that "despite the two texts' general equivalences, each religious tradition, the Judeo-Christian or Islamic, grants absolute primacy to only one."[30] Herein lies the "Principle of Privilege." However, as the following discussion of other traditionist *mufassirūn* indicates, the Islamic version of creation is not so univocal. Nevertheless, for all these commentators, the primary focus is on Iblīs and the angels, and only secondarily on Adam—and for Tabari, not on Adam as representing humanity, but only as a single prophet. Nor does the passage make explicit what Adam's role on the earth was to be (except to "settle" there), nor do the traditionalist commentators in this period ask the question. There might be one exception: Isma'il Imad al-Din Abu al-Fida ibn Kathir.

Ibn Kathir: Adam is Granted Favor
Despite the fact that Ibn Kathir (d. 774/1373) lived four centuries after Tabari, his approach to *tafsīr* is much like his. In fact, much of his material is simply copied from Tabari. Yet the sociopolitical situation had changed drastically since the days of his *maître à penser*. Baghdad had long been sacked by the Mongols and Ibn Kathir had witnessed the last throes of the Abbasid Empire. "In many ways, Ibn Kathir was a man of his time, aware of the vicissitudes of Muslim history, mildly polemical, but always fair and informative."[31] He is a "man of his times" in another respect: he is much more critical than Tabari in his use of

---

[29] Ibid., 176.
[30] Ibid., 177.
[31] Ibid.

ḥadith.[32] Further, is it not possible that out of a totally different historical
and cultural context he also comes to the qur'anic text in a different way?
At least he is the only classical exegete cited by Ayoub who sees in God's
announcement to the angels an indication of humanity's high status:

> Ibn Kathir sees in God's declaration to the angels, "Behold, I am about to
> place a vicegerent in the earth" an indication of the special favor of man
> with God. "God tells us in this verse of His favor toward the children of
> Adam by mentioning them in the highest company before creating them"
> (Ibn Kathir, I, p. 120). Ibn Kathir argues that the word *khalifah* here
> refers not only to Adam but also to his progeny. This argument is based
> on the angels' protest, "Would You place therein one who will spread
> corruption and shed blood?"[33]

My own reading of Ibn Kathir supports Ayoub's opinion. Despite
numerous quotes of commentators lifted from Tabari's *Tafsīr*, the atmosphere in and around his remarks is sharply different from the latter's.
Here are his opening words following the text of Q. 2:30: "God (not
'Allah' but *taʿālā*, 'may He be exalted!') announces the pouring out of
his favor (*imtinānihi*) upon humankind (*banī Ādam*) by making special
mention of them before the heavenly council (*al-malāʾa al-aʿlā*)."[34] He
goes on to emphasize that it is not Adam as a person who is designated
by this title, but rather humanity in general. How otherwise could one
explain the angels' objection? Surely, it is the sin of human beings that
they have in mind. Yet even here, there is some ambiguity: for, on the
one hand, the *khalīfa* is "a people succeeding one another century after
century, generation after generation" (quoting several plural instances of
the term in the Qur'an), while, on the other hand, citing the Andalusian
scholar al-Qurtubi (d. 671/1272),[35] he writes, "the angels understood
from the term *khalīfa* that he was the one who would sort out the wrongs
done to people and deter them from crimes and transgressions"—clearly

---

[32] One of the salient aspects of Ibn Kathir's commentary is the influence the anti-
intellectual and often polemical Ibn Taymiyya's work (d. 728/1328) had on him.
Andrew Rippin notes, "Fundamentally antagonistic to intellectual speculation of all
types, whether legal or exegetical, Ibn Taymiyya and Ibn Kathir stand in contrast to
the general tendency in *tafsīr* to allow for diversity. The latter champions dogmatism
in his attempt to juxtapose and reconcile the Qur'an and the *sunna*, both understood
as revealed books" ("*Tafsīr*," *EI*², vol. 10:87).
[33] Ayoub, *The Qur'an and Its Interpreters*, 74.
[34] *Tafsīr al-Qurʾān al-ʿaẓīm*, ed. Mustafa al-Sayyid Muhammad et al. (Giza, Egypt:
Muʾassasa Qurṭuba, 2000), vol. 1:336.
[35] Ibid., my translation. All quotes from Arabic texts in this and following chapters
are my translations, unless otherwise indicated.

a spiritual/political appointment.[36] On the whole, he tends to fall on the side of humanity. Thus, when explaining God's response to the angels ("I know what you do not"), Ibn Kathir explains, "I know what is the greatest benefit (*al-maṣlaḥa al-rājiḥa*)[37] from the creation of this species, which outweighs the detriments you mentioned."[38] Among the advantages of bringing forth humanity is the sending of prophets, messengers (*rusul*), righteous people, including saints, ascetics, scholars and martyrs. Here is a bird's-eye view on Islamic civilization that Tabari could not have entertained four centuries earlier.

Significantly, Ibn Kathir cites Ibn Ishaq's opinion, which Tabari had summarily dismissed: God placing a *khalīfa* on earth means "one who would dwell on it and cultivate it, causing it to flourish and filling it with inhabitants (*sākinan wa-ʿāmiran yaʿmuruhā wa-yaskunuhā*), a creature unlike you [i.e. the angels]."[39] Ibn Kathir's use of the root ʿ-m-r to describe *kh-l-f* comes up again in relation to his commentary on Q. 6:165 ("He is the One who made you *khulafāʾ* on the earth"). He writes, "He made you civilize it generation after generation, century after century, from ancestors to descendants." I translated *yaʿmurūnahā* as "civilize it" in order to indicate that al-Qadi's translation ("cultivate") is weak. This root (ʿ-m-r) is richer than that. It conveys the idea of thriving, abundance and success, as well as the fostering of civilization in its many aspects—hence the word *ʿumrān*, "prosperity," "luxuriance," or "civilization." Ibn Kathir then cites as parallel passages several other instances of *khulafāʾ*, as well as the Adam verse. The legends about *jinn* and angels continue to be passed on, but the interpretation of *khalīfa* as the human race called to populate the earth and cause it to flourish seems to have taken over in his mind. Thus his first comment on v. 34 ("We said to the angels, 'Bow down to Adam'") is the following exclamation: "This is great dignity and favor that God confers on Adam and on his descendants."

What is certain, then, is Ibn Kathir's recognition of the significance of the creation of humanity in the Qur'an. A unique feature of his thinking on this, perhaps, is that he seems to be setting aside the matter of

---

[36] Ibid., 337. Two pages later, he quotes Tabari's statement about the *khalīfa* being "the supreme ruler," that is, the caliph. But this is among many other opinions he duly reports and on which he does not follow up.

[37] *Rājiḥa* literally means "preponderant." It is a common term in Islamic jurisprudence used with *maṣlaḥa* for cases not covered by clear texts in the Qur'an and Sunna. When two possible rulings were considered, both equally permissible as far as the spirit of Shari'a was concerned, the jurist would choose the one that seemed most *rājiḥ* ("likely to him," or literally, "heaviest on the scale" of desirability).

[38] Ibid.

[39] Ibid., 340.

corruption and violence and saying that God is announcing a new and hopeful era with the placing of humans on the earth. This shows that even in traditional Islamic circles of the eighth/fourteenth century the atmosphere had evolved a great deal since the days of Tabari. Having said that, Ibn Kathir transmits many of the same *aḥadīth* that Tabari reported—particularly those that seek to explain why the angels imme-diately raised the objection of harm done to the earth and the spilling of blood. This was done through the presentation of different versions of the origin of the *jinn*, and various scenarios on how the *jinn* were the first inhabitants of the earth but then began to act with envy, violence and strife; usually some tribes of angels are sent down to annihilate the *jinn* and take their place. Such prophetic sayings are often traced back to Ibn 'Abbas, thus tapping into the common stock of Jewish-Christian lore (*isrā'iliyyāt*) available in the first couple of Muslim generations. A good example of such a *ḥadīth* attributed to Ibn 'Abbas is this version of God creating Adam from clay, which Ibn Kathir copies word for word from Tabari's commentary:

> So He sent Gabriel to earth to bring some clay from there, and the earth said: 'I seek refuge in God from you, lest you reduce me or disfigure me!' So he returned without taking anything. He said: 'Lord, it sought refuge with You, so I granted it refuge.' So God sent Michael, and it sought refuge from him... So He sent the angel of death, and it sought refuge from him, so he said: 'I too seek refuge in God, lest I return without executing His command!' So he took [clay] from the surface of the earth, and he mixed the [clay] and did not take it from only one place: he took red, white, and black earth, and that is why human beings came out different [colours]."[40]

Ibn Kathir, despite his celebration of human dignity, which he sees in the Adam verse and in the command for the angels to bow to the new crea-ture, feels obligated to relate traditions from Tabari and others that point in another direction. The above narrative, for instance, portrays the relationship between humans and their physical environment as skewed. Humankind's creation necessitates the angel of death's strong hand to scoop up the needed clay from the earth. There is no hint here of Adam's calling as God's *khalīfa* on earth, though at least a tentative explanation for human diversity is offered. By contrast, when Ibn Kathir attends to his own ideas, it is the concept of civilization that comes to his mind, with the notions of abundance and rationality rising to the surface.[41]

---

[40] Cooper, *The Commentary on the Qur'an*, 215.

[41] I comment on his interpretation of the "names" that God teaches Adam in the next section. Rationality might be an overstatement, but at least there is a recognition of the role of naming in human language, and implicitly the role of language in science.

Their collective placing on earth is reminiscent of the Genesis 1 account, "Be fruitful and multiply, and fill the earth and subdue it" (v. 28). The idea of civilization begins to make its appearance.

The Rationalist School of Tafsīr

By "rationalists" I mean Muslim commentators of the Qur'an who drew deeply from the well of Greek philosophy in order to construct their theological interpretations.[42] Gilliot describes these interpreters as "the dialectical/speculative theologians (*mutakallimūn*).[43] In this section I single out three commentators of distinction: Zamakhshari, Razi, and Baidawi. And then, building on the momentum of the first two traditions of *tafsīr*, I shall offer an evaluation of some of the relevant issues affecting these scholars' hermeneutics—starting with the three-fold hypothesis I put forward in the beginning,[44] and then reserving a sub-section to a particularly thorny, yet capital issue: the politics of the Caliphate.

---

[42] Hermeneutically speaking, they come to the text with Greek philosophical categories in mind, such as the duality of soul/spirit and body, or the concept of universals and particulars, for instance. An enlightening window into the fierce debate still blazing in eleventh-century Baghdad between traditionalists and rationalists (mostly Mu'tazilites, but also in some other areas, Ash'arites) is provided by George Makdisi's *Ibn 'Aqil: Religion and Culture in Classical Islam* (Edinburgh: Edinburgh University Press, 1997). Ibn 'Aqil, intensely imbued with Ibn Hanbal's thinking, is nevertheless a staunch supporter of a moderate use of *kalām* methodology: "He stands between two extremes: the ultra-Rationalists, advocates of the primacy of reason over revelation, and the ultra-Conservatives, who would deny some of reason's rights" (ibid., 93). It is only reason that can determine the authenticity of the Qur'an and the *hadith*. His greatest ire is reserved, not for the *mutakallimūn* (they sometimes succeed in removing the doubts in the believers' minds when reading the Scriptures) but the "sham Sufis," "those contemptible fools [who] have no regard for either reason or the revealed law" (ibid., 96). Yet, even as a person somewhat in between the Traditionalists and the Rationalists, he bases the totality of his theology (*uṣūl al-dīn* leading to *uṣūl al-fiqh*) on Aristotelianism—in a very similar way to Thomas Aquinas, more than a century later. Therefore, with regard to *tafsīr*, the functions of reasons are as follows: "freeing the creed from difficult passages dispersed in the Qur'an and the Prophetic Traditions, and diverting them from being interpreted as God's injustice, or from being taken in the sense of anthropomorphism" (ibid., 95). This could become, admittedly, a fairly heavy-handed hermeneutical grid.

[43] Gilliot, "Exegesis of the Qur'an," 114.

[44] By way of reminder: (1) the relationship between Creator and human creature; (2) the relative rank of angels and human beings; (3) the human relationship to the created order.

Zamakhshari: A Rational Khalīfa *but Angels First!*
The first great exegete in this tradition, 'Umar al-Zamakhshari (538/
1144), was born a century and a half after Tabari died.[45] He is known to
have been a Mu'tazilite in his theology, but compared to such Shi'a
authors as al-Tusi (d. 460/1067) or al-Tabrisi (d. 548/1153) his theology
barely shines through.[46] Gilliot notes that he continues to be read in
conservative Sunni circles for "his qualities as a grammarian, philologist,
and master of rhetoric and literary criticism."[47]

    When moving from Tabari and Ibn Kathir to Zamakhshari, one is
immediately struck by the difference in style and approach—no more the
lengthy quotes from all the predecessors and the panorama of vastly
different perspectives. Zamakhshari seldom quotes anyone's views
(except in footnote-like passages) and engages the reader in the style of
the expert that leads his students with both authority and concision. What
is this *khalīfa* that God is about to place on the earth? His first meaning—and the one he seems to prefer—is "someone who follows someone
else, meaning 'your successor, because they [the angels] had inhabited
the earth and Adam and his posterity replaced them on it.' "[48] This would
seem a clear-cut decision to follow Tabari in at least one of the
isrāʾīliyyāt options he proposes. That might be a premature conclusion,
however. Just a few lines further, Zamakhshari advances a second possibility, though with less confidence: "It possibly means that [God] was
saying, 'I want someone to be my *khalīfa*,' because Adam was God's
khalīfa on his earth—and the same goes for all the prophets—'for We
have made you a *khalīfa* on the earth.'"[49] And then, in good scholastic

[45] According to J. Cooper, "He belonged to the school of theology of the Muʿtazila,
the early rationalist and speculative school that was later eclipsed among the Sunnis
by the theology of the school named after al-Ashʿari, a contemporary of Tabari"
(Cooper, *The Commentary on the Qur'an*, xxiv). He is considered one of the most
authoritative qur'anic commentators, mainly because of his combination of *hadith*,
theological discussion, use of poetry, and lexicographical and grammatical exegesis.
Further, his Muʿtazili influence does not readily appear in the course of his
arguments.

[46] Rippin puts it this way: Zamakhshari "is distinctive primarily for his special
outlook and not for the presence of an overall theological argument per se, nor for
the quantity of such argumentation" ("*Tafsīr*," 85).

[47] Gilliot, "Exegesis of the Qur'an," 115. Ayoub has this comment on his alleged
Muʿtazilite slant: "[h]is Muʿtazili ideas are so subtle that a number of commentaries
have been written to determine where and how his theological bias has influenced
his work" (Ayoub, *The Qur'an and Its Interpreters*, 5).

[48] *Al-Kashshāf ʿan haqāʾiq al-tanzīl wa-ʿuyūn al-aqāwīl fī wujūh al-taʾwīl*, 4 vols.
(Beirut: Dār al-Maʿrifa, 1980), vol. 1:61.

[49] Ibid. The last quote is from Q. 38:26, when God addresses David in those terms.

style, he moves on to a hypothetical question a reader might ask, "Why did God inform the angels of this?"

Clearly, Zamakhshari is stuck with the "succession" idea, and follow-ing the angels is the one and only option he offers. Yet he leaves some room for the "deputy of God" as well, particularly as it pertains to prophets such as Adam and David. Thus in his commentary on the David verse, he explains that God deputized (*istakhlafa*) David over the king-dom (*mulk*) of the earth (al-Qadi's meanings 2, 3 and 5).[50] He then adds an explanatory comparison: "as one who is deputized by several rulers (or 'sultans,' *salāṭīn*) over part of the land and becomes ruler over it. From this [is derived] their saying, 'deputies of God on earth.'" The pro-noun "their" is a clear reference to the title claimed by the late Umayyad and early Abbasid caliphs—a topic I come back to in a further section. So, succession in his mind is also tied to rule and authority, at least in this context, and possibly in the Adam verse as well. Yet when it comes to humanity as a whole, Zamakhshari still prefers the version of Adam and his descendants being called by God to succeed the angels who preceded them on earth.

Further, the weakness of humanity's authority in this world is well illustrated in a passage commenting on Q. 38:75 in which God rebukes Iblīs for not prostrating himself "to one whom I have created with My hands." I offer here Helmut Gätje's translation of Zamakhshari's com-mentary:

> Thus it was on the basis of his higher rank that he was too proud to bow himself before this creature. In this Iblīs did not take into consideration the following: God gave the commandment to bow down (before Adam) to those among his servants who are dearest to him and closest to him in favour, namely the angels. These, rather than all others, would have been entitled to be too proud to humble themselves before the small man and to scorn bowing down before him… [Iblīs] should have known that, by bowing down before a lower being at the command of God, the angels devoted themselves more strongly to the service of God than if they had bowed down before God himself.[51]

Several points can be made following this paragraph. First, angels are of higher rank and dearer to God's heart. Second, humans are relatively insignificant. Third, the main point of this incident has nothing to do with

[50] *Al-Kashshāf*, vol. 3:326. See also W. Montgomery Watt's chapter, "God's Caliph: Qur'anic Interpretations and Umayyad Claims," in Watt, *Early Islam: Collected Articles* (Edinburgh: Edinburgh University Press, 1990).
[51] Helmut Gätje, *The Qur'an and Its Exegesis: Selected Texts with Classical and Modern Interpretation* (Berkeley: University of California Press, 1976), 167.

Adam but is rather to highlight the virtue of the angels and the arrogance of Iblīs. But this is not the whole story; there is more to Zamakhshari's understanding of *khalīfa* in this passage. One also has to take into account his interpretation of the phrase, "And God taught Adam all the names." Here, it seems, his Mu'tazilite leanings come to the fore. He has just finished stating that the previous phrase ("God knows what you do not know") means that God knows "the benefits (or advantages, *maṣāliḥ*) of this (presumably, his naming of Adam and his descendants as his deputies on earth), a matter which is hidden to them." Then, in the question and answer style of the *mutakallimūn*, he goes on: "Then if you said, 'Has He not made plain to them those benefits?' I answer, 'It is enough for humans ("the slaves" or "servants" of God, *al-ʿibād*) to know that all of God's works are good and wise, even if the content of that goodness and wisdom is hidden from them.'"[52] Then he offers this interesting transition: "However, He has shown them some of that [goodness and wisdom] in what follows." He then adds, "And God taught Adam all the names").

After a brief philological interlude (whence the word "Adam" came—it is definitely "foreign," *aʿjamī*), he dives into the heart of the subject. "All the names" means "the names of the things to which they refer" (*asmaʾ al-musammiyyāt*). The construct of "the names" (*al-muḍāf ilayhi*, here *al-musammiyyāt*) was simply elided, Zamakhshari informs us, because it is so obvious. But why, then, are the names more important than their referents? This is so, "because teaching must be connected to names, not to their referents." Note also that God said to the angels, "Tell me the names of these (*hāʾulāʾi*)."[53] "And if you say, 'What does the teaching of the names mean?' I answer: 'He showed him the names of the species (*al-ajnās*) He created, and taught him that the name of this was a horse, that the name of this was a camel, that this was called a such and such, and so on; and he taught him about each object's characteristics and what its benefits might be in this world and in the next." The text

---

[52] *Al-Kashshāf*, vol. 1:62.

[53] There is a grammatical problem alluded to here; but, as with the other commentators, it is never fully faced—much less solved. The plural of inanimate objects in Arabic is rendered by the feminine singular. Out of five pronominal instances relating to the names in this passage, only one (the first one: "And God taught Adam all the names, *al-asmaʾ kullahā*) in feminine singular, and that is because it refers to the word "names." All the rest are masculine plural (*hum*). Ayoub notes that this fact led Tabari to attach the names to "his [Adam's] descendants and those of the angels." This must be the case, he argued, "because in the Arabic language the masculine pronominal suffix *hum* indicates human beings and angels" (Ayoub, *The Qurʾan and Its Interpreters*, 80).

also says that God "presented them to the angels." For Zamakhshari, God presented to the angels the objects or creatures (*al-musammiyyāt*) of which Adam had been given the names. Such knowledge was, apparently, beyond them.

This all seems to make good common sense: the "names" God teaches Adam have to do with science, or the classifying of species on the basis of their characteristics. Yet there is more in this statement than meets the eye. In fact, Zamakhshari devotes a long paragraph to this question in a footnote-like section at the bottom of his page. This is necessary, he writes, because the *ahl al-sunna* (the traditionists and followers of al-Ash'ari, d. 935) believe that "the name *is* its referent."[54] This, he objects, is to ignore what the qur'anic phrase "O Adam, inform them [the angels] of their names" requires. Grammatically, the masculine plural possessive pronoun "their" refers to the referents of the names, and not to the names themselves. Further, to say that the name adds nothing to the referent (by saying that the name *is* the referent) is to deny the existence of universals. A name adds to the existence of a particular being by positing a real link between many particulars that have characteristics in common. That is precisely the link between the name and the referents this conversation between God, Adam and the angels teaches us. "This addition," explains Zamakhshari, "is justified by virtue of the differentiation between the general (or universal) and the specific." Without language, then, teaching would be impossible. The nominalism of the Ash'aris undercuts the possibility of teaching, he complains. If a name only covers a specific thing, then any kind of generalization would be tenuous, and science as we know it becomes impossible. In his words, "The central purpose [of this passage] is to teach the being [or ontological status] of the referents (*dhawāt al-mussamiyyāt*); to examine their realities (*haqā'iqihā*) and what God (may He be exalted) has placed in them, in terms of characteristics and secrets; and finally it is [to examine] their naming."[55]

[54] Ibid.

[55] Ibid. This debate is remarkably similar to the one that was then appearing in Europe. The French monk/scholar Peter Abelard (d. 1142), a contemporary of Zamakhshari, developed a coherent and comprehensive version of nominalism out of the arguments of his predecessors. His concern was to articulate a logical analysis of language that built on Aristotle, but at the same time safeguarded the sovereignty of God and kept faith above reason in theological discourse—a common concern of the Ash'arites. Michael J. Loux's definition of nominalism helps to underscore the issues at stake, both in the medieval Muslim and Christian contexts: "Whereas the Platonist defends an ontological framework in which things like properties, kinds, relations, propositions, sets and states of affairs are taken to be primitive and irreducible, the nominalist denies the existence of abstract entities and typically seeks to show that

At the same time, Zamakhshari tries to avoid controversy and to minimize the extent of disagreement between the two schools. Notice how he expresses the problem: "This is a summary of the issue of the name and its referent as it concerns this verse. God willing, it will suffice, even though this issue—despite its careful scrutiny by the *mutakallimān* with their science of *kalām*—turns out in the end to be a dispute over words (*masʾala lafẓiyya*)." The disagreement between Ash'arites and Mu'tazilites on this issue does not have much to do with reality.[56] If anything, this seems like an attempt to please both sides, while scoring his own points in the process. For in Zamakhshari's theology, the knowledge of the names has to do with the gift of rationality, which in turn is connected to the human trusteeship given to Adam from the beginning. His comment on God's word to the angels, "if you be truthful," is significant in this respect:

> It is God's response to the angels' declaration that He has deputized (*istakhlafa*) those who wreak havoc and spill blood. He is saying that those he has deputized are worthy [of that honor] by virtue of the scientific benefits, which are the foundations (*uṣūl*) of all benefits. He showed them (the angels) all of this and demonstrated to them some of the most beautiful advantages of their deputyship (*istikhlāfihim*) in His saying, "For I know what you do not know."[57]

Finally, in order to situate better Zamakhshari's theological stance, I will briefly compare it with Ibn Kathir, on the one hand, and with the Shi'ite and Mu'tazilite Abu Ali al-Fadl al-Tabrisi (d. 548/1154), on the other. The traditionist and Ash'arite Ibn Kathir, considering the abundance of material on all other aspects of Adam's creation (drawing from other passages as well), says little about "the names of all things." But what is clear is that, from the beginning (several Companions are cited), God's teaching of the names to Adam meant that he gave him knowledge of the universe, from the animals to the stars, including "their smallest particles (*dharrātiha*), their characteristics (*ṣifātiha*), and their actions"; also "the being (*dhawāt*) of the names."[58] Then, in his commentary on the phrase,

---

discourse about abstract entities is analyzable in terms of discourse about familiar concrete particulars" ("Nominalism," in *Routledge Encyclopedia of Philosophy*, vol. 7:17-23, at 7).

[56] *Al-Kashshāf*, vol. 1:63. This conclusion, it seems to me, sweeps under the carpet a difference in conception that he himself just declared of capital importance. The phrasing is purposely ambiguous (*lā yarjaʿu ikhtilafu al-ashʿariyya wa-l-muʿtazila fīhā ilā kabīr min ḥaythu al-ḥaqīqa*), which leads me to think it is more of a diplomatic gesture that a clear statement of his own position.

[57] Ibid., 62.

[58] *Tafsīr al-Qurʾān al-ʿaẓīm*, vol. 1:348.

"And he presented them to [the angels]," he quotes the early commentator Abd al-Razzaq (d. 211/827) as saying that Adam showed them "the
referents" (*al-mussamiyyāt*), presumably pointing to objects of creation
while giving their names. Yet Ibn Kathir offers no theory, or even any
further explanation for the word *al-mussamiyyāt*. On the other hand, he
plainly links knowledge of the names, the trusteeship of Adam and his
descendants, and God's command to the angels to bow down to Adam.
His first sentence in response to the verses about Adam and the names
(Q. 2:31-33) is this: "In this passage God [may He be exalted] mentions
Adam's honor above that of the angels by virtue of his knowledge of the
names of all things—something they did not possess."

Thus we must conclude that, although Ibn Kathir is uninterested in any
philosophical discussion about the meaning of the names (the issues of
both language and rationality), he is just as certain as Zamakhshari about
humankind's dignity, even with respect to the angels. Tabrisi, a younger
contemporary of Zamakhshari, is more open about his Mu'tazilite
theology. His ten-volume commentary, *Majmaʿ al-bayān li-ʿulūm al-
Qurʾān*, despite its clear debt to the work of his Shi'i predecessor, Abu
Ja'afar al-Tusi (d. 460/1067), has become a Shi'i classic until this day.
But after completing the *Majmaʿ*, he discovered Zamakhshari's *Kashshāf*
and decided to write a one-volume abridgment of it. Finally, near the end
of his life and at the request of his son al-Hassan, Tabrisi wrote a four-
volume *tafsīr* entitled *Jawāmiʿ al-jāmiʿ* (or, *al-Tafsīr al-wasīṭ*).[59] Completed in only one year, according to tradition, it represented a distillation
of his previous two works. It is from this last work that I will be quoting.

Tabrisi's first comments on the Adam verse are strikingly similar to
those of Zamakhshari (almost word for word), especially the declaration
at the start that *khalīfa* refers to Adam and his descendants as successors
to the angels on the earth. Tabrisi uses the same phrase, but with an addi-
tive: "It is possible that He wanted 'a *khalīfa* to succeed me,' because
Adam was the *khalīfa* of God on earth, and that is the correct option, if
one looks to His word, 'O David, We have made you a *khalīfa* on the
earth.' "[60] That confident choice of options was lacking in Zamakhshari,
but Tabrisi supplied it for his readers. His Shi'ite identity makes us wonder whether this rare use of *khalīfat Allāh* in the *tafsīr* literature is not in
fact linked to his Shi'ite convictions—a thought to which I will return.

Again, Tabrisi offers the same succinctness for the next section, the
commentary on v. 31: "That is, the names of all the referents" (*Ay asmāʾ*

[59] E. Kohlberg, "Al-Tabrisi," *EI²*, vol. 10:40.
[60] *Jawāmiʿ al-jāmiʿ fī tafsīr al-Qurʾān al-majīd*, 4 vols. (Beirut: Dar al-Adwa, 1985),
vol. 1:46.

al-musamiyyāt kullihā).[61] Then he summarizes in one paragraph what Zamakhshari took two long paragraphs to explain—the subject of names, their referents, and how this relates to the knowledge and science with which human beings have been entrusted.

Tabrisi also goes further than Zamakhshari when he comments on v. 34 (the angels are commanded to bow down to Adam). Whereas Zamakhshari found it difficult to say God esteemed Adam above the angels, Tabrisi writes, "And in this verse [we find] a clear indication of God's favor of Adam over that of the angels, because he puts Adam above the angels by ordering them to bow to him."[62] Then, in the next few sentences, he argues that the command to bow to Adam could only have been to show his preference for Adam (*ʿalā wajh al-tafḍīl*) and his desire to heighten his status (*ʿalā wajh al-taʿzīm li-shaʾnihi*). Why else, he contends, would Iblīs spitefully exclaim, "Seest Thou? This is the one whom thou has honoured above me!" (Q. 17:62).

No doubt, Zamakhshari had given more thought to the knowledge of the names than his predecessors and concluded that humankind is indeed an honored creature, but he cannot escape his own worldview in which angels live in the heavens and obey God without fault and, hence, are superior to humans who inhabitant the earth and are capable of many crimes. The next commentator draws that thought more systematically to its logical conclusion.

Razi: God's Wisdom Upheld

The truly great exegete in the tradition of the rationalists, however, is Fakhr al-Din al-Razi (d. 606/1209).[63] On the one hand, he was a careful student of *falsafa*[64] and *kalām* and his famous qur'anic commentary, *Mafātīḥ al-ghayb* (or more commonly, *al-Tafsīr al-kabīr*), contains by far the most extensive philosophical and theological discussions of any

---

[61] Ibid., 47.

[62] Ibid., 49.

[63] Ayoub gives a helpful summary of his style: "Razi was one of the most learned and brilliant men in Muslim history... Razi was a philosopher of high caliber and not primarily an exegete. He sets forth his opinions on verses in a complex and involved style with layer upon layer of arguments and counterarguments, although often without reaching any conclusion. He also digresses so far from his subject that one becomes lost in philosophical and theological arguments that are at best distantly related to *tafsīr*" (*The Qur'an and Its Interpreters*, 5).

[64] *Falsafa* is the name of the movement that developed philosophy from the Greeks, while adapting it to an Islamic framework. Razi at times commented on al-Farabi (d. 339/950), the first great Muslim philosopher, and he devoted at least two books to the works of Ibn Sina (or Avicenna, d. 428/1037).

other commentary. On the other hand, he was a firm partisan of al-Ash'ari's theology and spent his career refuting Mu'tazilite views.[65] Finally, in contrast to the other commentaries we have consulted, Razi's is encyclopaedic, in some ways like Tabari's, but whereas the latter sought to be comprehensive in his gathering of traditions, Razi attempts to cover all the theological and philosophical issues from every possible angle. Thus he divides his texts into questions (*maʾsala*, pl. *masāʾil*); then some of these are sub-divided into aspects (*wajh*, pl. *wujūh*), and sometimes into even smaller parts. The popular edition I have consulted comprises 32 volumes.[66]

Concerning his exegesis of sura 2, v. 30, Razi brings little new light on the subject of *khalīfa*. His opening statement is this: "I know that this verse points to the way in which Adam was created (upon him be peace) and to the way in which God (may He be extolled) exalted him. This also represented a general pouring out of [divine] favor on all of his descendants."[67] I agree with Ayoub: after enumerating a series of arguments and counterarguments, Razi then declares that the purpose of the announcement to the angels about placing humankind as his deputies on earth was "either that the angels should ask him and hear the answer, 'I know what you do not know,' or in order to teach men the virtue of consultation."[68] A large proportion of the 16 pages devoted to this verse is concerned with the angels—based on all the traditions, who are the major players, why and how they know about the potential evil of the human race, and the issue of their own nature (are they necessarily without sin?). For instance, much energy goes into the question of whether the angels were guilty of questioning God's will. One possible interpretation, he reasons, is that their purpose "was not to doubt God's wisdom, but to marvel at the perfection of His wisdom and its encompassing of all things."[69] This is an apt summary of the attitude of all the classical exegetes on the creation story. Understandably, the main theological focus is on God. In an article about early Islamic creation accounts, Roger E. Tim observes, "References to creation…serve primarily to focus on Allah and various attributes of Allah."[70]

---

[65] See G. C. Anawati's article, "Fakh al-Dīn al-Rāzī," in *EI²*, vol. 8:751-55.

[66] *al-Tafsīr al-kabīr* (Cairo: al-Maṭbaʿa al-Bahiyya al-Miẓriyya, 1934–62).

[67] Ibid., vol. 1:159.

[68] Ayoub, *The Qur'an and Its Interpreters*, 77.

[69] Ibid.

[70] Roger E. Tim, "Divine Majesty, Human Vicegerency, and the Fate of the Earth in Early Islam," in *Essays on Islam: Felicitation Volume in Honour of Dr. Muhammad Hamidullah*, ed. H. M. Said (Karachi: Hamdard Foundation, 1991), 4.

While the focus is first on God, it is secondly on the angels. That this is the case with Razi (the bulk of his material is devoted to them) is also illustrated in his second *mas'ala*. The problem is phrased in these terms: "Some people say: theological discussion (*kalām*) about the angels should take precedence over discussions about prophets."[71] He agrees, for two reasons: (1) the Qur'an tells us to believe in God, his angels, his books and his prophets (in that order); (2) "the angel served as an intermediary between God and the Messenger in the communication of the revelation and the *shari'a*." Another way of posing the problem is the following: "Some people say: theological discussion about prophecies has priority over discussions about angels, because we have no way of ascertaining the existence of angels through reason (*bi-l-ʿaql*), but rather through revelation (*bi-l-samaʿ*)."[72] This simply means that revelation is the origin of our knowledge about angels, and in that sense it takes precedence. "But it needs to be said that the angel comes before the prophet in honor and distinction (*bi-l-sharaf wa-l-ʿilya*) but after him when it comes to our reasoning and thinking (*fī ʿuqūlina wa-adhānina*), because of our reaching them (the angels) through our ideas." Then he adds what I consider to be a fundamental statement of not only his theology, but also his worldview: "And I know that there is no disagreement among the learned and wise (*al-ʿuqalāʾ*) that the honor of rank corresponding to the higher world comes from the presence of angels in it, and that the honor of rank corresponding to the lower world comes from the presence of humanity (*al-insān*) in it."

Despite the angels' higher rank, is Razi still impressed with the divine gift to humankind in the form of rational thinking? Indeed he is. God taught Adam the names of all things, and then he had Adam demonstrate his knowledge to the angels "in order to manifest the perfection of his gift of knowledge (*ʿilm*) and their lack thereof."[73] Then he launches into 37 pages of discussion (*kalām*) on the issue of the names God taught Adam, and then another 27 on the meaning of the angels' bowing to Adam. Though I have no space here to delve into the details of his argumentation, I will offer some illustrations of it.

The first paragraph sets the scope of all the issues (*masāʾil*) to be considered: God taught Adam the names of everything, and as he asked Adam to demonstrate his knowledge of them he was thereby pointing out to the angels their inability to do the same. "He [God] demonstrated for

---

[71] *al-Tafsīr al-kabīr*, 160.
[72] Ibid.
[73] Ibid., 175.

them Adam's favor, a fact that was still unknown to them."[74] His first issue (*masʾala*) is a semantic one. Here, surprisingly, Razi states the view of al-Ashari, which coincided with that of the famous Muʾtazilite, Abu al-Jubba'i (d. 915), namely that the names refer to a concordance of all the languages. He then proceeds to show that this would imply that God had imparted a "necessary" knowledge—imposing certain meanings on objects (*musamiyyāt*).[75] This would mean that both wise and unwise would inevitably come to the same conclusions. Without naming Zamakhshari here (which he had done several times before under the name, "author of the *Kashshāf*"), Razi makes the same distinction between language, teaching and learning. If this knowledge were given wholesale (the "necessary" between the signs/words, *alfāẓ*, and their meaning, *maʿnā*), then "God's character (*ṣifa*) would be known by necessity, in spite of the fact that his being (*dhātahu*) is known by deduction (*bi-l-istidlāl*)."[76] But since he has written about this at length in his work on *uṣūl al-fiqh* (theory of jurisprudence), he does not dwell on it here.

The second issue sets out the meaning of the names as "the description of things (*ṣifāt*), their attributes (*nuʿūt*) and characteristics (*khawāṣ*)." Here Razi continues his argument in the vein of Zamakhshari, asserting that languages can only be known through learning (the "names" are not really about words or language), but "only the mind (*al-ʿaql*) can attain the knowledge of the realities of things.[77]

On the fourth issue Razi takes the Muʾtazilites to task for their assertion that Adam's knowledge of the names was a miracle that pointed to his apostleship (*nubuwwatihi*) at that time. He therefore seeks to downplay the miraculous nature of the event. In the end, he substitutes the word *karāma* (an exceptional act of favor) for the word *muʿjiza* (an act that goes against the customary functioning of things, *nāqiḍ al-ʿāda*). Why the subtle change? He offers three reasons: (1) this is before his sin,

---

[74] Ibid.

[75] Roger Arnaldez explains that one of Razi's purposes in writing his commentary is to "show that Qur'anic Arabic is a model and that it only is capable of communicating to people true wisdom and science" (*Fakhr al-Dīn al-Rāzī: Commentateur du Coran et philosophe* [Paris: Librairie Philosophique J. Vrin, 2002], 85). Many Arab philologists held the view that language was purely conventional. On this view, some prior language must have existed, for how could God have communicated with the angels? God only taught Adam the meaning of the names, and hence the foundation of science. Razi argues the opposite, even stating that God communicated with the angels through another means than language (ibid., 86).

[76] *al-Tafsīr al-kabīr*, 175.

[77] Ibid., 176.

a grave sin in fact (*kabīra*); (2) an apostle is sent to a people, yet there is only Eve at the time; (3) from the narrative in sura 20, v. 122, we read, "And then his Lord chose him"—this is after his sin, not before.[78]

While the third and fifth issues are very brief, the sixth *masʾala* is the longest by far. "This verse points to the virtue of knowledge, for God (may He be exalted) has not revealed the perfection in this creation of Adam but for the sole purpose of demonstrating Adam's knowledge; and if there existed something more honorable than knowledge, then [Adam's] favor would have to be demonstrated by that thing and not by knowledge."[79] Then follows a lengthy demonstration of the excellence of knowledge, first from the Qur'an and Sunna (pp. 178-92);[80] then from the traditions (*āthār*, presumably less reliable and thus not qualifying as Sunna in his estimation; pp. 192-93); proverbs, or sayings (*nukat*, pp. 193-94); stories (*ḥikayāt*, pp. 194-98); "rational indications" that end up referring to the qur'anic as well (pp. 198-201).[81] Finally, while the seventh issue is philosophical (though based on al-Ashʾari—how a definition of "knowledge" can avoid circularity), the eighth is strictly philological, examining 30 synonyms of *ʿilm*, or knowledge (pp. 201-8).

The next section (Razi's commentary on the angels' bowing to Adam) is more homogenous than the preceding one.[82] There are seven issues, but they are treated more from a theological viewpoint, with the philosophical, semantic, philological aspects as ancillary to that perspective. This is so because the overarching question for Razi is whether Adam is higher than the angels or not. The issue of the relative position of angels, prophets and humankind in the hierarchy of being was much discussed in Islamic theological circles in the fourth/tenth and fifth/eleventh centuries. The author of the first-known work on the "roots of religion" (*uṣūl al-dīn*, following many such works on the "roots of jurisprudence"), ʿAbd al-Qahir al-Baghdadi (d. 429/1037), sets out the issues in the clearest manner:

---

[78] Ibid., 177-78.

[79] Ibid.

[80] This includes three enigmatic quotes, one from the "Psalms" of David (*al-Zabūr*, no indication where), and from the gospel (*al-Injīl*, one "in sura 17," and the other quoted by Muqatil b. Sulayman), none of which are to be found in the canonical writings indicated (ibid., 188-89).

[81] He also comes back to the Adam verse, arguing that it is the characteristic of knowledge that qualifies him as *khalīfa* of God "over the lower world" (ibid., 199).

[82] Razi has a short section in between commenting on v. 33, comprised of six issues; it deals mostly with the issue of whether the angels sinned and what was meant by God knowing the "secrets (*ghayb*) of heaven and earth" (pp. 209-11).

The large majority of our friends maintain the superiority of the Prophets over the angels. Some of them admit the possibility of the superiority of some of the Faithful as compared with the angels; they do not, however, mention anyone individually. The followers of Tradition do not teach the superiority of the angels over the Prophets, except al-Hasan ibn al-Fadl al-Bajali. The Mu'tazilites are divided on this point. The majority of them are of opinion [*sic*] that the angels are more excellent than the Prophets, they even consider the guardian angels of the fire as being superior to every Prophet... According to the Imamites the *imams* are more excellent than the angels. And the extreme Shi'ites say they are themselves more excellent than the angels.[83]

As he had stated before, Razi considers that the angels are bowing before Adam as the representative of a new species God is creating. This act of reverence is the fourth "grace" (*niᶜma*) God showers on humankind. This is his opening statement for the section:

I know that this is the fourth of four general graces [bestowed] on all of humanity (*jamīᶜ al-bashar*): He (may he be praised and exalted) had the angels bow before our father because, in first place, He (may he be exalted) he set Adam apart as the *khalīfa*; then secondly set him apart through great knowledge; then [thirdly] through his attainment of such knowledge that the angels were incapable of reaching such a level of knowledge; and now his being the object of the angels' bowing.[84]

Razi then fills up the next 26 pages with *kalām* about the degrees of worthiness between the angels, men and prophets, as well as issues related to Iblīs.[85] What I find particularly noteworthy is that his general anti-Mu'tazilite stance pushes him in a "modern" direction. By this I mean that, since the Mu'tazilites (and *falāsifa*, or philosophers) consider angels to be above humans,[86] and since Razi does not have to worry

---

[83] *Kitāb uṣūl al-dīn* (Stambul, 1928), 295, quoted and trans. A. J. Wensick, *The Muslim Creed: Its Genesis and Historical Development* (Cambridge: Cambridge University Press, 1932), 200-201.

[84] *al-Tafsīr al-kabīr*, 212.

[85] For an exposition of Razi's discussion on various Muslim views on whether or not prophets can sin and his own position on the matter, see Arnaldez, *Fakhr al-Dīn al-Rāzī*, 99-100.

[86] Razi states this in several passages. One of them reads, "The majority of the *ahl al-Sunna* say that the prophets are better than the angels, whereas the Mu'tazilites retort that no, it is the angels who are superior to the prophets—an assertion also made by a large body of the Shi'a" (*al-Tafsīr al-kabīr*, 215). This is not reflected in Gisela Webb's article ("Angels," *EQ*, vol. 1:84-92). Because her focus is mainly on Sufi commentators, Webb's generalization is misleading: "Muslim commentators interpret this qur'anic statement as a demonstration of a human capacity which the angels lacked, that of creative knowledge, the knowledge of the nature of things. By

about a philosophically coherent hierarchy of beings as in Neo-Platonism, he approaches the text about angels bowing before Adam more literally. If they are commanded to bow, he reasons, this must mean that they are in some way superior to the angels. This is "modern," as we shall see, in that, since the Renaissance in Europe, the focus of ideas has increasingly been on humankind's freedom of thought, moral responsibility and scientific abilities.

Take for instance, his twentieth "proof" (*hujja*) for the superiority of the prophets over the angels, from a *hadīth qudsī* (Muhammad's transmission of words from God not found in the Qur'an): "If my servant remembers me before the assembly, I will remember him before a better assembly." This means, Razi notes, that the heavenly assembly (*al-malā'a al-a'lā*) is superior to any earthly one, and thus by analogy that the council of angels is better than a human one. He goes on, "And a human assembly is comprised of notables, not of prophets, so the fact that angels are superior to humans in general (*'āmmat al-bashar*) does not mean that they are superior to prophets."[87] This is the last of the "textual indications" (*dalā'il al-naqliyya*) he notes. Then, in the same sentence (according to this editor, in any case), he transitions to another section: "And I know that the philosophers have agreed on the fact that the heavenly spirits called angels are better than the human rational spirits (*al-arwāḥ al-nāṭiqa al-bashariyya*), and that they have relied upon intellectual considerations (*wujūh 'aqliyya*)." Razi then offers twelve arguments advanced by the philosophers to prove their point (pp. 228-32). Each one is then countered by Razi's own objections, and, if that were not enough, he continues answering each one individually again in summary fashion (pp. 232-35).

Two points in particular prefigure the Muslim modern emphasis on humankind's *khilāfa*. The first is his emphasis on humankind's heavy burden of passions. Thus when the philosophers state that "spiritual substances are exempt from the passion that causes the spilling of blood whereas human spirits are bound to it,"[88] Razi objects: "No doubt, the

---

virtue of his knowledge of the names, Adam became master over created things... The story is also seen as an affirmation of man's vicegerency. God creates Adam as his vicegerent on earth (Q. 2:30) and ordered the angels to prostrate before him; hence the view that the human being (*insān*) is superior to the angels" (ibid., 86). As a matter of interest, Razi nowhere uses *insān* in these passages—only "sons of Adam," *bashar* or *ins*. By contrast, in contemporary literature, *insān* is used ubiquitously.

[87] *al-Tafsīr al-kabīr*, 228.

[88] Ibid. He states the philosophers' conclusion: "that which is free from a source of evil is more honorable than that which is tempted by it."

perseverance in one's service in spite of numerous hindrances and obstacles is a better indication of sincerity than such perseverance without hindrances and obstacles; and that shows that humanity's status with regard to love (*maḥabba*) is higher and more perfect."[89] Again and again, Razi highlights the cost of human obedience to God and therefore its greater merit before him. Angels are almost forced to serve God by their constant praises, except that they do share one passion with human beings, namely, pride. In the discussion about whether the angels sinned or not in this story, Razi believes they did; and certainly Iblis gave way to his arrogance and ambition. His second "modern" emphasis is related to the first. Because obedience is so difficult for people, they find greater pleasure (or delight, *ladhdha*) in it. Herein lies the privilege (*maziyya*) of humanity, and it may be that to which Q. 33:72 ("the *amāna* verse") is referring. The trust that God offered to the heavens, the earth and the mountains, which they refused, he offered it also to humanity, who accepted it. This may be the pleasure of drawing near to God through courageous service—like the delight one has in good things after having been deprived of them for a long period.

In the end, however, Razi remains ambivalent with regard to the human race as a whole. Sometimes he can be read as saying that humankind is higher than the angels because of its gift of knowledge—this is helped by his argument to the effect that Adam was not yet called as a prophet at this stage. Yet, at other points, he reverts to the idea that only prophets can be superior to angels. Thus, in a comment about Q. 33:33 ("For God chose Adam, Lot, Abraham and the family of Imran over all others," lit. "over the worlds"), Razi says "God chose them above all creatures, and no doubt angels are among those creatures; this verse inevitably leads to the conclusion that God (may He be exalted) chose these prophets above the angels."[90] This ambivalence is common to the whole period.

---

[89] Ibid. This is the only mention of love as a characteristically human trait in these passages.

[90] Ibid., 233. The seventh proof advanced by the philosophers is a good indication of his reasoning. Based on Q. 4:172 ("Christ disdaineth not to serve and worship God, nor do the angels, those nearest [to God]," Ali), it may be said that Jesus was inferior to the angels, but not Muhammad, Moses and Abraham (who were also higher than Jesus). Yet even this is to miss the point of the verse, he argues. For grammatical reasons, this is not a *mubālagha* structure in which the second term is added for emphasis (and therefore would make the angels superior to Jesus). In fact, he concludes, all the prophets are superior to the angels (ibid., 220-22).

Baydawi: Angels, Jinn *and Prophets*

The last exegete I will cite in this category is the famous Nasir al-Din al-Baidawi (d. 716/1316). His *tafsīr* (mostly an abridgment of Zamakhshari's *Kashshāf*, with added grammatical notes and an overall attempt to blot out traces of Mu'tazilism) has remained, over the years, the standard Sunni commentary and "one of the single most popular commentaries in the Muslim world.[91] Baydawi brings little new light in his discussion of Q. 2:30. In his introductory statement he emphasizes God's honoring the human race by bestowing three graces upon them: "Adam's creation, the honoring of his person [through his deputyship], and his preference over the angels, as they were ordered to bow before him, a bestowal of grace that includes his descendants."[92] This wide definition of "human exaltation," however, is progressively narrowed as the commentary proceeds.

What is a *khalīfa*? It is "someone who succeeds someone else, who acts as someone else's representative (*yanūbu manābahu*)."[93] This concerns Adam, he adds, "for he was God's *khalīfa* on earth, as well as every prophet God made his deputy (*istakhlafahum*) for the sake of civilizing the earth,[94] the management of people's affairs (*siyāsati l-nās*), the perfecting of their souls and the application of [God's] command among them." Yet *khalīfa* could also mean "one who succeeds those who inhabited the earth before them, or he [Adam] and his descendants, because they succeeded those before them, or they succeed one another." He had mentioned before that the *jinn* first occupied the earth, but that due to their corruption a group of angels (including Iblīs) was sent down to destroy them and take their place. Now it is humankind's turn. Part of the purpose of this announcement to the angels, hypothesizes Baydawi, is to teach them the art of "consultation (*al-mushāwara*), to elevate the status of the one made to be [*khalīfa*], in that He proclaimed the glad tidings (*bashshara*)[95] of people who were to inhabit His kingdom (*malakūt*), and granted [Adam] the title [of *khalīfa*] before creating him." Baydawi continues his sentence with a meditation on the profound theological implication of God's choice as "the demonstration of His

---

[91] Gilliot, "Exegesis of the Qur'an," 116; Cooper, *The Commentary on the Qur'an,* xxv.

[92] *Anwār al-tanzīl wa-asrār al-ta'wīl,* ed. Mahmoud 'Abd al-Qadir al-Arna'ut, 2 vols. (Beirut: Dār Sādir, 2001), vol. 1:53.

[93] Ibid., 54.

[94] The Arabic reads *fī ʿimārati l-arḍ,* a verb I have previously translated as "inhabit" or "cultivate" and as I mentioned previously, it is the root that gives us "civilization" (its synonym *ḥaḍāra* carries more the idea of occupation and settlement).

[95] The use of the verb *bashshara* here may be an intended pun on Baydawi's part: it is the same root as *bashar* ("humankind," like *insān*).

favor overriding the negative impact foreseen in their questioning; [God] answers and shows that wisdom demands the bringing to life of that which is preponderantly good, because to forgo a greater good for a lesser evil would be a great evil in itself."[96]

With regard to the names, God "inspired [in] him [Adam] the knowledge of the nature of things (or 'being': *dhawāt al-ashʾya*), their characteristics and their names, as well as the foundations of science (lit. 'sciences'), the laws behind the professions (or handicrafts) and the working of their instruments."[97] What is remarkable then, in comparison to Zamakhshari, is that his discussion of the names stops here. Gone are the philosophical speculations about names and their referents, language and the meaning of knowledge. Overall, Baydawi is impressed with the honor bestowed on Adam and agrees with Zamakhshari that part of it should be attributed to humankind's gifts of knowledge and ingenuity. Here his theological expression takes on noticeable mystical intonations:

> [God] made Adam the focus (*qibla*, as in the direction of ritual prayer) of their bowing in order to raise his status…as if He created him in such a way as to make him a model of all his innovations, in fact of all that is, a means by which the angels could fulfill the perfections assigned to them, and a link revealing to them the levels and degrees that were different than theirs.[98]

Baydawi's marveling about man's stature in the universe, however, seems to be mitigated by his ambivalence on the issue, as we saw in the opening paragraph. Adam is honored, and all of humanity as a result of their God-given knowledge and creativity. On the other hand, Baydawi also attributes Adam's privileged status to his prophethood and it is only this factor that, in the end, justifies God's command to the angels to bow before him.[99]

Here humans are seen as superior to angels—though not humans in general, only prophets. All said and done, this is no lofty theology of humanity. It does underscore a point, however, that can be traced in just about all we have seen so far among some notable authorities on *tafsīr*, which is that these commentators all assume a cosmology, in which angels and *jinn*, legends and stories about creation and the prophets play predictable roles. In this play, only prophets may hoist themselves above the angels.

[96] Ibid.
[97] Ibid., 56.
[98] Ibid., 57.
[99] Ibid., 58.

A First Evaluation
It may be that my hypothesis is substantiated here: these interpreters' low view of Adam's caliphate (with the traditionalists not seeing it as applying to humanity at all) seems to be connected to:

1. Their inability to see—or fear to admit—any connection between humanity's capacities of reason and free will and God's attributes as their Creator. Very likely this comes as a result of the great theological battles of those centuries. Anything that smacked of anthropomorphism was deemed anathema.[100] Also, with heated debates (backed with persecution, depending on the Caliph in power) on free will versus determinism, and especially while the Ash'arite side was gaining the upper hand, it would be difficult to imagine scholars emphasizing humanity as God's representatives on earth, at least in official *'ulamā'* circles.
2. Their worldview, which dictated angels as God's messenger par excellence, carrying out God's will on earth. Only prophets may rise above them as transmitters of God's messages to humanity.
3. Their view of the physical world: in spite of breathtaking advances in mathematics, astronomy, medicine, art and architecture, the human physical environment remained an often threatening, mysterious and awesome reality in the medieval world. It was not until the Industrial Revolution that people began to gain a solid sense of mastery over nature.

The Red Flag of Caliphate Politics

I see a fourth reason for the conscious (maybe unconscious in most cases) decision to avoid pouring into the word *khalīfa* any notion of sovereignty or dominion. This may well turn out to be the weightiest reason of all:

---

[100] Yet there are some intriguing *aḥādīth* on this subject. Dirk Bakker speaks of two *hadīth* using the word *ṣūra* ("form, image") for the image of God in man. "The one, that is, that Allah created man in the image of the Benefactor, *'alā ṣūrati l-Rahmān*, is considered to be apocryphal. The other, which reports that Allah created man in his (own) image, *'alā ṣūratihi*, is interpreted by the Muslim theologians as follows: Allah created man in the image of God, which existed previously in Allah in an intelligible form. The Mu'tazilites regard it as a confirmation of the existence of reason, *'aql*, in man, whereas some Ash'arites understand by it that Allah gave man understanding concerning the existence of his attributes" (Bakker, *Man in the Qur'an* [Amsterdam: Drukkerij Holland N. V., 1965], 26). Wensick gives several references for this *hadīth* (*fa'inna l-Allah khalaqa Ādama 'ala ṣūratihi*): Ahmad b. Hanbal, vol. 2:244, 251, 315, 323, 434, 463, 519; al-Bukhari, *isti'dhān*:1; Muslim, *birr* 115; *janna* 28 (*Concordance et indices de la tradition musulmane*, ed. A. J. Wensick and J. Mensing, vol. 4 [Leiden: Brill, 1962], 438).

the political history of the Islamic community since its inception, and the institution of the Caliphate in particular. The word *khalīfa* is a highly charged political term, starting with Abu Bakr's election right after Muhammad's death. T. W. Arnold quoted al-Shahrastani (d. 548/1153) in saying that "no article of faith has given rise to such bloodshed and contention in every period of Muslim history as this."[101] The Sunni doctrine in question was first expressed in the form of a *hadīth*. Arnold summarizes the pertinent details:

> [O]ne, that he must be of the tribe of the Kuraish...and the other, that he must receive unhesitating obedience, for whosoever rebels against the *Khalīfa*, rebels against God... This claim on obedience to the despotic poser of the *Khalīfa* as a religious duty was impressed upon the faithful by the designations that were applied to him from an early date— *Khalīfa* of God, and Shadow of God upon earth.[102]

It is not surprising to find the *'ulamā'* consistently resisting the appeal of the despotic Umayyad and Abbasid caliphs to Q. 2:30 and 38:26 to justify their hold on power. Some of the Umayyad claims, for instance, were outrageous: "The earth is God's; he has entrusted it *(walla-hā)* to his *khalīfa*: he who is head on it will not be overcome." Or: "We have found the sons of Marwan pillars of our religion as the earth has mountains for its pillars."[103]

---

[101] T. W. Arnold, *"khalīfa," EI²*, vol. 4:884. This debate goes on today. The Indonesian writer Harun Nasution argues in a consciously Mu'tazili way "that Islamic politics and *kalam* arose out of a single troubling event in nascent Islam, the assassination of the third caliph or successor to Muhammad as head of the Muslim community, 'Uthman ibn 'Affan (d. 656). That the civil war *(fitna)* followed had enormous implications for subsequent Islamic political and social history is not in dispute. The interpretation of that first *fitna* and the ones that followed, however, has been disputed throughout Islamic history" (Martin and Woodward, with Dwi S. Atmaja, *Defenders of Reason in Islam*, 9). Nasution's text argues that these civil wars raised profound theological disputes. In particular, he asserts "that Mu'tazilism argued for a middle ground that sought to preserve human ethical responsibility as well as social cohesion" (ibid., 10).

[102] Arnold, *"khalīfa,"* 884. While dealing with the origins of the imamate in his book, *The Themes of Islamic Civilization* (Berkeley: University of California Press, 1971), John A. Williams writes, "The origins of the caliphate-imamate have been the most troubled question in Islamic history" (p. 61). Williams also quotes Arnold (from his landmark book from 1924, *The Caliphate*), who notes that the majority interpretation of *khalīfa* in the Qur'an as "successor," but also that "other Muslim authorities interpret *'Khalīfa'* as meaning a vicegerent, a deputy, a substitute... It is obvious that such an interpretation could be used to enhance the dignity and authority of the Caliph" (p. 60).

[103] Watt, "God's Caliph," 62.

Patricia Crone and Martin Hinds' arguments are worth pondering at this juncture. Against much traditional scholarship which sees the religious claims of the Umayyads and Abbasids as a mere after-the-fact ruse for adding legitimacy to their power, they argue that *khalīfat Allāh* in the Qur'an and as used by the caliphs did originally mean "Deputy of God" (and continued to mean that throughout both Umayyad and Abbassid periods) and that from the very beginning the caliphate was considered both a political and religious institution. The suspicion that the *ʿulamāʾ* essentially rewrote early Islamic history to protest the abuses of power by the Damascus and Baghdad caliphs in order to find authoritative religious support for their view is confirmed for these scholars by three factors:

> First, those reported to have rejected the title *khalīfat Allāh* add up to Abu Bakr, 'Umar, 'Umar II and the *ʿulamāʾ*, or in other words the *ʿulamāʾ* and their favourite mouthpieces. Statements attributed to the first two caliphs and 'Umar II are usually statements by the *ʿulamāʾ* themselves, especially when the statements in question are of legal or doctrinal significance. Why should statements on the nature of the caliphate be an exception?[104]
>
> Secondly, our sources claim that Abu Bakr and 'Umar rejected the title of *khalīfat Allāh* for that of *khalīfat rasūl Allāh*, adding an apocryphal story about 'Umar intended to drive home the message that *khalīfa* means successor. In other words, *khalīfat rasūl Allāh* makes its appearance in a polemical context.
>
> Thirdly, if the caliphate was conceived as successorship to the Prophet, why did the title *khalīfat rasūl Allāh* more or less disappear? After Abu Bakr and 'Umar it is not met with until early 'Abbasid times, or in other words not until the *ʿulamāʾ* had acquired influence at court under a regime conscious of its kinship with the Prophet. And even then, it failed to acquire much prominence among the caliphs themselves.[105]

Besides adducing a wealth of evidence for this view (early documents, court poetry, and coinage of the period), Crone and Hinds also appeal to

---

[104] To this discussion should be added the evidence from the *ḥadīth* literature. For another serious treatment of the Caliphate, see Louis Gardet's *La Cité Musulmane: Vie Sociale et Politique*, 4th ed. (Paris: Librairie Philosophique J. Vrin, 1981 [1954]). On page 154 n. 2, Gardet presents the *aḥādīth* that are most often cited by the legal (*fiqh*) defenders of the caliphate, from al-Mawardi to Rashid Rida (last century) and concludes that (1) they are not well attested *ḥadīth* in general, and (2) there are many more which give the opposite view: "In fact, on no other subject have apocryphal *ḥadīth* been so abundant, according to the circumstances and the political necessities. They are nonetheless instructive of the mentality and common conceptions [of their times]" (ibid., 154).

[105] Patricia Crone and Martin Hinds, *God's Caliph: Religious Authority in the First Centuries of Islam* (Cambridge: Cambridge University Press, 1986), 22.

the sociopolitical dynamics of the founding of Islam. Speaking of God's spectacular intervention into Arab affairs through the Prophet, they make the following point:

> No sooner had He sent a prophet to the Arabs that He made the super-powers of the day collapse, enabling His adherents to leave their imme-morial life of "sand and lice" for incredible wealth and power in the Fertile Crescent and beyond: to the exhilarated participants in this venture, God was synonymous with success. Everything which happened was God's own handiwork on their behalf; and it was this which made it seem natural to them that He should have a representative on earth here and now, however the idea may have suggested itself to them in the first place.[106]

I would have to agree with Arnold, therefore, when he writes, "In none of these verses is there any clear indication that the word *khalīfa* was intended to serve as the title of the successor of Muhammad."[107] Muslim historians (Sunni of course) point out that the title *khalīfat rasūl Allāh* was first used by Abu Bakr. What is certain is that the expression *khalīfat Allāh* was consistently used by both Umayyads and Abbasids (with an appeal to Q. 2:30 and 38:26) and that it was hotly contested both by Kharajites and Sunni *ʿulamāʾ* throughout.

Muhammad Qasim Zaman, in his review of this book, calls the new interpretation thought-provoking, but remains unconvinced. Most of the evidence, he objects, is taken from the official sources censured by the caliphs.[108] His own conclusion is, "Even so, the title [*khalīfat allāh*] represents an effort for the enhancement of religious prestige, and must not be taken as a demonstration of the existence of religious authority."[109] Zaman may be underestimating the impact and authority of al-Mawardi's (d. 450/1058) *al-Ahkam al-Sultaniya* (Governmental Statutes), the first official exposition of the theory of the caliphate. Rosenthal notes that it was written at the height of the Abbasid empire, yet the striking fact is that even after it began to decline with the rising competition between caliph and sultan, the *ʿulamāʾ* "had to keep faith with the ideal *umma* headed by an imam with sole supreme authority, and to preserve the

---

[106] Ibid., 109.

[107] Arnold, "*khalīfa*," 881.

[108] Adel Allouche also points out that court poetry would be expected to embellish and flatter the rulers. Hence poetic license could account for much of the ideology surrounding the title *khalīfat Allāh* ("God's Caliph," review article, *Muslim World* 79.1 [1989]: 72).

[109] Muhammad Qasim Zaman, "God's Caliph: Religious Authority in the Early Centuries of Islam," *Islamic Quarterly* 32.1 (1988): 61.

inheritance of Muhammad and his *ummat al-Islām*."[110] This delicate
balancing act will be assiduously pursued by no less towering figures
than al-Ghazali and Ibn Khaldun. Ibn Taymiyya is the only exception.
Rosenthal offers a useful reminder of the established view that Ibn
Taymiyya was trying to counter—the view presumably held by most of
the (Sunni) *mufassirūn*:

> Only Ibn Taymiya ignored the political struggle and tried, by his concen-
> tration on the *Sharī'a*-government for the *umma*, to create the conditions
> necessary for the reconstitution of a Muslim community guided by the
> Sunna of the Prophet... But since his reforms tended to lead back to a
> golden but primitive past through his narrow interpretation of what con-
> stitutes Sunna and what constitutes *bid'a* (innovation), his plea for a
> *siyāsa shar'iyya* went unheeded, and his appeal to the *umma wasṭa* met
> with little or no response.[111]

In Martin Hinds's posthumously published *Studies in Early Islamic
History*, he argues that the murder of the third Caliph, Uthman, reveals
the presence of three distinct groups of opposition to him among the
Islamic leadership in Medina: the *muhājirūn* (who "emigrated" to Medina
with Muhammad in 622) and *ṣaḥāba* (Companions) who were incensed
by some of Uthman's "innovations" in Islamic law and the appointment
of his own family members to prominent positions in the newly founded
empire; the *anṣār* (early Medinan supporters of Muhammad) whose
position of influence in their home town had been considerably dimi-
nished by his policies; and finally, the *muhājirūn* like Talha and al-
Zubayr who objected to Uthman's favoring the Banu Umayya over the
rest of the Quraysh tribe. As a result, Ali, the next caliph, lost his way
and, in the end, his own life, trying to navigate the rapids between the
"Islamic" leadership and the tribal leadership.[112]

---

[110] Erwin I. J. Rosenthal, *Political Thought in Medieval Islam: An Introductory
Outline* (Cambridge: Cambridge University Press, 1958), 60.

[111] Ibid.

[112] Martin Hinds, Jere L. Bacharach, Lawrence I. Conrad and Patricia Crone, eds.,
Studies in Late Antiquity and Early Islam (Princeton, NJ: Darwin Press, 1995), 29-
55. Interestingly, the ex-Director General of the International Institute of Islamic
Thought, AbdulHamid AbuSulayman, seems to espouse a similar view: "During the
reign of the Righteous Caliphs, the leaders struck a harmonious balance between
religion and politics ("Islamization of Knowledge with Special Reference to Political
Science," *American Journal of Islamic Social Sciences* 2.2 [1985]: 267). True, he
does not discuss the *khalīfat Allāh* issue, yet he declares that Muhammad himself
established the caliphate (ibid., 265-66). The leadership split as a result of the
increasing influence of traditional tribal leadership in the Islamic community and

Whatever the conclusion one might adopt regarding the relation between Q. 2:30 and the caliphate (or the meaning that was actually ascribed to *khalīfa* in Q. 2:30 immediately after the Prophet's death), one cannot escape the fact that the politics of the next centuries weighed heavily on any *tafsīr* effort. Certainly in this kind of religio-political setting, it is not difficult to see why traditionalist (or traditionist, i.e., specialists in *hadīth*) *mufassirūn* would tend to shy away from ascribing to Adam the title of "God's Deputy." In this respect, it is significant that the only commentator to opt explicitly for the deputyship of Adam by God over humanity's replacing the *jinn* on earth was the Shi'ite Tabrisi. True, Tabari had made an overt connection between the Adam verse and the Baghdad caliphate of his time, and Zamakhshari (in spite of his Mu'tazilite sympathies) had seemingly linked the David verse with the official rhetoric of the caliphs.

However, partly because of the angels' objection and partly because of the plural verses in the Qur'an (*khalā'if* or *khulafā'*), from the beginning most commentators opted for all of humanity as the target of this title. This factor should caution us not to overstate the political factors that undoubtedly shaped the exegesis of the Adam verse. After all, Tabari's reference to the saying of the caliph 'Umar that all Muslims were God's *khulafā'* in order to be tested by God does not easily fit Crone and Hinds' thesis. If all are *khulafā'*, then what is so special about their ruler? Would Umar say that only he is God's deputy, perhaps as Adam was by virtue of being a prophet? We cannot know for certain. Yet Tabari's next quote—perfectly contradictory—making the same caliph cite the Adam verse in support of his own office, sounds more like a forgery, as al-Qadi alleges. In the end, we have to conclude that while the political controversies did affect the interpretation of *khalīfa* in Q. 2:30, more research is needed to determine its full extent. After all, much controversy still shrouds the history of the early Muslim community.

The Mystical School of Tafsīr

The Mystical School of *Tafsīr*, more than the preceding movements, elevates the status of humanity in its theology. Yet the concerns are different. Regarding one of the earliest Sufi commentaries, Gerhard

---

hence the progressive wedge driven between the intellectuals (*'ulamā'*) and political leaders. A possible rapprochement might be seen here between Sunni and Shi'a, since the latter have fervently maintained from the beginning the unity of the political and spiritual succession of Muhammad—though exclusively through 'Ali's line.

Böwering indicates that its author, Sahl al-Tustari (d. 283/896), "views God from the vantage point of a mystic's enlightened experience, not from the standpoint of a professional theologian."[113] God may be distant and unknowable (*ghayb*) by virtue of his infinite knowledge and power, yet he also draws intimately close to individuals who seek him as the "immanent secret (*sirr*)" of their innermost being. Two names stand out from later generations: Nisaburi, whose work reflects the more popular side of Sufi piety, and Ibn al-Arabi, the towering figure of Islamic mysticism. Before that, however, I will mention a word about al-Ghazali (d. 505/1111), a contemporary of Ibn Aqil, who bridged, not simply between Rationalism and Traditionalism, but between those two and Islamic mysticism. Then I will look at al-Jami's treatment of the Perfect Man, simply because he represents the climax of Ibn al-Arabi's school.

Ghazali: The Divine–Human Inner Relationship
Two centuries before Baydawi conceded that prophets such as Adam could be God's deputies on earth, Ghazali argued that by virtue of an "inner relationship" (*munāsaba bāṭina*) between Creator and creature, Adam and his descendants were indeed appointed by God to represent him on earth. Here is a useful summary by Fritz Steppat of Ghazali's *Iḥyāʾ ʿulūm al-dīn* on the subject:

> He also found in the inner relationship an explanation for the dictum that God had created Adam after his form: it did refer to God's form, but his inner form, not the outer form, implying among other things "man's closeness to his Lord in the properties which he is obliged to imitate, and in accepting the moral qualities of divine glory"; only sufis would be able to find and name other elements in the relationship. Al-Ghazali's conception of man obviously was far ahead of that of the majority of orthodox scholars in his time, but he did mark the way for them.[114]

Ghazali, in fact, goes one step further. This special endowment of humanity by God through Adam is to be identified with their mission as his *khulafāʾ* on earth. Lamb quotes from the *Iḥyāʾ* on this very point in a passage commenting on sura 15, v. 29 and God's breathing into Adam of his spirit:

---

[113] Gerhard Böwering, *The Mystical Vision of Existence in Classical Islam: The Qur'anic Hermeneutics of the Sufi Sahl At-Tustari (d. 283/896)*, ed. Bertold Spuler, Studien zur Sprache, Geschichte und Kultur des Islamischen Orients (Berlin/New York: W. de Gruyter, 1980), 145.

[114] Fritz Steppat, "God's Deputy: Materials on Islam's Image of Man," *Arabica* 36 (1989): 166.

For this reason the angels of God made obeisance to Adam. It is understood also from this verse: I have made you my successor in the world. The Prophet Adam would not have been fit for *khilafat* unless he was given the connection (*munasabah*) of spirit. The Prophet said, hinting at this connection: God created Adam according to His image.[115]

Clearly, Ghazali is here exegeting the Qur'an as a mystical thinker, one who has consciously distanced himself from the Greek-oriented philosophers/theologians of *kalām*. He is no doubt conscious, on the other hand, of the potential wrath of the traditional exegetes who cautiously avoided any link between the human and the divine. In two places in the *Iḥyā* he states that "the special endowment of humanity must not be spelt out in detail in a book."[116] Yet in the same passage quoted above, Ghazali cites a *hadīth* in which God rebukes Moses for not visiting him (God) when he was sick, a passage reminiscent of Mt. 25:31-46.[117]

Nisaburi: Humanity Reflects God's Light in the World
With Nizam al-Din al-Hasan al-Nisaburi (d. 730/1329) one takes a step further into Sufi thought. Besides his abundant borrowings for both Zamakhshari and Razi, Nisaburi in his *Gharā'ib al-Qur'ān wa-raghā'ib al-furqān* (Wonders of the Qur'an and desirable features of revelation) often engages in "spiritual exegesis" (*ta'wīl*)—a term Gilliot uses for the more esoteric, mystical interpretations gleaned from various Sufi commentators.[118] With regard to humanity's *khilāfa*, he expressly states that humankind is God's vicegerent, and, as such, has been put in charge over the whole created order: the physical and the spiritual, the earthly and the

---

[115] Lamb, *The Call to Retrieval*, 42.

[116] Ibid., 4.

[117] Ibid., 43. Here I bring up one of Cragg's articles, "Islam and Incarnation," in *Truth and Dialogue in World Religions: Conflicting Truth-Claims* (Philadelphia: Westminster Press, 1974), 128, emphasis original. I must agree here with Cragg that Christians entirely sympathize with the zealous Islamic iconoclasm and its uncompromising defense of God's unity. For now (I come back to him later), I present one side of his total argument, the assertion of the necessary "connection" between God's proscription of idols and the human prophets' innate capacity to cooperate with God's injunctions. Having cited the foundational *lā ilaha illā Allāh*, he comments, "The source of prophethood is within the unity. Thus it is exactly that urgent and total 'dissociation' of God from idols, and from human representation therein, that involves and requires his authentic 'association' with human servants and with the human scene, in order that the meaning of His lordship may be obeyed and confessed among men. We might almost say that there *are* prophets *because* there *are not*, and should not be, idols. The mission of the former and the insistent non-entity of the latter are reciprocal demonstrations of the divine unity."

[118] Gilliot, "Exegesis of the Qur'an," 113.

heavenly. Because humans possess divine qualities they are able to shine God's light in the world.[119] The individual's heart is a glass in which the oil of the spirit burns brightly. "Thus, when man's lamp is so illuminated with the fire of the light of God, he becomes God's vicegerent in His earth, manifesting the light of his attributes in this world through justice, well doing, compassion, mercy, kindliness, and domination."[120] These attributes, he concludes, are beyond the reach even of angels.

Ibn al-Arabi: Humans, Ontology and the Divine Form
Whereas Nisaburi relied heavily on Tabari and Zamakhshari, Ibn al-Arabi was a creative thinker, outstanding in his spiritual insights and powerful in his arguments. "He brought together mysticism, philosophy, and theology in a vast synthesis, reflected in a complexity of technical vocabulary which has engaged the attention of commentators of his works to this day…"[121] On the subject of Adam's caliphate Ibn al-Arabi declares: "Man is My vicegerent forming his character according to My character, and is known by My characteristics. He executes My command and rules over My creatures—managing their affairs, organizing their government, and calling them to obedience to Me."[122]

---

[119] Nisaburi comments on the knowledge of the "names" which God taught Adam. He relates that some commentators see those names as human languages. He then quotes Mu'tazili thinkers who reason that, since languages are a human invention, God only gave Adam the capacity to create new languages. "It is also possible that He taught him all the expressions of peoples who were before him. It may also be that He taught him the attribute of things, their descriptions, peculiarities, and religious and worldly uses" (quoted in Ayoub, *The Qur'an and Its Interpreters*, 80). Although this passage is more philosophical than mystical, it does reflect the centuries of Islamic progress in the physical sciences, linguistics and history. Again, the questions asked of the text are changing. Noticeably, a definite concept of culture is emerging—and the gifting to create it comes from God, as we saw with Ibn Kathir's concept of civilization.
[120] Ayoub, *The Qur'an and Its Interpreters*, 78.
[121] Cooper, *The Commentary on the Qur'an*, xxvii.
[122] Ayoub, *The Qur'an and Its Interpreters*, 79. Abdurrahman Badawi shows how Ibn Arabi sees humankind as placed by God in the center of the universe. He quotes from a chapter entitled "On Human perfection" in the ʿUqlat al-mustawfiz: "God has created humankind as a noble summary in which he has gathered the concepts of the macrocosm and of the names in the heart of God's presence. The Prophet (May God pray for him and bless him!) says of humanity: that God created him in his image. This is why we say that the world is formed in the image (of God)… As the Perfect Man is in the image (of God), the caliphate and deputyship on God's behalf are his full right and prerogative" ("L'humanisme dans la pensée arabe," *Studia Islamica* 6 [1956]: 84, my translation).

A doctrine of humanity, no doubt, must be linked to a corresponding doctrine of creation. In this regard, Annemarie Schimmel offers an enlightening analysis of the mystical tradition in Islam and its connecting humanity's creation, life on earth, and judgment.[123] All Muslims would follow the clear teaching of the Qur'an concerning the meaning of creation and say that "everything is created to worship God and to serve Him in veneration."[124] Yet Muslim mystics, particularly in Ibn al-Arabi's school, come to this point through the *ḥadīth qudsī*, "I was a hidden treasure and wanted to be known, therefore I created the world." God is seen not only as the totally transcendent God, but perhaps even more as the "longing and desiring God who in a certain way needs man and creation so that His names and attributes can be properly reflected."[125] Love, then, becomes the appropriate experience in the relationship between Creator and created. And it is in this connection that one notices the linking of humanity's "caliphate" and assuming of the "trust" (*amāna*), as Razi had done before (though more tentatively). Schimmel explains:

> The notion of "hidden treasure" is central to Sufism but since it is not found in the Koran it need not be accepted by the faithful as binding truth. The Koranic statement about the *amāna*, the 'entrusted good' (Sura 33:72) is more widely relevant: when God created the world he offered something in trust to heaven and earth. They refused to accept it, but man took it, yet proved ignorant and tyrannical; he did not properly know what to do with this treasure. The orthodox might understand the *amāna* as man's duty to believe and to surrender to God's will, a duty which man constantly forgets; the mystics might see here the secret of love which works through creation but can attain perfection only in man, for mountains would melt under its burden.[126]

This connection will become completely assumed by modern writers, as I point out in the next chapter. But the context is entirely different. Instead of consisting of a mystical relation with God for moderns, the human deputyship is either free will or a mandate to rule over the created order under God's headship.[127]

---

[123] Annemarie Schimmel, "Creation and Judgment in the Koran and in Mystico-Poetical Interpretation," in *We Believe in God: The Experience of God in Christianity and Islam*, ed. Annemarie Schimmel and Abdoldjavad Falaturi (New York: Seabury, 1979).

[124] Ibid., 155.

[125] Ibid., 156.

[126] Ibid., 157.

[127] Schimmel herself continues with this thought: "Others, particularly the modernists, speak of the trust of free will... Or the trust may be the position of Divine

A closely related issue between humanity and creation for the mystics is the idea of "the perfect man" (*al-insān al-kāmil*), but this philosophical/theological doctrine takes several forms. Not surprisingly, the person most often associated with this idea is Ibn al-Arabi. Arthur Jeffery presented a translation with notes on Ibn Arabi's *Shajarat al-Kawn* (*The Tree of the Universe*) in two successive articles showing how his original expression *al-insān al-kāmil* refers to the Prophet of Islam. He is identified as the "perfect man," "the firstborn of Allah and the archetypal man."[128] Jeffery argues that the four main points of the Christian doctrine of Christ as the Logos are transferred to Muhammad in this essay (without any notion of salvation or redemption, as one would expect). On the other hand, the identification of the *al-insān al-kāmil* with the Prophet does not essentially detract from his view that humanity as a whole serves as God's caliphs on earth. The light emanating from the World Tree is the light of Allah and the first creature to be formed from this light is the Light of Muhammad.

How are Adam and his descendants related to *al-insān al-kāmil* for Ibn al-Arabi?

> As the universe was brought into being to manifest the glorious nature of Allah, so mankind was brought into being that the nobility of Muhammad might be made manifest, and the first objective of Iblis is to spoil that manifestation… The Light of Muhammad, having fulfilled its function in creation, necessarily had a part to play in history after creation was completed, and so Ibn al-'Arabi brings it into connection with Adam, and Adam's conflict with Iblis, and with the succession of Prophets after Adam, in each of whom the prophetic light was derived from the Light of Muhammad.[129]

The peculiarity of his overall scheme, however, is that the universe is a monad—all is essentially one, and all that happens (including Adam's encounter with Iblīs) is predetermined and recorded in the Preserved Tablet.[130] This kind of extreme determinism seems to deny the possibility

---

vicegerency, which was granted to Adam (Sura 2:28), and which he did not always use to the best. This trust would have enabled him to perfect God's creation and to rule it; but he forgot this prerogative again and again and neglected his God-given duties" (ibid.).

[128] Arthur Jeffery, "Ibn Al-Arabi's *Shajarat Al-Kawn*: Introduction," *Studia Islamica* 10 (1959): 51 n. 2.

[129] Ibid., 56-57.

[130] After years of research and writing on Sufism, William C. Chittick has more recently come back to Ibn al-Arabi with one volume on the framework of his cosmology (*Principles of Ibn Al-'Arabi's Cosmology: The Self-Disclosure of God*

of free will for humanity and, in that sense, at least in our postmodern age, this must be seen as reducing the dignity and scope of our so-called vicegerency.

Nevertheless, at the risk of betraying his prodigiously fertile and complex thought, I will venture three more remarks on humanity and *khalīfa* in his system:

First, human beings are pivotal in God's scheme of creation. They are the link between the only true reality, Reality of Realities, or the Breath of the All-Merciful (God says "Be!" and that thing is engendered), or pure *wujūd*, and contingent existence, that is, everything outside of God that derives its existence from him. Strictly speaking, only God has existence. And when we speak of him we must at once enunciate two seemingly contradictory propositions: he is incomparable (God's *tanzīh*), and yet at the same time we experience *wujūd* in a way that enables us to make some sense of our existence. Thus we find in ourselves some commonality with God, and that Ibn al-Arabi calls God's *tashbīh*. In the middle, between existence and non-existence, is an isthmus or *barzakh*, which only humans can bridge through their God-given faculty of imagination.[131]

This leads to the possibility of human perfection, which is for William Chittick the core of Ibn al-Arabi's teaching. Notice again the Sufi use of the image (*ṣūra*, or "form") of God in humanity:

> Human beings are different from other creatures because they are forms of the whole, while other creatures are parts. "God created Adam in His form," and He likewise created the cosmos in His form. Both the cosmos and the human being are integral forms of God. But the cosmos displays the divine names and attributes in indefinite dispersion in all worlds and all space and time, so each individual thing in the cosmos can only be an

[Albany, NY: SUNY, 1998]). Two factors make a study of his thought an arduous task. The first is his methodology. As Chittick points out, Ibn al-Arabi is not a Western linear thinker. There is "no point to get to": "He is simply flowing along with the infinitely diverse self-disclosures of God, and he is suggesting to us that we leave aside our artificialities and recognize that we are flowing along with him. There is no 'point,' because there is no end" (ibid., xi). The second factor is closely tied to the first. His philosophical framework is Aristotelian—with a great debt to Ibn Sina—with touches of Neoplatonism and the whole being contextualized in a magisterial way according to the qur'anic worldview that held sway across a good part of the known world in the fifteenth century.

[131] Chittick explains, "As a faculty of the soul, imagination is able to perceive the self-disclosures of God, discerning his presence in each thing. Imagination, in other words, can perceive the words of the All-Merciful as God's self articulation. It sees the things as He even as reason recognizes that they are not He" (ibid., xxvi).

infinitesimal part of the whole. In contrast, human beings are called upon to display the entirety of the divine form. The degree to which they achieve this goal establishes their worth as God's servants and vicegerents and determines their situation in this world and the next.[132]

Second, on the surface, at least, it seems that humanity is called to be God's vice-regents on earth and that they will be held accountable for the extent to which they are faithful to this calling. Yet for Ibn al-Arabi, trusteeship is circular. In Q. 73:9 we are exhorted to take God as our Trustee (*wakīl*). All of being is rolled up into the concept of *tawḥīd* (the unicity of God) and nothing can escape the paradox of a human act accomplished freely in response to God's command; yet at the same time it was God's acting through this person. In this sense, both we humans and God himself—though in different ways—are vicegerents, or trustees. Consider this passage translated by Chittick:

> Hence the property of vicegerency and trusteeship—which is a vicegerency and a deputyship—never ceases to be perpetual and endless in this world and the last world. After all, the Real *Each day* [*sic*]—the "days" being the breaths—*is upon some task* [55:29] in which you have made Him your trustee, since for your sake He acts freely and for your sake He takes control of that in which He has taken you as vicegerent. Hence you act freely by command of your vicegerent, just as he is the King of the Kingdom through trusteeship.[133]

In spite of the "novelty and freshness" of Ibn al-Arabi's teachings "that proved irresistible to a large body of Muslim intellectuals,"[134] he was also a man of his time. I offer the following passage as a reminder that, even in the fifteenth century, it would have been impossible to totally extricate a doctrine of humanity as God's *khulafāʾ* on earth from its political overtones: "The *khalīfa*—caliph or vicegerent—is he who comes after someone else and succeeds him or represents him in his function. Thus a caliph is 'later' in relation to the person whom he succeeds. The political caliphs succeeded the Prophet in his legal and governmental functions." But for him as a Sufi, the politics of the *umma* only represent surface or secondary phenomena. At a deeper and more significant level, "The cosmic vicegerents stand in God's stead in ruling the cosmos, since they are His realized forms so they are God's vicegerents and Muhammad's inheritors."[135]

[132] Ibid., xxi.
[133] Ibid., 78, emphasis original.
[134] Ibid, xvi.
[135] Ibid., 204. It is also clear from the following that the concept of Adam and hence, the whole human race, as God's *khulafāʾ* on earth is not central in Ibn al-Arabi's

Jami': Adam and the Theophany of the Name

The theme of the Perfect Man was further elaborated by Ibn al-Arabi's disciples. According to Chittick, the person who best summarizes "a whole school of thought in himself and brings it to a climax is Abd al-Rahman ibn Ahmad Jami' (d. 898/1492).[136] Jami', the much lauded Persian poet, scholar and mystic of his day, represents the culmination of Ibn al-Arabi's school. In his writings on the Perfect Man, with regard to his being the "ontological prototype of creation," three points are made which relate directly to the discussion of humanity and *khalīfa*:

1. Adam is taught the "names" (Q. 2:31) and these refer to the attributes of God. This means that he was taught the name "Allah," the Name that embraces all the names. Adam represents all of humanity, though in fact only those who would reach the state of perfection. He is the first prophet and, "as the knower of Allah is the first locus of theophany for that Name in the world and the first corporeal manifestation of the eternal reality of the Perfect Man."[137]

2. Humanity provides the reason for the creation of the world, on the basis of the previously quoted *hadīth*, "I was a hidden treasure and I wanted to be known, so I created the world." In Neo-Platonic fashion, Jami' reasons that God, who is infinite and perfect, needs to add finite existence to the divine Essence in order to make it truly perfect. Hence creation is seen as process of emanation: "This finite existence itself adds a new dimension of knowledge to the non-manifest Essence, for the Names and Attributes which in the Essence are known only inasmuch as they are one with the Essence, are known in manifestation or creation as separate and distinct realities in the midst of multiplicity."[138]

thought. His focus is much more on God, *tawhīd*, and the process by which people ascend the stations of knowledge, first through the revelation of the Qur'an, *hadīth* and Sunna, then through the practice of *sharī'a*, and then finally through the esoteric path he is tracing himself, he the self-proclaimed "seal of the saints." Inevitably, the Sufi way breeds a form of elitism: "[He has lifted some of you above others in degrees], that some of them may take others in subjection [43:32] through the bestowal of the perfection of humanity—the [divine] form—on some of them. They are the ones whom God has lifted up. Those above whom they have been lifted are the animal human beings" (ibid., 77).

[136] W. C. Chittick, "The Perfect Man in the Sufism of Jami," *Studia Islamica* 49 (1979): 138.

[137] Chittick, *Principles of Ibn Al-'Arabi's Cosmology*, 144.

[138] Ibid., 149.

3. In a clearer fashion than Ibn al-Arabi, Jami' identifies the Perfect Man, the locus of the theophany of the Name "Allah" as Adam, referred to as God's *khalīfa* in Q. 2:30. As God's vicegerent, the Perfect Man is first and foremost the *barzakh* between this world and the next, and as keeper of the World of Imagination (*ʿālam al-khayāl*), the intermediary between the physical world and the spiritual world, and thus embracing the attributes of both. This makes for a substantive and comprehensive theology:

> It is precisely man's quality of being an isthmus which has made him worthy of being God's vicegerent. Since he is an isthmus, he comprises the attributes of both lordship and servanthood. Through his attribute of lordship, that is, his divine nature, he takes from God what the creatures demand. And through his attribute of servanthood he is able to establish contact with the other creatures and to see that they receive what they need from God.
>
> Expressed differently, the Perfect Man is the means whereby the world is maintained...
>
> It follows that without man, there would be no world.[139]

A Second Evaluation

At the beginning of this chapter I offered the following three-fold hypothesis as a working hermeneutical tool: a commentator's view of humankind and *khalīfa* is directly affected by his own belief in humankind's relationship

1. to God (what connection is there, if any?)
2. to other spiritual beings (e.g. are they higher or lower than the angels?)
3. to the physical world (is it a positive one? Is there any sense of human control over it?).

As seen in this chapter, the traditionalist and rationalist schools carefully refrained from positing any similarity between God and humans.[140]

---

[139] Ibid., 15. The tension in Ibn al-Arabi between humanity vs. the Perfect Man is still not resolved in Jami'. Chittick moves back and forth himself from one to the other. Besides the conceptual difficulty resulting from the monistic view of *wujūd* ("existence/being"), I would point out again the strain imposed by the mystical attraction to the "chosen few." This could not be the kind of theological foundation upon which a Muslim-Christian coalition for peace, justice and responsible earth-keeping could be built.

[140] Watt may be correct in stating that "in the Qur'an and in early Muslim thinkers no use was made of the conception of human freedom." This would have been a declaration of rebellion of the slave (*ʿabd*, the most common epiteth of man in the

But this is where the mystical tradition comes in, with an appeal to the *hadīth* about God's creating humans according to his *ṣūra*/form. Thus Al-Ghazali spoke of the *munāsaba bāṭiniyya* (the inner connection) as a link between the divine and the human. But he does so gingerly—after all, it is not in the Qur'an, and as a respected *ʿālim* he cannot afford to be branded a Muʿtazilite. Ibn al-Arabi is much bolder, as are his followers, hence incurring the condemnation of the *ʿulamāʾ* on this and other issues.[141] For these mystical thinkers humanity is the peak of creation, the *barzakh* that holds being together with non-being, the eternal forms in God's Essence, and the fleeting appearances of this-worldly things. And when humans attain to the level of perfection for which they were created, they truly function as God's trustees, his *khulafāʾ* on this earth.

At the same time, a discernable evolution in worldview seems to have taken place: across the board, by the seventh/thirteenth century, commentators are more impressed with the achievements of human research and science, creativity and cultural artifacts in a more urban environment. Thus Ibn Kathir marvels at the favor God bestows on Adam as the progenitor of the human race, which is called to inhabit and cultivate (with the added notion of civilization, al-Qadi's fourth meaning of *khalīfa*); and Razi ponders the weight of moral responsibility that sets humans apart from angels (and above them when it comes to prophets).

In the end, however, only the mystics seem to have been able to rise above the fascination that medieval people had with angels and *jinn* and their tendency to subordinate everything to them, including humans. The traditional attitude tends to lead to a demeaning of the natural realm—an attitude shared by the mystics, however, partly because of their strong Aristotelian biases, and partly because of the prevailing worldview in both the East and West at the time. Thus, what is lacking in all three

Qur'an) against his Master. He goes on, "Thus freedom could be in no sense an ideal to be striven for, but only a disaster to be avoided. The chief way in which the modern Muslim rises above the status of slave is when he becomes God's agent or steward in this world" (*Islam and Christianity Today: A Contribution to Dialogue* [London: Routledge & Kegan Paul, 1983], 127). Then he explains Q. 2:30, which parallels the thought of medieval Christian thinkers. Only the modern period, with its characteristic turn to the autonomous self, signals a change in this way of thinking about humans and freedom.

[141] Chittick mentions one of the main reasons for the opposition he often stirred up: "One was certainly his challenge to the received authority of the *ʿulamāʾ*. Not that he denied the authority of transmitted knowledge, but he was harshly critical of all claims to have codified and rationalized Islamic knowledge with finality, and it was upon such claims that religious authority was built" (*Principles of Ibn al-'Arabi's Cosmology*, xv).

schools of thought in the medieval period is a positive engagement with the created order. Undoubtedly the natural sciences had leaped ahead in Baghdad and Almoravid Spain. Yet, as mentioned previously, there was not a sense of human beings being capable of mastering the physical world and harnessing its resources for the fulfillment of their needs—and much less so in Medieval Europe. For this kind of worldview one must skip from Ibn al-Arabi in the thirteenth century directly to the eighteenth. India's Shah Wali Allah appears to be "the bridge between medieval and modern Islam in India," and beyond.[142]

I close this chapter with two more comments on hermeneutics. Here I lean on Peter Heath's analysis of hermeneutical styles in his overview of the *tafsīr* work of al-Tabari, Ibn Sina, and Ibn Arabi, with special reference to their attitude toward the qur'anic text. Keeping in mind that these three men represent the traditionalist, rationalist and mystical schools respectively, his assessment is worth quoting here:

> In the first stance, the interpreter accepts the privileged status of the text and consciously relegates interpretation to a lower position. The latter's subservient position is never questioned. In the second, the interpreter rejects, implicitly or explicitly, the privileged nature of the text and attempts to displace it by reading it in terms of other texts or modes of discourse granted higher privilege. In the third, texts (or modes of discourse) are accepted as equal, varied but congruent expressions of the truth. Here the hermeneutic of one discourse confirms the other; the other, the one."[143]

A traditionalist Muslim scholar might quip, "This confirms what I have always thought. The rationalists privileged Greek philosophy above the sacred text and the Sufis imposed their mystical interpretations on the text at the expense of its plain meaning." Yet in the light of Chapter 5, I would answer that we all come to any text, and especially one that we consider to be the fountainhead of our faith, with an "agenda." Our culture, our socioeconomic and political situation, the academic discipline(s) we privilege in our study, and our personal issues—all these reality domains and their corresponding validity claims are lenses through which we all read the text. All three schools of *tafsīr* consider the Qur'an to be God's eternal Word. They are simply using different perspectives through which to view and determine its meaning for them.

---

[142] Aziz Ahmad, "Political and Religious Ideas of Shah Wali-Ullah of Delhi," *The Muslim World* 52 (1962): 22.

[143] Heath, "Creative Hermeneutics," 203. By way of clarifying the last category, Heath had just shown that Ibn al-Arabi affirmed both the "inner" and the "outer" meanings of the Qur'an. So, as a Sufi, he could hold simultaneously to a literal and an esoteric interpretation (ibid., 202).

My second and last comment relates to the first reality domain in Francis Fiorenza's hermeneutical model: language. When it comes to the Arabic of the Qur'an for Muslims, scholars have consistently noted that language in interpretation is much more than just symbols guided by rules of grammar and syntax. It has a life of its own because of how and where the text was revealed. For this reason translations of the Qur'an, till today, are only tolerated as paraphrases, and thus carry little spiritual weight in the Islamic community. Ibn al-Arabi's high regard for the Arabic language may not be an extreme example. To those advocates of a strong version of postmodernism—"reality is determined by language"—he would reply that they are correct, except that the source of language is not the self but God. Chittick explains how language, self, the world and God fit together for Ibn al-Arabi:

> The self is nothing but a word articulated by the Essence, and the Essence remains always and forever beyond articulation. So also, every language, whether meta-, cosmic, human, or infrahuman, is an articulation of Unarticulated *Wujūd*... In one word—the speech of God is articulated in the Koran. This means that the Koran, as Chodkiewicz has amply illustrated, is his constant point of reference. Over and over again, his basic intention is to bring out the meaning inherent in this divine speech, whose articulation is aimed specifically at establishing human awareness of self, cosmos, and God.[144]

Admittedly, his view of language is heavily imbued with Neo-Platonic philosophical categories, but his "mystical" respect for the language of the Qur'an is still shared by Muslims generally today. Nonetheless, many other elements of one's worldview, which, as I have shown in this chapter, contribute to differing readings of the text. And in particular, as I hope to show in the next chapter, humanity and *khalīfa*, as a gloss of Q. 2:30, can take on vastly different meanings in the modern period.

---

[144] Chittick, *Principles of Ibn al-'Arabi's Cosmology*, xxxiii-xxxiv.

Chapter 8

TAFSĪR OF Q. 2:30: MODERN PERIOD

From the height of the classical period before the Mongol seizure of
Baghdad (656/1258) to the eighteenth century, a good deal of political
reshuffling, consolidation of the Islamic sciences, and phenomenal
growth of the Sufi brotherhoods took place in a vast Islamic realm
stretching from Spain and North Africa to the Indonesian archipelago.
Yet apart from the work of commentators and theologians in the school
of Ibn al-Arabi and a growing interest in *falsafa* even in ʿ*ulamāʾ* circles,
little new ground was broken in these areas. Hodgson, in his description
of the intellectual traditions in the "Later Middle Period" (1258–1503),
isolates two main characteristics in the general trend toward conserva-
tism in Islamic education. First, knowledge that was worth transmitting
was conceived as a finite amount of true sayings and memorizable
formulas from the past. "Hence not only was knowledge, in principle, a
fixed corpus of statements; its authenticity was made to depend on the
word of a limited number of great men, whose authority was not to be
questioned, at least not by the student."[1]

The second fact flowing out of the conservative worldview that ruled
supreme in premodern times, both East and West, was that education was
seen as primarily the inculcation of norms—how people were meant to
behave according to their station in society. Even the philosophers and
Sufis, who naturally could engage in more innovative content than their
ʿ*ulamāʾ* counterparts, used the method of apprenticeship (as in all the
crafts) in order to pass along not just facts but the guiding values of the
good life proper to their social standing. By the eighteenth century,
however, with the expansion of the Western powers now being seriously
felt in the heartland of Dar al-Islam, several Muslim thinkers began to
see the need to rethink and reformulate their Islamic legacy.[2]

---

[1] Marshall G. S. Hodgson, *The Venture of Islam: Conscience and History in a World
Civilization*, 3 vols. (Chicago: University of Chicago Press, 1974), vol. 2:439.
[2] Charles Kurzman lists five major threats posed by the advance of the "Christian"
West: (1) *military*: the superior firepower and technology of Western armies;

Indeed, the eighteenth century was a pivotal time for Muslims world-wide. Though it may have been a period of decline militarily and politically, it was also a period of invigoration, growth and expansion for the Sufi *ṭuruq* (pl. of *ṭarīqah*), and with them, for the increasing influence of Muslim merchants and *'ulamā'* in Africa and Southeast Asia. But even more than that, according to John O. Voll,

> The eighteenth century is also a time of laying the foundations for Muslim experiences in the modern era… A number of groups of modern importance have direct roots in eighteenth-century renewalism. The best known is the Wahhabi movement which provides the base for the Kingdom of Saudi Arabia… Other organizations with inspirational roots in the eighteenth century which have helped to shape twentieth-century political concepts can be seen in North Africa, Nigeria, the Sudan, and Somalia. In South Asia, a major eighteenth-century Muslim thinker, Shah Wali Allah, has been credited with providing the Islamic intellectual foundations for most Islamic thought in the modern era.[3]

Yet if this bubbling of qur'anic interpretation, flurry of Sufi organization and flocking to Islamic institutions in general is to be called "revivalism," then it could only be so in a conservative sense. In itself, it hardly represented new directions in Islamic thought and practice, but rather a call for purification and a strict return the teaching of the Qur'an and Sunna. Though he no longer uses this term, Voll used to call this return to the sources a "fundamentalist" mode of religious activism: "In each age of Islamic history there are people who try to create effective links between the vision and the reality. It is, in many ways, a critical element in the emergence of fundamentalist or revivalist movements: how does the individual revivalist become inspired?" The linking, he continues, varies according to the particular situation. In the eighteenth

---

(2) *economic*: the generation of new products and commodities that were now desired by the elites in the Muslim world and the West's vast industrial capabilities; (3) *intellectual*: the advance of Western science created a gap that is still present today; (4) *political*: the Muslim intelligentsia also attributed the West's superiority to its constitutional form of government—social peace fosters advancement in all other areas; (5) *cultural*: Western mores including the consumption of alcohol and the mixing of the sexes in public threatened to undo the traditional Muslim social fabric (*Modernist Islam, 1840–1940: A Sourcebook*, ed. Charles Kurzman [Oxford: Oxford University Press, 2002], 6-7).

[3] John O. Voll, *Eighteenth-Century Renewal and Reform in Islam*, ed. Nehemia Levtzion and John O. Voll (Syracuse, NY: Syracuse University Press, 1987), 18-19. See also Voll's chapter, "The Foundations of the Modern Experience: Revival and Reform in the Eighteenth Century," in *Islam: Continuity and Change in the Modern World* (Syracuse, NY: Syracuse University Press, 1994), 24-83.

century, however, the vehicle for this was "a special combination of *Hadith* studies and social reform-motivated *tariqah* affiliation."[4] This amalgamation afforded both the vision of what the reform should be and the community framework for carrying it out. This is the context of Shah Wali Allah's contribution.

Shah Wali Allah: Transition to Modernity

Though Shah Wali Allah (1703–1762) and Muhammad ibn Abd al-Wahhab studied at the same time in the Hijaz, probably with some of the same teachers of *hadīth*, and while both acknowledge their debt to the revivalism of Ibn Taymiya, their contribution to the future development of Islamic thought would be very different. The Wahhabi movement is often considered the spiritual father of contemporary islamism, whereas Wali Allah's irenic disposition and comprehensive philosophy is seen as the foundation for "liberal" Islam.[5] Yet, of the two, it was the latter who truly laid the foundation for the revitalization of Muslim thought, already in the eighteenth century, which continues (albeit in diverse streams) until today. And much of the turning point, I will argue, revolves around his theology of humanity.

Wali Allah was perhaps, more than anything, a reconciler of seeming contradictory and opposing movements. As a South Asian his worldview already integrated the holistic and reconciling approach of Hinduism, and though he was also a strict follower of orthodox Sunni belief and practice, in his own person he was able to harmonize many loose and disparate strands of the Islamic practice of his day. From his father he inherited the leadership of a Naqashbandi Sufi *ṭarīqah*. Wali Allah kept as a goal throughout his life the completion of the work initiated by Sheikh Ahmad Sirhindi a generation before, "the channeling of the streams of Sufi spiritual heritage into Islam, reoriented on the basis of the Prophet's traditions."[6] Further, as a Sufi thinker, he looked to Ibn al-Arabi for inspiration in constructing a comprehensive philosophical framework.[7] In this

---

[4] Ibid., 88.

[5] Kurzman in the Introduction to the first edited book, *Liberal Islam: A Sourcebook* (Oxford: Oxford University Press, 1998), also considers him the founder of what he calls "liberal" Islam. It is the view—with many shades of meaning and positions— that "Islam, properly understood, is compatible—or even a precursor to—Western liberalism" (ibid., 6). As such, it stands against both "customary Islam" and "revivalist Islam," the other two main streams of contemporary Islam in his analysis.

[6] Ahmad, "Political and Religious Ideas of Shah Wali-Ullah of Delhi," 23.

[7] Levtzion and Voll confirm this influence on eighteenth-century scholars: "Some of the major figures in the eighteenth-century revivalist networks of scholars…were

tradition he saw himself as the *qāʾim al-zaman*, "the religious pivot of the age," who, in exemplifying the internal (*bāṭinī*) nature of *khalīfa*, "is entrusted with the responsibility of giving direction to the *ʿulamāʾ*, the rationalists, the legists and the Sufis."[8] In order to respond to the challenges of the day, Aziz Ahmad argues, Islam had to be "liberal, resilient, tolerant and composite":

> As the *qāʾim al-zaman* of his own particular age, he considered it to be a part of his mission to restore the solidarity of the ummah by emphasizing a formula of compromise based on whatever was commonly accepted by the various sects of Islam, and by force of conciliatory logic to blur the dividing line between the mystic and the theologian,[9] between the Muʿtazilite and the Ashʿarite; but even more specially between the four orthodox schools of law in Sunni Islam.[10]

In relation to his view of humanity I see three contributions made by Wali Allah that others were able to build on, and thereby expand the borders of Islamic thought. The first is his call to all believers to use their right of *ijtihād*. In his own words, "Time has come that the religious law of Islam should be brought into the open fully dressed in reason and argument."[11] His overriding concern, according to Freeland Abbott, was to find "some way by which his society could be strengthened, and could, as a theologian, once again be reconciled with Islam."[12] Whereas most

still under the influence of Ibn al-Arabi. There was, however, a shift away from the approach and mood of Ibn al-Arabi. This shift was not usually taken to the extreme of the positions of Ibn Taymiyya, but toward the more moderate style and thought of al-Ghazali" (*Eighteenth-Century Renewal*, 9).

[8] Ahmad, "Political and Religious Ideas," 24.

[9] Sirhindi had denounced all the Sufis as pantheist mystics. He responded that "[b]oth views were based on true revelations…and that of Shaikh Ahmad actually confirmed that of Ibn ʿArabi. 'If real facts,' he wrote, 'are taken into account and studied without their garb of simile and metaphor, both doctrines will appear about the same.' It has been said that what Waliullah did was to give a philosophical and Sufistic base to Islamic orthodoxy" (Freeland Abbott, "The Decline of the Mugul Empire and Shah Waliullah," *The Muslim World* 52 [1962]: 120-21). He also set out to reconcile rival Sufi orders: "He was too much the Indian himself to reject mysticism, but he did hope to promote co-operation among members of the mystic orders. Initiated into each of the four primary Sufi orders on the subcontinent himself, he started the practice of initiating novices into each of the orders simultaneously—a practice that has been continued by the Muslim theological school started at Deoband in the wake of his influence" (ibid., 121).

[10] Ahmad, "Political and Religious Ideas," 24.

[11] Ibid., 26.

[12] Abbott, "The Decline of the Mugul Empire," 116.

Muslim jurists and scholars considered that the "gates of *ijtihād*" had been closed in the sixth/twelfth century, Wali Allah agreed with Sirhindi and 'Abd al-Wahhab that a slavish and blind following of the opinions of the medieval authorities would keep society from changing in a progressive and healthy way. In the following, notice his awareness of hermeneutics, and even the contextualization of theology, particularly in his redefinition of the word *sharīʿa* and his anticipation of sociological themes:

> The prophetic method of teaching, according to Shah Wali-Ullah is that, generally speaking, the law revealed by a prophet takes special notice of the habits, ways and peculiarities of the people to whom it is specifically sent. The prophet who aims at all-embracing principles, however, can neither reveal different principles for different peoples, nor leave them to work out their own rules of conduct. His method is to train one particular people, and to use them as a nucleus for the building up of a universal *shariʿat*. In doing so he accentuates the principles underlying the social life of all mankind, and applies them to concrete cases in the light of the specific habits of the people immediately before him. The *shariʿat*-values (*ahkam*) resulting from this application...are in a sense specific to that people; and since their observance is not an end in itself they cannot be strictly enforced in the case of future generations.[13]

Here is a recognition of the necessary shaping of revelation in a particular culture and the need to continue its interpretation in ever changing settings.

A second area of pioneering in Wali Allah's thought is his holistic view of religion. In a philosophical system clearly reminiscent of Ibn al-Arabi (and to a lesser degree, Ibn Tufayl and al-Ghazali before him), he conceives of "the totality of existential phenomena as a manifestation of the Divine creative and sustaining order (*tadbīr*) pervading the whole cosmic scheme."[14] Thus he is able to relate "ontology and cosmology with teleology, law, ethics, psychology, politics and socio-economics under a comprehensive and systematic framework of thought derived essentially from Revelation."[15] From an epistemological standpoint, he thus brings together revelation, empirical observation and the structuring provided by human reason as complementary ways of discovering and unfolding reality. Though this may not be, strictly speaking, "critical realism," his inductive analysis of human society and his appeal to the

---

[13] Ahmad, "Political and Religious Ideas," 25.
[14] Muhammad al-Ghazali, "Holistic Trend in Islamic Thought: Pioneering Contribution of Shah Wali Allah," *Hamdard Islamicus* 18.4 (1995): 43.
[15] Ibid.

necessary translatability of revelation to specific human contexts is a bold step in that direction.[16]

This is perhaps Wali Allah's most original contribution. Anticipating anthropological methods, he starts with an inductive study of human societies and shows that there are common societal patterns in all of them. All the best of their cultural, intellectual and moral achievements are upheld and promoted in the sharīᶜa. But Islam goes one step further, he argues. It teaches a universal paradigm for human society—humanity's khilāfa is based on its fiṭra—which develops into four levels of usefulness (irtifāqāt): (1) people's normal pursuit of instinctive needs physically, culturally and economically; (2) expansion of the first level gives rise to a diversification in vocations and professions which secures a modicum of social cooperation; (3) as human society further expands it feels the need for political organization in the form of a city state, which allows for the double protection from outside aggression and internal oppression and injustice (a form of monarchy); (4) the pinnacle of human socio-cultural, economic and political organization is a higher level of civilization which he calls khilāfa, the main function of which is to coordinate all of human concerns in a just and equitable way. And because it is linked with humankind's God-given nature (fiṭra), it is common to all men—beyond religion, one might say: "Thus khilafah in Shah Wali Allah's philosophy emerges as a natural outcome of a free and healthy pursuit by man of his socio-cultural aims as dictated by his natural and temperate disposition, provided he is unhampered by moral perversion or any other deviation from the natural course."[17]

This leads to his third contribution. In the center of the cosmic divine tadbīr stands the human being, who ranks above the animals, not so much because of his capacity to reason, but because of his ability to make moral choices (not unlike Razi). What defines people is their taklīf (their being endowed with moral responsibility and accountable to God's law). And then, as he goes to the Qur'an to define the nature of humanity, Wali Allah links the two passages that continue to be linked to this day (as Razi and Ibn al-Arabi had done): "I will create a khalīfa on earth"

---

[16] Though this is not specified in al-Khalifa's article, Aziz Ahmad spells out the nature of this khilāfa ("Political and Religious Ideas," 27). It is a worldwide application of sharīᶜa law, a vision recaptured later by Jamal al-Din al-Afghani, and especially Rashid Rida, Muhammad Abduh's disciple, in his Al-khilāfa aw al-imāma al-ᶜuẓma ("The Caliphate or the Supreme Imamate") in 1923, as the Ottoman Caliphate was breathing its last. See also Erwin I. J. Rosenthal's fourth chapter, "For and against the 'Khilāfa'," in Islam in the Modern National State, 64-102.

[17] Al-Ghazali, "Holistic Trend," 51-52.

(2:30) and, "We did indeed offer the Trust to the Heavens and the Earth and the Mountains; but they refused to undertake it" (33:72). The Pakistani scholar Muhammad al-Ghazali notes that for Wali Allah this verse was to be taken metaphorically. Only humanity is given this natural ability to make moral judgments. At the same time, their choice can go either way, as the end of the verse indicates: "He was indeed unjust and foolish." As al-Ghazali puts it, "Since it is man alone who is given the dual capacity of good and evil, virtue and vice, erudition and ignorance, he alone is fit by nature to take-up the onerous office of the vicegerency.[18]

The Period of the Reformers

In the nineteenth and twentieth centuries these were the Muslim intelligentsias who sought to articulate a constructive response to European colonialism. Kurzman's recent anthology of reformers from 1840 to 1940 amply demonstrates (1) that they were widely represented throughout the Muslim world, from North Africa to the Middle East, and from Central Asia to South and East Asia; and (2) that, despite varying attitudes toward the Western powers themselves, they all strove to incorporate in their theological understanding certain key values they saw in Western modernity, hence his reference to them as "modernists."[19] Their dominant ideology was nationalism—already a conscious (perhaps unconscious in some cases) borrowing from foreign values. Certainly the more vocal anti-imperialists among them, in their bid to reduce the hegemony of the West, deliberately appropriated other Western ideals and methods as well. Bruce Lawrence notes that on the top of this list was an emphasis "on science and technology in education, on constitutional structures and parliamentary democracy in politics, and on the revised role of women in social life."[20]

In this section I examine Muhammad Iqbal, considered one of the founding fathers of Pakistan. Then, turning to Egypt, first, the struggle between two polar opposites, Ali Abd al-Raziq and Rashid Rida, during the 1920s as the Caliphate was being dismantled; then I conclude with a recent female authority on *tafsīr*, Bint al-Shati.

---

[18] Ibid., 47.
[19] Kurtzman, *Modernist Islam.*
[20] Bruce B. Lawrence, *Shattering the Myth: Islam Beyond Violence* (Princeton, NJ: Princeton University Press, 1998), 45.

Muhammad Iqbal: Humanity as God's Co-workers

An outstanding poet, Iqbal (d. 1938) was one of the great reformers of India, who in time became a symbol of the new state of Pakistan, though he died before it saw the light of day. His book, *The Reconstruction of Religious Thought in Islam*, is a handbook for the revitalization of Islam in the first decades of this century. In his chapter on "The Human Ego—His Freedom and Immortality," Iqbal announces the three cardinal points taught by the Qur'an about humanity. The first is that "man is the chosen of God," and he quotes Q. 20:114: "Afterwards his Lord chose him [Adam] for Himself and was turned towards him, and guides him." The other two points are, in fact the connecting of the *khalīfa* and the *amāna*:

> (ii) That man, with all his faults, is meant to be the representative of God on earth: [he quotes Q. 2:30], and then Q. 6:165: "And it is He Who hath made you His representatives [*khalāʾif*] on the Earth, and hath raised some of you above others by various grades, that He may prove you by His gifts."[21]
>
> (iii) That man is the trustee of a free personality which he accepted at his peril: [he quotes the "*amāna* verse," 33:72].[22]

Iqbal finds it surprising that "the unity of human consciousness which constitutes the centre of human personality never really became a point of interest in the history of Muslim thought."[23] Perhaps he realized it, but chose not to state it: a focus on the consciousness of the individual was one of the leading themes of Western modernity. As a modern himself, educated in the West, this would naturally have been a concern to him in reading the qur'anic text. Yet not all Western thinking is helpful to the Muslim, he objects, particularly the "dead metaphysics" of the mind–matter dualistic perspective introduced into Islamic thinking through Greek philosophy. The despising of sense perception was a fatal flaw:

---

[21] Q. 6:165. For Iqbal, the names God teaches Adam, which the angels know nothing of, are the capacity of humans to form concepts: "Thus the character of man's knowledge is conceptual, and it is with the weapon of this conceptual knowledge that man approaches the observable aspect of Reality" (*The Reconstruction of Religious Thought in Islam* [Lahore: Shaikh Muhammad Ashraf, 1960], 13). He continues with some verses from the Qur'an that speak of the marvels of nature which God gives to humans as signs. They are to read these signs and ponder them. From here, Iqbal asserts, we see that the Quran calls us to the empirical study of the world around us.

[22] Ibid., 95.

[23] Ibid., 96.

How unlike the Quran, which regards "hearing" and "sight" as the most valuable Divine gifts and declares them to be accountable to God for their activity in this world. This is what the earlier Muslim students of the Quran completely missed under the spell of classical speculation. They read the Quran in the light of Greek thought.[24]

The "trust" (*amāna*) offered by God to humans is, in Iqbal's view, two-fold. The first is their calling to exert mastery over nature, and by this task they become "co-workers" with God:

It is the lot of man to share in the deeper aspirations of the universe around him and to shape his own destiny as well as that of the universe, now by adjusting himself to its forces, now by putting the whole of his energy to mould its forces to his own ends and purposes. And in this process of progressive change God becomes a co-worker with him, provided man takes initiative:

"Verily God will not change the condition of men, till they change what is in themselves" (13:12).[25]

The second component of the *amāna* is humanity's burden of freedom, at once an exhilarating and daunting experience: they must choose between good and evil, and in that choosing they seal their eternal destiny, whether in hell or in paradise.[26] This theme will be picked up again.

Iqbal, finally, deals head on with the issue of the Caliphate. Commenting on Turkey's abolition of it, he writes that this is sound *ijtihād*.[27] He advocates both independence for Muslim nations and the formation of a "League of Muslim Nations" to preserve the unity of the Muslim *umma*.[28]

Al-Afghani and Abduh: Roots of Reformism and Fundamentalism
No doubt, the father of Egyptian Islamic reformism is Muhammad Abduh. As was the case with Sayyid Ahmad Khan in India and Ismail

---

[24] Jagan Nath Azad, "Iqbal, Islam and the Modern Age," *Islam and the Modern Age* 9.1 (1978): 41.

[25] Iqbal, *The Reconstruction of Religious Thought*, 12.

[26] Ibid., 88.

[27] Ibid., 157.

[28] Quoting a close associate of Iqbal, Azad confirms that the poet-thinker in his latter years became "a severe critic of the Congress, Muslim Conference and the Muslim League." He then cites one of Iqbal's poems, addressed to his son Javed, and through him directed to the younger generation as a whole. Iqbal's vision soars magnificently above political, nationalistic and religious politics to embrace "the brotherhood of man": "What is humanity? To respect a human being / Be aware of the dignity of human being. / Human being survives by mutual contacts; / Put your foot on the path of friendship. / A man of love follows the path of God, / And is compassionate both to the infidels and the faithful" ("Iqbal, Islam and the Modern Age," 65).

Gasprinskii in the Russian empire, Abduh positioned himself between the secular intellectuals, who, though educated in the West, lambasted Western interference in the name of nationalism, and the traditional *ʿulamāʾ*, who rejected any shade of Western epistemology. For this reason, Esposito and Voll consider both al-Afghani and Abduh "modernists" (agreeing with Kurzman).[29] Nevertheless, J. Dudley Woodberry pointed to the fact that Abduh's theological work, *Risalat al-Tawhid* (*The Treatise on Oneness*), was often quoted in the works of Hasan al-Banna and his Muslim Brotherhood colleagues.[30] Moreover, al-Banna quotes a footnote by Muhammad Rashid Rida on Abduh's text on two occasions in his creedal work (*ʿAqāʾid*).[31] Woodberry's following observation serves as a useful introduction to this and the next section: "al-Banna considered himself in the reforming tradition which may be traced from the activist al-Afghani, to the educationally-oriented Abduh, and then to the more traditionalistic Rida."[32] Indeed, al-Banna, arguably the founder of twentieth-century islamism, traces his spiritual roots to the these reformers—as do Bint al-Shati and a host of other contemporary reform-minded Muslims.

Recent assessments of his role concur with Woodberry.[33] Johannes Jansen, a recognized specialist on modern Egyptian Islam, declares that "[t]he history of fundamentalism in modern Islam starts with al-Afghani's activities in Cairo during his stay there in 1871–9."[34] Very possibly an Azerbaijani of "Iranian Shiʻi descent," al-Afghani's pan-Arab and Islamic revivalist vision included a political program in which he would play a prominent role. Those dreams remained unfulfilled, yet he inspired a

---

[29] John L. Esposito and John O. Voll, *Makers of Contemporary Islam* (Oxford/New York: Oxford University Press, 2001), 18-19.

[30] J. Dudley Woodberry, "Hasan al-Banna's Articles of Belief" (Ph.D. dissertation, Harvard University, 1968), 49.

[31] This footnote referred to "a tradition that, in variant forms, encouraged reflection on the creation but denied it on God and his essence" (ibid., 84). Woodberry adds further on that al-Banna's father had studied under Abduh (ibid., 89).

[32] Ibid., 50.

[33] Lawrence represents the majority view, which sees "two parallel but divergent groups" protesting and resisting British colonialism in Egypt. The first is led by Abduh, "the leading light of Islamic reformism and founder of the opposition Wafd party." The second group, naturally, was the Muslim Brothers, who "advocated nothing less than the restoration of Islamic identity to all levels of Egyptian society, but particularly the government" (*Shattering the Myth*, 64). This is true, as far as it goes, but in fact Abduh laid the foundation for both movements.

[34] Johannes J. G. Jansen, *The Dual Nature of Islamic Fundamentalism* (Ithaca, NY: Cornell University Press, 1997), 26.

whole generation of young intellectuals in the many countries to which he ceaselessly traveled, advocating a strong new Muslim identity that bridged wisely between the best of the past and the best of modern civilization.[35]

Jansen's second chapter is entitled "The Century of Al-Afghani and Ibn Taymiyya," in which he argues that both figures (one from the nineteenth century and one from the thirteenth century) set in motion the intellectual ferment of the twentieth century.[36] Erwin Rosenthal too, in his classic volume, *Islam in the Modern National State*,[37] sees the contemporary discussion among Muslim intellectuals regarding the respective roles of politics and religion as aligned with either Ibn Taymiyya or Ibn Khaldun (fourteenth-century North African). Rosenthal's book (published in 1965) anticipated much of what was to happen from the 1970s on. He claims that the parameters of the modern debate were fixed in the uproar following the establishment of the Kemalist Republic in Turkey. According to Jansen, a look at that situation

> convinces us that the question "Islamism or Westernism," a religious ordering of life or a lay state within Western civilisation, agitated many intellectuals and still does in other Muslim lands... If we refer once more to Ibn Khaldun, it would appear that Islam has always been able to absorb within its realm several cultural streams that flowed side by side most of the time. Sporadic attempts at integration, at harmonisation, were made, notably by the *falāsifa*, the Muslim philosophers, who tried to resolve the dichotomy of faith and reason. Though such attempts never affected the mainstream of Islam as a faith and a way of life, they greatly enriched Islamic civilisation, and their echo can be heard today among certain Muslim intellectuals.[38]

Indeed, it is not only the issue of *ijtihād* that is at stake, but more fundamentally, what is to be understood about the relationship between the various political arrangements under centuries of *dār al-Islām* and the new nation-state *fait accompli* of the modern era. To answer this question, traditional Islamic theology has to be reworked within the new setting. And this is the process that al-Afghani began.

Significantly, he joined forces with Abduh in founding a journal called the *Al-ʿUrwa al-Wuthqā*, "which argued repeatedly that the West was not as strong as it looked and could be defeated."[39] The central methodology

---

[35] Ibid., 28.
[36] Ibid., 26.
[37] Cf. n. 16.
[38] Ibid., 50.
[39] Jansen, *The Dual Nature*, 28.

they proposed was a reinterpretation of Islam to meet the requirements of the new age. For Abduh especially, this meant going back to the right-eous forbears (*al-salaf al ṣāliḥ*),[40] in order to distill the authentic values and norms of the first generation of Islam and then to reinterpret them in the new framework of modern rationalism. With the reality of decoloni-alism on the horizon, this *Salafi* project seemed all the more urgent from Abduh's perpective. By then he was an influential professor at Al-Azhar University in Cairo and, as Grand Mufti of Egypt, he was in a position to propose several reforms that were actually carried out—both at Al-Azhar and in Egyptian law at the national level.

How, then, can this member of the *ʿulamāʾ* class of Al-Azhar, only mildly anti-Western and a skeptic with regard to many alleged miracles held up by the Islamic tradition, become the unlikely "founding father of modern fundamentalism"? Jansen explains this thesis, now widely accepted:

> In traditional Islamic political literature it is a time-honoured assumption that just rule is assured by knowledge of the law and consultation with the *ʿUlamāʾ* on the part of the ruler. Abduh gave a modern, Mediterranean version of this assumption when he argued that only a just dictator, *mustabidd ʿādil*, could modernise the Muslim East. Moreover, since Abduh is the author of a Koran commentary which is nowadays widely read, his reputation established the Sunni credentials of what was to become modern Islamic fundamentalism.[41]

However, this link would not have been so clear cut had Rashid Rida not left Syria to join Abduh, become his closest associate and gently pres-sured him to embark on his now famous qur'anic commentary, *Tafsīr al-Manār* (*The Lighthouse Commentary*). Founded in 1898, the *Tafsīr* appeared weekly until Rida's death in 1935. *Al-Manār* was not only the "first sustained attempt to use the modern media systematically as a vehicle for the progress of Islam," but the decisive factor in rendering the movement started by al-Afghani "irreversibly identified with Sunni Islam."[42] Rida is the figure who rediscovered and popularized Ibn Taymiyya (d. 728/1328) at about the time when the Caliphate was offici-ally abolished (1924). He set himself to re-edit the scholar's works in 1925, and for Jansen it is no coincidence that Ibn Taymiyya is the most quoted author "in recent fundamentalist writings."[43] Rida took up the

---

[40] Hence the name "Salafism" generally given to the stream of thought that came out of Abduh through Rashid Rida.

[41] Ibid., 31.

[42] Ibid., 32.

[43] Ibid.

Salafi mantle of his mentor, and, against the backdrop of Kemalist Turkey, took definite political positions, which from that period on became increasingly anti-Western.

Yet Rida represents only one strand of thought that came out of the modernist "lighthouse." The other built on the side of Abduh that was greatly impressed with Western science and rational thought and which took this emphasis into a more secularist direction. Intellectuals such as Qasim Amin and Taha Hussein "stressed the modern elements of this tradition and essentially became secular intellectuals and participated in their triumph in the interwar era.[44]

This brief historical sketch was necessary for two reasons. First, the *Tafsīr al-Manār* sets the tone for all subsequent commentaries of the Qur'an, particularly in its doctrine of humanity. Second, both the *Tafsīr* and Rida's landmark book, *Al-Khalīfa aw al-imāma al-ʿuẓma* (*The Caliphate, or the Supreme Imamate*, 1923), serve as foundation for both contemporary islamist movements, at least of the Sunni variety, and for the renewed interest in reformist theology—a movement that has gained momentum of late.

Abduh and Rida: Tafsīr al-Manār. Rashid Rida composed the *Manār* on the basis of notes taken from Abduh's copious notes and lectures. Until his death in 1905, Abduh read and commented on at least all of the exegetical articles of the journal, but by that time the journal had only reached v. 165 of the fourth sura (*Sūrat al-Nisāʾ*).[45] Thus, in the following discussion on the second sura, we can be certain that this is Abduh's teaching.

What about the *Manār's* exegesis of Q. 2:30?[46] Here is the opening statement:

> God informs the angels that he is placing on earth his *khalīfa*, and they understood from this that God had implanted within the nature (*fiṭra*) of this species—which he is making a *khalīfa*—the capacity of absolute will of unlimited choice in his work, and that the discernment between the works that present themselves to him should be made according to his knowledge; and that this knowledge, if it is not endowed with the qualities

[44] Esposito and Voll, *Makers of Contemporary Islam*, 19.
[45] Jacques Jomier, *Le commentaire coranique du Manar: tendances modernes de l'exégèse coranique en Égypte* (Paris: Éditions G. P. Maisonneuve, 1954), 51. Jomier indicates that at Rida's death in 1935 the *Manār* had just begun its twelfth volume, while still commenting on Sūrat Hūd (the eleventh sura).
[46] I have commented on this elsewhere at greater length: "The Human *Khilāfa*," particularly 39-44.

of welfare and utility (*al-maṣāliḥ wa-l-manāfiʿ*), is corruption (*fasād*). This is a necessary requirement, because perfect knowledge can only belong to God—may He be exalted![47]

Abduh acknowledges that Muslim commentators generally held two different interpretations on the meaning of *khalīfa*. The first is, as might be expected, the "successor" meaning—with the accompanying lore of *jinn* and angels who lived on earth before humanity. Yet it is soon dismissed, in light of the better of the two "ways" (*ṭarīqatāni*): *khalīfa* as God's representative on earth. In this perspective God is saying, "I am placing a deputy for myself (*khalīfatan ʿannī*), and because of this the conviction spread that humanity was God's deputy on his earth."[48] This affirmation is to be placed side by side with the David verse (38:26), yet one should not thereby gain the impression that God's deputies are only prophets, kings and caliphs. Rather, "the meaning of this deputyship extends to all people through all the qualities with which God has set apart humankind from the rest of the creatures."[49] All living creatures have a limited amount of knowledge and have been assigned specific functions—including the angels whose calling is limited to praising God and executing his commands. But to be called "the deputy" of the One God who knows no limits in wisdom and power is to be granted a higher function still. No doubt, humans are prone to weakness, ignorance and rebellion at times, as revelation makes plain. Yet the secret of their secret endowment lies with their power of reason: "Through this power humankind is endowed with limitless potentiality, limitless aspirations, limitless knowledge, limitless action. Hence the human race, despite the weakness of individual members, behaves in the universe with actions that have no limitation, by God's leave and decree."[50]

In order to measure the difference in worldview exemplified in these pages compared to the classical commentaries, we need only point to Abduh's turn toward rationalism. Fritz Steppat notes that, for Abduh, humans, unlike other created beings, however "weak and ignorant," have the capacity to develop their sense perception (*iḥsās*), their ability to know (*shuʿūr*), mainly because of their God-given capacity to reason (*ʿaql*). It is the realization of these potentialities that enables people to exercise dominion (*sulṭān*) over all other created beings. Particularly instructive is Abduh's comment on God teaching Adam the "names":

[47] *Tafsīr al-Qurʾān al-ḥakīm al-shahīr bi-Tafsīr al-Manār*, 12 vols. (Beirut: Dār al-Maʿrifa, 1990), vol. 1:255-56.
[48] Ibid., 258.
[49] Ibid., 259.
[50] Ibid., 260.

> Through this story he lets us know our own worth and what was put into
> our nature, distinguishing it from the other creatures. Now we must strive
> for perfection by the sciences (*al-ʿulūm*) for which he prepared us at our
> creation, before the angels and the rest of the creatures, so that God's
> wisdom would become visible to us.[51]

One would also have to notice here, at least for Abduh, a pre-World
War I worldview in which unlimited progress was assumed—propelled
by scientific discoveries and maintained by a humanity imbued with
reason and harmony. Together with this unmitigated optimism, Steppat
observes two more corollaries to Abduh's and Rida's understanding of
humanity. One has to do with *sharīʿa*,[52] which only figures indirectly in
this discussion as divine regulations and laws (*aḥkām wa-sharāʾiʿ*): "Certainly man, when exercising his dominion, is following God's design;
but if he remains subject to the divine law this plays a rather secondary
role as a means of orientation."[53]

The second corollary, related to the first, is that religion is not a factor.
Humankind as a whole is called to be God's deputy on earth. This reading of the Qur'an would seem to indicate a clear-cut reformist position
for the authors. Yet even in this text now under consideration, Rida parts
ways with his mentor. In the few pages that precede the commentary on
Q. 2:30, he discusses his own definition of Salafism. This entails a distinction between the *salaf* and the *khalaf* (normally "ancestor" and
"descendant"). For Rida, the distinction is mostly theological: the *salaf*
follow reason to the extent that anthropomorphisms (*tanzīh*) in the Qur'an

---

[51] Steppat, "God's Deputy," 167.

[52] On Abduh and Rida's view of *sharīʿa*, see Johnston, "An Epistemological and
Hermeneutical Turn," 256-66.

[53] Ibid. Michel Hoebink disagrees somewhat here. Abduh's key hermeneutic was the
distinction between the *ʿibadāt* (rules in the Quran dictating humankind *muʿāmalāt*
(qur'anic rules governing human relationships in this world) which are general principles which must be reinterpreted constantly as new problems arise within the ever-
flowing and changing human situation. Thus, "The notion of monotheism (*tawḥīd*) in
such modernist thinking became a dynamic principle referring to a continuous
human effort to re-unite the eternal ideal with the evolving reality" ("Thinking about
Renewal in Islam: Towards a History of Islamic Ideas on Modernization and Secularization," *Arabica* 46 [1998]: 49). This same thought is found today among such
"modernist islamists" such as Muhammad Amara and Rashid Ghannushi. Yes, an
Islamic state must be established on the basis of Islamic law, but its substance "is
largely for humans to determine, according to the circumstances in which they find
themselves." Hoebink rightly adds, "Some of the later modernists have endowed
humanity with such a degree of moral and social autonomy, that one could ask
oneself what the practical difference is between their Islamic state and the sociopolitical order advocated by the Muslim secularists" (ibid.).

are not taken literally, but reason should never take one beyond the limits prescribed by revelation. By contrast, the *khalaf* interpret the text on the basis of *taʾwīl*[54]: "they say the bases of the Islamic religion were laid on the foundation of the mind (*al-ʿaql*) and nothing in the whole structure is exempt from reasonableness (*al-maʿqūl*), and if the mind discerns something clearly and it contradicts scripture, then the definitive (*qāṭʿī*), reasonable judgment is evidence that a literal meaning is not intended."[55] Hence, the need to resort to *taʾwīl* (symbolic interpretation), says the Shaykh—meaning Abduh. Rida is uncompromising here: "But I follow the path of the *salaf* in claiming the need to surrender and entrust matters that pertain to God, his attributes and the knowledge of the unseen." Then follows this programmatic statement from Rida, showing both his respect for Abduh and his desire to make this journal a truly collaborative effort: "And thus we embark on our understanding of the verses according to both ways, because theological discourse (*kalām*) must be shown to be useful, for God—to Him power and majesty!—does not address us except in a way that is meaningful."

Though Rida shares much of his mentor's vision, he moves one step beyond the al-Afghani legacy, however, and deals with the Caliphate issue head-on. Already with the events of World War I, but especially with the dissolution of the Ottoman Empire and the caliphal system, Rida turns more anti-Western and returns to the pan-Islamic themes of al-Afghani. He attempts to revive the political dimension of *khalīfa*, which was so prominent a factor in the medieval interpretation of Q. 2:30, and which still haunts Islamic thought today, albeit in a different way.

The 1920s Debate on the Caliphate. I have just mentioned Rida's keen interest in the thirteenth-century Islamic revivalist Ibn Taymiyya. In order to understand the debate on the relative roles of religion and politics in Islam, one would also have to know about the writings of fourteenth-century Muslim political scientist and first philosopher of history, Ibn Khaldun.[56] Rosenthal explains how this thinker's distinction between

---

[54] This is similar, though not to be confused with the Sufi meaning of *taʾwīl*, which represents a more esoteric and mystical approach to the scripture (with a distinction between the surface or literal meaning and the deeper or *bāṭinī* meaning).

[55] *Manār*, vol, 1:252.

[56] Rosenthal continues, "Muhammad Abduh was a keen student of Ibn Khaldun's *Muqaddima*, as was Rashid Rida, who nevertheless violently disagreed with Ibn Khaldun (as did Ali Abd al-Raziq, though for opposite reasons). Nor should we forget that Taha Husain, one of the foremost contemporary Egyptian thinkers, devoted a penetrating study to Ibn Khaldun's social philosophy, and that Arab

khilāfa and *mulk* has become so essential to our contemporary discussions. *Mulk*, for Ibn Khaldun, or a "government by authority based on power," represents a natural state of human politics.[57] As people historically moved from their rural settings to found cities, political power as such was born. The quintessential human drive, he argued, is the will to exercise power and domination. Thus it is crucial that a ruler's evil inclinations be held in check by one of two sources of wisdom. The first is the highest and most effective, because it not only brings harmony but it also allows the flourishing of all human potentialities. This is divine revelation, and when it is applied, contends Ibn Khaldun, we have *siyāsa dīniyya* ("religious rule"), or the *khilāfa* ("Caliphate"), as it was exercised by the first four "rightly guided caliphs."

The second avenue of wisdom capable of limiting humanity's tendency to use power to oppress—in the absence of prophetic revelation—is *siyāsa dīniyya* (the rule of reason, much better, he maintains, than Plato's *siyāsa madaniyya*, the ideal state of the philosophers). "Their rulers exercise authority by power and/or ʿaṣabiyya, which unites their supporters."[58] This common bond is necessary for the functioning of any state, religious or otherwise. The passing of time brings about an inevitable evolution, even in the case of the most ideal states, like that of the *khulafāʾ al-rāshidūn* ("the rightly guided caliphs"). *Sharīʿa* ruled unopposed in the golden age of Islam, with the supreme leader standing under its authority as all other members of the *umma*. Every Muslim was equally a servant of Allah. Yet, for Ibn Khaldun, the *khilāfa* quickly degenerated into *mulk*, starting with the advent of the Umayyad dynasty. "With the eyes of a political scientist," notes Rosenthal, Ibn Khaldun considered that "the decline of the original *khilāfa* as a political entity was attributed to the decline of religion, and he saw altogether a close connection between religion and politics.[59]

The crucial point here is that, however inferior the Umayyad and Abbasid dynasties might have been, for Ibn Khaldun they were still Islamic states. There are degrees, certainly, in a state's responsiveness to the divine laws. The more a ruler lays aside his *qawānīn siyāsiyya* (laws regulating the administration of the *mulk*), the better to heed the advice of the ʿulamāʾ, the more Islamic the state will be. But he is not calling for the clock to be turned back and for Muslims to return to the "golden

---

nationalist writers often discuss his ideas with approval or disapproval" (*Islam in the Modern National State*, 17).

[57] Ibid., 18.
[58] Ibid., 18.
[59] Ibid., 19.

age." As a precursor of modern historiography, Ibn Khaldun observes and describes the facts, and then lays down a minimum of acceptable religious components (though still vague), in order to qualify a state as practicing *"siyāsa dīniyya."* In a section of his *Muqaddima* (*Introduction*) entitled "The Transformation of *Khilāfa* into *Mulk*," he traces the degeneration process from the unity of prophethood and political power to its rightful expression in the first four successors to the prophet to the gradual "secularization" of the Islamic state (i.e. pure politics take over the religious dimension undergirding the Caliphate at its inception) under the Umayyads and Abbassids. For Ibn Khaldun, there is inevitability in the way societies evolve. Here are the closing words of this section: "From the following it is clear that the *khilāfa* at first existed without *mulk*, then their respective meanings overlap and became confused, and finally the *ʿaṣabiya* of *mulk* became separated from the *ʿaṣabiyya* of *khilāfa*."[60] Yet, in the articulation of the distinction between *khilāfa* and *mulk* and in the positing of their inevitable separation, Ibn Khaldun was parting company with the classic theorists of the Caliphate.

Rashid Rida strongly disagreed with this viewpoint and explained himself in *The Caliphate or the Supreme Imamate*, which was "the classical formulation of what an Islamic state is and should be in the twentieth century and beyond."[61] This book, far from being a treatise on theology for *ʿulamāʾ* consumption only, was essentially a "battle-cry for action." It was to serve as an ideological guide to the political party he founded, the Salafiyya, and it was first published by installments in the *Tafsīr al-Manār*. Its expressed goal was to resist the negative influences of the West and to retrain a whole new generation in view of the re-establishment of the Caliphate under the undisputed rule of *sharīʿa* law. This new generation of *ʿulamāʾ* would be able to exercise badly needed *ijtihād* in order "to interpret and apply the *Sharīʿa* in the spirit of early, pure Islam and in conformity with the requirements of the age."[62]

---

[60] Abd al-Rahman b. Muhammad b. al-Khaldun, *Muqqadimat Ibn Khaldūn*, edited, annotated and introduced by Ali Abd al-Wahad Wafi, 2 vols. (Cairo: Lajnat al-Bayjān al-ʿArabī, 1965), vol. 1:548, cited in Rosenthal, *Islam in the Modern National State*, 21 (my translation, not Rosenthal's).

[61] Ibid., 67. He adds, "The significant fact is that, apart from a few details and perhaps a slightly different emphasis here and there, no contemporary advocate of an Islamic *Sharīʿa*-state has gone beyond Rida. This is not surprising, since Rida is a rigid adherent of the classical theory of *khilāfa* and firmly believes in the need to re-establish the caliphate in the best interest of Islam and the umma of believers: the *ahl al-sunna wa-l-jamāʿa*, who are governed by the *siyāsa sharʿiyya* as expounded by his great examples, Ibn Taymiya and the latter's disciple Ibn al-Qayyim al-Jawziya" (ibid.).

[62] Ibid., 69.

By contrast, a professor and *shaykh* at Al-Azhar University published a response to Rida's book two years later (1925): *Al-Islām wa-uṣūl al-ḥukm* (*Islam and the Principles of Government*). In it, Ali Abd al-Raziq applauds the direction taken by the Turkish political leadership and proceeds to exegete the Qur'an in such a way as to show that Islam is first and foremost a call (*daʿwa*) to people from every race and tribe, and that Muhammad only set out to found an *umma* (religious community, as opposed to an Islamic state).

> Again and again he dissociates religion from politics. Islam is a call to God; the prophet issued a call to religious unity by faith in Allah. Universality belongs to Islam as a *waḥda dīniyya*, a religious unity, and a *waḥdat al-īmān wa-l-madhab al-dīnī*, a unity of faith and a religious way (and direction), but not as a *waḥddat dawlatin wa-madhāhib al-mulk*, a political unity and temporal-royal institutions (and directions). Islam is a *daʿwa dīniyya*.[63]

In his separation of *khilāfa* and *mulk*, *siyāsa dīniyya* and *siyāsa ʿaqliyya*, Abd al-Raziq shows himself an ardent disciple of Ibn Khaldun. After a brief summary of the classical formulations of what the Caliphate represented for the Muslim community, he quotes from the latter, in fact from the long paragraph preceding the conclusion (quoted above) from Ibn Khaldun's section, "The Transformation of *Khilāfa* into *Mulk*." This graphic portrayal of the gradual disappearance of religious fervor, Arab solidarity and moral rectitude serves Abd al-Raziq's purposes well. He then goes on to say that the official ideology of the Caliphate would have us believe that the institution itself was directly called for in the sacred sources. In fact, he maintains, Muslims followed two views on the legitimacy of the caliph. The larger group, represented by all the scholars commissioned by the caliphs to draw up the orthodox version (as al-Mawardi), taught that the caliph was directly commissioned by God to follow in the Prophet's stead. He then cites a number of sources, from the early poetry of the *umma* (e.g. al-Farazdaq) to later commentaries and super-commentaries, which elevate the status of the caliph to fanciful and frankly heretical degrees. To listen to them, he remarks, they were the incarnation of divinity on earth![64]

---

[63] Ibid., 99.
[64] Ali Abderraziq, *L'islam et les fondements du pouvoir*, new translation and Introduction by Abdou Filali-Ansary (Paris: Éditions La Découverte, 1994), 60. The Arabic version I have is *al-Islām wa-uṣūl al-ḥukm: bahth fi al-khilafa wa-l-hukuma fi al-Islam* (*Islam and the Foundations of Political Rule: Research on the Caliphate and Government in Islam*), ed. and comments by Mamdouh Haqqi (Beirut: Dār Maktabat al-Ḥayāh, 1966).

The other view, held by some scholars throughout the centuries (and particularly inscribed in the latest documents of Ottoman Empire), emphasizes the consent and commissioning power of the *umma*. He agrees with Ibn Khaldun that every Muslim empire in history was more *mulk* than *khilāfa*: "We do not doubt that brute force has always been the support of the institution of the *khilāfa*."[65] Yet at the same time he seriously parts company with Ibn Khaldun, and indeed with the consensus of Islamic tradition since the beginning:

> It is important to realise that the *ʿālim*'s [singular of *ʿulamāʾ*] argumentation is scholastic, medieval, the result of a critique of the religious and historical sources of Islam by means of the rational, empirical method employed by Ibn Khaldun, who deeply influenced him even where he disagreed with this North African Muslim thinker of the fourteenth century...
>
> The theory of the *khilāfa* or *imāma* is the work of theologians and jurists and is strictly separated from the history of the *khilāfa*. While he rejects the theory altogether, he subjects the history of the caliphate to a searching criticism, guided by reason and experience. Implementing Muhammad's religious message has, he avers, nothing to do with politics, which is left to man and his reason exclusively... The jurists and Ibn Khaldun are wrong when they claim the religious and political unity of Islam.[66]

That the reaction to his work was immediate and vehemently condemnatory is not in the least surprising. By dismissing the qur'anic basis for *khilāfa*, Abd al-Raziq was at the same time, in fact, dispensing with *sharīʿa*: "It may be assumed that this was his primary purpose in demolishing the orthodox theory, since the *khilāfa* is practically synonymous with the Sharīʿa, and that for this reason he had to pay the penalty for his unorthodoxy."[67] Not only is there no stipulation for any specific kind of government in the sacred texts, argued Abd al-Raziq, but the califate itself "was and still is a catastrophe (*nakba*) for Islam and for Muslims, a source of evil and corruption."[68] He was, unsurprisingly, deposed from his position at Al-Azhar, and his book was banned.

The de-politicization of the interpretation of *khilāfa* in the Qur'an naturally opens the way for fresh hermeneutical ventures. While commenting on Q. 2:30, Cragg sees Abd al-Raziq's interpretation as groundbreaking:[69]

---

[65] Abderraziq, *L'islam et les fondements du pouvoir*, 90.

[66] Rosenthal, *Islam in the Modern National State*, 86-87.

[67] Ibid., 101.

[68] *al-Islām wa-uṣūl al-ḥukm*, 83.

[69] Cragg, *The Privilege of Man*, 29-30. On page 30, n. 2, he comments on his own sentence, "It is a perversion to regard the Islamic privilege of man as somehow capable of being interpreted, still less achieved, in this external, political expression."

> There can be no question that 'Ali 'Abd al-Raziq was right in his empha-
> sis in 1925, shortly after the final demise of the Caliphate at the hands of
> Atatürk, that it had been a de facto creation of the Arab genius, authentic
> and vital through long centuries but nevertheless lacking any explicit
> Qur'anic basis. For that perspective helps to bring into unclouded promi-
> nence the deeper concern of the Qur'an for the inclusive "caliphate" of
> man himself.[70]

In concluding this section on the debate that raged in the 1920s in
Egypt, I wish to draw the reader's attention to the interpretive dynamics
at work. The trigger for the hermeneutical storm was the political fact of
the 1923 Turkish constitution and the subsequent cancellation of the
Caliphate. By now the debate included the issues of nationalism, pan-
Arabism, and post-colonial political, economic and cultural reconstruction. The foundation of a more inclusive theological (as opposed to a
political) concept of *khalīfa* had been laid by Wali Allah in India and
Abduh in Egypt. But now, for the first time in Islamic history, the question of the relationship between Islam as a faith and Islam as a political
reality had been poignantly posed. Certainly in this respect Rida went
further than his mentor. Though few in the Muslim world today would
advocate the re-establishment of the Caliphate as a political institution,
both Shi'i and Sunni advocates of the theological necessity of the political embodiment of Islam abound.

On the face of it, Abd al-Raziq's work was a milestone in qur'anic
interpretation, a foundation that others would build on. Heath ends his

---

He explains: "This was the main thrust of 'Ali 'Abd al-Raziq's study. Though something of a *tour de force* at the time of its publication, *its arguments have now been
widely agreed* [sic]. We still urgently need his emphasis that continuity in Islam has,
and can have, no external guarantee in any form of government, but only in living
conformity to its meaning" (ibid., 30, emphasis added). Here Cragg sides with liberal
Muslims.

[70] For a similar argument, see Tamara Sonn's chapter, "The Islamic Call: Social Justice and Political Realism," in *Islamic Identity and the Struggle for Justice*, ed.
Nimat Hafez Barazangi, M. Raquibus Zaman and Omar Afzal (Gainesville, FL:
University Press of Florida, 1996), 64-76. She argues against the islamist view that
religion and politics must be united. The postcolonial borders and nation-state structures are a fact of life. Islamists "ignore the primary requirement of national stability: a source of social solidarity logically congruent with the nation's geographic
borders." She concludes that "there is nothing in secular nationalism that conflicts
with Islamic sociomoral goals" (ibid., 74). Hasan Askari also argues that the idea of
a theocratic state is an Islamic contradiction in terms ("Religion and State," in *Islam
in a World of Diverse Faiths*, ed. Dan Cohn-Sherbok [New York: St Martin's Press,
1997 (1991)], 178-87).

article on "Creative Hermeneutics" with the "Principle of Particularism." By this he means that abstract hermeneutical principles must be accommodated to specific cases and circumstances. He defines three such levels of accommodation (similar to Fiorenza's "reality domains"): historical influences, individual influences, and hermeneutical influences. He believes that most historians have some sensitivity to the first two levels. According to Heath, however, the last one is often overlooked:

> Interpretation is a dialectically creative process. What emerges is often very different than what enters. This becomes important from a theoretical perspective when textual particularities which to an external observer appear of minor or idiosyncratic importance spark the interest and imagination of an individual interpreter. It is not unusual that such seemingly particularistic sparks give birth to new intellectual currents.[71]

To sum up, Wali Allah and Abduh saw in Q. 2:30 God's mandate for the Islamic *umma* to wake up from its sleep and to use its God-given inner resources to rebuild its dying civilization on the basis of *ijtihād* and the modern sciences. But this stream of interpretation, now firmly in motion, meets the historical fact of Atatürk's lay republic and the demise of the Caliphate. Abd al-Raziq musters his vast talent and skills as a *mufassir* to challenge the notion that Islam was ever meant to express itself in a particular political institution. True, his bold reinterpretation is condemned, yet he initiated a new direction, one which his contemporaries could not ignore. Bint al-Shati is one qur'anic scholar, young enough to be Abd al-Raziq's daughter, who reworks his position, consciously or unconsciously. I now turn to her exegetical work.

Bint al-Shati: Unfettered Tafsīr
A woman of many talents, Aisha Abd al-Rahman (1913–1998), or, as she is better known in the Arab world, Bint al-Shati ("daughter of the seashore"), began her writing career in the summer of 1925 with a series of articles on rural Egypt on the front page of the Egyptian daily *Al-Ahram*. The next year, while still a first-year university student, those articles were published in a book that took Egypt by storm. She eloquently depicted the plight of the Egyptian *fellah* and called for a series of reforms (including land reform) to address the cruel injustices that beat down incessantly on the farmers.[72] Following the book's publication,

---

[71] Heath, "Creative Hermeneutics," 210.
[72] For more details, see Yunan Labib Rizq, "Rural Start," *Al-Ahram Weekly Online*, A Diwan of Contemporary Life (553), no. 698 (July 8–14, 2004), http://weekly .ahram.org.eg/2004/698/chrncls.htm. In citing her literary achievements, this

she launched into a successful career as a professor of Arabic Literature (Ain Shams University and several years in Morocco), with numerous literary contributions as a literary critic, novelist and poet. She was also a respected qur'anic exegete who wrote a column on the Qur'an and current issues for several decades in *Al-Ahram*.

As a commentator, Bint al-Shati stands in the line of Muslim reformers, owing much in particular to Abduh.[73] Of all the possible choices, Cragg selected her (somewhat randomly, he confesses) as a representative of contemporary qur'anic exegesis and examined her treatment of sura 93 (*al-Ḍuḥā*) from her *Al-Tafsīr al-Bayānī li-l-Qur'an al-Karim*.[74] Isa Boullata, in an article published a year later, deepens Cragg's analysis. Concerning the "trust" verse in *Sūrat al-Aḥzāb* (Q. 33:72), Boullata explains that Bint al-Shati's word study of *amāna* and *ḥamala* reveals the following:

> *Amāna* according to her does not refer to material things entrusted to man, or to the religious duties required of him, nor is it the intellect of man or his obedience as interpreted by some exegetes, but it is his free will, his responsibility of choice, his accountability.[75] To carry this *amāna* means to assume its results as man takes his position of "viceroy in the earth" (S. 2:30) and achieves the fullness of his humanity by being free and respon-

colleague of hers at Al-Ahram never once mentioned her work as a Qur'an commentator.

[73] For a fascinating account of a personal encounter with her, see Joyce M. Davis's *Between Jihad and Salaam: Profiles in Islam* (New York: St. Martin's Press, 1997). In this work, Davis presents interviews with seventeen opinion leaders in the Arab world, including Hassan al-Turabi, Rachid Ghannouchi, and Khurshid Ahmad. As a respected journalist she has also engaged in serious research into contemporary Islam and offers insightful comments on each interview. In this respect her work parallels Mary Anne Weaver's *A Portrait of Egypt: A Journey Through the World of Militant Islam* (New York: Farrar, Straus & Giroux, 1999). This is how Davis opens her chapter on Bint al-Shati: "Aisha Abdel Rahman is known to millions of Muslims as Bint al-Shati, 'daughter of the coast,' and columnist in the Egyptian newspaper Al-Ahram. She is one of the best-known proponents of women's rights in Egypt and a respected Muslim scholar. Bint al-Shati described herself as 'a type that is not repeated,' and for a woman born into a religious family in rural Egypt in 1913, the intensity of her lifelong quest for education and independence certainly had been unique" (*Between Jihad and Salaam*, 167).

[74] Cragg, *The Mind of the Qur'an*, 70-74.

[75] This is precisely why such an interpretation would have been difficult to defend in the classical era. It was a verse of choice for the proponents of the Muʿtazilite position (cf., for instance, J. Woodrow Sweetman, *Islam and Christian Theology: A Study of the Interpretation of Theological Ideas in the Two Religions*, 2 vols. [London: Lutterworth, 1947], 185).

sible, in contrast to all other creatures, though he is foolish and does not know the extent of the burden undertaken and does not realize the difficulty in handling it.[76]

So as better to explain her contribution to *tafsīr* in general and to the current Muslim understanding of humanity and *khilāfa* in particular, I will first look at her hermeneutical methods and then in more detail at her book *Al-Qurʾān wa-qaḍāyā al-insān* (*The Qur'an and Issues Touching Humanity*).

Bint al-Shati's Hermeneutical Method. Here is how Cragg characterizes Bint al-Shati's hermeneutical method:

> There is directness of interpretation about *Al-Tafsīr al-Bayānī* and a refusal to be deterred by the pious overgrowth of sacrosanct authority. There is a refreshing freedom from inventive subtlety and a will to see the text steadily and naturally... Some may feel that its careful disengagement from traditional niceties is too lengthy and deferential. But the logic of the writer's emphasis does point to an imaginative reckoning with Qur'anic meanings."[77]

Before going into more detail, I first turn to Jansen for a helpful overview of the *tafsīr* landscape of Egypt in his work, *The Interpretation of the Koran in Modern Egypt.* According to Jansen, all modern commentators of the Qur'an can be pictured as working together in a large room in the center of Cairo, using the same reference library of classical commentators, dictionaries, Al-Wahidi's work on the *asbāb al-nuzūl* (occasions for the revelation of specific passages), and Al-Suyuti's introduction to the Qur'an. Yet they come to the Qur'an from mainly "three different viewpoints: natural history, philology and the day-to-day affairs of the Moslems in this world."[78] This classification would be somewhat in line with the reality domain I added to Fiorenza's list: "academic disciplines." But he also includes the exegetes' purposes. Perhaps this would best be labeled "desired output":

1. Some exegetes have as their main concern "to prove that the modern sciences are not in contradiction to the Koran, or even that they can be deduced from the Koran."[79] Their focus, understandably, is on the natural sciences, both as a discipline to come

---

[76] Isa J. Boullata, "Modern Qur'an Exegesis: A Study of Bint Al-Shati's Method," *The Muslim World* 64 (1974): 112.

[77] Cragg, *The Mind of the Qur'an*, 74.

[78] Johannes J. G. Jansen, *The Interpretation of the Koran in Modern Egypt* (Leiden: Brill, 1974), 6.

[79] Ibid., 7.

to the text with, and as a purpose for writing about the Qur'an. Jansen labels this approach the *tafsīr ʿilmī* ("scientific interpretation").

2.   Another group of exegetes express their main purpose as the uncovering of the meaning of the text as it would have been understood by the Prophet's contemporaries. This calls for some historical study as well as the heavy use of philology.

3.   The third group of interpreters seeks to bring contemporary society in line with society as it was in the original *umma*. Though Jansen calls this orientation "practical exegesis," with the perspective we now have three decades later, this category would more properly be labeled "islamist exegesis." He comments, "They are not sure whether contemporary Egyptian society is 'Islamic enough.'"[80]

Bint al-Shati, as it turns out, follows her husband, the late Amin al-Khuli (d. 1967), and would clearly fall into Jansen's second category. Al-Khuli taught qur'anic exegesis at the Egyptian University in Giza and, though he never published a commentary, he "developed a theory on the relation between philology and Koran interpretation that has exerted some influence in Egypt."[81] He insisted that a commentator should delve into the historical background of the Qur'an, the people and society in which it was revealed, but also the history of its genesis. He often quotes from Theodor Nöldeke's *Geschichte des Qorans*[82] and adopts what Jansen calls "the 'traditional-historical approach' to a written text as opposed to the modern formalist, psychological or sociological approaches."[83] He takes the traditionalists to task by arguing that the Qur'an "came to humanity in Arab garb (*fī thawbihi al-ʿarabī*) and that we, therefore, in order to understand the Koran, should know as much as possible of these Arabs and their time."[84]

In terms of a practical guide to the task of *tafsīr*, al-Khuli's advice is rather straightforward:

[80] Ibid., 8.

[81] Ibid., 65.

[82] Theodor Nöldeke, *Geschichte des Qorans*, 2d ed. (Leipzig: Dietrich'sche Verlagsbuchhandlung, 1926).

[83] Ibid.

[84] Ibid., 66. His most controversial stance, however, was to agree to supervise the thesis of Muhammad Ahmad Khalafallah on the art of story telling in the Qur'an. By implying that the stories were not necessarily true historically, but that their main value lies in their teaching religious values, the published work caused quite a commotion in Egypt and the novice professor was quickly deposed (ibid., 68).

Firstly, he urges the scholar who intends to write a Koran commentary to take notice of all verses in which the Koran talks about a subject, and not to limit himself to the interpretation of a single passage neglecting other Koranic statements on the same topic... Secondly, Amin al-Khuli stresses the need for a careful study of the meaning of every word, not only with the help of the classical dictionaries but in the first place with the help of the Koranic parallel occurrences of the same word or the same root. Finally, the Koran interpreter should analyse how the Koran combines these words into sentences, and attempt to explain the psychological effect the language of the Koran has on its hearers.[85]

I have presented an outline of al-Khuli's method of *tafsīr* because, according to Jansen, "We may safely assume that we have a reliable picture of what a Koran commentary by al-Khuli would have looked like in two books that are dedicated to him and published by his widow, Aisha Abd ar-Rahman."[86]

---

[85] Ibid., 67. Notice how he is moving toward a dynamic speech-act conception of hermeneutics, before it was popular in the West.

[86] Ibid., 68. *Al-Tafsīr al-bayān ī li-l-Qurʾān al-karīm*, vol. 1, deals with suras 93, 94, 99, 79, 100, 90, 102 (in this order); and vol. 2, Suras 96, 68, 103, 92, 89, 104, 107. All of these are put in the first Meccan period by Th. Nöldeke (Jansen, *The Interpretation of the Koran*, 68 n. 61). The assumptions behind her method as presented by Boullata are based on a naïve realist epistemology: a combination of good philology, historical research and internal literary analysis will yield the desired result. As yet there does not seem to be an awareness of the crucial role in the extraction of meaning played by the interpreter herself. She does, however, consciously engage in a "contemporary *tafsīr*" and devotes a good fifty pages, for instance, to the theme "Contemporary Humanity Between Religion and Science" (with a subsection "Humanity and the Moon") in her book about humanity from a qur'anic perspective, *Al-Qurʾān wa-qaḍāyā al-insān*. Yet she is not consciously "doing theology," but rather letting the Qur'an speak for itself to the questions posed by our contemporary situation. Just the same, her own historical, contextual and personal orientations determine much of how she links passages together and finds meaning for them. For instance, when explaining the qur'anic idea of the *ghayb* (the unseen/unknown), she digresses into a commentary on the spiritist movement in nineteenth-century England. Instead of quoting classical commentators (she does, but very little), she offers a lengthy quote from Bernard Shaw (against Marxism, p. 217) and the late Egyptian poet, Muhammad 'Abd al-Wahab (p. 236). At one point she criticizes at length Ibn Sina's dualistic Greek worldview, and then in the next chapter comments on Pope Paul VI's visit and signed agreement with Soviet Premier Gromyko. This kind of *tafsīr muʿāṣir* needs to be recognized as "doing theology in context." The outcome is strongly shaped by the lens used to examine the qur'anic text.

Humanity's Khilāfa *and* Amāna. Bint al-Shati begins the Foreword to her book, *Al-Qurʾān wa-qaḍāyā al-insān*,[87] with the following admission, "My own sorrow about the worries, preoccupations and tragedies of humankind today (*insān al-ʿaṣr*) directed me to present in the first instance the research in this book under the title 'The Qur'an and the Issues of the Age.'"[88] But then she changed her mind, realizing that "contemporary thought" is inevitably understood by her readers as the discussion of foreign ideas, the "schools of thought," either expressed in political terminology (from the "right" or the "left"), or economic theories or sociological orientations.

By contrast, she rejects all those foreign intrusions. "My membership," she confesses, "is within humanity in its most absolute sense, and my loyalty is to my own conviction through which I judge, and to my *umma*, the only 'school' I recognize." In the next section she explains the content of her preoccupation with humanity: "In the context of my humanity I am concerned with the issues (or "problems," *qaḍāyā*) that have and will always exist, the preoccupation of humankind wherever and however they find themselves: the mission (or "trust": *amāna*) of their humanity, the demands of their existence, the worries of this world and the concerns for the hereafter."[89]

Then follows a sentence which could well serve as a subtitle to the book, though she rarely brings it up directly again: "Our age has bestowed upon us [*yaminnu*, the same root for the word 'manna'] the Declaration of Human Rights, announced by the United Nations about twenty-five years ago." What an irony this is, she continues. This is the time span of the generation of our children who "breathed, while still as embryos in the womb, the horrific dust of Hiroshima and Nagasaki, and welcomed, the same year as the Declaration of Human Rights, the crime of the century which amputated a portion of the homeland of Arab humankind…"[90]

---

[87] Bint al-Shati, *Al-Qurʾān wa-qaḍāyā al-insān* (Beirut: Dār al-ʿIlm li-l-Malāyīn, 1982). This is its fifth printing, an indication of its wide circulation in the Arab world. Bint al-Shati taught for several years in Morocco and her two-volume commentary was reprinted at least three times in her native Egypt. This book is also dedicated to her late husband, Dr. Amin al-Khuli.

[88] Ibid., 5. Since Dr Al-Rahman is first and foremost a literary critic, poet and a professor of Arabic Literature, she writes in an elegant Arabic style, rich in vocabulary and unusually eloquent. For this reason I confess my own inability to do justice to her writing in English.

[89] Ibid., 6.

[90] Ibid.

The tale of irony and tragedy then moves from Palestine to Algeria, where the war of independence cost the lives of more than one million civilians, and the continents of Asia and Africa, which have witnessed several famines and genocides. Referring to the Cold War, she writes that all this suffering has been in the name of "equalizing the balance of forces." Without mentioning either side (the word "West" does not even appear), she follows up with this powerful image: "And in the show of masks, the mantle of the saint resembles the Devil's cloak."[91] The enemies of humanity themselves announce the good news of human rights. Where are the genuine values of humanity in evidence, she asks? "[T]he slaves of centuries past found their hands and feet in chains and shackles but their conscience was intact, and their hearts bespoke a sanctity which could not be violated and would not submit to any fetter or control."[92]

The scope of her vision is twofold, then: first, the *umma*, but at the same time, humanity as a whole. Perhaps a better way of putting it is that she is addressing her Muslim audience with the message of the Qur'an, which reaches beyond their borders. Thus she ends her Foreword: "A question persists in my mind: what is willed for my *umma*?[93] I see ourselves torn apart by the schools [*madāris*, not the traditional *madhāhib* for schools of Islamic law and various theological tendencies], the circumstances and the systems, in divisions, parties and sects; we have gone our way, dispersed in many directions."[94] Yet much of the book concerns humanity in general.

In order to gauge what place these concepts have in her thinking, a short summary of the book is necessary. Significantly, it is divided into two parts: "Humankind and the Present Age" (*Al-insān wa-l-ʿaṣr*) and "The *Umma* and the Present Age" (*Al-umma wa-l-ʿaṣr*).[95] The first part consists of two fairly even halves: "The Story of Humanity from Beginning to End" and "The Destiny of Humanity: Existence and Non-Existence" (*al-wujūd wa-l-ʿadam*). Whereas the second part of this work is devoted mainly to the defense of a "contemporary reading of the Quran" (*tafsīr ʿasrī*), with examples given in such areas *al-ghayb* (the qur'anic

---

[91] Ibid., 7.

[92] Ibid.

[93] I take that passive voice to mean, "What does God want for my *umma*?"

[94] Ibid. There seems to be a deliberate word play here: *fa-dhahabna ṭarāʾiq qidadan* means literally, "we have gone in many groups." The verb *dhahaba* picks up the central theme of the "schools" that have divided the *umma* and at the same time expresses her feeling that because of its internal divisions the *umma* has "gone" astray, left the straight path set out for her in the Qur'an and Sunna.

[95] The first part is slightly longer, with 240 pages.

term for the unknown that only God knows), human freedom vs. determinism, being and non-being, the Qur'an and modern science, and more, the first part lays the qur'anic foundation for a doctrine of humanity.

The chapter titles of the first part on the "story of humankind" are indicative of her convictions: "The Caliph on Earth"; "Bow before Adam"; "He created Humankind and Taught them *Al-Bayān*"; "The Trust (*amāna*) of Humankind"; "The Freedom of Humankind"; "Freedom and Bondage"; "Freedom of Religion"; "Freedom of the Mind and Thought"; "Freedom of the Will." On the one hand, *khalīfa* and *amāna* are her two pillars for a qur'anic theology of humanity, and on the other, the greatest concern arising out of that fact is how to understand the relationship between humanity and freedom from a qur'anic perspective. Indeed, this is one of the central purposes of this work.

The whole first section of *Al-Qurʾān wa-qaḍāyā al-insān* has an introductory chapter which seeks to define the term *al-insān* (humankind) from within the Qur'an.[96] Humans are not *bashar*, "that Adamic substance that eats food and walks in the markets"[97]—a word she says that simply designates humans as a biological species nor are they "people" (*nās*)—an indication of our species or of the category of beings we represent; nor are we *ins*, which in the Qur'an is always used in contrast to the *jinn*; so what is *al-insān*? What are the qualities in the qur'anic concept of *al-insāniyya*, which set *al-insān* apart from other earthly creatures? She answers:

> In [the term] humanity [*al-insāniyya*] is the elevation to the rank that enables them to fulfill the caliphate [*khilāfa*] on earth, to shoulder the responsibilities of their calling and the trust [*amāna*] of humankind, for they are the ones whose specialty is science and revelation [*al-bayān*, or

---

[96] Her methodology here is consistent throughout the book: first, what the questions are; second, what the qur'anic terms are, how often they are found in the Qur'an and, through the examination of several representative passages, how they come to be defined. She seldom turns to classical commentators, except when the topic is controversial (like the question of freedom vs. determinism). But even then she quotes sparingly compared to traditional, including contemporary, commentators. She seems more inclined to quote poetry, but more for homiletical purposes than for exegetical ones. This is no disinterested *tafsīr* work. She is consciously building on what she considers sound and solid exegetical work in order to edify. Her message is that Islam must be renewed on the basis of the ethical values of the Qur'an, applied to our contemporary society, and informed by the achievements of science and technology. Though she studiously avoids the word *ijtihād*, this is exactly what she is advocating: a renewed vision of her readers' God-given calling as Muslims and secondly as humans, keeping in mind the questions of our contemporary culture.
[97] Ibid., 15.

"Qur'an"], the use of their minds and discernment, together with all that exposes them to the tribulation [*al-ibtilāʾ*] of good and evil, the temptation and deception caused by their need to feed themselves and expend energy.[98]

At the same time, scores of qur'anic passages remind us humans that we were made of dust, or of sperm, or of a bloodclot, or of gushing water, and furthermore, that our nature is weak and forgetful of God's wondrous works. This is so that humans "will not overstep their power and rebel, or become conceited."[99]

Here is her opening statement for the chapter "Caliph on the Earth": "The story of humankind begins with the creation of Adam, the father of humanity" (*al-bashariyya*). After several pages of qur'anic commentary on creation, and a tribute paid to the "illiterate Prophet" (according to the traditional meaning of *ummī*), she comments on the angels' bowing to Adam, God's deputy about to placed on earth. There is in this announcement something that appears to be new. She explains whence the novelty: "since only humankind is singled out among creatures for their ability to think and discuss, and for their responsibility to choose..."[100] In other words, Bint al-Shati sees in humanity's *khilāfa* a combination of 'Abduh's vision of knowledge and science, and Iqbal's second emphasis on humanity's "burden of personality which gives him freedom to choose between good and evil and holds out to him the prospect of achieving personal immortality by personal effort."[101]

There is a difference, however. Bint al-Shati distinguishes between *al-bashar* and *al-insān* specifically in this respect: the truly human person (*al-insān*) "shoulders the requirements of the trust [*amāna*], the covenant, the commandment, the burdens of his charge [*taklīf*], responsibility and painful endurance [*mukābada*], as well as being the one who specializes in science, thinking and revelation [*al-bayān*]."[102] The *khilāfa* and *amāna* of humankind, in other words, are intimately related.[103] She

---

[98] Ibid., 19.

[99] Ibid., 22.

[100] Ibid., 25.

[101] Steppat, *God's Deputy*, 167. We also saw this in Ibn Kathir and even more so, in Razi.

[102] Bint al-Shati, *Al-Qurʾān wa-qaḍāyā al-insān*, 172.

[103] As mentioned above, this connection is assumed today. I offer two more examples. In an article entitled "The Islamic View of Freedom" (*Islam and the Modern World* 6.2 [1975]: 41-60), A. R. Doi begins with the statement that "the whole universe is a manifestation of God's will in which Man's place is the highest." God granted man "freedom of choice...and gave him authority to acquire and make use of the things around him. In short, He granted him a kind of autonomy and appointed

defines the trust that the heavens and earth and mountains refused but which humans chose to accept as "the testing [or 'tribulation'] through the demands of their mission [or 'charge': *taklīf*], the freedom of their will, and the responsibility of choice."[104]

Bint al-Shati goes deeper than Iqbal in the matter of human free will. One could almost read a Kierkegaardian attitude to her depiction of the human struggle. Existence is a fight between good and evil, with the only consolation being our hope for the hereafter. "Life passes by and does not stop…and the human being continues his endless fight (*niḍālahu*) so that life might overcome." One's willingness to take on this costly struggle is what elevates one from the level of *bashar* to *insān*. In her words,

> [S]o we perceive that our earthly expedition over the bridge between life and death is, according to the quranic text [*al-bayān al-qurʾānī*], nothing less than painful testing [*ibtilāʾ*]—an echo to the bearing of the trust which the heavens and the mountains and the earth shunned and the responsibility of which they were forcefully exempted. The meaning of testing [*ibtilāʾ*], as far as my research has allowed me to feel confident, is the endurance [*mukābada*, with the idea of suffering] along the way in order to achieve the ideal human existence, the defiance of the obstacle in the way of keeping a firm grasp on the horizons of truth and goodness, and the intrepid fight [*mujāhada*] to resist the tendencies of selfishness and evil and the attractions of temptation [*fitna*] through the lusts of this world and their fleeting appearance: "…who created life and death to try you (that He may see) which of you is best in deeds? (Q. 67:2)."[105]

I conclude with another comment on Bint al-Shati's use of the word *insāniyya*. As previously quoted in her Preface, she offers her readers a key to her thinking on the issue of humanity and *khilāfa* (though it is not tackled head on): "My membership is within humanity in its most absolute sense…" Then she confesses her Islamic faith, "and my loyalty is to

---

him as His *Khalīfa* (vicegerent) on the Earth and instructed him to live according to his Guidance." He then comments on the reason for God's command to the angels to bow before Adam. Interestingly, one of the reasons for humankind's superiority over the angels is its emotional nature. Consequently, the angelic hosts cannot understand "the whole of God's nature, which gives and asks for love." In a similar article ("Qur'anic Humanism"), Syed Vahiduddin considers this a central quranic concept: "Man is considered the vice-regent of God on earth and endowed with a trust (*amāna*) which no one dared to carry… Here it is that man is scolded and scolded lovingly; he is ever trying to overdo himself and yet it is he alone who seems to be capable to attempt what is beyond his reach" ("Qur'anic Humanism," *Islam and the Modern World* 18.1 [1987]: 3.

[104] Bint al-Shati, *Al-Qurʾān wa-qaḍāyā al-insān*, 2
[105] Ibid., 172-73.

my own conviction through which I judge, and to my *umma*, the only school I recognize."[106] Two pages later she confirms the priority of her belonging to the human family, "From my humanity (*insāniyyatī*) I gaze intently at my *umma* in its tribulation at the hands of humanity's (*al-insān*) enemies: in a matter of hours, her most powerful army was led in the heart of the Arab fatherland and the Islamic world, from the war in Yemen to the cemetery of Sinai."[107] Together with the tragedies of Hiroshima and Palestine, she condemns oppression of the weak and violence to people's dignity in the name of *insāniyya*. This is no way to treat a fellow human being, whatever his/her ethnic or religious background. Might does not make right.[108]

Good deeds and high moral values are no doubt the foundation for a common theology of humanity in all of the three monotheistic religions. This impression is confirmed by another passage in the section devoted to the tension between science and faith. Again, the crucial word is *insāniyya*, much more than "humanity" in the English sense, but comprising also moral and spiritual values that raise our species above the rest of the living creatures:

> As humanity (*al-insān*) greets the age inaugurated by the landing on the moon she has a right to question what this age has to contribute to her peace of mind after it has afflicted her with a splitting apart of the material and the spiritual [*al-maʿnawī*] and wore her out with the futile struggle between science and faith…[109]
>
> With all these sufferings this age must also grant her, if nothing else, healing of the soul and spirit, and after that, recovery from the effects of the doubling of her personality—the material and the spiritual—and bestow on her a balance between the gravity of the earth into which the human roots have sunk deeply from time immemorial, and those heavenly horizons where one wears belts to compensate for the weightlessness.[110]

---

[106] Ibid., 5.

[107] Ibid., 7.

[108] While dealing with the destiny of humankind she quotes several verses along these lines: "Every living soul shall taste death; and We try you with evil and good as a trial. And to Us you shall return" (Q. 21:35). She comments: "And because of this trial [*ibtilāʾ*] humanity's journey through this life is not empty or vain, but as the Adamic creature [*bashar*] dies, his high values, his kind words and good deeds remain as a treasure to humanity [*insāniyya*] in the course of time and as peaceful lanterns to light the way [of those who follow]… The human values [*al-qiyam al-insāniyya*] alone are that which is eternal and remains" (ibid., 173).

[109] Here I chose to keep the Arabic feminine for "humanity," *insāniyya*.

[110] Ibid., 215. The image of gravity contrasted to the "belts of anti-gravity" (or "weightlessness") is not only pleasingly poetic but pungently theological. In this age of astronauts on the moon, we perhaps are in a better position to grapple with the

Finally, as a way perhaps to emphasize the universal application of this vision, Bint al-Shati widens the scope of her remarks: "On the wide horizon of our new world has begun to appear initiatives which reveal a lucid realization of the futility of any attempt to find a substitute for religious faith..."[111] She then recounts the official visit of a delegation of the Vatican to the Soviet Union followed by Andrei Gromyko's visit to Pope Paul VI in 1966. During that same year, diplomatic relations were established between the Vatican and Czechoslovakia, after prior successful attempts in the same direction with Hungary and Yugoslavia. Her point is well made: Rome acquired the right to care for her flock in those countries without any opposition from the political authorities and thereby exposed the failure of Marxism to eradicate the spiritual longings of the people it had tried to control.

The door is wide open in Bint al-Shati's view of humanity in the Qur'an to a fruitful dialogue with believers of other faiths and particularly where good works, thoughtful economic development and the pursuit of justice and human dignity is concerned. Humanity's *khalīfa/amāna* is to care for the created order and express solidarity for the whole human family, making use of human intelligence and keen moral sensitivity. In the meantime, she has helped to inspire a new generation of Muslim women scholars, who seek to tackle issues of deep concern to them in the light of qur'anic norms of social justice, compassion and human values prized by all.[112]

The reformist vision, then, never died during fourteen centuries of Islamic life. For all the pressure, humiliation and oppression occasioned by European colonialism in the Islamic world, these same forces acted as a catalyst for renewed theological efforts to interrogate the sacred texts and find answers to the new questions raised by a quickly changing world. Some drew inspiration from a combination of Sufi, reformist mechanism in Islamic law and modern ideas such as those promoted by

---

fact that the attraction of this world can be pernicious and therefore needs to be balanced with the weightlessness of the world to come. In any case, it is the classical qur'anic contrast *dunyā/ākhira*, and its contextualization is all the more inspiring.
[111] Ibid., 216.
[112] See the Muslim Women's Studies website, connected to the Zahira Abdin Chair for the Study of Women and Gender (Cairo), and the proceedings (in Arabic only) of the conference held under their auspices in March 2000, "Aicha Bint al-Shati (1913–1998): Women's Discourse or Contemporary Discourse: A Genealogy of Ideas." The larger umbrella is a non-profit organization founded in Egypt in 1999, called Association of Women and Civilizational Studies. See http://www.muslimwomenstudies.com/English.htm.

Shah Wali Allah in the eighteenth century. Others, including al-Afghani, Abduh and Bint al-Shati, sought to appropriate Western rationalism and press it into the service of a more Mu'tazili-like reformism. Rida continued this tradition, but as he pondered both the fall of the Caliphate and Abd al-Razzaq's version of a depoliticized Islam, opted for a more ideological version of the Islamic faith. In the early 1930s he realized that it was Hasan al-Banna who was succeeding to do this on a grand scale. Political Islam and new streams of reformism are the topic of the next chapter.

Chapter 9

TAFSĪR OF Q. 2:30: POSTMODERN PERIOD

Hasan al-Banna left his small town in the Nile delta in order to attend the best institute for teacher training at the time, the Dar al-'Ulum in Cairo. Arriving in the capital city (still sixteen years old) in 1923, Banna was shocked both by the moral laxity of the youth and by the political turbulence of Egyptian politics, which he attributed mostly to the British occupation. One of the mentors he sought out was Rashid Rida. According to Richard Mitchell, he often visited Rida, and after the latter's death in 1935, he and his newly founded Society of the Muslim Brothers (by now growing at a breath-taking pace)[1] took over the editing of *al-Manār*.[2]

Whereas a sharp distinction has often been drawn between Islamic "reformism" and "fundamentalism," in this chapter I seek to emphasize the continuity of motivations and objectives (if not strategies) on the part of Muslim thinkers and activists, starting with Banna, Sayyid Qutb, the chief ideologue of his movement in the two decades following Banna's assassination (1949), the South-Asian scholar/activist al-Mawdudi, two Palestinian leaders of Hamas, as well as current writers who might be labeled "progressives" or "liberals," moving as they do toward a postmodern epistemology. As I look at their exegesis of Q. 2:30, it appears that they all share a renewed sense of human dignity that increasingly prizes the values of human rights and democracy.

I begin, however, with a contemporary *ʿālim*, a Syrian member of the Al-Azhar faculty, Abd al-Majid al-Najjar. From one angle, his views present an easy transition from the reformism of Bint al-Shati to the more exclusivist exegesis of many islamists. But they are also important to

---

[1] See Brynjar Lia, *The Society of the Muslim Brothers in Egypt: The Rise of an Islamic Mass Movement 1928–1942*, with a Foreword by Jamal al-Banna (Reading: Ithaca, 1998).

[2] Richard Mitchell, *The Society of the Muslim Brothers*, with a Foreword by John O. Voll (Oxford/New York: Oxford University Press, 1993 [1969]), 5, 23.

highlight because of the context from which they are uttered: the oldest university in the world, an Islamic icon, yet also an institution that has been co-opted for government purposes since the revolution of 1952. President Gamal Abd al-Nasir succeeded in wresting the control over family law from Al-Azhar (1956) and in adding secular departments, thus initiating sweeping changes in its structure (1961). Still under both Nasir and Sadat, the Egyptian state could claim its legitimacy in terms of pan-Arab nationalism. After the latter's assassination in 1981, it became clear that Egyptian society was taking a turn toward a more pious and religiously conservative stance on issues of politics and society. What is more, the state was locked into a violent struggle with groups that splintered off from the Muslim Brotherhood and aimed to take power. President Hosni Mubarak, then, had to shore up the religious credentials of his government—a situation, argues Steven Barraclough, which has enabled Al-Azhar to reassert its authority.[3] As a bridge between the state, the islamists and a population that had become more conservative in the last two decades, Al-Azhar has been able to play the role that ʿulamāʾ have for centuries in various Islamic state configurations—that is, act as a counterweight to state authorities. In the 1990s, for instance, Al-Azhar was granted wide powers that enabled it to control Egypt's mosques, censure the media and issue important legal decisions allowing sharīʿa considerations to trump the civil code.[4] This context should be remembered while examining al-Najjar's views.

A Traditionalist: al-Najjar's Khilāfa Methodology

Abd al-Majid al-Najjar, a professor of uṣūl al-dīn (theology as opposed to law), is a good example of a traditionalist theologian who consciously places himself between liberalism and fundamentalism. Is he a reformist? The question remains open, though I will argue for a "conservative" or "traditionalist" label.[5]

---

[3] Steven Barraclough, "Al-Azhar: Between the Government and the Islamists," *Middle East Journal* 52.2 (1998): 236-49.

[4] The most famous ruling was in relation to one of its own professors, Nasr Abu Zayd, who on the basis of his exegesis of the Qur'an, was declared an apostate and saw his marriage dissolved (forcing him into exile).

[5] He began his career with research on the reformist Berber scholar Ibn Tumart (d. 524/1130), wrote several books on issues related to faith and reason, then more recently on the environment and on questions related to culture from a theological perspective (*Muqārabāt fī qirāʾat al-turāth* [*Approximations on the Reading of Cultural Legacy*] [Beirut: Dār al-Badāʾil, 2001]).

Najjar was interested in bridging the gap between various historical factions in Islamic theology in the 1980s, and among his writings from that period is a short book entitled *The Vicegerency of Man: Between Revelation and Reason*.[6] Humankind is an eminent species among all of creation, he writes, for the human person "is a combination of material and spiritual ingredients," and thus "in Islamic literature, man is referred to as the microcosm, the core, the kernel, and the gist."[7] Since he was declared God's deputy from the start, and since he "was fashioned [in due proportion]" while God breathed his Spirit into him (Q. 38:72), there can be no room for Darwin's evolutionary theory.[8] In fact, *khilāfa* sums up the high calling of humanity, which "means implementing Allah's intent on Earth and practicing His rules."[9] It is another way of saying that people are to "enjoin the good and prohibit evil." Human beings are first called to worship: "I have only created Jinns and men, that they may serve Me" (Q. 51:56).[10]

Though Najjar never quotes the Trust verse directly (Q. 33:72), the linking of *khilāfa* and *amāna* runs throughout. Human vicegerency is a matter of free will and of moral obligation:

> Only man is entrusted with the divine assignment, as the sky, the mountains, and the Earth found it difficult to undertake. The assignment means explaining to man, in his capacity as *khilāfah* [*sic*], what is expected of him and then letting each individual decide freely whether or not, after contemplating the consequences, to fulfill the corresponding obligations.[11]

This choice to follow God's path is a decision to fight one's inner passions and selfish desires (*hawā*). "It is a kind of psychological jihad

---

[6] Abd al-Majid al-Najjar, *The Vicegerency of Man: Between Revelation and Reason: A Critique of the Dialectic of the Text, Reason, and Reality*, trans. Aref T. Atari (Herndon, VA: International Institute of Islamic Thought, 1999). The original is *Khilāfat al-insān bayna al-waḥī wa-l-ʿaql* [*The Vicegerency of Man between Revelation and Reason*] (Beirut: Dār al-Gharb al-Islāmī, 1987).

[7] Najjar, *The Vicegerency of Man*, 17. Atari's translation is in fact an abridgment of the original: whole paragraphs and sentences are cut out, and sentences are simplified to the greatest extent. The result is a more readable version, which loses some of its original pungency however. Here the sentence reads in the original: "As an expression of humanity's constitutional polarity (*quṭbiyya takwīniyya*, i.e., its material-spiritual composition), he has been described as a microcosm, the core of the world, its quintessence and its gist" (*Khilāfatfat al-insān bayna al-waḥī wa-l-ʿaql*, 41).

[8] Najjar, *The Vicegerency of Man*, 14.

[9] Ibid., 21.

[10] "Serve" translates the root ʿ-*b*-*d*, which gives us ʿ*abd*, "servant or slave," and "worship," ʿ*ibāda*.

[11] Ibid., 23-24.

leading to gradual growth and perfection through interacting with the universe, during which human beings observe Allah's injunctions by enjoining right and refraining from wrong. This jihad climaxes with the realization of *khilāfah*."[12] Either people choose to draw closer to God by means of their voluntary obedience (*ṭāʿa*), or, as they are inclined to do naturally, they refuse the assignment and inevitably become unjust, thereby trampling upon human dignity.[13]

The only source of guidance in the path to fulfill this vocation is "the Qur'an and *Hadīth*." On the one hand, "[t]he revealed text clearly states the ultimate end to which man's life should be geared." This includes regulations applying to areas of human endeavor that are precise and unambiguous. On the other hand, since the revealed norms pertain to "categories of deeds," with the passing of time and changing social conditions the categories will need to be reworked.[14] Herein consists the role of "reason"—"that discerning human faculty" at the origin of the human *khilāfa*. "Revelation is a divine means of revealing truth, while reason is a human faculty encompassing the human faculties of perception and discernment."[15] The Qur'an itself often calls its audience to use this capacity of reflection, which is essential if people are to apply the norms of the text to their everyday lives. Yet reason is limited and its judgments will always remain relative and tenuous. Not so with the clear injunctions of the *sharīʿa*. If the language is unequivocal, then the command or prohibition must be followed to the letter. A thief's hand must be cut off (Q. 5:38).[16] Revelation is always purposeful; it seeks to "achieve the broad interests of man and secure human happiness now and in the hereafter."[17] It may be that we cannot always fathom the

---

[12] Ibid., 24.

[13] Najjar occasionally adopts an apologetic tone, as he does here: "In contrast, human beings are inclined to decline, resign from the inherent assignment, be unjust, and encroach upon human dignity when they feel that their own life has become useless, that existence has become vain, and that aims have been exhausted. This was experienced by early societies, and also by contemporary societies, particularly in the West, which suffers from such symptoms as a result of misunderstanding man's task on Earth" (ibid., 24-25).

[14] Ibid., 26.

[15] Ibid., 31.

[16] Ibid., 49.

[17] He quotes the Granadan jurist al-Shatibi (d. 790/1388) more than any other classical jurist. The purposes of law (*maqāṣid al-sharīʿa*), or the "divine intent," as it is translated here, is an expression that gained currency only after the sixth/twelfth century in Islamic jurisprudence. As a methodology, it was systematically incorporated within the range of acceptable legal strategies by al-Shatibi, but did not gain prominence until recently (cf. Johnston, "An Epistemological and Hermeneutical

purpose of some injunctions. Yet Islamic law states that the divine intent covers five main areas: the protection of religion, the self, human property, progeny, and reason.[18] Men and women guilty of adultery must be flogged (one hundred lashes, Q. 24:2) in order to protect human dignity, and here, specifically, human progeny. Hence, sexual intercourse is strictly limited to marriage.[19]

At the heart of Najjar's procedure is a faithful allegiance to the traditional Islamic welding of theology and law—much like Judaism. Revelation in daily life is about categorizing human acts, and Islamic law over the centuries has evolved a complex, yet coherent and somewhat flexible system. I cannot here offer further details, except to say that when new situations arise that are not covered by the explicit texts, then reason proceeds cautiously on the basis of analogy (*qiyās*), past consensus (*ijmāʿ*), the common good or public utility (*maṣlaḥa*), or blocking the way to evil (*sadd al-dharāʾiʿ*).[20] What is noteworthy in this book is Najjar's insistence that this path should continue to be followed rigorously. He castigates "the so-called liberal Muslims," or "the so-called Islamic left," who take divine intent as a license to cancel the clear commands of the text.[21] They reason that the circumstances in which the Qur'an was revealed were vastly different from those of today. Further, Muslims now "live in a world having a universal value system and new covenants, such as human rights, all of which reflect modernization, progressiveness, humanism, and the *zeitgeist* (*rūḥ al-ʿasr*, the spirit of the times)."[22] Hence, polygamy should be abolished, thieves' hands should no longer be amputated, and banks may charge interest.[23]

Turn"). Yet many use this method by highlighting "public benefit" (*maṣlaḥa*) with very conservative results (as Najjar), while others use it as a means to dramatically revamp Islamic law.

[18] Ibid., 48.

[19] Ibid., 50.

[20] This is an over-simplification, naturally. There are other criteria, and various schools and individual scholars emphasize certain juridical tools over others. There has been a tendency in the last century to deemphasize the four traditional schools of Islamic jurisprudence and to pick and choose from all the schools according to the issue at hand. For an accessible (and thus not comprehensive) and accurate introduction to this field, see Bernard G. Weiss, *The Spirit of Islamic Law* (Athens/London: University of Georgia Press, 1998).

[21] Ibid., 59-68.

[22] Ibid., 59.

[23] Najjar quotes the Egyptian philosopher Hasan Hanafi's book *al-Turāth wa-l-tajdīd* (*Legacy and Renewal*) in order to illustrate the arguments of the "liberals" (though it does not appear in the bibliography of the translated version).

This is to commit three errors, counters Najjar. First, revelation reveals not only the purpose of the law but also its methodology (the *"Khilāfa* methodology," *minhāj al-khilāfa*). These people ignore the fact that clear texts are to be universally applied. Second, to assert that commands only applied to a specific time is to deny the universality of the texts. Finally, this approach reveals an unhealthy fixation with contemporary norms. "Our argument," he writes, "is that what is new is not necessarily better."[24] Along with the remarkable scientific achievements of our era, he continues, we have witnessed the egregious attacks on Muslim nations and culture by Western colonialism and neo-colonialism, with as a result "poverty, dependency, and cultural alienation."[25]

How, then, is Najjar countering those "literalists" who want to apply the rulings of classical *sharīʿa* (and especially the *ḥudūd*, "limits," or "penalties") *in toto* today? Only twice does he criticize those who "call for a mechanical implementation of the Shari'ah."[26] His argument against them is that in order to apply revelation's plain rulings, one has to study carefully (1) the divine intents behind them and (2) the realities of life in each context (using all the social sciences). When this is done—and the process is never completed—then a new *ijtihād* will emerge, one which at the same time draws from the juridical legacy of the centuries and builds its application on a careful appreciation of the realities of today's world. When human acts have been assiduously analyzed and categorized in this new (changing) context, the next task is to ascertain whether the conditions for its application are present or not (one such condition is that it should not lead to undue hardship for the *umma*). Thus it is quite possible that a clear ruling in the text should be suspended for the time being.[27] This does not in the least invalidate the universality of that ruling, Najjar is careful to emphasize. Nevertheless, Muslims have much work ahead of them in several areas. Here is his reasoning about *ribā* (interest, or usury):

> Applying the prohibition of *ribā* to the existing financial domain requires an in-depth investigation of various transactions to distinguish those which fit in this category and those that do not. After *ribā* has been singled out, they should be thoroughly studied to determine their nature, volume, immediate and remote causes, international dimensions, and their short-

[24] Ibid., 65.
[25] Ibid., 66.
[26] Ibid., 75; see also 80. These attacks amount to very little, compared to his invectives against the "liberals."
[27] An example, which he brings up here, is very common in the literature: the second caliph, Umar, suspended the penalty for theft during a year of famine.

term consequences. On the basis of such considerations, the ruling of *ribā* may be temporarily waived for some transactions in this category. A grace period may be deemed necessary to prepare for an interest-free financial system designed to replace the existing interest-based one.[28]

Quoting again from eighth/fourteenth-century jurist al-Shatibi, Najjar describes this meticulous and ongoing legal reasoning as "accomplishing the trust" (*taḥqīq al-munāṭ*).[29] In theological terms, the human person is called to be God's trustee on earth (*khalīfa*), and in order to fulfill this trust (*amāna*), he or she must fulfill the requirements God has revealed in the Qur'an and Sunna—thus obeying the specific commands and prohibitions therein, and following the specific revealed methodology. Put in these terms, the "caliphate of humanity" is a calling that only Muslims can fulfill. That would be the inevitable consequence of Najjar's position—a very different viewpoint from the one expressed by Bint al-Shati, who is more interested in people following their conscience, irrespective of religious affiliation, and in upholding the dignity of all human persons rather than obeying legal precepts.

This brings me to a shortened section on Islamic "fundamentalism" and the human *khilāfa* —shortened because of space, but also because it is so widely discussed in the literature. Najjar is a traditionalist scholar, one who has been entrusted with a living tradition, which he handles with great respect and caution. At the same time, he is a politically conservative scholar, who, by teaching at Al-Azhar University, walks a tightrope between subservience to a regime that is more Islamic in name than in fact and outright condemnation of it in synch with a population that for the most part identifies with the Muslim Brotherhood. Banna's movement, though still officially prohibited from functioning as a political party in Egypt (though it renounced violence in the early 1960s), nevertheless has many of its members active in the parliament. Najjar too, whether officially or not one of the Brotherhood's ranks, joins with them in calling for incremental changes in a more "Islamic" direction. For this reason, I see him more as a traditionalist than as a reformist or an islamist.[30] In contrast to Abduh, Najjar presents no reformist agenda—only a cautious proposal to "postpone" certain literal injunctions of *sharīʿa*.

---

[28] Ibid., 81-82.

[29] Ibid., 85. The word *munāṭ* here ("that which is entrusted") is a synonym of *amāna*.

[30] A prime example of a traditionalist is that of Kashmiri scholar Ghulam Nabi Ganai, who lectures in Islamic studies at the University of Kashmir (Srinagar). In an article devoted to the topic of the caliphate, he nowhere indicates that the concept of *khilāfa* might apply even to Adam, much less to humankind—he simply rehashes arguments and counterarguments about the role and qualification of the *khalīfa* or imam, and in what cases a people might rebel. Mawdudi is quoted in one sentence on the definition

1954–2009: Islamism or Continued Reformism?

I seek to question the term Islamic "fundamentalism" in this section— and not so much because, as many Muslim writers rightly point out, it was first applied to a Christian movement in the USA in the early 1920s. In fact, other Muslims themselves from within the ranks of "political Islam" and from without have often used the term *uṣūlī* (fundamentalist). My only argument with the term is that the reality it seeks to denote is far too complex to fit under its label. The 1970s and 1980s witnessed a resurgence of religious fervor on a worldwide scale, among Christians, Muslims and Hindus on the one hand, and in Western societies in general (including the spiritualities of the east like Buddhism and hybrid forms like "New Age" movements). The Islamic revival should be seen in that context,[31] though conditions proper to Muslim societies did undoubtedly play a major role as well—first and foremost the Arab humiliation following the "Six Day War" of 1967 and later the Iranian Islamic revolution of 1979. Yet it was the rising tide of religious enthusiasm in general that prompted the "Fundamentalism Project" spearheaded by Martin Marty and Scott Appleby.

In the first volume of that series, Voll tackles the subject of "Islamic fundamentalism," admitting that it is not an easy concept to define, mainly because of the variety of individuals and groups associated with it. He notes the reticence on the part of some Muslims scholars to use the term "fundamentalism" that originally applied to Christian movements. Yet he, along with other authors, still chose to use the label to refer to "the complex cluster of movements, events, and people who are involved in the reaffirmation of the fundamentals of the Islamic faith and mission in the final decades of the twentieth century." In Islamic terms, this means:

> All Muslims affirm the truth of the revelation in the Qur'an and they have an obligation to implement the fundamentals of that truth in their lives and societies. However, those commonly referred to today as "funda-

of the term. Even so, there is no hint as to how this might apply today ("Muslim Thinkers and their Concept of Khilāfah," *Hamdard Islamicus* 24.1 [January–March 2001]: 59-72).

[31] The German specialist Gudrun Krämer puts it nicely: "Since the late 1970s, Islam has come to renewed prominence in the Muslim world as the guiding principle of individual behaviour and public life" ("Visions of an Islamic Republic: Good Governance According to the Islamists," in *The Islamic World and the West: An Introduction to the Political Cultures and International Relations*, ed. Kai Hafez, trans. from the German by Mary Ann Kenny [Leiden: Brill, 2000], 35).

mentalists" adopt an identifiable approach to this common obligation, an approach marked by an exclusivist and literalist interpretation of the fundamentals of Islam and by a rigorist pursuit of sociomoral reconstruction.[32]

Yet the "exclusivist and literalist" approach is common to all traditionalists or conservatives—Najjar is a good case in point. Accordingly it must be the political element that sets them apart, the "rigorist pursuit of sociomoral reconstruction." Here again, all sincere Muslims of a conservative bent are committed to shaping their society into a more Islamic mold (whatever that might mean to them). So, if we are talking about a gradualist approach, this is the agenda of reformism (from the Arabic term *iṣlāḥ*)—the legacy of al-Afghani, Abduh and many before and after them. If, on the other hand, we mean "overthrowing governments by violent means," then we have "political Islam," "Islamism," or "radical Islam"—terms used interchangeably with "Islamic fundamentalism" in the media as well as academic publications. But again, if it is violence that marks off "fundamentalists" from "reformists," then how does one account for the fact that many movements that went through violent phases (the Muslim Brotherhood is a good example), joined the political process at a later stage? The use of violence has to do with the means a group chooses in order to reach its political goals.[33] "Fundamentalism" and "political Islam" are not helpful terms in that respect.

This is why in the 1990s the term "islamist" became most widely used, by Muslims and non-Muslims alike.[34] No doubt, language mirrors

---

[32] John O. Voll, "Fundamentalism in the Sunni Arab World," in *Fundamentalisms Observed*, ed. Martin E. Marty and R. Scott Appleby (Chicago, IL: Chicago University Press, 1991), vol. 1:347.

[33] As I see it, the best sociological analysis of "three decades of Islamism" is Gilles Kepel's *Jihad: The Trail of Political Islam* (Cambridge, MA: The Belknap Press of Harvard University Press, 2002). His thesis is that the movement to overthrow existing regimes and install "Islamic" governments in the Muslim world had petered out by the late 1990s, and that spectacular terrorist operations by the loose international networks of hard-line islamists are actually the sign of desperation of a movement that has failed. Terrorism will not disappear quickly, but neither will these networks succeed in converting the masses to their worldview.

[34] Najib Ghadbian stated that "'Islamist' (*'Islamiyyūn'*) is what people belonging to Islamic movements call themselves, while 'fundamentalist' is what their opponents derisively call them in a foreign tongue... 'Islamist' is distinct from 'Muslim' in that the former refers to people with a conscious activist agenda while the latter is a nominal identity for the people of a gamut of ideological views. The majority of the Arab world is Muslim, while only those with ideologies that call for the implementation of Islam in the public as well as private realms are Islamists (*Democratization and the Islamist Challenge in the Arab World* [Boulder, CO: Westview, 1997], 7).

trends (that evolve constantly), and since labels are often used to casti-
gate one's opponents, the emotional packaging of a word will affect
different people very differently.[35] I agree with two French specialists,
Gilles Kepel and Olivier Roy,[36] that a corner has been turned. "Political
Islam" has failed to capture the imagination of the masses, at least if
they were counting on them to rise up and overthrow all the existing
"un-Islamic" regimes in the Muslim world. In fact, islamist movements
vary enormously in their ideology and tactics from country to country,
though the trend has been toward greater participation in local or national
governments and toward eschewing violence.[37]

British political scientist Salwa Ismail, for her part, disagrees with
Roy's thesis of islamism's failure, which sees the deployment of Islamic
signs, idioms and symbols in ever-widening fields of civil society as
a diluting of their potency (Roy's term is "re-Islamization"). "This
assumes," she retorts, "that there are practices that are inherently Islamic
and practices that must be intrinsically un-Islamic. What this ignores,
however, is that meanings are not given, but are invested socially and
historically."[38] In the 1970s, the islamist movement was primarily univer-
sity-based, though with time it became a popular groundswell sweeping
across all the major professional syndicates in Egypt in the 1980s. In
many other states as well, this renewed religious zeal expressed itself in a
flurry of new NGOs, neighborhood groups, sports associations, and the
like. This massive "societal investment" may be a different strategy from
the direct takeover of a state, but it should not in any sense be seen as a
"failure."

---

[35] In July 1998 I made the mistake of using the word "extremist" (*mutaṭarrif*) while
referring to groups such as Hamas and Islamic Jihad in front of the mayor of Yatta, a
town and collection of villages south of Hebron. He raised his voice in protest and
gave me a passionate ten-minute lecture on how these people were simply Muslims
trying to apply their religion to their life situation. His reaction might have been just
as negative had I used *uṣūliyyūn* (the closest term translating "fundamentalists"). Nor
did he use the term *islāmiyyūn* (whence we get our word 'islamists'), which I have
heard used consistently by people inside and outside of the movement.
[36] Gilles Kepel and Olivier Roy, *The Failure of Political Islam*, trans. Carol Volk
(Cambridge, MA: Harvard University Press, 1994).
[37] See, for instance, Gundrun Krämer, "Cross-Links and Double Talk? Islamist
Movements in the Political Process," in *The Islamist Dilemma: The Political Role of
Islamist Movements in the Contemporary Arab World*, ed. Laura Guazzone (Reading,
PA: Ithaca, 1995), 39-67; also John L. Esposito, ed., *Political Islam: Revolution,
Radicalism or Reform?* (Boulder, CO/London: Lynne Rienner, 1997).
[38] Salwa Ismail, *Rethinking Islamist Politics: Culture, the State and Islamism*
(London/New York: I. B. Tauris, 2003), 170.

Ismail's call for an interdisciplinary approach that deploys a number of methodologies common to the social sciences is greatly needed, if only because the issues are so complex. Some authors rightly point out that the turn toward a more conservative mood on the Muslim streets could still be like gunpowder awaiting a single spark to explode. Former CIA Vice-Chairman Graham Fuller warns that US policies could provide that fateful spark.[39] Israeli scholar Emmanuel Sivan bemoans the listless appeal of Islamic liberals in the face of islamism's growing popularity.[40] There is much truth in these analyses, yet the bigger picture might simply reveal a process that is following its natural course: religious discourse (which I call "theology") is intimately linked to socioeconomic and political events. In fact, in times of great change—postmodernity since the 1970s—religion in its social and historical dimensions "undergoes redefinition." As religious symbols are appropriated and redeployed in new social spheres, they carry with them modified meanings. "This process must be understood in terms of the interaction of religion with the social as entailed in the mobilization of particular religious traditions and their reworking and re-insertion into new domains."[41] Perhaps this is only the outward manifestation of a reformation process—inevitably a "messy" project. As I have argued elsewhere, the boundary between islamism and reformism is more often than not "fuzzy."[42]

This is the thesis of two recent edited books on the issue. One particular chapter summarizes well the thesis I recommend here, "Inside the Islamic Reformation." In this piece, Dale Eickelman seeks to describe "the immense spiritual ferment taking place today among the world's nearly one billion Muslims"—a complex scenario that cannot be captured by the buzzword "fundamentalism" or "the catchy phrases such as Samuel Huntington's 'West versus Rest' or Daniel Lerner's 'Mecca or mechanization.'"[43] Mass education and mass communication (with a proliferation of new media) may be the single most influential factors behind this bubbling of theological inquisitiveness. From Indonesia to Iran to Turkey, Morocco and France, Eickelman sees Muslims engaged

---

[39] Graham Fuller, *The Future of Political Islam* (New York: Palgrave Macmillan, 2003). Fuller was Vice-Chairman of the CIA's National Intelligence Council.
[40] Emmanuel Sivan, "The Clash within Islam," *Survival* 45.1 (2003): 25-44.
[41] Ismail, *Rethinking Islamist Politics*, 173.
[42] David L. Johnston, "Fuzzy Reformist-Islamist Borders: Malek Bennabi and Rachid Ghannouchi on Civilization," *The Maghreb Review* 29.1-2 (2004): 123-52.
[43] Dale F. Eickelman, "Inside the Islamic Reformation," in *Revolutionaries and Reformers: Contemporary Islamist Movements in the Middle East*, ed. Barry Rubin (Albany, NY: State University of New York Press, 2003), 203.

in a "reconstruction" of religion "from below."[44] Authoritarian regimes can no longer hope to control information and belief while media such as the Internet, cell phones and satellite TV encourage a "civil society of dissent." When, for instance, the secular Syrian thinker Sadiq Jalal al-Azm engaged the media-savvy Shaykh Yusuf al-Qaradawi in 1997 on Qatar's al-Jazira channel, many viewers found the secularist more convincing.[45] While the expansion of civil society does not automatically translate into greater democracy, cautions Eickelman, it seems certain that for Muslims a new configuration of public space is taking shape that is "discursive, performative, and participative, and not confined to formal institutions recognized by state authorities."[46]

Barry Rubin, for his part, notes two main victories of Islamism since 1979: the trappings of the Iranian revolution are still in place and "revolutionary Islamist doctrine and groups have become the principal opposition force throughout the region" (Middle East).[47] Meanwhile, they have also suffered major setbacks—their prospects for seizing power anywhere are fast diminishing and their level of popular support is dwindling. While sociopolitical conditions vary greatly across the Middle East, Rubin denotes three main types of islamist groups: (1) the revolutionaries engaged in an armed struggle to overthrow existing regimes in order to create a "true Islamic state";[48] (2) national liberationists, such as Hizbullah in Lebanon and Hamas or Islamic Jihad in Palestine; (3) reformists, as seen in the dominant islamist movements in Kuwait, Egypt, Jordan, Israel, Morocco and Pakistan, which have renounced violence and are working within the political systems in place. Within each movement one can witness two main tendencies: either moderation is a means to achieve political power or the gradual transformation of society in a more Islamic direction *is* the goal. Though this second trend (for Rubin, "reformism") seems on the rise, events could lead again to a hardening of positions. Yet Rubin's thesis is worth pondering: these organizations should be put in parallel with the role Christianity and

---

[44] Ibid., 204.
[45] Ibid., 205.
[46] Ibid., 206.
[47] Barry Rubin, "Islamist Movements in the Middle East," in Rubin, ed., *Revolutionaries and Reformers*, 207.
[48] He lists six such movements in Bahrain, Saudi Arabia, Algeria, Egypt, Syria and Iraq. Out of those, four are also ethnic-national liberation movements: "Shi'ite in the case of Bahrain, Saudi Arabia and Iraq; Sunni in the case of Syria" (ibid., 214). Since Rubin wrote this, Iraq has become the prime training ground for militant Sunni groups as well (with Afghanistan not too far behind).

social movements (aligned either with it or against it) played in the West since the Renaissance. The agendas of islamist groups seek to provide viable responses to modernity and postmodernity while engaging in nation building and development. In sum, islamist groups "are part of the broader history of nationalism."[49]

The second book takes a similar tack, as the title indicates: *Shaping the Current Islamic Reformation*. The editor begins his own chapter with the words, "[i]t is an uncertain business understanding the impact of political Islam and salafism in today's Middle East."[50] While exact prognoses are not possible, argues Roberson, one can reasonably state that the tolerant and inclusive Islamic way of life that has been the rule by and large since the first/seventh century will continue to prevail. First, Islamic law was always the prerogative of the *ʿulamāʾ*, not the political rulers of Islamdom—hence the notion that the *sharīʿa* was a "private law," not the law of the state. Second, five main schools of law (counting the Shi'ites) are considered orthodox, despite their many disagreements in details. Third, because of this built-in diversity and the mechanism of *ijtihād*, Islamic jurisprudence evolved a great deal over time and within a variety of political arrangements, even into modern times. Indeed, over the centuries, the picture looks like "the passage of a long series of intermittent 'mini-reformations.'" Roberson's contention is that a new period of reformation began in earnest in the eighteenth century and continues to this day, in the "modern and postmodern era."[51]

My point here is not to analyze his thesis in detail but to suggest that in the wider context of three decades of self-questioning and spirited theological debates, what has been dubbed Islamic "fundamentalism," "radical Islam" or "Islamism" represents both a short-lived phase (in its political intensity) and one option that will continue to remain on the table. Rashid Rida vehemently disagreed with Abd al-Raziq about the political implications of an "authentic" Islamic society, and he recognized in Banna's reformation of society from the bottom up a modern way of achieving an ancient goal: an *umma* that was vitally committed to the embodiment of Islamic values in all spheres of its existence. Yet the question of the human caliphate remains: Is it about the human person or, as Najjar believes, about Muslims only? The answer to this question

---

[49] Ibid., 217.
[50] B. A. Roberson "The Shaping of the Current Islamic Reformation," in *Shaping the Current Islamic Reformation*, ed. B. A. Roberson (London/Portland, OR: Frank Cass, 2003), 1.
[51] Ibid., 5.

takes us again beyond the traditional perimeters of the debate about "political Islam." It raises the issue of theological reform at a fundamental level.

Mawdudi: A Reinterpreted Caliphate

Sayyid Abul-A'la Mawdudi (1903–1979) is undeniably the cornerstone of South Asian Islamic revivalism. A descendant of his sixteenth-century namesake, the founder of the Chishti Sufi order in India, Mawdudi was also related to the influential modernist Muslim thinker Sayyid Ahmad Khan.[52] In order to understand his ideas on the caliphate one must first look to his life and the circumstances surrounding it.

His education (in near seclusion), meticulous as it was at the hands of his father, was anything but Western. At the age of eleven he completed his first book, a translation of Arabic into Urdu of a recent Egyptian book. By the time he turned fifteen, his father had passed away and he had moved with his brother to Delhi, embarking on a journalistic career. In 1921 he resumed his studies, this time in Arabic: *tafsīr*, *fiqh*, *adab* (literature), *mantiq* (logic) and *kalām* in a seminary associated with Deoband Sufism. By 1926 he received his degree, but though he was now recognized (and functioned in part) as a ʿ*ālim*, by then his political ambitions and his determination to influence educated Muslims led him to downplay this title.[53] And, in fact, increasingly he used his pen to give direction to the Muslim community caught in the turmoil caused by the double challenge of British colonialism and Hindu/Muslim communal tensions.

The significant turning point came for him in 1937 when, on the verge of Indian independence, he discerned that the direction taken by the loose coalition of Hindu leaders was a grave threat to the Muslim community of India. He took up his pen and addressed the political situation from a qur'anic perspective in a series of essays published between 1937 and 1941 in the journal he had bought in 1932, the *Tarjumān al-Qurʾān* (*Qur'anic Interpretation*).[54] If the future government of India becomes

---

[52] Seyyed Vali Reza Nasr, *Mawdudi and the Making of Islamic Revivalism* (Oxford/ New York: Oxford University Press, 1996), 9.

[53] Ibid., 18.

[54] Charles J. Adams, "Mawdudi and the Islamic State," in *Voices of Resurgent Islam*, ed. John L. Esposito (Oxford/New York: Oxford University Press, 1983), 101. Nasr reports that 1933 was, from Mawdudi's perspective, the turning point. He called it a "conversion": "In reality I am a new Muslim" (*Mawdudi*, 31). This is when he realized that Islam called for political organization and that he, therefore, was called to propagate a more political vision of Islam. This vision, nonetheless, did not

secular and democratic, he reasoned, then in effect the Muslim minority would have to submit to the dictates of the Hindu majority. For Indian Muslims it was not race, geography or even culture but faith that was their rallying point. In fact, Islam was their nationality. "Hence he urged the Muslims not to participate in the freedom struggle being led by the Indian National Congress and its nationalist Muslim supporters."[55]

The 1940 Lahore Resolution enacted by the Muslim League called for autonomous states to be established in the subcontinent's areas of Muslim concentration. From then on, mainstream Muslim opinion rallied behind the founding of Pakistan. But this too Mawdudi considered a danger. Pakistan would only be the "Muslim" counterpart of India, and just as secular. He began to view Muhammad Ali Jinnah, the leader of the Muslim League, as his direct rival.[56] In August 1941 Mawdudi founded his own party in Lahore, the now famous Jama'at-i-Islami. His career as an ideologue had now shifted to politics, though he continued to edit and write for the *Tarjumān al-Qurʾān* till his death in 1979.

Mawdudi is first and foremost a Muslim theologian. In the next section I first offer some thoughts about his concept of *tafsīr* and then turn to his view of humanity and *khalīfa*.

Mawdudi's Rethinking of Islam. Mawdudi was a firm believer in *ijtihād* —the independent interpretation and application of *sharīʿa* by qualified scholars to suit evolving sociopolitical conditions. At first sight, he was proclaiming nothing but traditional Islamic theology. The starting point is the absolute sovereignty of God,[57] and the corollary truth, the unicity of God, *tawhīd*. This means that the law that God has revealed through the Qur'an and Sunna must be obeyed. In fact, the prime characteristic of humanity's relationship to their Creator is submission (*islām*) and obedience. True to the qur'anic text in which the most often repeated word in this context for humans is the word *ʿabd* (servant/slave), "he viewed absolute obedience to God as a fundamental right of God about which man had no real choice."[58] So far, nothing unusual, except that, in

crystallize until 1937, when he left Hyderabad for Delhi. This is where he addressed the issues now being debated by the Indian National Congress.

[55] Adams, "Mawdudi and the Islamic State," 103.

[56] Nasr, *Mawdudi*, 40.

[57] His Urdu term for this was translated in the Arabic versions of his works by *hākimiyya*—a neologism which was to become one of Qutb's foundational concepts (Yvonne Y. Haddad, "Sayyid Qutb: Ideologue of Islamic Revival," in Esposito, ed., *Voices of Resurgent Islam*, 89). Through Qutb this word has become a key term in islamist discourse.

[58] Nasr, *Mawdudi*, 58.

harmony with contemporary islamists, he asserted that this could not faithfully be lived out except in an Islamic state. And thus his use of the word *daʿwa* took on a different meaning from that of orthodox Islam.[59] For him, it was absolute submission to God, which necessarily included the commitment to strive for the establishment of an Islamic state. To say the least, according to Seyyed Vali Reza Nasr, he was not supported in this interpretation by the majority of Indian Muslims:

> Herein lay the source of Mawdudi's break with traditional Islam, for the doctrine of absolute obedience to God in lieux [*sic*] of human choice and volition in matters of faith subtly but surely challenged the traditional Islamic position. Its ostensible "radical orthodoxy" rejected the prevalent norms and institutions of Islamic life. It was exactly on this point that ulama such as the rector of the *Nadwatu'l-ʿUlama* in Lucknow, Mawlana Sayyid Abu'l-Hassan ʿAli Nadwi, echoing the sentiments of the Islamic establishment of the Indian subcontinent, criticized Mawdudi's rigid interpretation in the strongest terms, accusing him of parting with the fundamental tenets of Islam. Their debate elucidated the extent of Mawdudi's challenge to traditional Islam.[60]

Significantly, Nasr entitles the second half of his book on Mawdudi's revivalism "Islam Reinterpreted." My only point in this section is to reemphasize the formative events in Mawdudi's life: his childhood, education, mingling with prominent politicians throughout the turbulent years leading up to independence, and increasing personal political involvement, first in the Khilafah Movement, then, after 1924, increasingly through his pen, in giving leadership to the Indian Muslim community and advocating a separate state for them. This background, together with the dramatic events that shaped his nation's life, first in India and then in Pakistan, must be seen as influential in his qur'anic hermeneutics.

In fact, all five reality domains in the hermeneutical model presented in Chapter 5 can be seen to bear on Mawdudi's hermeneutic:

1. *Language:* his position of influence and his writings were facilitated and shaped, to some extent, by his phenomenal skill and grasp of both the Urdu and even the Arabic language.

---

[59] *Daʿwa* is "the call," or the spreading the faith: "Mawdudi's *daʿwa*, as the embodiment of his reinterpretation and revival of the Islamic faith, ultimately became a 'movement' directed at regimenting the lives of all those who had accepted Islamic ideals and molded their lives accordingly, erecting an Islamic order, and, eventually, revolutionizing human thought by instilling Islamic values into it. His scheme was holistic and all-inclusive; it began with the individual Muslim and culminated in a new universal order" (ibid., 56).

[60] Ibid., 58-59.

2. *External reality/epistemology:* Mawdudi, besides his staunch
   traditionalism, is also a modernist. I agree with Nasr that
   Mawdudi, in order to engage with and rebut certain claims of
   modernity, adopted several of its assumptions, "especially those
   involving scientific truths, which he saw as value neutral." He
   goes on:

> His views also involved a process of modernization, but under
> the guise and in the name of Islam. This modernizing impetus of
> Islamic revivalism was not limited to the use of tape recorders,
> facsimile machines, and other instruments of the modern world,
> as some observers of this phenomenon have contended, but
> encompassed values, ideas, and institutions. Revivalists are not
> only moderns but modernists.[61]

In essence, Mawdudi assumes a "naïve realist" epistemology.
While he is happy to exercise his right of *ijtihād*, he would not
have been prepared to see this as simply "one possible reading"
of the Qur'an, but rather condemned the other *ʿulamāʾ* for being
un-Islamic.

3. *Social reality domain:* it would be difficult to deny the role
   played by the sociopolitical effervescence in India in the emergence and formulation of Mawdudi's ideas. In fact, his whole
   orientation "was informed by the urgency of sociopolitical
   exigencies, and he sought to base the historical and spiritual
   significance of the Qur'an on temporal contingencies.[62]

4. *Academic reality domain:* here again, Mawdudi's study of *fiqh*
   in the Sufi context no doubt gave him more freedom in seeking
   solutions beyond those advocated by the traditional and official
   ʿulamāʾ. Further, his growing interest in politics pushed him to
   read widely, and together with his own involvement, political
   science and economics came to bear more directly on his
   writings.

---

[61] Ibid., 50-51. Krämer calls attention to this common distinction made by islamists
between "techniques and values": "Islamists hold that techniques are entirely neutral
from a religious perspective, and provided that Islamic values are preserved intact,
they can be adopted from other civilizations without jeopardizing Islamic authenticity. This applies not only to scientific discoveries and modern technology, but also to
methods, instruments and institutions of economic, political and social organization"
("Visions of an Islamic Republic," 37). Naturally, she notes, this is just as problematic as the distinction made in the same breath between "a fixed a stable 'core' of
Islam and its time and place dependent 'variables.'"
[62] Nasr, *Mawdudi*, 62.

5. *Personal reality domain:* it could plausibly be argued that his early sheltered education and the influence of his mystical and profoundly religious father helped to instill in him the idea that he was to give leadership to the Muslims of India. Certainly his innate gifts for articulating these ideas in writing helped to propel him into the limelight.

Though the scope of the study does not allow me to apply all of these reality domains to each of the Muslim thinkers I have presented, my doing so for Mawdudi hopefully illustrates the powerful impact of these factors in the hermeneutical process.

Mawdudi's View of Khilāfa. Adams lists four basic principles behind Mawdudi's Islamic vision:[63]
1. The sovereignty of God, already mentioned (*hākimiyyat Allāh*).
2. The authority of the Prophet of God: the Sunna, along with the Qur'an, then, is the ultimate basis of law—itself the foundation of the political sovereignty of God embodied in the Islamic state.
3. Qur'an 24:55, though speaking of David, applies to all heads of Islamic states. They are God's *khulafāʾ*, his vicegerents on earth. In one sense the Islamic state is like other states in that it exercises sovereignty over the territory it controls. In another sense, it cannot disregard God's law, and, in fact, must uphold it completely. "An Islamic state should properly, therefore, be called a caliphate for such is its nature."[64] At this level Mawdudi vacillates, and his overall view of *khalīfa* according to Q. 2:30 is more fuzzy: in one breath the state itself is the *khalīfa* of God, and therefore is embodied in its supreme ruler or imam, in another he wants to speak about democracy originating in Islam. The *khalīfa* therefore becomes the body of Muslim citizens forming the Islamic state that collectively carry out the *khilāfa* of humanity. But in either case the *khilāfa* of Adam becomes restricted to the Muslim community and must be expressed politically through the functioning of a state based on *sharīʿa* law.
4. The fourth principle is that of *shūrā*, or mutual consultation among the Muslim citizens.[65] "The practical meaning of this

---

[63] Adams, "Mawdudi and the Islamic State," 115-19.

[64] Ibid., 116.

[65] Mawdudi may not have been the first to propose *shūrā* as the basic "democratic" procedure in the modern Islamic state. In the 1930s he was likely to have run across the writings of the Muslim Brotherhood in Egypt. Mitchell remarks that Banna only touches on this question briefly in his writings (*The Society of the Muslim Brothers*,

popular viceregency is that the government of the Islamic state
can be formed only with the consent of all the Muslims, or at
least a majority of them, which can remain in office only so long
as it continues to enjoy their confidence."[66] The Qur'an and
Sunna do not specify the exact form this consultation must take,
but it does *ipso facto* rule out any form of dictatorship, despotism or even monarchy. Mawdudi writes that in effect he is talking about the "Kingdom of God" on earth, or a theo-democracy,
resting on the twin principles of the sovereignty of God and the
caliphate of humankind.

For Nasr, this is far from a Western understanding of democracy. The
sovereignty of God trumps any volition of the people, be it individual or
collective. The state embodies the divine and the popular will, thus
becoming the "sole political actor": "The individual would have to relinquish his own vicegerency to the Islamic state, which is the expression of
a collective vicegerency. The individual would be bound by the writ of
the state, backed by the full force of religious law and the more paramount power of the collective vicegerency."[67]

My concern here is not so much the outworking of this collective
khilāfa as it is the perspective his scheme offers on a theology of
humanity. For this reason a word about human rights is in order. The
continuity of the state, remarks Nasr, is guaranteed by the Islamic code
of law that in turn requires obedience. Dissent, in most instances, would
be considered apostasy. What is missing here are "guarantees for demo-

with a Foreword by John O. Voll [Oxford: Oxford University Press, 1993 [1969]).
But it was expanded on by both Abd al-Qadir Awda (*al-Islām wa-awḍāʿunā al-
siyāsiyya*, 1951) and Sayyid Qutb (*al-ʿĀdala al-ijtimāʿiyya fī'l-Islām*, 1949) (ibid.,
243). The term is found in Q. 42:38 ("those…who [conduct] their affairs by mutual
Consultation," Y. Ali). This is the only verse where the noun *shūrā* appears in the
Qur'an. Bassam Tibi is dismissive: "Historically, this precept conjures the pre-Islamic
system of intertribal consultation among the leaders of ethnic groups. In following the
Qur'anic commandment and in keeping with this tradition the Prophet Muhammad
consulted with his close contemporaries and followers, foremost among whom were
Abu Bakr and Omar. The four Righteous Caliphs of early Islam maintained this
tradition, the caliph Omar increasing the number of counselors to six. None of the
Umayyad or Abbasid caliphs practiced the *shūrā*" (*The Challenge of Fundamentalism: Political Islam and the New World Disorder* [Berkeley: University of California
Press, 1998], 174). After Mawdudi, to invoke *shūrā* as Islam's answer to Western
democracy has become an almost universal reflex among mainstream islamists.
[66] Adams, "Mawdudi and the Islamic State," 116.
[67] Nasr, *Mawdudi*, 90.

cratic procedures, protection of individual rights, and, most important, a mechanism for translating popular interest into policy."[68]

Yet to read Mawdudi, one gains the impression of a Muslim scholar attempting to "contextualize" Islamic dogma for a contemporary readership. In a long paper entitled "Human Rights in Islam," he lists "the right to participate in the affairs of the state" as the last of fifteen "rights" guaranteed by Islam. This right is based on the qur'anic principle of *shūrā*, which he considers "the legislative assembly" and which guarantees the following:

1. The executive head of the government and the members of the assembly should be elected by free and independent choice of the people.

2. The people and their representatives should have the right to criticize and freely express their opinions.

3. The real conditions of the country should be brought before the people without suppressing any fact so that they may be able to form their opinion about whether the government is working properly or not:

---

[68] Ibid., 91. This article was reprinted (eight years after his death) in one of the official Iranian journals of the Islamic Republic of Iran. This fact alone speaks highly of Mawdudi's influence, even among Shi'i Muslims. There are, nonetheless, some reservations expressed in a footnote on the first page—and it has to do with his view of *khilāfa*: "As on the occasion of the Ten-Day Dawn Celebrations, on completion of 8 years of Islamic Revolution of Iran [*sic*], the Islamic Propagation Organization selected the theme of 'Human Rights in Islam' for the 6th Islamic Thought Conference to be held in Tehran from January 29–31, 1987, the IPO is taking a privilege [*sic*] of reprinting the paper on the subject written by the late 'Allamah Abu al-A'la Mawdudi. However, we would like to mention that the late 'Allamah's views regarding *khilafah* as the representation of the Creator are controversial. We believe that the representative of the Creator, i.e, the Caliph who is responsible for the affairs of the Muslims should necessarily possess full knowledge of Islamic *Shari'ah* and be a just person having full control over his desires. Such a person is known as *al-Wālī al-Faqīh*" ("Human Rights in Islam," *al-Tawhid* 4.3 [1987]: 59). From a strict Shi'i perspective, of course, the *khilāfa* cannot be collective but can only be entrusted by God to one person. It may be more accurate to say that the above mentioned "*Wālī*" is the deputy (*khalīfa*) of the hidden Imam: "As described by one of the major Shii political theorists at the time of the Iranian revolution, Ayatollah Baqir al-Sadr (who was executed by the Iraqi government in 1980), in this structure there is a jurist who holds the position of final religious authority and formally is the 'Deputy General of the Imam' (the divinely selected messianic leader in Shii theology), the position held by the Ayatollah Khomeini following the Iranian revolution in 1979" (John L. Esposito and John O. Voll, *Democracy in Islam* [Oxford/New York: Oxford University Press, 1996], 24).

> ...There should be adequate guarantee that only those people who have the support of the masses should rule over the country and those who fail to win this support should be removed from their position of authority.[69]

Esposito and Voll consider Mawdudi's "reconceptualization of Islam in the contemporary context" as foundational and therefore common with other views, both Shi'i and Sunni: the "political system of Islam has been based in three principles, viz.: *Tawheed* (Unity of God), *Risalat* (Prophethood) and *Khilafat* (Caliphate). It is difficult to appreciate different aspects of the Islamic polity without fully understanding these three principles."[70] They further quote him as saying that the caliphate is carried out by the community as a whole, which fulfills its mandate in harmony with the principle of *tawḥīd*: "This is the point where democracy begins in Islam. Every person in an Islamic society enjoys the rights and powers of the caliphate of God and in this respect all individuals are equal."[71] Though many Muslim writers, like Nasr, would question the extent to which such a view guarantees democracy in practice, Esposito and Voll are more sanguine about the possible harmonization of 1990s Islamic revivalism and Western notions of democracy. Notice here the key terms Bint al-Shati uses, albeit with a different application:

> The absolute sovereignty and oneness of God as expressed in the concept of *tawhid* and the role of human beings as defined in the concept of *khilafah* thus provide a framework within which both Sunni and Shi'i scholars have in recent years developed distinctive political theories that are self-described and conceived as being democratic. They involve special definitions and recognitions of popular sovereignty, and an important emphasis on the equality of human beings and the obligations of the people in being the bearers of the trust of government. Although these perspectives may not fit into the limits of a Western-based definition of democracy, they represent important perspectives in the contemporary global context of democratization.[72]

At the end of this section, therefore, it may be possible to say that with Rida and Mawdudi Islamic thought has come full circle. Though the

---

[69] Mawdudi, "Human Rights in Islam," 84.
[70] Esposito and Voll, *Democracy in Islam*, 23.
[71] Ibid., 26. Khurshid Ahmad, a current islamist leader in Pakistan, is also expressing Mawdudi's position when he writes, "Secular democracy, as it has evolved in the post-enlightenment era, is based upon the principle of the sovereignty of man, conceptually speaking. Islam, on the other hand, believes in the sovereignty of God and vicegerency of man, the difference being that man is God's *Khalifah*, or vicegerent on the earth" (ibid., 26-27).
[72] Ibid., 27.

latter had much more opportunity to flesh out his ideas on an Islamic state, both write after the dissolution of the Caliphate in Turkey and reinterpret the traditional qur'anic concept of humanity as God's *khalīfa* in similar ways. On the one hand, they insist that from the beginning *dīn wa-dawla* (religion and state) have always been one in Islam; on the other hand, they veer away from the traditional monarchical view and selectively incorporate some Western values associated with democracy. Though in practice they would not extend the *khilāfa* of Adam to all of humanity (unless all were led to embrace Islam), at least they see it as applying to Muslims collectively and, in the case of Mawdudi, begin to wrestle with the rights of individuals as well.[73] Qutb, to whom I now turn, while taking some inspiration from Mawdudi, displays an even more confrontational stance with the West and existing Muslim regimes.

Sayyid Qutb: A Narrowed khilāfa

So much has already been written on Qutb that I will resolutely confine my remarks to his view on the *khilāfa* of Adam.[74] Executed in 1966 for

---

[73] Esposito and Voll confirm this viewpoint, writing that Islamic political thought over the centuries was dominated by the Caliphate and that as a result the "successor" meaning of *khalīfa* dominated. Yet, after the First World War and the dissolving of the Caliphate along with the Ottoman Sultanate, the "dominion" meaning began to creep in: "However, there is a profoundly different meaning of the term that has received increasing attention in the second half of the twentieth century. In addition to the connotations of 'successor' that the Arabic term *khalifah* involves, there is also a sense in which *khalifah* is a deputy, representative, or agent. It is possible to interpret some sections of the Quran as identifying human beings in general as God's agents (*khalifahs*) on earth, and human stewardship over God's creation as the broader cosmic meaning of *khilafah*. Mawdudi utilized the concept of *khilafah* defined in this way as a basis for his interpretation of democracy in Islam" (ibid., 26). Though I agree with their analysis of the general trend (it is the centerpiece of my own thesis), I would question their assessment of Mawdudi's position. His caliphate of humanity is inexorably linked to *tawḥīd* and *nubuwwa* (prophethood). In other words, you would have to be a Muslim, and a Muslim subscribing to his political interpretation of Islam, to participate legitimately in the divine *khilāfa*.

[74] Called by historian R. Stephen Humphreys "the most important ideologue of the [Muslim Brotherhood] movement after Hassan al-Banna's death," he estimates that his book *Milestones* "at once became and remained (to use an odd but useful image) the Bible of Islamic activism. His call for *jihad* against the forces of tyranny and moral corruption within the Islamic world—in effect, for an Islamic revolution—bore fruit in Egypt by the mid-1970s... Ultimately [the Islamic movement] came to dominate [Egypt's] universities, even that once unassailable stronghold of secular liberalism, the University of Cairo... By the end of the 1970s, Sayyid Qutb's call had found an echo in Syria..." (*Between Memory and Desire: The Middle East in a Troubled Age* [Berkeley: University of California Press, 1999], 194-95).

plotting to overthrow the Egyptian state along with other leaders of the *Ikhwān al-Muslimūn* (Muslim Brotherhood), "[h]e stands with costly passion for a verdict about Islam which must be heeded."[75] Explaining that Qutb found mentors and colleagues in the likes of Mawdudi and the Indian scholar Ali Nadwi, Cragg traces even further back some of his inspiration: "With them he shared a deep discipleship to the medieval champion that the doctrine of *Jihad* may have to be pursued against ostensibly Muslim rulers who behave untruly, the redoutable [*sic*] Ibn Taimiyyah of the seventh Muslim century."[76]

Perhaps just as influential as his writing was the evolution of Qutb's life. Beginning as an Egyptian intellectual enamored with the West, he slowly underwent a change of mind in the 1940s as he watched Great Britain use its colonial power to manipulate its Arab protégés in Egypt and betray them in Palestine. From 1949 to 1951 he was granted a research fellowship in educational administration in the United States. This experience marked the actual turning point: he came back from the United States disgusted with the materialism, lack of moral values and the profound anti-Arab bias in the American media.[77] This is when he joined the Muslim Brotherhood, following which he soon became their voice through his prolific writings.[78] But more than just through his pen, Qutb dove into the fray of growing Brotherhood opposition to King Faruq's regime, which led him to be arrested with many others in 1954.

Before turning to his specific view of humanity as *khalīfa*, I begin with a comment on his worldview—how it was shaped by the events he experienced in Egypt in the aftermath of World War II, and how this affected his hermeneutical approach to Q. 2:30. To be sure, Qutb's "conversion" in 1951 was more than a rediscovery of "Islamic theology." This would be to misunderstand the holistic nature of Islam to begin with. Indeed, the convergence of growing nationalism, increasing Westernization with (ironically) the phasing out of direct colonialism and

---

[75] Cragg, *The Pen and the Faith* (London: George Allen & Unwin, 1985), 60.

[76] Ibid. Qutb considered the current leaders of Muslim countries as apostates, as well as the majority of its citizens. They had in effect reverted to the state of paganism that preceded the arrival of Islam in seventh-century Arabia, or *Jahilliya*. For information on the cleavage between the two factions of the Muslim Brotherhood (for or against Qutb's hard line, or *takfīr*, i.e. declaring other Muslims "infidels"), see David L. Johnston, "Hassan al-Hudaybi and the Muslim Brotherhood: Can Islamic Fundamentalism Eschew the Islamic State?," *Comparative Islamic Studies* 3.1 (2007): 39-56.

[77] Haddad, "Sayyid Qutb," 69; Robert D. Lee, *Overcoming Tradition and Modernity: The Search for Islamic Authenticity* (Boulder, CO: Westview, 1997), 83-84.

[78] Haddad cites 24 books besides his qur'anic commentary, *Fī ẓilāl al-Qurʾān* (*In the Shade of the Qur'an*) ("Sayyid Qutb," 69).

the trauma in the Arab psyche occasioned by the founding of the Israeli state (and Hasan al-Banna's assassination the next year as well)—all these factors favored the emergence of a radical and confrontational version of Islam. Qutb came back to Egypt in 1951 with a profound disgust for Western moral decadence and the staunch American support for Israel. Yet, according to Stephen Humphreys, his theology was a more original blend than most would recognize:

> In spite of all claims to the contrary by Muslim activists and many West-ern commentators, this version of Islam was a radically new interpretation of the faith, with few real precedents from earlier centuries, and it was aimed not at recovering the past but at controlling the future. This purified Islam was sharply confrontational in its rhetoric and manner, for it had been constructed precisely to challenge the religious status quo, to rid the faith of superstition and corruption, to compel Muslims to reject the entice-ments of the West and live in accordance with God's revelation.[79] As the reformers saw it, Islam was a religion of action, and to that end it had to be stripped down to its essentials.[80]

Thus most *tafsīr* traditions of the past must be laid aside—some because of their love for Greek philosophy and others because of their penchant for Christian and Jewish fables. What is needed, he writes—and Qutb's life is a model of this—is an activist and dynamic hermeneutic. Some-what closer to Rida's *salafī* tendencies in the *Manār*, Qutb adds a new dimension, according to French scholar, Olivier Carré:

---

[79] For instance, Qutb reads into the battles of the Qur'an the contemporary struggles of the *umma*: "the Muslim world has experienced a 'Uhud' [first Islamic defeat after the spectacular victory of Badr] at the hands of the dominating West, but which is actually a Muslim victory for the sake of their souls through the restoration of the *shura*, which will make possible an ulterior military victory on the model of the conquest of Mecca" (Olivier Carré, *Mystique et politique: lecture révolutionaire du Coran par Sayyid Qutb, frère musulman radical* [*Mysticism and Politics: A Revolu-tionary Reading of the Qur'an by Sayyid Qutb, Radical Muslim Brother*] [Paris: Les Éditions du Cerf, 1984], 278). In his worldview, Islam is pitted against the Christian West, which continues to "crusade" against them, especially through the effort of the orientalists and international Zionism.

[80] Humphreys, *Between Memory and Desire*, 189. Carré summarizes Qutb's herme-neutics in these words: "In his exegesis Qutb quotes very little from the classical commentators, and then only from the traditionists: Ibn Ishaq's *Sira* (as transmitted through Ibn Hisham and al-Tabari) for the Prophet's life, Ibn Hanbal and Malik for questions of jurisprudence, some of the 'Jalalayn' and Zamakhshari for extra com-ments on other passages. But only Muhammad's own *tafsir* (through well-attested *ahadith*) can claim 'decisive value' (*qāṭʿī*) and important exegesis for Qutb's entire doctrine, for his notion of *ḥākimiya* in particular (neologism formed on *ḥukm*)" (*Mystique et politique*, 267).

> Islam is a dynamic way of life which only the activists, the militants, the fighters may truly understand and interpret. Books in themselves are useless, they are cold and lifeless. The men on the battle front (at the side of, or following Muhammad's example as commander in chief)—those are the ones who understand, not those who stay behind. This is of course the heritage of the Muslim Brotherhood, expressed by Hassan al-Banna when he was situating himself in relation to Afghani, 'Abduh, Rida, non-militant intellectuals, in his estimation.[81]

Within this black and white, almost apocalyptic worldview, how is the qur'anic narrative of creation interpreted? Here is Qutb's *Fī ẓilāl al-Qur'ān*'s opening word on Q. 2:30:

> The context—what precedes this—presents the procession of life, indeed, of all of reality. Then it speaks of the earth—in a picture of God empowering people...and creating everything for them. So in this passage comes the story of Adam's mandate as trustee (*khalīfa*) on the earth, his being handed its reins of power, according to God's covenant and its attending conditions, and his being granted the knowledge which allows him to fulfill his trusteeship. This story in turn opens the way for the people of Israel's mandate to become vicegerents on earth through God's covenant; then the cutting off of their trusteeship and the passing on of the reins of power to the Muslim community.[82]

On one level, this interpretation is very similar to Iqbal's. Humanity is empowered by God, and indeed mandated to put their God-given intellectual and moral gifts to work for the task of administrating the created order—or so it would seem. Actually, at a deeper level, because he has accepted Mawdudi's starting point as his own (the sovereignty of God and humans at his service), submission to God as revealed in the Qur'an is the authentic nature of humanity (its *fiṭra*). This means that individuals must discover freedom by overcoming their animal nature and entering into covenant with God. Further, such a "conversion" can only take place within the context of a radical, believing and Islamically engaged community. "Since no such society exists today, human fulfill-ment thus demands the formation of a revolutionary elite, the overthrow of existing social arrangements, and the construction of a society regulated by divine rules."[83]

Carré shows that while even laying aside the radical political implica-tions of such a view, its very conception of human nature is tied to God's successive covenants in history which culminate in the revelation of the Qur'an and the Prophet's establishment of the Islamic *umma*:

[81] Ibid., 267-68.
[82] *Fī ẓilāl al-Qur'ān*, 6 vols. (Beirut: Dār al-Shurūq, 1973), vol. 1:56.
[83] Lee, *Overcoming Modernity*, 88.

This promise, this testament, this pact goes back to humanity's origins: Adam is instituted "lieutenant of God on the earth" by a primordial pact which includes the conditions of this trusteeship on both God's side and man's side.[84] He (God) grants him the knowledge of the things which will enable him to carry out this trusteeship, and Adam, for his part, commits himself to worship the only one God and live according to His law... The "covenants" enacted in the course of sacred history are nothing but that same primordial covenant (adamic) and final (Qur'anic), and it comprises also a prescribed order for society.[85]

One must therefore posit for Qutb two levels of *istikhlāf* (mandate for trusteeship): the first is that all of humanity has been empowered by God to use their knowledge in managing the earth, and the second is that the full mission of *khilāfa*, which includes obedience to the conditions of the primordial covenant, is reserved for the Muslim community. This kind of elitism—meant in a good sense—would also be present in much of Christian theology. Yet my concern is for a Muslim–Christian dialogue in the activist mode of advocacy for justice and a clean environment. How much common ground is there in our theologies of humanity?[86]

---

[84] Qutb is not only connecting *khilāfa* and *amāna* but he is linking them to a third qur'anic concept, usually referred to as the "primordial covenant" (7:172): "When thy Lord drew forth from the Children of Adam, from their loins, their descendants, and made them testify concerning themselves, (saying): 'Am I not your Lord (who cherishes and sustains you)?' They said, 'Yea! We do testify!' (This), lest ye should say on the Day of Judgment: 'Of this we were never mindful.'" Sweetman, in the course of his monumental work, mentions this as a quick aside, too obscure really to comment on: "In some passages there seems to be a hint of pre-existence, and the day of Alast has passed into the stock of common ideas in Islam. The passage is Sura 7:171. Alast is the primeval covenant of the pre-existing souls with God" (*Islam and Christian Theology*, vol. 1:185). Not surprisingly, Arkoun sees this as the moment of unity for all three monotheistic faiths: "The complex nature of the contents, functions, ends and possible future to be found in the Qur'an is such that societies based on this phenomenon (and the same goes for the Biblical and evangelical phenomenon) are going to organise their space according to *existential* differences (which give shape and add dynamism to human existence) in the ontological framework of the primordial Alliance (the *mithāq* through which God delegates to man part of his power by creating him 'God's vicar on earth,' because man is the only animate being to have accepted his Creator without condition)" ("Religion and Society: The Example of Islam," in *Islam in a World of Diverse Faiths*, 136-37, emphasis original).

[85] Carré, *Mystique et Politique*, 106.

[86] See Hoebrink's discussion on Mawdudi and Qutb, describing them as "moderate islamists," because, while they emphasize human fallibility and poor judgment, the work of interpretation is nonetheless recognized. "This becomes clear where they use the notion of human vicegerency (*khilāfa*) as complementary to God's exclusive sovereignty (*ḥākimiyya*)... In such cases humans, as the vicegerents of God on earth,

Apart from the political dimension of Islam Qutb espouses, one could convincingly argue that his doctrine of humanity is entirely orthodox, and possibly, inescapable. Turning to a Western Muslim for comparison, the Sufi-leaning Charles Le Gai Eaton, we find similar conclusions regarding human nature based on a qur'anic worldview. Islam, he reminds his readers, is the *dīn al-fiṭra*, "the religion of primordiality," or maybe better, "the original religion." According to the Qur'an, *fiṭra* exemplifies the standard from which humanity has fallen away. Taken from the root of one of the verbs used for God's creating activity, it points to what humanity was originally created to be. "It follows that the image of human perfection (or, quite simply, of human normality) lies in the past, not in the future, and the way of attainment lies not in an aspiration focused on a distant goal or in any miraculous redemption from inherent sinfulness but rather through the removal of accretions and distortions that have both corroded and twisted a perfection that is, in essence, natural to mankind."[87] *Tawḥīd*, or the "unitarian perspective of Islam," recognizes no distinction between sacred and secular.[88] This is why humans as God's *khulafāʾ* are also his *ʿibād* (slaves/servants/worshippers). I believe this also accounts for Bint al-Shati's moving directly from the foundation of the caliphate of humanity to the idea of the Qur'an and its role in sealing their gift of knowledge. As in the beginning of *Sūrat al-Raḥmān* (55): "He taught the Qur'an; he created man; he taught him *al-bayān*."[89]

I conclude that by following Mawdudi's linkage of *khilāfa* with *nubuwwa* and *sharīʿa*, Qutb leaves very little room for meaningful Muslim–Christian cooperation in the socio-political and economic realms. And, in the end, whereas Mawdudi played the democratic game till the end in Pakistani politics, Qutb declared all Muslim governments apostate.

can rely on their own fallible and subjective interpretations of the divine Will" ("Thinking about Renewal in Islam," 53). However, when confronted with an explicit text (*naṣṣ*), "there is no room for interpretation."

[87] Charles Le Gai Eaton, "Man," in *Islamic Spirituality: Foundations*, ed. Sayyed H. Nasr (New York: Crossroads, 1987), 366.

[88] Ibid., 371.

[89] What facilitates Muslim–Christian dialogue at this point is that Islam recognizes the prophethood (*nubuwwa*) of Moses and Jesus, because of their divinely inspired books, the *tawrāt* and *injīl*. We can disagree on the nature of *fiṭra* (a denial of original sin), but we can agree on most of the ethical values enshrined in *khilāfa*. This can be seen from the interviews mentioned in the next section.

Two Palestinian Islamist Leaders on Khilāfa

While in the West Bank and Gaza Strip on a research trip in January and February 1999, I had the opportunity of asking some of these questions to the Sheikh Ahmad Yasin (1934–2004), the founder and spiritual leader of the "Movement for Islamic Resistance" (Hamas).[90] In the course of the 45-minute taped interview he granted me, the following statements are some of the highlights concerning the topic at hand.

My first question was about his interpretation of Q. 2:30. Here is an excerpt from his lengthy answer:

> So God wanted to place on earth a vice-regent [*khalīfa*] who would execute what he wanted him to execute on earth. And God's will requires that the earth be a land of plenty [*khayr*], of cooperation, of development [*binā'*] and investment so that people would experience comfort and ease, justice and a kindness that befits human relationships [*insāniyya*]…
>
> To summarize, what I am trying to say is that God wants to place on earth one who will act in His stead [*man yakhluf 'anhu*], obeying Him by managing it, by developing it, by bringing goodness to it, by forbidding and resisting evil on it [when it appears]. For whoever does what is good prospers and whoever does what is evil will be punished.[91]

My next concern was to make sure who exactly is the *khalīfa*: "Adam clearly received this mission from God. Does Adam in this verse represent all of humanity?" His answer was to the point:

> Adam represents all of humanity. He is the father of humanity, its very origin. The message that was addressed to Adam is addressed at the same time to the apostles, and to their followers. So when God speaks to Adam he is speaking to us as well. What he reveals to him he is revealing to us because he is the original one, the father, the first human God created in the beginning.

Finally, the question I was most eager to pursue: "Doing good covers a lot of ground. There is then the possibility of us working together as humans—Christians or Muslims—for the sake of the good. And part of doing good is to forbid injustice, which unfortunately is everywhere." Here is just the beginning of a long exchange:

---

[90] A paraplegic, he had been tortured while incarcerated by the Israelis from 1988–1997 and was killed by an Israeli missile as he came out of a mosque (in his wheelchair) after morning prayers on March 22, 2004.

[91] "*Khalīfa*, Culture Change, and Muslim–Christian Cooperation in Hebron District Community Development," unpublished paper, 1999, 43-44. The interview was entirely in Arabic. I later transcribed and translated the taped interview that took place in his residence on February 1, 1999.

I understand what you are saying. Opportunities for doing good and working at it are numerous. The main thing is that whether humanity is able to accomplish this caliphate or not is entirely up to them: according to their nature, their principles, the extent of their faith. In our Islamic way of thinking all people are equal before God. There is no difference between this one or that one. The Lord said, "[A]nd help one another in goodness and piety, and do not help one another in sin and aggression" (Q. 5:2). That is the basic rule. As Muslims we are ready to cooperate with people of all stripes for the sake of the good. But with regard to evil, no. By this I mean Muslims, Christians, Jews, even Buddhists or any others—we can work together for the good. However, when it comes to evil, we are not prepared to cooperate with anybody.[92]

Shaykh Nizar Ramadan was one of the 420 deportees who were brutally arrested by the Israeli authorities on a cold December night in 1992, bused and left stranded on the Israeli–Lebanese border, near the village of Marj Al-Zuhur. Their ordeal lasted a whole year. Ramadan is a leader now in his early forties, but early on specialized in qur'anic *tafsīr* and then turned to issues of society and politics. Author of three books, he founded the Islamic Cultural Center of Hebron in 1997. I met him in his Hebron office for three separate interviews.[93]

---

[92] This quote is taken directly from my complete transcript of that conversation. To put this and the next interview into their proper perspective (which I cannot do in the present context), one would need to refer to Nüsse's excellent work, *Muslim Palestine: The Ideology of Hamas* (Amsterdam: Harwood, 1998). From a careful study of Hamas publications, the 1988 Mithāq (Charter), but also the British journal *Filasṭīn al-Muslima*, Nüsse demonstrates the amazing ideological flexibility of this movement and the creativity of its reinterpretation of traditional Islamic symbols and dogmas. Also useful is Khaled Hroub's *Hamas: Political Thought and Practice* (the recent English translation of the original Arabic version of 1996) (Beirut: Institute of Palestine Studies, 2000). For an interesting attempt to document the impact of the transnational islamist movement on Muslim culture, and in particular its creation of a very large literature, see Fedwa Malti-Douglas's chapter, "Postmoderning the Traditional in the Autobiography of Shaykh Kishk," in *Tradition, Modernity, and Postmodernity in Arabic Literature*, ed. Kamal Abdel-Malek and Wael Hallaq (Leiden: Brill, 2000), 389-410. She offers a helpful definition of the new Islamic literature: "From phases of social realism, stream of consciousness, etc., modern Arabic prose has evolved into new formal domains. One of these is a distinctive brand of metafiction and postmodernism that exploits and manipulates the rich textual tradition, the *turāth*. This highly evocative word, which literally means 'heritage,' englobes works ranging from the theologico-philosophical through the literary and the historical to the biographical and philological... As the creation of a civilization that was fundamentally Islamic in ethos, these works stand not only for tradition but also, to a considerable degree, for Islamic tradition" (ibid., 391).

[93] I returned for a visit in the fall of 2002. In his office I was told that he had been arrested and incarcerated by the Israelis about six months prior to this. Like most

When asked about Q. 2:30, with no hesitation Ramadan answered that God chose Adam after he had been sent from heaven to live on earth. As such, Adam represents all of humanity. *Khilāfa*, the caliphate of humanity, is

> the leadership and active responsibility of humanity on the earth in every aspect of their existence. But however we define this, our application of this principle must not deviate from what is revealed in the Quran. God's rulership (*ḥākimiyyat Allah*) through humanity represents God's method on earth (*minhāj-Allāh fī-l-arḍ*). Among the missions of this caliphate: to spread the religion [i.e. Islam], which calls for social relationships that bring development, which also includes human solidarity (or 'bonded-ness': *tarābuå*) in society. This solidarity has three elements:
> 1. love (*maḥabba*) between people
> 2. a common commitment to applying and promoting Islam
> 3. the building of an Islamic society which resolves all human problems in an Islamic way.[94]

In my interviews with Ramadan I was eager to raise the question of hermeneutics. It was he that brought it up. He had told me in his second interview that Hasan Turabi most epitomized the position and attitude he espoused, so I followed up with a question on Qutb in my last interview, on February 6. The following is a paraphrase of his answer; for security reasons I was not able to tape these interviews.

> Sayyid Qutb? No, he was too confrontational—that was because of the context of the sixties! Turabi's ideas are still new—something like 10 years old.[95] But Muslim circles in this part of the world are increasingly influenced by his ideas. Now, one has to say that different people inter-pret Islam in different ways. When the Crusades came, there was a spirit of violence that pushed us to respond in kind. So for many centuries now the feeling of Muslims for Christians (at least Western ones) has been that of animosity. The colonial movement only reinforced this prejudice. Algeria is still struggling, partly due to this legacy.
> Since the end of colonization a new reality has taken hold: Islam must be rethought in a spirit of dialogue and mutual understanding. Hamas has

of the 7000 prisoners in his situation, Ramadan was not given the right of a lawyer or a trial.

[94] Johnston, "*Khalīfa*, Culture Change, and Muslim–Christian Cooperation," 53. This is from my first interview of Ramadan, January 14, 1999.

[95] For a useful article on this by Turabi, see "Principles of Governance, Freedom, and Responsibility in Islam." In this article, Turabi emphasizes the equality of all human beings: "In principle, all believers, rich and poor, noble or humble, learned or igno-rant, men or women, are equal before Allah, and they are His vicegerents on earth and the holders of His trust" (*The American Journal of Islamic Social Sciences* 4.1 [1987]: 4-5).

been shaped almost exclusively by the reality of Israeli occupation. But even in Palestine, Turabi's influence is being felt. In the Muslim world at large, perhaps 65 percent believe in a progressive, open ended interpretation of the Qur'an and Hadith. The percentage here is lower but it is growing.[96]

Can we cooperate for the sake of the good? "Certainly," he answered. In fact, Ramadan and I talked in detail about what we hoped would become a cooperative venture with a Christian non-governmental organization (NGO). Though this is not breaking new ground,[97] it is a unique experience because of its intentional dialogue perspective. Whereas Shaykh Yasin saw cooperation mostly as a Western Christian effort of advocacy for the cause of justice and human rights on behalf of Palestinians, Ramadan was interested in the practicalities of income generation and microenterprise lending schemes.[98] Yet both assumed the *khilāfa* of Adam involved all of humanity, a position that has profound implications for peace and justice in the twenty-first century global context. My feeling is that many others in Ramadan's generation are looking for new ways of reading the Qur'an today, who in like fashion would rather see a spirit of dialogue replacing the old confrontations—on the condition, of course, that their counterparts display the same spirit of understanding.

Two Muslims on Postmodernity

In this section I compare two others contemporary scholars I have mentioned above, Akbar Ahmed and Ziauddin Sardar. I then close the chapter with a brief look at Muslims who move toward a postmodern hermeneutic.

---

[96] From my personal field notes.

[97] Thomas Neu, at the time Palestine Director for American Near East Refugee Aid (ANERA), said to me during a visit in July 1998 that most aid organizations in Gaza preferred, when possible, to work with Hamas charitable societies (as opposed to PA sponsored ones) because the funds were always better managed.

[98] Maurice Borrmans incorporates both aspects—development and advocacy for justice: "In many places the struggle against underdevelopment mobilizes Christians and Muslims alike, autochthones and foreigners, to unite in labor movements, for example, or in other ideological campaigns, based on commonly held economic and political principles. Such cooperative efforts, having as their goal the values of economic justice and the common good, are they not opportunities for a dialogue of faith between Christians and Muslims, whether on the national level or the international (the United Nations Organization, for example)? We may call this the dialogue of professional, economic and political values" (*Guidelines for Dialogue between Christians and Muslims*, 30).

Ahmed's *Postmodernism and Islam* seeks to strike a balance between what he calls "occidentalism" (a reaction of African and Asian scholars to "orientalism") and "orientalism."[99] Both sides—the West and Muslims—have developed (over many centuries) erroneous perceptions of each other and, as a result, have embarked on a collision course. For this reason Ahmed ferrets out the philosophical, theological, cultural and political dimensions of this clash (one chapter is entitled "Confrontation and Clash"), and posits that in the end the greatest danger for today's Muslim community lies in the all-pervasive pressure and corrosion of the Western media.

For Ahmed the essence of Islam is *ʿadl* and *ihsān* (balance and compassion), *ʿilm* and *ṣabr* (knowledge and patience), that quintessential Islamic balance between *dīn* (religion) and *dunyā* (world). Nothing is said anywhere about humanity as *khalīfa*, but his worldview assumes the human being's nature as that of a free agent called by God to mark out his or her destiny according to God's revealed purposes. The book ends, therefore, with a double challenge. The first is addressed to the Muslim community:

> At the threshold of the twenty-first century the confrontation between Islam and the West poses terrible internal dilemmas for both. The test for Muslims is how to preserve the essence for the Quranic message, of *adl* and *ahsan*, *ilm* and *sabr*, without it being reduced to an ancient and empty chant in our times; how to participate in the global civilization without their identity being obliterated. It is an apocalyptic test; the most severe examination. Muslims stand at the crossroads. If they take one route they can harness their vitality and commitment in order to fulfill their destiny on the world stage, if the other, they can dissipate their energy through internecine strife and petty bickering: harmony and hope versus disunity and disorder.[100]

His challenge to the West is uncannily close to the purposes of the present study: "to expand the Western idealistic notions of justice, equality, freedom and liberty beyond their borders to include all humanity and without appearing like nineteenth-century imperialists; to reach out to those not of their civilization in friendship and sincerity." Finally, his last phrase captures the necessity of increased cooperation in an increasingly

---

[99] His definition of "occidentalism": "This is as much a rejection of colonialism, with which orientalism is associated, as it is an expression or revolt against the global civilization dominated by the West. No survey of the Muslim intellectual landscape would be complete without a comment on this little-discussed phenomenon" (*Postmodernism and Islam*, 177).

[100] Ibid., 264.

fragmented world: "In both cases a mutual understanding and working relationship are essential."[101]

In spite of the hope offered for dialogue, Bryan Turner's assessment of Ahmed raises some questions. Though "he provides a reasonably full account of postmodernism, he fails to resolve an important question: is postmodernity after modernity or against modernity? Alternatively, is postmodernity in fact a form of high modernity?"[102] Ahmed has not tackled the hermeneutical issue. In fact, he never specifies how post-modernism represents a threat to Islam—only that a consumerist society with pluralistic values erodes traditional Islamic belief. His personal summary of Islam under the headings of *ʿadl* and *iḥsān*, *ʿilm* and *ṣabr* is certainly an attempt at "doing theology in context." Yet he does not recognize this fact, nor does he spell out any rules for pursuing such a course.

This is not the case of Sardar. Also a Pakistani living in England, his academic training, however, was in the sciences, and it is in that field that he has mostly worked as a journalist. Yet this fact hardly does justice to the breadth of his concerns and achievements. He has published extensively on Islam in international journals and served as consultant for the Hajj Research Centre at the King Abdulaziz University in Jeddah, and "director of the Center for Policy and Future Studies at East-West University in Chicago."[103] Though both Sardar and Ahmed come to similar conclusions, the former is more interested in theology. I will indicate three aspects of Sardar's work, *The Future of Muslim Civilization*, that pertain to the present discussion:[104]

1. In a section entitled "the epistemology of Islam," Sardar declares, "It is the epistemology of a people that gives unity and coherence to the body of their sciences—a unity which is the result of critical examination of the sciences in the light of their beliefs, convictions and value system. There is no such thing as a non-aligned truth."[105] This sounds like the postmodern slogan, "all facts are value-laden." But lest anyone wonder whether he has thrown out objective truth, he counters that the Qur'an and

[101] Ibid., 265.

[102] Turner, *Orientalism, Postmodernism and Globalism*, 14.

[103] Tomas Gerholm, "Two Muslim Intellectuals in the Postmodern West: Akbar Ahmed and Ziauddin Sardar," in *Islam, Globalization and Postmodernity*, ed. Akbar S. Ahmed and Hastings Donnan (London/New York: Routledge, 1994), 192.

[104] Ziauddin Sardar, ed., *Islamic Futures and Policy Studies* (London/New York: Mansell, 1987).

[105] Ibid., 24.

Sunna are the "Absolute Frame of Reference" of Islam. "The very definition of what constitutes knowledge and error in Islam solves the problem of source and validity. The absolute source of knowledge and the absolute judge of validity is, of course, the Qur'an and Sunnah."[106] Does this mean that hermeneutics are dispensed with?[107] To the contrary, following the paradigm of the scientific method, the consensus in all disciplines is that in doing research one solves problems. Collecting facts or seeking after truth is elusive. There is no such thing as a bias-free observation. Even facts, "reliable statements about the world—come to be recognized as such by social process... How the evidence for a particular solution is regarded and presented depends on the cultural temperament of that society."[108]

2. Theology must be formulated. In Islam, right after the foundational principle of *tawḥīd* comes humanity's *khilāfa*: "Man is the best of Allah's creation. He has been gifted with the capability to adorn himself with the divine attributes. But he has the choice to use his capability or to ignore it. He has his destiny in his own hands."[109] To be more specific, "*Khilafat* signifies man's vicegerency of Allah's attributes. Man is charged with the responsibility of sustaining himself and other creatures of the globe

---

[106] Ibid., 25.

[107] This is certainly true of the late Isma'il al-Faruqi's understanding—one of the founding fathers of the contemporary "Islamization of Knowledge" movement (from the first conference in Switzerland in 1977, the subsequent founding of the International Institute of Islamic Thought, the *American Journal of Islamic Social Sciences* and in Pakistan, the journal *Islam and the Modern Age*, among others). In his article entitled "On the Nature of Islamic Da'wah," Faruqi writes that the hermeneutical problem with regard to the Qur'an is inexistent because "Arabic is the only language which remained the same for nearly two millennia, the last fourteen centuries of which being certainly due to the Holy Qur'an" (p. 39). In a footnote he goes on to say, "the Qur'anic text is not bedeviled by a hermeneutical problem. Differences of interpretation are apodictically soluble in terms of the very same categories of understanding in force at the time of revelation of the text (611–632 A.C.), all of which has continued the same because of the freezing of the language and the daily intercourse of countless millions of people with it and with the text of the Holy Qur'an" (ibid.). This is hardly congruent with modern linguistic theory. Further, a call to *ijtihād* today cannot be divorced from an epistemology of critical realism and a solid understanding of the hermeneutical process in approaching any text from the past (Esack, "Qur'anic Hermeneutics: Problems and Prospects"; Arkoun, *Rethinking Islam*; "Religion and Society, The Example of Islam").

[108] Sardar, *The Future of Muslim Civilization*, 229.

[109] Ibid., 28.

faithfully according to the Divine characteristics of *al-Rabb*."[110] It follows that political theory in Islam flows "from the sovereignty of Allah and the vicegerency of man. Two outcomes of the theory of Divine sovereignty are absolute equality before the law and the limitations of man's power of legislation."[111] In the economic realm, Islamic values "are derived from the pragmatic value of Divine ownership and the universal brotherhood of man. As such, every creature of Allah has a natural right to draw its sustenance from the fruits of mother earth."[112]

3.   More consciously postmodern than Ahmed, Sardar sees history as a narrative constantly in the process of being unfolded:

> The realization of the state of Islam is a constant state of becoming. The development of the personality of a *mumin*, the perfect man, is a continuous striving. Each step forward requires adjustment; new problems have to be tackled from epoch to epoch so that the state of Islam can be reached and its true dimensions realized. Only in the constant process of becoming and in the continuous state of striving can we implement the dream that comes to us from the depths of our historical consciousness. This dream constitutes an inspiring challenge for future generations of Muslims as we move toward the Medina state with complete trust in Allah.[113]

---

[110] Ibid., 29-30.

[111] Ibid., 30.

[112] Ibid., 31. S. Manzoor wrote an article, quoting heavily from postmodern literature and in particular from the field of eco-philosophy. He explains how human ecology can serve as a metacritique of modernity and operate as a foundation for a holistic and integrative vision, which could solve the problems of contemporary civilization. Surprisingly, Islam only makes its appearance in the last paragraph: "The sacred text of Islam, it has been recognized even by outsiders, contains a 'theology of ecology.' The Qur'anic notion of *Khilafah*, which fosters a uniquely comprehensive ethic of custodianship and responsibility, anticipates all the seminal insights of human ecology. Today, when some of the leading ecological minds of our civilization are pleading for the replacement of our image of the universe as a machine by that of the Garden, they are simply reiterating ecological guidelines enunciated by the Qur'an and accepted by other revealed religions" ("Human Ecology and the Quest for a Universal Science," *MAAS Journal of Islamic Science* 7.2 [1991]: 107). Though the article does not present the author's full perspective on humanity's *khilāfa* and ecological responsibility, it certainly represents an intriguing attempt to theologize in a way that is very compatible with the present study.

[113] Sardar, *The Future of Muslim Civilization*, 260. For an evocative analysis of development discourse in a Muslim context from a postmodern perspective, see Toine Van Teeffelen's chapter "Development Discourse: The Case of Palestine," in *Changing Stories: Postmodernism and the Arab-Islamic World*, ed. Inge Boer,

Learning from Postmodernism

In my discussion of postmodern hermeneutics in Chapter 4, I mentioned that some evangelical scholars were beginning to interact with these ideas and that some were willing to incorporate new insights.[114] I also indicated in that context that Muslim scholars such as Esack and Arkoun were consciously reflecting on postmodern epistemology in a theological way.[115] With the present backdrop of Muslim interpretations of Q. 2:30 behind us I will add a few more remarks, as a catalyst for further research.

Contrasting Attitudes toward Postmodern Thought
A comparison between articles in the conservative journal *Muslim World Book Review* may prove useful as an introduction.[116] In a review of the edited book, *Changing Stories: Postmodernism and the Arab-Islamic World* (cf. n. 113), the editors adopt a derisive and even sarcastic attitude toward social scientists "of a postmodernist bent." For scholars in that category the book may not "represent a disappointment," but for Muslims it does. This is for two reasons: (1) all authors "are equally preoccupied with their *Übervater*, Jean-François Lyotard" (especially his distinction between local and grand narratives—Islam itself becomes one the grand narratives that now disenchant); the result, they write, is "a sterile, self-contained, incestuous pseudo-scientific monologue, blurring rather than

Annelies Moors and Toine V. Teeffelen (Amsterdam: Rodopi, 1996), 37-52. Her goal is to put the "development discourse" used with only slight variation by foreign NGOs and by Palestinian and official Palestinian Authority channels. On the one side, the particular Dutch NGO, NOVIB (which she uses as case study), assumes that quasi-technical language framing "an abstract development story can be transposed upon a highly politicized and conflict-ridden domain" (ibid., 45)—a shrewd compromise aimed at pleasing both donors and partners at the same time, but not in touch with the reality "on the ground." From an analysis of the official Palestinian development discourse she isolates the metaphor of "nation building" which implies "agency and control" as central values. Yet the problem here is that "[a]lthough the schema incorporates a diversity of groups and viewpoints, the small stories are fully subordinated to the compelling order of the master story" (ibid., 49). This kind of analysis in sociology/anthropology is extremely helpful in teasing out many of the internal tensions of peoples, nations, and sub-groups (and organizations, for that matter).

[114] I mentioned in Chapter 4 Merold Westphal's edited book, *Postmodern Philosophy and Christian Though*; and David Naugle's study on worldview.

[115] See the first section, "Hermeneutical Rumblings in Muslim Circles."

[116] It is published by the Leicester Islamic Foundation in conjunction with the International Institute of Islamic Thought in Herndon, VA.

elucidating reality"[117]; (2) its obsession with deconstruction; of course, "phenomena can be viewed from different perspectives. Big deal!" One study brands the feminist "discourse" of Egypt as an "imported narrative." Then comes this rhetorical question on the part of the editors: "What have we gained when reading that 'the grand narrative of Islam... can be countered by the diversity of smaller narratives like feminism'?" There is no desire here to dialogue, understand or learn from another viewpoint—a simple dismissive conclusion will do: "Sometimes the emperor has no clothes, and we might as well say so."[118]

Despite this dismissive attitude toward contemporary Western thought, we find that in the previous volume, the consulting editor, S. Pervez Manzoor (cf. n. 112), a scholar hailing from Stockholm, engages in a serious debate with a work I have discussed in Chapter 2, Habermas's *Between Facts and Norms*. Though he also reviews two other Muslim works in his articles (Wael B. Hallaq's *A History of Islamic Legal Theories* [1997], and Mohammad Hashim Kamali's *Freedom of Expression in Islam* [1994]), most of his effort is devoted to Habermas's work. In the end, he judges Habermas's project flawed for divorcing law from morality. In fact, he finds his theory of communicative action is "founded upon a circular logic":

> Citizens qua political community make laws through the exercise of political sovereignty; laws, enforced by the coercive sanction of the state, provided them with a system of rights; hence citizens qua legal subjects submit to the authority of the law and accept the legitimacy of the legal order they themselves have created. However, citizens qua moral community get disenchanted with existing laws; they don their legislative robes and frame new laws which they regard as binding, at least for the time being![119]

On my view, this is valid argument on the part of any religious person, particularly those of the monotheistic faiths who believe in creation and God's call for humanity to be his stewards/caliphs on earth. But Manzoor has come to this point by seriously engaging Habermas's arguments. He

[117] "Changing Stories," *Muslim World Book Review* 19.1 (1998): 41. Though certainly identifying with the reviewers' complaint, I wonder whether their caustic terminology is warranted: "The vicious impact of this approach is due to the suggestiveness of the key concept 'story'... To reduce Islam to a meta-story is to treat it at the level of a nursery rhyme. The current vocabulary of cultural anthropology may itself be a language game, but one with deadly relativistic, agnostic consequences" (ibid.).
[118] Ibid., 42.
[119] S. Pervez Manzoor, "Faith and Law: At the Crossroads of Transcendence and Temporality," *Muslim World Book Review* 18.3 (1998): 6.

not only credits Habermas as "the most cogent defender of the tradition of European reason today," but also as "the last of the philosophers committed to the project of modernity who is neither a Hegelian historicist nor ascribes [*sic*] to the transcendent rationality of the sovereign subject of Enlightenment."[120] Further, he declares his great admiration for Habermas's discourse theory, which "smoothly balances and mediates such notorious dualities as individual and community, law and state, freedom and coercion, validity and facticity, rights and responsibilities and all else that have plagued, and continue to plague, modern thought." However, this "splendid work," comparable in some ways to that of Thomas Aquinas in another era, accomplishes its feat at a great cost—the divorce of law and morality.

What is most remarkable in this essay is not Manzoor's philosophical expertise and wide knowledge of Western thought. Rather, it is his introduction to the central issue of the essay—the state of Islamic legal theory today. This is where his own epistemological assumptions shine through. "*Sharīʿa*," he begins, "the cardinal concept of Islam, presents human understanding with a most baffling challenge and an insoluble paradox. For it incarnates norm/doctrine, authority, exegesis, praxis, tradition and much else at the same time."[121] Based on revelation and on the authority of the divine will, *sharīʿa* nevertheless calls for people's efforts to interpret and reason, thus transforming it into legal norms for social consumption. Hence, to construe it as "law" (in and of itself) "is to fall prey to the positivistic irreverence of modernity and reduce it to a punitive code. *Sharīʿa* in this sense straddles "transcendence and immanence, faith and existence, eternity and temporality" and is both "overpowering and alluring, punitive and healing, retributory and redemptive."[122]

This said, what is the state of Islamic jurisprudence today? At a time when "the authority of tradition is confronted by the challenge of history," when legal claims clash with moral ones and when the rich heritage of Islamic jurisprudence "is put under the magnifying glass of modern skepticism," contemporary Muslim writings on Islamic law are "a veritable cacophony."[123] One searches in vain for any coherence of systems or epistemological vantage points: "Having renounced the unity of transcendence for the polymorphism of temporality, it has degenerated

---

[120] Ibid., 4. This seems to open the door to some kind of weak version of postmodernism.

[121] Ibid., 3.

[122] Ibid.

[123] Ibid.

into an idolatrous cult of nihilism. Little wonder there is not a ghost of meaning left in its epistemological mansion."[124]

On the other hand, Manzoor is most enthusiastic about Wael Hallaq's work. He is, undoubtedly, "the most eminent among the academic scholars of Islamic law and jurisprudence. His research and writing demonstrate "an originality of vision" and has thereby created a "paradigm shift" in the field.[125] Nevertheless, while Kamali's *Freedom of Expression in Islam* is truly "a labour of love and a work of devotion and piety," he opines that "a sharper intellectual vision of modernity and a more vigorous encounter with its polemics would have enhanced the already considerable worth of this work."[126] He then develops his argument. In a nutshell, since Kamali fails to develop an Islamic vision of the state, his discussion of "freedom of expression" simply assumes (and therefore supports) the modern secular "dialectics of individual and state, conscience and society, public and private." The alternative would be to consider Muslim societies as cultural and political islands untouched by modern secularism—a claim difficult if not impossible to substantiate. Perhaps the most "grievous" shortfall of this book, in Manzoor's eyes, is his "reified perception of the pivotal concept of Sharia: it is used such that it can be totally identified with the extant corpus of fiqh." But surely, an interpretation of the text is not the text itself, and even according to traditional Islamic norms, *fiqh* (in five main schools) is considered a fallible human attempt to understand God's Law. Manzoor's recommendation, therefore, is that in such a work "the philosophical and conceptual analysis of the key Qur'anic terms be given priority over the atomistic and literal approach of fiqh."[127] And this has yet to be done, he concludes. I wholeheartedly concur.

Abdolkarim Soroush: Epistemology Systematically Constructed
Whereas the Iranian thinker Abdolkarim Soroush[128] agrees with Manzoor, Arkoun and others, he has devoted much of his career to working out

---

[124] Ibid., 4.

[125] Ibid., 7-8.

[126] Ibid., 8-9.

[127] Ibid., 11.

[128] I was tempted to use Tariq Ramadan instead of Soroush, as they end up with similar positions on the issues of human rights and democracy. In fact, Ramadan is explicit about the human caliphate, whereas Soroush is not: "It is the role of humankind to manage the world on the basis of an ethic of respect for creation not only because people do not own it but, more deeply, because it is in itself an eternal and continual praise addressed to the Most High" (*Western Muslims and the Future of Islam* [Oxford/New York: Oxford University Press, 2004], 18). Despite the fact that

what an Islamic epistemology would look like in our postmodern context. I offer a summary of this thought, including a brief history of how it developed in his thinking, as an illustration of the kind of work I have been proposing throughout this book to conservatives, both Muslims and Christians. The exercise will also prove to be a useful summary of the previous chapters.

The Genesis of an Idea. Born in 1945, Soroush developed a passion for poetry while still in elementary school. As an adolescent, he attended extracurricular classes in qur'anic exegesis taught by an expert in Islamic law, who had a degree in physics and knew a great deal about Eastern philosophy. Though Soroush was unimpressed by this man's answers to his queries, his life-long interest in the intersection of faith and science had begun in earnest.[129] During his six years at the University of Tehran, where he earned a doctorate in pharmacology, Soroush took private lessons in Islamic philosophy (including the work of such contemporary thinkers as Tabataba'i and Morteza Motahhari), studied Marxism (the 1960s in Iran saw the growing political opposition to the Shah, especially in student circles), attended numerous qur'anic study sessions of a revivalist group ("Qur'anic Muslims"), and deepened his knowledge of Persian poets Jalal al-Din Rumi (d. 1273) and Hafiz of Shiraz (d. 1389). He then went to England for postgraduate studies in analytical chemistry. After a year, however, he switched his specialty to philosophy and the history of science.

The common thread in Soroush's intellectual odyssey throughout these years was the hermeneutical question. While still at the University of Tehran, he threw himself into the study of both Shi'ite and Sunni *tafsīr* literature and remembers being "fascinated" with "the details and intricacies of the differences in interpretation."[130] This he combined with his passion for Rumi's *Mathnavi* and Hafez's *Divan*, and the result was a growing desire to unlock the mysteries of "the art of textual interpretation."[131] In fact, it was the hermeneutical quest that led him to explore

---

both his theology and social ethics are more developed than they are for Soroush, Ramadan does not make explicit the theoretical questions that interest me here: epistemology and hermeneutics.

[129] The biographical information here is taken from an interview with Soroush by one of the two editors (unspecified) of Soroush's only book in English so far, *Reason, Freedom and Democracy in Islam: Essential Writings of 'Abdolkarim Soroush*, trans., ed. and with a Critical Introduction by Mahmoud Sadri and Ahmad Sadri (Oxford/New York: Oxford University Press, 2000), 3-38 (Chapter 1).

[130] Ibid., 6.

[131] Ibid., 7.

Western theories on the philosophy of science—a quest that could not be divorced from the epistemological issue of what we can know as human beings and how we come to know what we know. Interestingly, he notes, 1974 was the year he began his studies in the philosophy of science—the same year Thomas Kuhn's ideas began to meet with greater acceptance.[132]

The very notion that science might not be cumulative and value-free, but that it was rather the result of people making research decisions—influenced by their particular worldviews and biases—was both revolutionary and unsettling to Soroush. Looking back on this period, he states that this course of study "was a true revelation" for him. He found himself in a constant state of meditation and intellectual rumination:

> I ate, drank, slept, and walked philosophy. I was bombarded by challenging questions and stimulating insights. I was constantly at work sifting, revising, synthesizing, reconciling, and distinguishing different components of my education and knowledge. Particularly, I was grappling with the questions of the relationship between science and philosophy, that is, science and metaphysics. No single waking minute would pass, whether walking, riding on a subway, sitting at home, or working in the library, unless I was struggling with some serious and grand problem.[133]

Before Soroush returned to Iran in September 1979 (a few months after the revolution), he had already written four books (in Farsi) on these related topics.[134] It must also be stated that the context of Islamic revolutionary fervor played a part in the writing and the tone of these books. While uncomfortable with the more extreme statements of religious certitude, Soroush was actively involved with many of the leading architects of the revolution. Upon his return to Tehran, he was named chair of the department of Islamic Culture at the Teacher's College. Then he served for four years as a member of the Advisory Council on the Cultural Revolution. His next appointment was as a member of the Academy of Philosophy, following which he served until the late 1990s on the board of the Research Center for Humanities and Social Sciences.[135] Those "eighteen years of postrevolutionary thought," he says, represented "a period of unabated intellectual struggle." What was

---

[132] Ibid., 9.

[133] Ibid., 10.

[134] Besides a major pamphlet entitled "Epistemology," he wrote *What is Science, What is Philosophy?*, *Philosophy of History*, *Science and Value*, and *The Dynamic Nature of the Universe* (ibid., 11; no dates are given in the text).

[135] It was in 1997 that he was relieved of all of his official positions, so great had the rift become between his views and those of the Guardian Council. He has held teaching positions at Harvard, Yale and Princeton in the last six years.

the result? In the next two sections, I focus on two aspects of his thought: epistemological/hermeneutical and ethical/political.

The Theory of the Contraction and Expansion of Religious Interpretation. Muslim reformists throughout the history of Islamic societies sought to renew the core of Islamic beliefs and apply it afresh to changing circumstances. What they lacked was a comprehensive epistemological theory, argues Soroush. What they failed to distinguish was the difference "between religion and religious knowledge."[136] No religious revivalist can ever be a lawgiver. That was the role of revelation, which stopped when the Prophet of Islam passed away. Religion—read "the sacred texts"—is perfect, because it originates in heaven. But as soon as people (however well trained they might be) attempt to understand it, articulate it and apply it to specific contexts, they are displaying "religious knowledge," by definition limited, incomplete and flawed. Reformers, then, are only exegetes, never lawgivers. As such, human understanding of revelation can only proceed on the basis of human reason:

> Reason does not come to the aid of religion to complement it; it struggles to improve its own understanding of religion. The sacred *sharīʿah* never sits parallel to human opinions, so there is no possibility of agreement or disagreement between the two; it is the human understanding of religion that may be congruous or incongruous with other parts of human understanding.[137]

True religious reform and revival consists of patching up and reworking the knowledge of revelation passed down to a particular generation, in such a way that it can better be translated and fleshed out for those faithful of that time. It is precisely, according to the present book, "doing theology in context." Hence, it makes no sense to complain (as Muhammad Iqbal did) that "religious knowledge has remained under the tutelage of Greek thought for centuries." That was but one attempt to make sense of God's revelation, which comes to us from beyond the corridors of space and time. Yet the knowledge and articulation of our own understanding of that revelation will of necessity be bound by specific cultural and historical factors. Recall our discussion of Farid Esack's theological clashes with the South African *ʿulamāʾ*, who saw no reason to question the apartheid status quo. The same theological debates were going on among Christians. The result for Esack was an Islamic "liberation theology," which a decade later might already look somewhat different—the

---

[136] Ibid., 30.
[137] Ibid., 31.

world has become even more globalized and his own thinking has now evolved with more writing and interacting with others. This is what Soroush means by "contraction and expansion": theology is constantly taking on board new insights while at the same time discarding others.

From an epistemological viewpoint here, Soroush is affirming a critical realist position, namely, that all human knowledge (including that of divine revelation) is limited, fallible, and in constant need of revision. Though he never makes use of the notion of *khilāfat al-insān*, he assigns nevertheless a high role to reason, though a role chastened by the insights of what Seyla Benhabib characterized as the "weak version" of postmodernism:

> We human beings are now expelled from heaven and deprived of revelation. We are profane and fallible. To speak and act like prophets does not suit us. Apropos of our limited reason, we acquire a faint scent of the truth and act accordingly. We are *sharihan* [interpreters of religion], not *sahrīʿan* [initiators of religion]. We are the enticed, not the infallibles. Let them who deem their words above the mere understanding of religion beware: their hubris may at long last tempt them to don the mantle of the prophets.[138]

That last reference, unmistakably, was to the clerical rulers of revolutionary Iran. It will be clear from the next section how far his thinking is from a traditional Shi'ite political view.

On the other hand, Soroush is no relativist. He disagrees with those who say that "if religion proves to be time-bound," then why not "immerse ourselves in the concerns of our time altogether?" Yet this would confuse secular and sacred worldviews, answers Soroush. In fact, the Qur'an and Sunna are not open to an infinity of interpretations.[139] What is more, though the boundaries between "orthodoxy" and "heterodoxy" are often difficult to draw, truth itself is not relative. Faith, certitude and free will are main elements of true religion—hence *reasons* can be advanced for believing this over that. Debate is not only possible but necessary, with the expectation that these issues can be sorted out rationally. At the same time, false beliefs arise on the wings of multiple *causes*, but no amount of *reasons* can be adduced to defend them.[140]

---

[138] Ibid., 37.

[139] Ibid., 36.

[140] Ibid., 139-40. For Soroush, "reasons have to do with the proof of the truth or falsehood of beliefs, but causes deal with the process of their development. False beliefs have causes, but not reasons" (ibid., 214 n. 12). In other word, one has to pay attention to the historical, sociopolitical and cultural factors that led to people formu-

Here the context is a discussion of tolerance within the wider purview of democracy.[141] It is precisely here that the distinction between reasons and causes becomes plausible, at least as it concerns the issue of religious pluralism. Recall Paul Griffiths's distinction (cf. Chapter 5) between religious truth claims and claims relative to "the proper end of human persons," or "salvation." Religious discourse makes claims about metaphysical reality in relation to sacred texts. Soroush and I would agree on an "inclusivist" view, arguing that while we may find much overlap between various religious views (*contra* the "exclusivists"), some systems are truer than others (*contra* the "pluralists"; cf. Appendix D).

With regard to the proper end of human existence, each religion proposes its own path. Partly influenced by Rumi's view of truth as one, yet with multiple perspectives, and partly by his own epistemology, which finds more affinity with the Christian theologian John Hicks than with Karl Barth, Soroush follows the pluralist option. All religions have a single "essence" and their differences reside more at the level of degree than of kind.[142]

This is to say too that Soroush is a critical realist largely along the lines drawn by the physicist/theologian Ian Barbour (cf. Chapter 3). In fact, his discussion in Chapter 5 ("Doctrine and Justification") of the justifiability of religious doctrines ends up at a similar point. In Soroush's words, a religious doctrine (here Qur'an and Sunna) is necessarily tested in the flow of history. The fratricidal wars, doctrinal divisions and political oppression on the part of Umayyads, Abbasids and other states must be weighed against movements of revival and reform, saints and scholars who inspired the masses, and great cultural achievements: "All of these are part and parcel of the history of a doctrine and should enter its final evaluation... Factional conflicts and abuses of principles in Islam and Christianity owe as much to human volition as to the teachings of these religions."[143] For Soroush, it is human nature that is at stake here. No doubt, Barbour would agree. Recall that, for him, any scientific or religious theory must pass four tests: (1) agreement with the data ("critical" realism, but "realism" nonetheless); (2) coherence (parts relating to the

---

lating or espousing certain beliefs. To become aware of these factors would rid us of much unnecessary conflict.

[141] I believe Soroush is still overly optimistic about the capabilities of human reason. See my discussion of McIntyre, Naugle and Smart in Chapter 4. Rationality is in part determined by one's "worldview"—that rock-bottom core of beliefs, stories and myths that make sense of the world, which is at the heart of human culture.

[142] Ibid., 72.

[143] Ibid., 84-85.

whole, etc.); (3) scope (ability of a theory to cover the widest ground); (4) fertility (its ability to generate new research, and, in religion, to produce good results in individuals and society). All of the above flows logically from Soroush's epistemological theory.

Soroush on Ethics and the Politics of Religious Democracy. Now we come back to the issue of religious pluralism in a democratic context. Soroush's starting point builds on his previous distinction between religion and religious knowledge. This leads him to differentiate further between "intrareligious issues," such as rites, revelation, miracles, Satan, faith and the like, and "extrareligious issues," such as "free will, human rights, meaningfulness of religious propositions and practical verification of religion."[144] I have no space to discuss these views in detail, but only to point out some of their implications for my own project in this book. To be sure, exploring the doctrine of creation common to Muslims and Christians through the lens of the qur'anic "caliphate" of Adam is a venture into comparative "intrareligious" issues. Yet the preparatory chapters that developed a critical social and historical theory (response to postmodernity), coupled with background material in epistemology and hermeneutics (response to postmodernism) was surely an exercise in "extrareligious issues," or philosophy of religion. My contention—along with Soroush—is that one comes to the task of theology with the assumptions and methodologies of our time and culture, and that the more we lay bare those assumptions and examine those methodologies, the more honest, authentic and effective our theology will become.

Soroush's previous chapter, "The Sense and Essence of Secularism," had already laid out an ethico-political foundation. In it he argued that religious texts employ the language of duties, while modern law speaks only of human rights. Secularism gradually appeared on the heels of a still very Christian Renaissance, particularly at the point where the scientific method of empiricism was first applied to the ethics of con-temporary societies (beginning with Machiavelli). The newfound pride in harnessing the forces of nature and inventing new instruments "gave people the courage to revise social conventions and to initiate deliberate reforms in the world of politics."[145] Yet, for Soroush, the desacralization of politics, which eventually led to the French Revolution, was primarily an application of the scientific method to society, not primarily a rebel-lion against God. Thus, just as science and God need not be at opposite

[144] Ibid., 69.
[145] Ibid., 59.

poles, so the modern secular view of government as accountable to the people, and therefore submitted to checks and balances, might also be compatible with a more religious rule. After all, "[h]uman beings can remain spiritual and religious while enjoying the benefits of rational administration of their affairs."[146]

Notice how Soroush traces the accomplishments of modern secularism (starting with the Greeks and ending with European science and political philosophy) while painting them in neutral terms. When God was taken out of the picture (along with his "rights" for human beings), then a new humanity was erected, "autonomous, self-reliant," within a "causally integrated system that lacked the concept of just deserts." He goes on: "The experts of social, political, and economic sciences inherited this philosophical system. Thus the concept of rights and duties were abandoned." But the key phrase follows: "We have inherited this latter system."[147] This is a statement of fact—"the story of the secularization of humankind." But the good news, reasons Soroush, is that behind secularism is a "nonreligious reason," and therefore not necessarily an antireligious one.

For Soroush, then, secularization is a mixed blessing. The result of the swift takeover of Western science in the last two centuries, along with new philosophical theories both modern and postmodern, is that "no latitude has been left for stability and certitude." For people of all religions, the new tolerance in the area of beliefs has brought in its wake an erosion of epistemic confidence.[148] This difference between a traditional society basking in certainty and a global community now groping in the shadows of uncertainty has enormous ethical and political consequences. For one, human rights count more than beliefs. Only three centuries ago Europeans killed one another for their beliefs—Catholics against Protestants, capital punishment for heretics and witches. Though Soroush does not mention it, the Islamic conquests, the Crusades, the Inquisition and Reconquista, are all examples of religiously sanctioned killing. The law of abrogation (*naskh*), which for the majority of classical Muslim jurists meant that qur'anic verses enjoining the killing of idolaters (and even "people of the book" if they refuse to pay the poll tax), trumped any verses about peaceful relations with these people or even any verses about defensive fighting.[149] That consensus has eroded of

---

[146] Ibid., 61.
[147] Ibid., 68.
[148] Ibid., 125.
[149] See Rudolph Peters, *Jihad in Classical and Modern Islam* (Princeton, NJ: Marcus Wiener, 1996).

late: "Nowadays, killing people for their beliefs is deemed unacceptable and a breach of human rights."[150]

This brings up a second consequence. The Universal Declaration of Human Rights (UDHR, 1948) is often interpreted to say that "democratization of religious government means washing one's hands of convictions and surrendering to skeptical or secular ideas and to people's demands, to the exclusion of God."[151] For Soroush, however, this secular interpretation is not warranted, and here his category of "extrareligious knowledge" becomes crucial. On the age-old debate between the respective balance between reason (*ʿaql*) and revelation (*sharʿ*), Soroush falls clearly on the rationalist side. Religious scholars must not "shirk the responsibility of balancing the knowledge inside and outside religion, since many basic religious values such as truth, justice, humanity, public interest, and so on are integral to nonreligious value systems as well."[152] Why is this? Because religion in the present context must be "rationally acceptable." Hence, the ethical values of justice, the dignity of the human person, the search for the common good, and truthfulness, are not primarily products of religious dogma, but rather the common inheritance of all peoples in conditions of modernity.[153]

By the same token, moderns have reached a consensus on the desirability of democratic government: powers in the three branches of government are separated and equally weighted through checks and balances; policies are reached through a transparent mechanism in all three branches; elected rulers remain accountable to the people; freedom of expression and all other human rights are guaranteed; education is mandatory and the press is independent and free. By no means a comprehensive list, this much of a definition demonstrates the coherence of Soroush's overall epistemological theory. Starting with religion and religious knowledge as both equally based on the rational principles of

---

[150] Soroush, *Reason, Freedom and Democracy in Islam*, 125.

[151] Ibid., 126.

[152] Ibid., 127.

[153] Soroush contends that "justice is a metareligious category, and the right and acceptable religion should, inevitably, be just" (ibid., 132). He is adopting a position that the Islamic rationalists, the Mu'tazilites, held long ago—ethical objectivism (justice or goodness exists in and of itself) and its epistemological corollary: the human mind can grasp ethical values apart from revelation. The Shi'ites in general have held onto that rationalistic strand of Islamic theology, much more so than the Sunnis. See Majid Fakhry, *Ethical Theories in Islam* (Leiden: Brill, 1991). For a detailed argument about how modern Sunni legal theory began to move in this direction, see my essay "An Epistemological and Hermeneutical Turn in Twentieth-Century *Uṣūl al-Fiqh*."

justice and human dignity, it is an easy step to accept the fruits of the modern evolution toward democratic procedures in government and international law.[154] On the one hand, a religious society is defined as "sober and willing—not fearful and compulsory." On the other, "[s]uch religiosity guarantees both the religious and the democratic character of the government." This standard of rationality, however, is only possible "when the innerreligious and outerreligious domains are harmonized."[155] This also means that a "democratic religious government" is not a contradiction in terms, because it draws from the common well of human wisdom in the present age—and harmonizes nicely with his epistemological theory:

> This rational sensibility permits the transformation and variation of religious understanding. The acknowledgement of such varieties of understanding and interpretation will, in turn, introduce flexibility and tolerance to the relationship of the ruling and the ruled, confirm rights for the subjects, and introduce restraints on the behavior of the rulers. As a result, the society will become more democratic, humane, reasonable and fair. Expansion and contraction of knowledge, its constant renewal, the perception of truth as an elusive labyrinthine path, the recognition of man as a tarnished, slothful, and fallible creature who, nevertheless, possesses an array of natural rights have all been among the necessary prerequisites for and epistemological and anthropological foundations of democracy. If these same principles are included in religious knowledge and respected by religious people, the result will be religious democracy.[156]

As an evangelical theologian seeking to build common strategies with Muslims for a more just, peaceful and democratic society, I find in Soroush a welcome partner. On the one hand (unsurprisingly), he affirms the values of justice, human dignity, freedom of expression and the rule of law in a democratic body politic. On the other hand, Soroush points to the frailty of human knowledge and reason: "the perception of truth as an elusive labyrinthine path." Again, I agree—hence all the discussions about weak and strong versions of postmodernism. Reason here plays a role very much akin to Habermas's communicative action. The legitimacy of the state, as we have seen, is grounded in a discursive form

---

[154] "Evolution" here is used deliberately. Soroush for all of his emphasis on human fallibility, I believe, remains too optimistic about this evolutionary process: "The root of democracy is a novel insight that humanity has gained about itself and the limitations of its knowledge. Wherever the seed is allowed to germinate, the external manifestations of democracy will, inevitably, bloom" (*Reason, Freedom and Democracy in Islam*, 133).

[155] Ibid.

[156] Ibid.

of rationality—people hashing out different viewpoints while listening to each other. With Soroush we also rejoin Benhabib's contention that discourse ethics require two deontological moral principles from the start: (1) universal moral respect, and (2) principle of egalitarian reciprocity. Her Kantian universalism, fortunately, is combined with a communitarian concern for the embodied self and the necessity for individuals to work out their ethics in harmony with the narratives that inform their particular communities. This too is the point of Soroush's "religious democracy." There will have to be room at the table for everybody, the secular and the religious of all stripes.

A final comment on Soroush's two-tiered theology is in order. I find it crucially important that people come to agreement on basic democratic principles. The notion of human rights is inseparable from these principles, yet in the vast body of literature that has arisen around them there is not as much of a consensus as one might think. First, as we saw in my discussion of indigenous peoples and their rights, the notion of culture is a particularly knotty issue. Cultural relativism militates against hard and fast, universal definitions of human rights.[157] Second, the philosophical foundation underpinning human rights is tenuous at best. The originator of human rights discourse, John Locke, assumed these rights were grounded in God and revelation. The UDHR in 1948, precisely because of its universal claims, could not invoke theological principles to ground the said rights. The document simply sidestepped the issue.

An international relations specialist, Katerina Dalacoura, asserts that the human rights concept passionately interacts with the tenets of Islam in the political life of Egypt and Tunisia.[158] Drawing from social science and philosophy, she concludes that the notion of human rights has a metaphysical dimension—only some kind of faith could guarantee the dignity of the human person. Whether the context is secular or religious, human rights cannot be attributed to reason. Yet this concept tends to be more compatible with religious faith.

---

[157] For useful discussions about the impact of cultural relativism on the idea of human rights, see Michael Freeman, *Human Rights*, especially his Chapter 6, "Universality, Diversity and Difference: Culture and Human Rights"; also Michael Ignatieff, *Human Rights as Politics and Idolatry*.

[158] Katerina Dalacoura, *Islam, Liberalism and Human Rights: Implications for International Relations*, rev. ed. (London/New York: I. B. Tauris, 2003). Her approach is necessary and complementary to other approaches, but her limitations in the area of theology must be stated: she is no specialist in Islam and has to rely solely on French and English sources.

I find Dalacoura persuasive on this issue. I also believe Soroush would concur, though he firmly states that it is "reason" that dictates the necessity of a general respect for the rights of people *qua* human beings. If both could start with "discursive rationality," then at least the conversation would continue. What is certain, however, is that faith of some kind is involved, and this is why thoroughly secular persons find it difficult to subscribe to the concept. Yet in the light of genocides, environmental disasters, nuclear build-ups and the threat of deadlier wars to come, a majority of secular and religious people around the globe agree that human persons are to be protected everywhere, if our race is to survive on this planet. My contention in this book has been that an intentional theology of creation with humans as God's trustees of the earth would go a long way in mobilizing Muslims, Christians and Jews for the daunting tasks ahead.

A Three-Chapter Summary

At the end of this diachronic case study of Qur'anic hermeneutics around the theme of Adam's caliphate, I first want to confess the shallowness of much of what was "unearthed" and presented here from the wealth of Islamic writings over the centuries. This is due, no doubt, partly to my own limitations and partly to the necessary limits imposed on such a study. Apart from some passing remarks, I have not explored how Sufism today might illuminate the human caliphate.[159] On the other hand, my

---

[159] The most influential Muslim scholar in this regard would be the prolific Seyyed Hossein Nasr. In a paper presented in 1967, Nasr declares that Islam is the religion of unity (*tawḥīd*) but that only Sufis have realized the true integration of Islamic rituals and law with the experience of the oneness of God in the human soul—a "state of purity and wholeness" ("Sufism and the Integration of Man," in *God and Man in Contemporary Islamic Thought: Proceedings of the Philosophy Symposium held at the American University of Beirut, February 6–10, 1967*, ed. with an Introduction by Charles Malik [Beirut: American University of Beirut Centennial Publications, 1972], 145). The human caliphate is mentioned early on, but it functions more as a philosophical underpinning than as a doctrine of any practical importance: "Man, being the vice-gerent of God on earth (*khalīfah*) and the theatre wherein the Divine Names and Qualities are reflected, can reach felicity only by remaining faithful to this nature, or by being truly himself" (ibid., 144). In a recent work (*The Heart of Islam: Enduring Values for Humanity* [New York: HarperSanFrancisco, 2002]), he elaborates more on the philosophical aspect mentioned above (pp. 12-15), and then on its implications for ecology: "Islam sees men and women as God's vicegerents on earth. Therefore, in the same way that God has power over His creation but is also sustainer and protector, human beings must also combine power over nature with responsibility for its protection and sustenance" (pp. 142-43).

hope is that by presenting a sampling of Islamic *tafsīr* on this topic of Adam and *khalīfa* side by side with a working model of the hermeneutical process, at least the following points have stood out:

1. The Qur'an is the self-declared event of God's dynamic interaction with the prophet Muhammad and his people and, as such, is an intricate "speech-act complex." There is much work ahead of interpreters, Muslim or non-Muslim, to sort out validity claims, reality domains and their mutual intertwining.

2. This sorting out cannot happen in an epistemological vacuum. Traditional interpretation was done from a naïve realist (or naïve idealist) perspective, and certainly much of contemporary Muslim apologetics as well. The same applies to some evangelical biblical interpretation as well. Meaning must be connected to truth (how a statement connects to reality), as well as truthfulness, intelligibility, appropriateness and rightness.

3. The interpretation of Adam and *khalīfa* over the centuries clarified the crucial role played by the interpreters themselves— the lens through which they examined the Qur'an. As a result, the questions they asked of the text were shaped by their worldview, religious convictions, socioeconomic and political contexts, the disciplines and respective methodologies chosen, and, undoubtedly, their own personal issues as well.

4. There is a renewal of interest in the hermeneutics of the Qur'an, not so much in the traditional context of *tafsīr* itself, but in the crucible of sociopolitical stirrings and ethnocultural strivings. This has at least two consequences: first, the dynamic reformism launched by Shah Wali Allah and Abduh has continued unabated, even in so-called islamist quarters; and second, continued dialogue on theologies of humanity might well be fruitful avenues for a truly constructive Muslim–Christian dialogue, both in the context of the new sociopolitical, economic and cultural landscapes of postmodernity, and in the intellectual landscapes affected by postmodernism.

Manifestly, my own conclusions amply substantiate Fritz Steppat's assertion that "now the conception of man as God's deputy on earth has achieved general recognition."[160] Beside the interesting fact of this verse's long and varied career lies the widespread realization by Muslims today that humanity's empowerment by God is seen as strengthening a core value in our contemporary civilization. And yet, for Adam's *khilāfa*

---

[160] Steppat, "God's Deputy," 166.

to be a constructive tool for human solidarity and peace, it must be unabashedly applied to humanity as a whole and not too closely linked to its *fiṭra* (depending on how that is interpreted).

It will also require from the Christian side a willingness to commit to a holistic approach that will not only deal with relief and development issues, but also include a strong stand for issues of economic and political justice and a commitment to work for peace. Further, holism will require an examination of local situations in their global setting and ask the difficult questions about Western hegemony, the limits of democracy in a world increasingly run by a few transnational companies, and the issues of true power-sharing and responsible management of the earth's resources.[161]

---

[161] See Ziauddin Sardar's more recent book on this, *Postmodernism and the Other: The New Imperialism of the West* (London: Pluto, 1998). Often strident in tone, he nevertheless makes a good case for postmodernism (my "postmodernity") as simply a continuation of the West's expansive hegemony under modernity—it is "but an extension of the grand western narrative of secularism and its associated ideology of capitalism and bourgeois liberalism" (ibid., 273). Hence much Western relief work in the south, especially under the sign of the cross, follows a cynical path: "self-aggrandisement, promotion of western values and culture, including conversion to Christianity, inducing dependency, demonstrating the helplessness of those they are supposedly helping and promoting what has been aptly described as a 'disaster pornography'" (ibid., 78). Though overstated (in my view), it is a needed warning to Christians currently involved in relief work in Iraq and Afghanistan.

Chapter 10

A BIBLICAL TRUSTEESHIP OF HUMANITY

This chapter concludes the second section of this work on interpreting scriptural texts—the human caliphate as seen in the Qur'an, and now, as viewed from the Bible. First, I present some of the salient exegetical questions relevant to the dominion of humanity in the Genesis 1 and 2 creation narratives. It entails, among other things, taking into serious consideration Gen. 1:26-31 as a literary unit and a discussion relative to its central concept, the "image of God." Second, I take a historical look at this passage's interpretive "career" among Jews and Christians. This leads me, thirdly, to engage the theological discussion, a process that will call to mind not only other passages, both in the Old and New Testaments, but even more, a framework of thought discernible through the whole sweep of biblical revelation. The conclusion shows both the extent and the limit of a common Muslim–Christian theology of humanity and creation.

Exegetical Issues of Genesis 1 and 2

Just as qur'anic scholars are beginning to appreciate the literary appropriateness of interpreting a sura as a whole, so biblical experts have begun to distance themselves from strict methods of form, source and redaction criticism and are turning to more holistic perspective: How does the received text appear as a literary unit?

The Literary Unit as a Whole
The first two chapters of the Bible offer two parallel narratives on the origins of the world and humanity. Ostensibly, their purpose is to answer the theological questions of the ancient Near East, not twenty-first-century scientific ones.[1] Though many commentators would say that the

---

[1] Jewish scholar Jon D. Levenson argues compellingly for the post-exilic authorship of the Gen. 1 creation account by tracing the uncanny parallels between older

second is more ancient, I will simply sidestep those critical questions here and focus on the received text, the opening of the Torah, which Jews and Christians consider to be, in some sense, "the Word of God." I take the text at face value and start out by asking, "what is affirmed here?"

Genesis 2:4 ("This is the account of the creation of the heavens and the earth," NLT) forms the hinge between the two stories. The second story includes the disobedience of Adam and Eve and their expulsion from the garden of Eden, whereas the first one is much briefer and carries no negative comments about humanity.[2] While the first story is more general ("Adam" as male and female, v. 27) and keeps to the chronological sequence of the six days of creation and comments on the Sabbath

passages dealing with creation and the Babylonian creation story, *Enuma elish*, in which Marduk is declared sovereign leader over all the gods after his defeat of the rebellious forces allied with the sea, Tiamat (*Creation and the Persistence of Evil: The Jewish Drama of Divine Omnipotence* [Princeton, NJ: Princeton University Press, 1988]). Marduk then creates the world as we know it, by splitting Tiamat's body in two: one half becomes the heavens and the other half the oceans (cf. the parallel in Gen. 1:7). Nowhere does the Bible describe creation out of nothing—not even Gen. 1 (where creation in fact begins with God's Spirit brooding of the formless chaos of primordial waters). Then in passages like Ps. 74, creation happens as God drives back the threatening waters and defeats the great sea monster Leviathan (vv. 13-14). The Flood story in Genesis ties this kind of creation as victory over forces of chaos to covenant (God will not allow another such flood for the sake of human beings). The post-exilic "Second Isaiah" recalls this divine deliverance of days past and calls God to action in his day in order to deliver his people according to his covenant with Moses (Isa. 51:7-13; 54:7-10).

[2] Donald E. Gowan, following the now traditional form-critical assignment of the first narrative to the "P source" (dated to the exilic period) and the second narrative to the earlier "J source" (early years of the monarchy, possibly during Solomon's reign), speculates that human sin needed highlighting much more at the height of Israelite hegemony in the region than when gloom and despair had set in after the fall of Jerusalem and the realities of exilic life. Whatever the reliability of these theories (divine authority does not rise or fall with Mosaic authorship), I would certainly rejoin him when he affirms the complementarity of the two narratives: "The sober presentation of J is needed again, whenever life seems firmly under human control, as is the message of P, when the world seems to have gone completely out of control" (*From Eden to Babel: A Commentary of the Book of Genesis 1–11*, ITC 1 [Grand Rapids, MI: Eerdmans, 1988], 32). He then quotes a wise Hasidic teacher, Rabbi Bunam, "A man should carry two stones in his pocket. On one should be inscribed, 'I am but dust and ashes.' On the other, 'For my sake was the world created.' And he should use each stone as he needs it" (ibid.)—a necessary reminder that, just as the main emphasis in the Qur'an is on human submission to God as servant/slave (ʿbd), in the Bible humanity is totally dependent on divine grace for salvation.

rest of the seventh, the second goes into the details of how Eve was created and presents the garden of Eden in some detail. In my exegesis I am assuming the theological complementarity of both narratives:

> The first narrative begins with chaos (1:2), the second with a barren desert (2:5); the first associates humankind with the animals, both being created on the sixth day (1:24-25); the second speaks of the man's being formed from the dust of the ground (2:7), the same ground that brought forth the animals; the first speaks of dominion over the creatures of the earth (1:26-27), the second of naming the animals (2:19).[3]

Importantly, both narratives specifically mention humankind's God-given dominion over the earth. In ch. 1 it is actually mentioned twice: "and let them have dominion over the fish of the sea, and over the birds of the air, and over the cattle, and over all the earth, and over all that creeps upon the earth" (1:26, RSV). And: "Be fruitful and multiply, and fill the earth and subdue it; and have dominion over the fish of the sea and over the birds of the air and over every living thing that moves upon the earth" (1:28, RSV). The concept of "dominion" is a royal term implying that human beings were placed by God on earth to rule it—precisely the qur'anic concept of humanity's caliphate. A more recent version (NLT) reads, "They will reign over the fish in the sea, the birds in the sky, the livestock, all the wild animals on the earth, and the small animals that scurry along the ground."

The second creation narrative finds manifest echoes in the Qur'an: "Then the Lord God formed man of dust from the ground, and he breathed into his nostrils the breath of life; and man became a living being" (2:7, RSV). This is followed by a description of the habitat God prepared for Adam, the Garden of Eden. Then v. 15 picks up again the creation story: "The Lord God took the man and put him in the garden of Eden to till it and keep it" (2:15, RSV). The text had already mentioned God's positioning human beings in the garden (v. 8), but now it explains why: "to tend and watch over it" (NLT). The first Hebrew verb, *'abad*, is a cognate of the Arabic *ʿ-b-d*, with the same meaning, "to serve," and thus, "to worship." The second verb, *shamar*, "carries a slightly different nuance. The basic meaning of this root is 'to exercise great care over,' to the point, if necessary, of guarding."[4] In fact, it is used in the next chapter

---

[3] Paul K. Jewett, *Who We Are: Our Dignity as Human: A Neo-Evangelical Theology* (Grand Rapids, MI: Eerdmans, 1996), 28. I lean heavily on the late Professor Jewett in this chapter, a professor at Fuller Seminary who I admired, particularly for his contribution on the doctrine of creation.

[4] Victor Hamilton, *The Book of Genesis: Chapters 1–17*, NICOT 1 (Grand Rapids, MI: Eerdmans, 1990), 171.

to describe the cherubs' guarding the access to the tree of life with their flaming swords (3:24). The other obvious parallel with the Qur'an is the mention of Adam naming the animals. However different the two versions may be,[5] they both underscore humanity's mission to rule over the created order.

Beyond the last verse's unequivocal assertion that work (here, physical labor) is hallowed and therefore not a result of sin, it also points to the empowerment of humanity to manage wisely (and "carefully") the natural resources of the earth.[6] It is gift, and therefore needs to be cherished, "something to be protected more than it is something to be possessed."[7] Naturally this reverence for that which has been given to us from God means that to discharge our duty to "care for" is a deliberate act of worship. This obvious parallel to the Qur'an can be seen, not only in the use of the Semitic root "to serve/worship," but also in the many depictions of events in the natural realm as *ayāt* ("signs" from God).[8]

However luminous the qur'anic and biblical kinship may be at this level, the empowering of the human being to rule over the created realm cannot be understood in the Genesis 1 narrative apart from his or her creation "in the image of God."[9] But first, I offer a comment on the literary construction of the passage.

Stylistic Remarks Concerning Genesis 1:26-31
From a glance at the first narrative as a whole (1:1–2:4) one has to notice that in the seven-day creation narrative the second part of the sixth day (creation of humans) stands out from a literary perspective:

---

[5] Cf. again n. 14 in Chapter 5.

[6] Victor Hamilton remarks on the gender of "it" in "to tend and care for it." "Garden" in Hebrew is masculine yet the "it" here is feminine, as is the word for earth. He sees this as indicating the author's deliberate widening of the task to that of the earth and not simply the original garden (*The Book of Genesis*, 171 n. 1).

[7] Ibid.

[8] These passages almost always include a call to the reader to discern this fact and react appropriately, either by the expression of thanks to God (*shukr*) or by careful meditation (*tafakkur*). As Cragg notes, these injunctions serve to temper humanity's divine calling to trusteeship on earth. Recall this quote from him: "They clearly belong with the double vocation to mastery and to worship which constitutes man as the *khalifah* or viceroy. The sustained emphasis they receive in the qur'anic doctrine of the human standing is to be understood as the claim that dignity makes and the due condition of its exercise (*The Privilege of Man*, 34).

[9] In this sense, Jewett is perfectly justified in naming the last section of his monograph dedicated to the dignity of humanity in the Bible as "The Divine Image and the Dominion of Humankind" (*Who We Are*, 351-61).

[26] Then God said, "Let us make[10] man in our image, after our likeness; and let him have dominion over the fish of the sea, and over the birds of the air, and over the cattle, and over all the earth, and over every creeping thing that creeps upon the ground according to its kind. And God saw that it was good. [27] So God created man in his own image, in his own image he created him; male and female he created them. And God blessed them, and God said to them, [28] "Be fruitful and multiply, and fill the earth and subdue it; and have dominion over the fish of the sea and over the birds of the air and over every living thing that moves upon the earth." [29] And God said, "Behold, I have given you every plant yielding seed which is upon the face of all the earth, and every tree with seed in its fruit; you shall have them for food. [30] And to every beast of the earth, and to every bird of the air, and to everything that creeps on the earth, everything that has the breath of life, I have given every green plant for food." [31] And God saw everything that he had made, and it was very good. And there was evening and there was a morning, a sixth day. (RSV)

God's work of creation on the sixth day is retold in much greater detail than the creative acts of the other five days. The general literary pattern that precedes is respected: "Each creative act begins with an announcement ('and God said'), followed by a command ('let there be…'), a report ('and it was so'), an evaluation ('God saw that it was good'), and a temporal framework ('the *n*th day')."[11] Yet there is one glaring exception. Instead of a command (in the jussive), the reader encounters either self-deliberation or consultation (the cohortative "Let us make man…").[12]

The next noticeable feature is that the rhythm of the narrative changes. John Skinner notes that "[a]s the narrative approaches its climax, the style loses something of its terse rigidity, and reveals a strain of poetic feeling which suggests that the passage is moulded on an ancient hymn."[13]

---

[10] The Hebrew verb used here (*br*ᵓ) is different from the one used to denote God's creation of all the animals (*ᶜsh*)—a fact that further underscores the higher status of humankind in this narrative (Cohen, *"Be Fertile and Increase,"* 12). Incidentally, the Hebrew here is a cognate of one of the Arabic verbs used for creation in the Qur'an (*br*ᵓ; one of God's 99 names is *al-Bāriᵓ*).

[11] Edward M. Curtis, "Image of God (OT)," in *The Anchor Bible Dictionary*, vol. 3, ed. David Noel Freedman (New York: Doubleday, 1992), 390, emphasis original.

[12] More on this below.

[13] John Skinner, *A Critical and Exegetical Commentary on Genesis*, vol. 1 (Edinburgh: T. & T. Clark, 1910), 30. Curtis lists other factors in the text which indicate the monumental importance of the creation of humans along with the greater volume of material and the slowing down of the rhythm: "the threefold repetition (in 1:27) of the word *bara*ᵓ, "to create (a word reserved in the Hebrew Bible for God's creative activity), the fact that humanity is given dominion over the rest of creation, and the evaluation 'very good' that follows the creation of man and

Though Skinner's guess about the ancient hymn might be hard to prove, the poetry of v. 27 is manifest. "The three clauses are in apposition. The first two are arranged chiastically and emphasize the divine image in man, while the third specifies that women also bear the divine image."[14]

It is clear that the Hebrew *adam* in the first clause means "humankind," while in the next clause we find a Semitic chiastic parallelism, "in the image of God he created him," thus powerfully reinforcing a concept that to the strictly monotheistic Hebrew mind would have seemed shocking. The last phrase, then, unpacks the *adam* with relation to the image, asserting that the resemblance of humanity to God is somehow connected to their sexual differentiation. The other implication is simply that male and female equally participate in this divine connection, which at the same time separates them from the animals. All people, therefore, both male and female, equally receive the commands to have dominion. This is made clear as well by the next verse, which presents God's blessing of fertility.

Further, the two verses containing the dominion command are on either side of the verse affirming humanity's creation in the image of God. Already the first mention of humanity's creation in v. 26 is related to the "image": "Let us make man in our image." What reinforces even more this sense that the "image" is connected to the calling and empowerment of humans to take charge over creation is that vv. 26 and 28 both record the voice of God whereas the verse in between brings in the narrator's voice—a dramatic commentary in the middle of this cosmic narrative indicating that what is being enacted here is indeed the climax of creation and the key to its interpretation.[15]

The Problematic Plural: "Let us make…"
There has been much discussion about this first person plural of the first verb announcing *adam*'s creation ("let us make"). Gordon J. Wenham

woman… It is man and woman alone who are said to be created in (or as) God's image, and this appears to account for humanity's preeminent position in the created order" (ibid., 390).

[14] Gordon J. Wenham, *Genesis 1–15*, WBC 1 (Waco, TX: Word, 1982), 33.

[15] Hamilton brings attention to this fact and then offers the following explanation: "Perhaps the use of the third person singular pronominal suffix is deliberate and undercuts the possibility of any misunderstanding of the 'our' in v. 26. May this be the writer's way of saying that when man was created in the image of *elohim*, he meant 'God' and not 'divine counsel'? If the narrator had meant the latter, then we would expect, 'so God created man in *their* image'" (*The Book of Genesis*, 138, emphasis original). In what follows I show how Moltmann takes this one step further.

lists six positions taken by commentators over the centuries. At the outset, I will dismiss the interpretation of some of the early Church Fathers who read into this text a hidden formulation of the Trinity—a tempting hermeneutical short-cut.[16] However, the last view in Wenham's list is that of D. J. A. Clines, presented in his landmark essay, "The Image of God in Man" (1967, but still quoted in the literature). It relates somewhat to the above view but only has the Spirit involved—a view, I might add, he suggests "with hesitation":

> [W]e suggest therefore that God is addressing His Spirit, who has appeared in verse 2 in a prominent though usually little understood role (it is not simply a 'mighty wind'), and has curiously disappeared from the work of creation thereafter. In other Old Testament passages, however, the Spirit is the agent of creation, e.g. Job 33:4: 'The spirit of God has made me, and the breath of Shaddai gives me life'; Psalm 104:30: 'When thou sendest forth thy spirit they (animals) are created'; cf. also Ezekiel 37 (valley of dry bones and the recreating spirit).[17]

Though I do not necessarily retain this view about v. 26, Clines is certainly correct concerning the role of the Holy Spirit in creation. I would suggest, however (and he would agree), that the writer did not have a developed theology of the Spirit of God as a distinct person in the Godhead.[18] This is only discerned from Scripture as a whole. As seen before,

---

[16] According to Wenham, "Christians have traditionally seen this verse as adumbrating the Trinity. It is now universally admitted that this was not what the plural meant to the original author" (*Genesis 1–15*, 27). Here is one good reason to get "behind" the text and use the tools of historical criticism.

[17] D. J. A. Clines, "The Image of God in Man," *Tyndale Bulletin* 19 (1968): 69. In a classic study, Thomas O'Shaughnessy traces the different uses and meanings of the word "spirit" (*rūḥ*, akin to *ruaḥ* in Hebrew) in the Qur'an and finds that the evolution in the meanings closely parallel the origin of the suras considered. It is precisely in suras of the Second Meccan Period (based on Blachère's chronology) that *rūḥ* is used in connection with creation (and the virgin birth of Jesus in Mary's womb), for example, "Thy Lord said to the angels: See, I am going to create mankind from clay: so when I have formed him and breathed into him some of my spirit, fall down to him in obeissance" (Q. 38:72). "From a personal being set above the angels [in the First Meccan Period], the spirit is now transformed into an impersonal thing, a breath of life, originating with Allah and animating the human body" (*The Development of the Meaning of Spirit in the Koran* [Rome: Pontifical Oriental Institute, 1953], 25).

[18] This view cannot be dismissed out of hand, however, since some scholars, including Hamilton, have adopted it: "The best suggestion approaches the trinitarian understanding but employs less direct terminology... True, the concept may not be etched on every page of Scripture, but hints and clues are dropped enticingly here

theology rightly impacts one's hermeneutic, but only in allowing the totality of revelation to inform one's interpretation of earlier passages— the *sensus plenior*.

I will mention three other views, mainly because of the possible parallels they raise concerning the Qur'an. In the end, it will become apparent how this is connected to the caliphate of humanity. The first posits a self-deliberation in God, or even a "self-encouragement" (parallel to Gen. 11:7, and Ps. 2:3).[19] Following W. H. Schmidt, Jürgen Moltmann calls this God's "resolve."[20] This view, now widely held among commentators, offers some close parallels with the qur'anic discourse that has God speaking sometimes in the first person singular, sometimes the third person singular, and sometimes in the first person plural. This position is assumed by a theologian like Paul K. Jewett: "Although one among many creatures, humans are somehow different. When God made the human species, he began by taking counsel with himself—'let us make humankind.'"[21] And for him this "pause" in the divine project of creation is a way of showcasing the human dignity:

> As a diamond cutter contemplating a priceless gem pauses before the initial stroke, so the Creator paused, as one especially engaged in what he was about to do (1:26). Furthermore, he did not simply command the dust to bring forth but stooped to gather it in his hands that he might form the man (2:7). Then, taking a rib from his side, God made the woman as man's counterpart, the "helper fit for him" (2:18-22). By such studied deliberation and intimate involvement, the Creator commends to us the dignity of our nature as human.[22]

The second view involves some form of "plural of majesty"—somewhat akin to the English "royal we." Though this suggestion has been largely discounted among Bible commentators, it might well be the most popular in Muslim scholarly circles.[23] Most intriguingly, it is adopted by

---

and there, and such hints await their full understanding 'at the correct time' (Gal. 4:4)" (*The Book of Genesis*, 138).

[19] Wenham, *Genesis 1–15*, 28.

[20] In fact, Moltmann offers a very useful outline of Gen. 1:26-30: "introduction ['And God said'], resolve [rest of v. 26], creation of human beings [v. 27], blessing and commission [v. 28], provision [vv. 29, 30]" (*God in Creation: A New Theology of Creation and the Spirit of God* [San Francisco: Harper & Row, 1985], 217).

[21] Jewett, *Who We Are*, 28.

[22] Ibid.

[23] Wenham writes that this view, which shows that Hebrew verbs cannot be used in this fashion, has now been discredited (*Genesis 1–15*, 28). Concerning the Qur'an, see Robinson's chapter, "The Dynamics of the Qur'anic Discourse," in his

the author of the first Hebrew grammar and the first full (extant) transla-tion of the Hebrew Bible into Arabic, the Jewish theologian, philosopher and philologist of tenth-century Baghdad, Sa'adya b. Yosef: "It says 'We shall make man' in plural for this is for aggrandizement and honor as is the practice of the language of the Arabs that the king and the official and the distinguished person say: 'We commanded,' 'We said,' and 'We did.'"[24]

The third view is also one of the oldest and still probably the most widely held. "From Philo onward, Jewish commentators have generally held that the plural is used because God is addressing his heavenly court, i.e., the angels (cf. Isa 6:8)."[25] This view does not entail the notion that in some measure the angels are co-creators with God, but simply that like the cherubim guarding the Garden in 3:24 and the "sons of God" in 6:2, the contemporary Hebrew worldview contemplated God making deci-sions in the presence of a "heavenly host."

This idea is present in the first chapter of Job, in the Psalms, but also in the Revelation of John in which world history unfolds in the presence of the "four living beings," the "twenty-four elders" (presumably the leadership of Israel and the Gentile church), "thousands of millions of angels," and finally "a vast crowd, too great to count, from every nation, tribe and people and language, standing in front of the throne and before the Lamb" (Rev. 7:9, NLT).

My conclusion in this section is not to opt firmly for one position or another—probably the first and third both apply at different level—but to

---

Discovering the Qur'an. For him, the first person plural "emphasizes His majesty and power," the explanation Wenham says has now been dismissed for such instances in the Hebrew Bible (where these occurrences are much rarer than in the Qur'an) (ibid., 246). I would think that a deliberative view of this discourse could also be advocated, at least in some instances.

[24] Michael Linetsky, *Rabbi Saadiah Gaon's Commentary on the Book of Creation* (Northvale, NJ/Jerusalem: Jason Aronson, 2002), 107-8. Linetsky was born in the region just south of Cairo in 882, but moved to Baghdad in 921, where he became the "Chief Scholar" (Gaon) of the Jewish academy of Sura.

[25] Wenham, *Genesis 1–15*, 27. Concerning the "we" discourse in the Qur'an, Robin-son argues that he can only find two passages in which the "implied speaker" is the plurality of the revelatory angels (19:64; 37:164-66). He then says, "Attractive though is may seem, the suggestion that the Qur'anic 'We' should be interpreted in the light of the Hebrew Bible, as the self-designation of God and His angels speak-ing in unison, does not stand up to scrutiny... The Qur'an thus appears to correct biblical theology and angelology rather than to adopt it uncritically" (ibid., 237). On the other hand, it is clear that God is addressing the heavenly beings in assembly in the key passage of this study, Q. 2:30.

point to the necessary fluidity and therefore creativity of the hermeneutical task. On the one hand, in the context of the original writing of
this creation narrative the heavenly court was in view. But something
more powerful "took over" the text itself so that both Clines's suggestion
about God addressing the Spirit and that of God's self-deliberation are
both imbedded in the text. Consider this remark by Moltmann on the role
of the word of God in this narrative. It will serve as a useful introduction
to the discussion of the *imago dei* ("image of God") in the next section:

> When light is created, we read: "He said...and there was" (Gen. 1:3).
> When the animals are created, we read "He said...he created" (Gen.
> 1:20f.). But when human beings are created, the passage reads: "Let us
> make... So God created." Human beings come into being, not through
> God's creative word but out of his special resolve. The word which pre-
> cedes the resolve is addressed by God to himself. It is self-exhortation.[26]
> In a resolve, the author of the resolve acts on himself first of all. He
> resolves 'for himself' before he acts on anyone or anything else. In the
> self-exhortation we have here, God designates himself to be the Creator
> of his image before he creates that image. 'God resolves for himself.'
> Inherent in this resolve is God's *contraction* to this single possibility, and
> already inherent in this contraction is also a first self-humiliation on
> God's part. This is apparent from the fact that God 'implants' his image
> and his glory in his earthly creation, the human being, which means that
> he himself is drawn into the history of these creatures of his.[27]

The Imago Dei

I have no space to present such a wide topic except in its briefest form.
Yet I must mention it, for at least two reasons. First, it is intimately connected to the idea of humanity's dominion over the created order (Gen.
1:26); and second, it is inevitably discussed in the context of Muslim–
Christian dialogue.[28] And, surprisingly, there is much more convergence

---

[26] As noted earlier, this is not the majority view. Most still prefer the "heavenly
court" hypothesis. Gowan, for one, strongly opposes Karl Barth's understanding of
the divine plural as pointing to "a concert of mind and action in the divine being
itself," which he sees as having "no support in the OT" (*From Eden to Babel*, 29).
This does have a direct bearing on the meaning of humanity being created in the
image of God. Moltmann and Jewett, for instance, following Barth and Dietrich
Bonhoeffer, make much of the "I" and "Thou" of the human as male and female in
community (Gen. 1:27) reflecting the "I" and "Thou" in the divine self-consultation
in the previous verse. Jewett also sees Gen. 2:18 as a divine self-deliberation (*Who
We Are*, 161).

[27] Moltmann, *God in Creation*, 217, emphasis original.

[28] Briefly put, Islam's strict monotheism naturally recoils from any connection
between Creator and creature. Yet, as noted in Chapter 7, mystical writers such as al-

(at least potentially) on this issue between the two faiths than is generally believed—as shown by Jewett's astute observation:

> Thus we are taught that ours is a privileged place in the created order, a place that brings with it extraordinary responsibility. We are the stewards of creation, vested with authority as the vicegerents of the Creator. This dignity, worth, and responsibility with which the biblical story of creation invests the human creature comes to its sharpest focus, for theologians, in the concept of the image of God (the *imago dei*). It is the gift that grounds the I–thou relationship we have with God and with one another.[29]

The first appearance of the *imago dei* is in v. 26, which offers two different words: "Let us make man in our image, after our likeness" (RSV). The Hebrew word translated here by "image" is *tselem*, a difficult word to define, mainly because its root is never used as a verb in the Hebrew Bible. Besides ten occurrences that refer to a physical image and two others in the Psalms that compare humanity to an image or shadow (39:7; 73:20), the other five instances are in Genesis.[30] The only two references outside the creation account are 5:3 ("Adam became the father of a son in his own likeness," RSV), and 9:6 ("Whoever sheds the blood of man, by man shall his blood be shed; for God made man in his own image," RSV).

The word translated by the RSV by "likeness" is *demut*, from a root meaning "to be like," "to resemble." From Irenaeus on, Patristic exegesis tended to differentiate between the image and the likeness, the Latin terms, *imago* and *similitudo*.[31] But that such a distinction is artificial can easily be shown by quoting Gen. 5:1, which summarizes creation just before the Flood in these terms: "When God created man, he created him in his likeness (*demut*)." The two words are plainly interchangeable.[32]

Ghazali and Ibn al-Arabi made much of this image, largely because this doctrine is explicitly taught in the *ḥadīth* (cf. Chapter 7, n. 100) and thus makes possible the quest for union with God. In the following discussion I am proposing that this doctrine need not be an embarrassment to Muslim theologians, in that it actually underscores and amplifies the theme of humanity's caliphate without injuring God's transcendence.

[29] Jewett, *Who We Are*, 29.

[30] Wenham, *Genesis 1–15*, 29.

[31] "The image refers to the natural qualities in man (reason, personality, etc.) that make him resemble God, while the likeness refers to the supernatural graces, e.g., ethical, that make the redeemed godlike" (Jewett, *Who We Are*, 29).

[32] The second phrase "after our likeness" has often been interpreted as a toning down of what must have seemed very shocking (humanity as God's *tselem*) to the Hebrew monotheistic ears. Moltmann is unusual in his distinguishing the two: "The first of these terms is used for the concrete representation, the second is used for the

Wenham cites four main views of the *imago dei*:

1. The image consists of one or several faculties in human beings that differentiate them from the animals: reason, free will, self-consciousness, intelligence or personality. The difficulty here is that this concept is nowhere else addressed theologically in the Old Testament. Wenham quips, "In every case there is the suspicion that the commentator may be reading his own values into the text as to what is most significant about man."[33]

2. The image refers to a physical resemblance between God and man. If Adam's son is said to be in the image of his father, one could argue that there must be a physical element to this image. Several commentators have also held that the human upright posture is what marked them from the beginning as distinct from the animal realm. Of course this idea represents a very problematic interpretation for all three monotheistic faiths. The Hebrew Bible certainly does stress the invisible and incorporeal nature of God. Another argument against this comes from the near certainty today of this idea as originating in Egyptian and even possibly Mesopotamian culture of ancient times, in which the king was said to be in the likeness of his god. This of course had nothing to do with physical appearance but rather established the king's identity and defined his function.[34]

---

similarity. The first expresses more the outward representation, the second rather the reflexive inward relationship." Yet he immediately rejoins the majority opinion in the following phrase: "Both terms have probably been borrowed from Egyptian royal theology" (*God in Creation*, 219).

[33] Jewett, *Who We Are*, 30.

[34] Ibid. The first part of this sentence has to be somewhat qualified. Clines comments on the issue of anthropomorphisms in the Old Testament (a problem that has preoccupied Muslim commentators on the Qur'an as well): God's "hands," "ears," "eyes" and the like. He notes, "Such anthropomorphisms cannot easily be dismissed as merely metaphors, since everywhere else in the Ancient Near East these terms were understood to be literally true of the gods, and it is difficult to believe that Israel would have run the risk to faith of using such terminology if she had believed that Yahweh was pure spirit, without parts or passions. Nevertheless, it is significant that the anthropomorphisms used of Yahweh in the Old Testament do not enable us to construct an identi-kit picture of Yahweh's physical appearance, as is the case, for example, with Greek deities described in Homer, but rather concentrate attention on the personhood of Yahweh. Yahweh is depicted in human terms, not because He has a body like a human being, but because He is a person and is therefore naturally thought of in terms of human personality" (Clines, "The Image of God in Man," 70-71). God does appear to several people (prophets and non-prophets, some women included) in the Hebrew Scriptures and so takes on a human appearance. Here Clines

3. The image is related to humanity's vocation to represent God on earth. This in fact was the common perception of the ancient Near Eastern king in the second millennium BCE. In his seminal essay, Clines outlined this position that is still debated today, four decades later.[35] He began by noting that the preposition "in" ("in our image," v. 26) at first sight seemed to call for a meta-phorical meaning, but then shows at length (especially in com-parison with Accadian texts) how concrete the term must be taken. "It is not impossible that we should have here a vivid metaphor unparalleled elsewhere, but the linguistic evidence would suggest that it is most unlikely that *tselem* means anything but a form, figure, object, whether three- or two-dimensional.[36]

He then shows that a more satisfactory way of handling the *beth* ("in") would be to consider it a "*beth* of essence, meaning 'as,' 'in the capacity of.'"[37] After answering four main objections to this view, he concludes,

> [O]ur conclusion is that Genesis 1:26 is to be translated "Let us make man as our image" or "to be our image," and the other references to the image are to be interpreted similarly. Thus we may say that according to Genesis 1 man does not have the image of God, nor is he made in the image of God, but is himself the image of God.[38]

On this basis, then, he asks, what can this image have meant in the ancient Near East? He answers, first of all, that "the primary function of the image was to be the dwelling-place of spirit or fluid which derived

---

follows James Barr in saying, "when Jahweh does appear in a form, the human form is the natural and characteristic one for Him to assume" (ibid., 73). Christians will naturally point out the precedent for the incarnation. Yakub Zaki makes a surprising parallel between the 99 names of God, or his attribute, and the Christian doctrine of incarnation: "The Attributes are the clue to the Islamic understanding of God, for a god that was completely transcendent would be an agnostic god. Therefore the Attributes supply the *analogia entis* [analogy of being] otherwise supplied in Christianity by the Incarnation" ("The Qur'an and Revelation," 52).

[35] Levenson informs us that a crucial piece of research had been published two years earlier (Hans Wildberger, "Das Abbild Gottes, Gen. 1:26-30," *Theologische Zeit-schrift* 21 [1965]: 245-59) demonstrating from an impressive array of Egyptian and Mesopotamian sources that "[t]he link between the creation of humanity 'in the image of God' in Genesis 1 and their status as royalty can be clearly seen in ancient Near Eastern inscriptions in which it is the king who is described as the 'image' of the deity" (*Creation and the Persistence of Evil*, 114).

[36] Clines, "The Image of God in Man," 75.

[37] Ibid.

[38] Ibid., 80.

from the being whose image it was."[39] This fact is of primary interest to the biblical and qur'anic texts:

> As a bearer of spirit, the image is consistently regarded and treated as a living being. After it has been completed by the workman, the image is ritually brought to life by touching mouth, eyes, and ears with magical instruments. In Egypt the day begins with the call of the priest to the image "Wake in joy!" The little chapel in which the image has been shut up for the night is opened. In Babylonia also the images are awoken, dusted and washed, sometimes bathed in the sea; then a large breakfast is brought to the image, and so the day continues.[40]

The second parallel to be drawn (mentioned above) is that kings were themselves considered the image of their god.[41] Edward M. Curtis concurs: "There are indications in Gen. 1:26-28 that the 'image of God' terminology perhaps had its origins in the royal ideology of the ancient Near East." He follows this up with a conclusion espoused by many scholars today: "The idea of dominion and the idea of subduing are most appropriate in the context of kingship."[42] Wenham sees the royal character of humanity's appointment beautifully reflected in Psalm 8, "which speaks of man as having been created a little lower than the angels, *crowned* with glory and made to *rule* the works of God's hands."[43] Jewett is attracted by this hypothesis, but offers a cautionary note. God indeed placed human beings on earth as his representatives and "vicegerents,"[44] much as earthly rulers used to command "images of themselves to signify their authority in provinces where they did not personally appear."

---

[39] Ibid., 81.

[40] Ibid., 82. Curtis, in his Anchor Bible article on the "Image of God (OT)," follows this line: "The significance of the image did not lie in the way it described or depicted the god (though that was not totally unimportant); rather, it lay in the fact that the statue was a place where the deity was present and manifested himself. Thus, the presence of the god and the blessing that accompanied that presence were effected through the image. It was the function of the image rather than its form that constituted its significance" (p. 390).

[41] Clines quotes a seventh-century Assyrian king, Esarhaddon, as being addressed in the following letter as the image of Bel: "The father of the king, my lord, was the very image (*salmu*) of Bel, and the king, my lord, is likewise the very image of Bel" ("The Image of God in Man," 83).

[42] Curtis, "Image of God (OT)," 391. One has to note, however, that a great widening has now taken place in the biblical text—a democratization of sorts. In Genesis, God proclaims the whole human race to be his image bearers, not just an elite or simply a king.

[43] Wenham, *Genesis 1–15*, 30, emphasis original.

[44] This is the only instance I found of this term being used by a Christian theologian. I still find it an awkward English term, however.

Nonetheless, continues Jewett, "we would prefer to say not that this responsibility and privilege constitutes the image, but rather that it rests upon it. That is, that the divine image is gift; the dominion of humankind is the exercise of the powers with which they are vested by God when he created them in his own image."[45] Could this not have been written by a Muslim? I am bound to point out here that Muslims and Christians together share the same calling: the divine image imprinted in them also impels them to fulfill their solemn caliphate/dominion over creation. At the same time, the emphasis is on the accountability of humans to God for the way in which they discharge this responsibility.

Two last comments by Clines demonstrate the richness of the cultural background to the *imago dei*.[46] The first is that it underscores the consistency of the Hebrew theological anthropology: "Man according to the Old Testament is a psychosomatic unity; it is therefore the corporeal animated man that is the image of God. The body cannot be left out of the meaning of the image; man is a totality."[47] Creation is the basis for understanding the hope of the resurrection to come. It is not that humans are the image of God because he himself has a body, but that they represent him "in a place where he is not. If God wills His image to be corporeal man—union of physical and spiritual (or psychical)—He thereby wills the manner of His presence in the world to be the selfsame uniting of physical and spiritual."[48] This not only opens the way for the incarnation, but also closes it to all dichotomies between secular and sacred.

The second implication Clines sees in this borrowing of meaning from its wider cultural context is that the image was to bring the god's presence wherever it went, including in enemy territory. Likewise, "The king puts his statue in a conquered land to signify his real, though not his physical, presence there."[49] Thus in Genesis God's transcendence is clearly established, but at the same time his immanence is guaranteed through the work of his representatives. God freely creates the world and stands totally above it and distinct from it. The human being, on the other hand, is totally creature and therefore cannot be divinized. Though there is no contradiction between God's transcendence and immanence in the

---

[45] Jewett, *Who We Are*, 351.

[46] Here is a plain case of theologizing within a particular worldview. The content of the theology is modified, to be sure, but the cultural forms are used to communicate divine truth. This hermeneutic is an indication for us who live in a very different world to rethink the implications of this concept for today.

[47] Clines, "The Image of God in Man," 86.

[48] Ibid., 87.

[49] Ibid.

text, there is nevertheless a tension, similar to that which is inevitably felt in Islamic theology in the relationship between God, his earthly deputies, and his eternal Word, the Qur'an (deeply steeped as it is in the culture of the Prophet's Arabian milieu).[50] Clines quotes Bonhoeffer here: "The only continuity between God and his work is the Word. But from the sixth day of creation onward man, the image of God, becomes the continuity."[51] Might we not say, "And the Word became flesh"? It is certainly an intriguing pointer in that direction.[52]

4. Wenham lists a fourth option for the *imago dei*. This view, argued most eloquently by Barth, sees in the image the human being's capacity to hear, respond to, and make a covenant with God.[53] Hence, it is similar to the first view in that humankind is taken to resemble God in a certain way. But now it is not so much a quality but a capacity to be in relationship with the Creator. Ray S. Anderson, for example, in a very Barthian way, sees "being in the image of God" in three dimensions. The first is "freedom in dependence": "Human freedom is not a freedom from that which binds the self, but a freedom for that which

---

[50] This is according to what in fact became orthodoxy, the Ash'arite doctrine. The Mu'tazilites, on the other hand, argued that the Qur'an was created.

[51] Ibid., 88.

[52] William A. Dyrness makes this point even more strongly. Following other recent theologians who are beginning to realize the limits of a narrative theology that is simply *diachronic*, he calls for the wedding of time and space which gives theology its *synchronic* dimension. Thus God's commitment to creation calls his human creatures to participate in his design by shaping their culture (including art) in a way that reflects his purposes. In other words, God's purposes for his creation were embodied in particular ways, creating the logic and beauty which already foreshadowed his project of incarnation—regardless of the problem of sin. In his words, "the account of the embodiment of God's purposes in Genesis 1, which is itself a kind of progressive incarnation leading to God's own image in the man and woman, is evaluated by God as 'very good.' Throughout the Old Testament God continues to indwell creation in various ways, supremely in the Temple of Zion—which is clearly intended to be a microcosm of the whole of creation—where God dwells in righteousness. But now, because of the rebellion of the creature, God's presence must be intentionally directed toward atonement" (*The Earth Is God's: A Theology of American Culture* [Maryknoll, NY: Orbis, 1997], 22).

[53] Dutch theologian Johannes S. Reinders argues this, leaning on Greek Orthodox theologian John D. Zizioulas. Ontology is not rooted in "being in itself," but rather in the fatherhood of God, thus "being in communion" ("*Imago Dei* as a Basic Concept in Christian Ethics," in *Holy Scriptures in Judaism, Christianity and Islam: Hermeneutics, Values and Society*, ed. Hendrik M. Vroom and Jerald D. Gort [Amsterdam: Rodopi, 1997], 202).

determines the self... It is the freedom to be for God... Adam is also free for creation, and for his own creatureliness as the object of the Creator's determination."[54]

Anderson's second dimension is the "responsibility in hearing." The word of God addressed to Adam not only presupposes that Adam can hear and respond, but in a real sense elevates him as the only creature who can do so. Just like Lazarus, who upon hearing his name comes back to life, Adam becomes a living, fully human person as he hears the word of God and responds.[55]

The last dimension is the "differentiation in unity."[56] This phrase, "male and female he created them," is not simply to state that under the genus "human" come two sub-divisions, "male" and "female," or merely to set the stage for the upcoming blessing, "Be fruitful and multiply!" (as many commentators would have it). This shift from singular ("man") to plural ("male and female") is rather, as Moltmann puts it, because "to be human means being sexually differentiated and sharing a common humanity; both are equally primary."[57] Humanity's shared caliphate, then, takes on a richer theological meaning—from sexual differentiation to community:

---

[54] Ray S. Anderson, *On Being Human: Essays on Biblical Anthropology* (Pasadena, CA: Fuller Seminary Press, 1982), 80.

[55] This is where I would part ways with the Barthian approach. The *imago dei* is tied too closely to sin and redemption and therefore looses its moorings in a positive doctrine of creation that is maintained even after the fall. In this perspective there is no possible dialogue with a Muslim or a Jew. I also believe that Jesus Christ is the perfect image of God in humanity as the Second Adam. Yet the calling of humanity as a whole to function on earth as God's deputies is not vitiated by the Fall. Jewett's approach, however Barthian it may be, seems more sound to me: "We shall seek to understand the doctrine of the image by working from creation to redemption... It is not just that sin is the logical *prius* of salvation, and creation in the image the logical *prius* of the Fall; but that Scripture always supposes the three—creation, fall, and salvation—to be related in terms of *temporal* succession... Our approach, therefore, will reflect this temporal structure" (*Who We Are*, 60-61, emphasis original).

[56] It is not only that the human person becomes aware of her*self*, but she does so in seeing her reflection in the male person standing across from her, and in the ensuing dialogue discovers more about herself: "This means that the differentiation must necessarily be found in the creaturely form of male and female sexuality... Male and female expresses the polarity in terms of co-humanity" (Anderson, *On Being Human*, 86).

[57] Moltmann, *God in Creation*, 222. This is also a common creation theme in the Qur'an, though without the explicit reference to the image of God (e.g. Q. 49:13; 53:45; 75:39; 92:3).

> Sexual difference and community belong to the very image of God itself; they are not merely related to human fertility. So this community already corresponds to God, because in this community God finds his own correspondence. It represents God on earth, and God "appears" on earth in his male–female image. Likeness to God cannot be lived in isolation. It can be lived only in human community… The isolated individual and the solitary subject are deficient modes of being human, because they fall short of likeness to God. Nor does the person take priority over the community. On the contrary, person and community are two sides of one and the same life process.[58]

A similar point is made by Yale theologian Miroslav Volf. An individual becomes a human person by being addressed by God, his or her Creator, a person at the same time unique and necessarily "embedded in a network of multiple and diverse social and natural relationships." A human being cannot exist in isolation. Volf adds, "This may be the anthropological significance of the peculiar transition in Gen. 1:27 from singular to plural: 'So God created humankind in his image, in the image of God he created *him*; male and female he created *them*.'"[59]

But such a theology which can bring together the individual dignity of the person and its essential social vocation is only possible because through the lens of the whole sweep of the Christian canon one may discern the triune God who calls humanity to be his image on earth. The eternal and dynamic interrelationship of love between Father, Son and Holy Spirit is the wellspring of life and love between persons. It is at once the model and the power that enables humans to live out their calling as God's trustees in the world. Male and female are not only equal and complementary; rather, people of all races, classes and abilities are to embody together God's image on earth: "they rule over earthly creatures as God's *representatives* and in his name; they are God's *counterpart* on earth, the counterpart to whom he wants to talk, and who is intended to respond to him; and they are the *appearance* of God's splendour, and his glory on earth."[60] This means also that issues of justice, peace and

---

[58] Ibid., 222-23.

[59] Miroslav Volf, *After Our Likeness: The Church as the Image of the Trinity* (Grand Rapids, MI: Eerdmans, 1998), 183. Volf comments, "On the basis of personhood as grounded in God's creative relationship with human beings, human beings are in a position not simply of having to submit passively to their social and natural relations, but of being able to integrate them creatively into their own personality structure. Without this interactive–integrative activity of being a subject, human beings would exist merely as the reflection of their relationships" (ibid., 184).

[60] Moltmann, *God in Creation*, 221, emphasis original. Jewett offers a moving account of the mentally disabled in this regard. He distinguishes between personhood

equality, are right at the core of the biblical view of the human person. The dominion of humanity on earth is a gift from God, and people are therefore accountable to him for the way in which they treat both the physical environment and one another.[61] Striking a parallel note, the Qur'an says, "O people, we have created you from a male and female and divided you into nations and tribes so that you would get to know one another. Truly the most righteous among you is the most honored. For God is the most wise and knowledgeable!" (Q. 49:13, my translation).

Having gone through Wenham's list of possibilities for the interpretation of the *imago dei*, I must say that I cannot dispense entirely with any one of them, though the last two seem more compelling. This is partially

---

and personality: "As we see it (and we can do no more than give our opinion), those who are retarded are *persons* in God's image who, in this life, are unable to respond to their environment in a way that would manifest and develop *personality*. In the technical language of theology, they cannot actualize the ontic image; therefore they live their lives apart from any manifestation of the image in the dynamic sense. That is, they possess the image as given by the Creator but not as actualized by the creature… We are all but pilgrims and strangers on our way to a better land and a better life. They whose eyes have shown no human awareness will in death, we believe, cross the narrow sea with us into the life beyond. Then we who have known the Lord's compassion will also know the gratitude of these hapless ones to whom we have shown compassion" (*Who We Are*, 67, emphasis original).

[61] It is beyond the purview of this work to deal with the effect of the Fall on the image of God in humankind. Suffice it to say that on the subject of God's image, texts in Genesis "make it clear that humanity even after the Fall is still in the image of God" (Curtis, "Image of God [OT]," 390). Certainly a fundamental break takes place with Adam and Eve's disobedience. They are chased from the garden and the tree of life; and their sin brings a lasting curse on themselves and the natural world (Moltmann, *God in Creation*, 233). St Paul, himself a rabbi, sees it in these terms: "When Adam sinned, sin entered the entire human race. Adam's sin brought death, so death spread to everyone, for everyone sinned" (Rom. 5:12, NLT). It is the fundamental disorder of the human race that called for the radical solution represented by God giving up his Son on the cross for the redemption of humankind. Salvation, as a result, is, in Paul's terms, "the necessity of turning from sin and returning to God, and of faith in Jesus Christ" (Acts 20:21, NLT). Though sin has marred and greatly weakened humanity's ability to rule in the earth in God's stead, their vocation to rule in the earth has not been cancelled. It is now through grace— "the grace of the God who holds fast to his relationship to human beings in spite of their opposition" (ibid., 233). This is why the *imago dei* leads to the *imago christi* (the original image is found perfected in Christ), which points to the end of history when the image will be fully restored in the redeemed community and humankind will truly become the *gloria dei* (the glory of God) (ibid., 225-29). This provides an even greater incentive for Christians to work with all people of good will in order to fulfill this collective calling in a way that best represents the Creator's wishes.

due to the fact that the doctrine is only introduced in the creation narrative, but then left undeveloped in the rest of the Jewish canon. But more importantly, theology is always done in context. Thus, the New Testament picks up the *imago dei* theme in the light of the Christ-event and gives it great prominence. My context, however, is Muslim–Christian dialogue and my intent is to discover common ground. I can naturally assert that humanity's dominion on the earth (Latin: *dominium terrae*) will only be fully restored when Jesus Christ comes back to establish the final kingdom of God in the "new heavens and the new earth" (Rev. 21; 22).[62] But God's will for us humans beings to rule with righteousness on the earth today is not in question, for Muslims or for Christians. As always, though I usually do not articulate it, this also goes for Jews. For instance, Jewish scholar Jeremy Cohen concludes his survey of the literature on the command in Gen. 1:28 to multiply and rule, and comes to a similar conclusion:

> Mastering the earth and ruling over the animals, man and woman fashion a distinctly human civilization, the affirmation of whose intrinsic value sets the Bible apart from other ancient Near Eastern religions and cultures. These responsibilities bespeak a third realm of relationship—namely, God himself. Alone of all his works, God fashions humans in his own image, which they in turn must exemplify by meeting the charges of Gen. 1:28.[63]

As mentioned above, Psalm 8 constitutes an important parallel to Gen. 1:26-28. I will highlight it at this point as an introduction to a few comments I will make on ecology. The psalm opens and closes with a hymn of praise, "Yahweh, our Lord, how great your name throughout the earth!"[64] Then follows a paragraph on God's majesty in the heavens and his complete victory over all foes (v. 2). This leads the psalmist to raise his eyes to the heavens—the moon and the stars God has made—but it is not only to praise God for his greatness seen in creation, but especially to point out a contrast: "ah, what is man that you should spare a thought for

---

[62] To a large extent I would have to agree with Moltmann: "In the messianic light of the gospel, the appointment to rule over animals and the earth appears as the 'ruling with Christ' of believers. For it is to Christ, the true and visible image of the invisible God on earth, that 'all authority is given in heaven and on earth' (Matt. 28:18). His liberating and healing rule also embraces the fulfillment of the *dominium terrae*—the promise given to human beings at creation. Under the conditions of history and in the circumstances of sin and death, the sovereignty of the crucified and risen Messiah Jesus is the only true *dominium terrae*" (ibid., 227).

[63] Cohen, *"Be Fertile and Increase,"* 64.

[64] I will be using the Jerusalem Bible here.

him, the son of man that you should care for him?" (v. 4). This opens the way for a meditation on the lofty position God has granted humankind in the order of creation:

> Yet you have made him little less than a god,
>> you have crowned him with glory and splendour,
> made him lord over the work of your hands,
>> set all things under his feet,
> sheep and oxen, all these,
>> yes, wild animals too,
> birds in the air, fish in the sea
>> traveling the paths of the ocean. (8:5-8)

The psalm ends with the same phrase with which it began. Bernhard W. Anderson speculates that both this Psalm and Gen. 1:26-28 originated from "a cultic legend used in the Jerusalem Temple" and that the psalm might predate the creation narrative. Certainly the two passages present striking similarities:

1. They both seem to presuppose a belief in a heavenly council ("god" here refers to the assembly of heavenly beings) "within which, as in the Babylonian *Enuma Elish*, the decision to create humankind is announced."[65]

2. In both passages humans are given a status very close to that of the heavenly beings (the "sons of God" or *elohim*). But whereas in Genesis 1 they are created in God's image, the psalmist circumvents the delicate phrasing (perhaps because it is so shocking) and writes that humankind is slightly inferior to the *elohim*.

3. Both texts teach that humanity's high position carries with it the responsibility and power to rule over the animals.

Anderson points to a difference between the two passages, however: in Psalm 8 people are crowned by God very much in the way kings were crowned at that time. This crowning elevates their position over the rest of creation. In Genesis 1 the divine blessing empowers humans to fill the earth and subdue it. In light of the preceding, and in particular Clines's contribution, I would make less of the difference between the two passages.[66] In the end of his essay Clines himself comes back to the close

---

[65] Bernhard W. Anderson, *From Creation to New Creation: Old Testament Perspectives*, OBT (Minneapolis, MN: Fortress Press, 1994), 121. Notice the same context for the Q. 2:30.

[66] Levenson, in commenting on Ps. 8, writes, "Here again, the language is that used elsewhere for God's mastery over the world, and the assumption is that he has appointed humanity to be his viceroy, the highest ranking commoner, as it were,

parallel between the two formulations of humanity's vocation, and to the interpretive process of those who, under divine inspiration, coined it:

> In summary, Psalm 8 and the related passage in Genesis 1 are evidences of the new situation that prevailed in Israel when, with the rise of David, Israel accepted the alien institutions of temple and king and came under the influence of the royal theology of the ancient Near East.
>
> ...The coronation of humankind is seen in the context of Yahweh's rule in the heavenly palace from which he comes out to rout his foes—the powers of chaos who manifest their uncanny influence in the threats of disorder... *Adam* is Yahweh's viceroy on earth, having a status only slightly inferior to divine beings. Through human dominion Yahweh's name becomes glorious on earth.
>
> The situation is different in Gen. 1:26-28. Here the democratization of royal theology has been carried to its conclusion, leaving only vestigial remains, especially the motif of the image of God that entitles humanity to have dominion over the earth. *Adam* is the collective whole of humanity, differentiated according to male and female.[67]

How does this human dominion resonate with our current preoccupations about ecology? Before we explore this issue as it unfolds today, I must make a brief digression into the historical career of the Genesis 1 command: "Be fruitful and multiply, and fill the earth and subdue it." It affords us the opportunity to examine some of the Jewish and Christian concerns about humankind at creation, from about 200 BCE to 1500 BCE.

Classical Jewish and Christian Commentary on Genesis 1:28

Jeremy Cohen begins his exploration of this topic by nicely summarizing the present discussion: "Of all God's creatures, humans alone bear the image of God. They alone engage in direct conversation with their Maker. And they alone receive authority over the natural world: a responsibility to populate it, to control it, and thereby to civilize it."[68] God's apparent desire to include people in the continuation of his creation project through their own civilizational designs is one reason this verse has had so much impact on Western civilization, argues Cohen. Yet the theme of dominion over creation, so prevalent in today's commentaries, was almost totally absent in the commentaries from antiquity to the late

---

ruling with the authority of the king. The human race is YHWH's plenipotentiary, his stand-in" (*Creation and the Persistence of Evil*, 114).

[67] Anderson, *From Creation to New Creation*, 128-29, emphasis original.

[68] Cohen, *"Be Fertile and Increase,"* 1.

medieval period. What then monopolized the attention of Jewish and Christian commentators?

The first phrase of this "primordial blessing," "Be fruitful and multiply," directly raised the issue of human reproduction, and indirectly posed the question of covenant—especially when the second phrase was connected to the first: "fill the earth and subdue it." On the one hand, how could people deal with their sexuality so as to qualify for the reward of earthly dominion? On the other, both Jews and Christians found the universal scope of this blessing rather troubling. How might this square with other passages that seem to indicate that God had chosen a specific people, whether defined ethnically or religiously? "In the world of biblical antiquity, these concerns—not worry over the ecological implications of technology—gave Gen. 1:28 its meaning and significance, a conclusion that is confirmed by centuries of Jewish and Christian commentary on the Bible."[69]

Jewish Commentary on Gen. 1:28

I start out with the exception to the rule cited above. Plainly, in the intertestamental period and in the first two centuries of the Christian era, Jewish commentators focused much more on the second half of the verse (human dominion), than on its first half (reproduction).[70] Ben Sira (second century BCE), for instance, noted that despite humanity's short existence, God "granted them authority over the things upon the earth. He endowed them with strength like his own, and made them in his image."[71] Besides the five physical senses, God added two non-physical ones as well: "the gift of mind," and reason, "the interpreter of his operations."[72]

As Jews evolved in their mostly Hellenistic environment, they naturally came to use Greek philosophical categories to make sense of the biblical text. The most sophisticated philosopher in this regard was Philo of Alexandria (d. 50 CE), who, while rarely citing Gen. 1:28, expresses nevertheless a keen interest in the human dominion. In fact, he does so while appropriating the closest Greek word for "vicegerent": "So the creator made man after all things, as a sort of driver and pilot, to drive and steer things on earth, and charged him with the care of animals and plants, like a governor subordinate (*hyparkos*) to the chief and great

---

[69] Ibid., 39.

[70] Cohen indicates, however, that clear allusions in this period to Gen. 1:28 "are strikingly few indeed" (ibid., 69).

[71] Ibid., 69.

[72] Ibid., 70.

king."[73] This implies that for Philo Gen. 1:28 is both blessing and commandment, and that the human *imperium* means "an administrative responsibility of management on behalf of a superior officer"—precisely the notion of *khalīfa* in Arabic.

This specific focus on Philo's part also meant that he wrote little or nothing about the first part of the verse, which deals with sexuality and procreation[74]—a noteworthy point considering his tendency to interpret the Bible allegorically. While this "literal" meaning of dominion can apply to both Adam and later to Noah ("increase and multiply and fill the earth and dominate it," Gen. 9:1), Philo instinctively preferred the more spiritual meaning: people are called to dominate the beasts (exerting will power against evil in its many forms), over the birds (arrogant and vain thoughts), and over the reptiles (the noxious passions of the soul).[75] At the same time, as a Jew, Philo never lost sight of the literal meaning, for the Torah must necessarily speak to every day life.

The emphasis changed radically in the following centuries, the period of rabbinic *midrash 'agadah*, or the discourse of nonlegal works of edification. Living as they did in a Christian or Muslim environment that could often be hostile, the rabbis emphasized procreation as an act of faith and generally connected the duty to multiply in the present with the hope of future redemption. Part of this, naturally, could refer to the desire of a minority to survive. Yet the greater motivation was theological in nature: "The messianic king will never come until all those souls intended for creation have been created."[76] In other words, a certain number of souls to be born was stipulated by God at creation, and until they are actually born, the messianic era will have to wait. Procreation, then, becomes paramount. Yet it represents only one part of the threefold blessing of Gen. 1:28: procreation, settlement and human civilization. The overall objective is to define human nature against the tapestry of cosmogony (the origins of the world) and "the fulfillment of God's cosmic plan."[77] Privilege goes hand in hand with responsibility. Further, obedience merits reward, and the greatest reward is that one be ushered into the messianic kingdom of the end times.

---

[73] Ibid., 72.

[74] Ibid., 74.

[75] Ibid., 75.

[76] J. Theodor and C. Albek, eds., *Midrash Bereshit Rabba: Critical Edition with Notes and Commentary* (Hebrew), ed. 3 vols. (Jerusalem: Sifre Vahrman, 1903–36, repr. 1965), 1:233, cited in Cohen, *"Be Fertile and Increase,"* 117.

[77] Ibid., 122.

So much for the Aggadah, or Judaism's theological discourse. As it is for Muslims, Jews are equally (if not more so) concerned about rabbinic law, the Halakhah, by which Talmudic Judaism is translated into specific laws for daily living. Without going into details, marriage (and marriage to "a woman who was ostensibly capable of bearing children")[78] was ranked by some sages as having the same priority as studying Torah. Yet the linking of marriage to procreation had been a staple of Greco-Roman society as well. The Stoics displayed much the same attitudes, "and even Plato had called for female officials appointed by the state to supervise conjugal relations, like the Mishnah ensuring that no childless couple would remain married for longer than ten years."[79] In the same way as the Greeks, writers of the Talmud were concerned that procreation resulted in increasing their own community. Yet this theological position—connected as it was to their messianic hope—put pressure on the plain meaning of the text, which clearly speaks to humanity as a whole. In fact, this tension has always persisted, the reasoning of exegetes like Bahya b. Asher (d. 1340) notwithstanding. Bahya distinguished between "settling the earth" and procreation in the service of God:

> It is known that marriage has two dimensions. One seeks to maintain and expand the line of descent for the sake of settling the world, in the sense that it is written "He did not create it a waste but formed it for habitation"; and this intention pertains to the nations of the world. The second seeks to direct descendants to the service of the Lord, may he be blessed, and to know him and to recognize him, and this intention applies solely to Israel.[80]

"Still," avers Cohen, "the Talmud eventually bent over backward to exempt Gentiles from the mandate for procreation, a process that demanded considerable prowess in exegesis and casuistry."[81] For instance, he cites an argument between two rabbis in the Babylonian Gemara of the third century. A man converted to Judaism (becoming a "proselyte") after fathering several children. One rabbi ruled that he had already fulfilled his duty to bear children while the second rabbi esteemed he had not, since in becoming "like a newborn child" he was in effect starting over. Yet one could also find dissenting views. In the late medieval period, for example, the *Sefer ha-Qanah* outright contradicts the Talmud: "You should know, my son, that the Gentiles also are

---

[78] Ibid., 135.

[79] Ibid., 139-40.

[80] Bahya b. Asher, *Kad ha-Qemah*, in *Kitve Rabbenu Bahya*, ed. C. B. Chavel, 3 vols. (Jerusalem: Mosad Ha-Rav Kuk, 1969), 183, cited in Cohen, *"Be Fertile and Increase,"* 152.

[81] Ibid., 145.

commanded with respect to procreation; for the Gentiles too are included in every commandment which was uttered before the revelation of the Torah and was not repeated at Mount Sinai."[82]

Regardless of the status of Gentiles, the consensus of the Talmudic rabbis was that "be fertile and increase" was a commandment on a par with any other in the Mosaic law. In fact, to neglect this commandment was to commit a transgression akin to murder.[83] More importantly, it was the sign of blessing—that of God's gracious covenant with and election of the Jewish people. Already in the third century, the Mishnah considered "be fertile and increase" as an obligation, and several scholars date its origin to the reign of Caesar Augustus (at the time of Jesus' birth), when Roman policy officially encouraged and rewarded large families.[84] This Jewish emphasis on procreation, originally (perhaps) to comply with official decrees, only increased with the Roman wars that led to the destruction of Jerusalem in 70 CE and the brutal crushing of Jewish self-assertion in 135 CE. As mentioned above, the Jewish communities of the medieval period had just as much motivation to ensure the propagation of their own people.

Though the kind of decisive evidence that might link the origin of the interpretation of Gen. 1:28 as a commandment is lacking, Cohen sees another sociological factor that might explain it. The phenomenology of religion teaches us that a society's way of classifying roles and statuses within its ranks is directly related to its worldview, and in particular to how people remember and envision their origins.[85] Hence, doctrines of creation are crucial to the shaping of these taxonomies. Undoubtedly, muses Cohen, to interpret the primordial blessing of Gen. 1:28 as statutory law was to do violence to the text. On the other hand, the rabbis of the Talmud were simply claiming elite status in a world shaped like a pyramid: with them at the top, direct heirs to the covenant transacted by God with their forefathers at Sinai, they ranked above women, slaves, and in relation to the rest of humankind, above all non-Jews in general.[86]

Jews of the Middle Ages continued to honor the basic directions of Talmudic teachings (both aggadic and halakhic) as they sought to direct

---

[82] *Sefer ha-Qanah* (1894; repr. Jerusalem, 1973), 105a, cited in Cohen, *"Be Fertile and Increase,"* 153.

[83] Ibid., 158.

[84] Ibid., 159.

[85] He quotes two works: Ralph Linton, *The Study of Man* (New York: D. Appleton–Century, 1937), and Howard Eilberg-Schwartz, "Creation and Classification in Judaism: From Priestly to Rabbinic Conceptions," *History of Religions* 26 (1987): esp. 374.

[86] Cohen, *"Be Fertile and Increase,"* 164.

their second-class lives in the ambient ocean of Muslim or Christian societies. As was the case with Islamic jurisprudence, a study of the leading rabbis' *responsa* (Latin for "answer," sing. *responsum*)[87] in the litigation of marital disputes in the area of childbearing should shed light on how general rulings of the Halakhah are actually enforced (or not) in practice.[88] Yet the implementing of legislation in this area is likely to face formidable obstacles: "The intimacy of the marriage relationship, the intensity of the emotions accompanying all human efforts, successful or otherwise, to have children, and the truly universal character of the desire to reproduce bore directly on the ability of rabbinic legislation to control behavior."[89] Thus we can only sympathize with rabbis—the legal obligation emanating from "be fertile and increase" notwithstanding—who decided that the Talmudic law of divorce for a couple childless after ten years of marriage could not be enforced in reality. Additionally, this more relaxed attitude could be attributed to changing social conditions. According to Isaac b. Sheshet Perfet (d. 1408),

> If the courts took it upon themselves to enforce the letter of the law and exert compulsion in the selection of mates, they would have to exert compulsion in all such instances. And most wives nowadays would be divorced and would collect their dowries and the settlements stipulated in their marriage contracts. And since there is no such settlement which would go undisputed, strife and dissension would abound.[90]

Perhaps taking this unenforceability of the law into account, rabbis would often be willing to relax laws, which in other contexts would place limits on the number of marriages contracted, or at least the number of potential children born. One could permit widows or widowers to remarry more quickly, or one might provide dispensations for those who asked to be married on a Sabbath or religious feast. A Jewish man might also be allowed to marry a Karaite woman,[91] or a woman who had made an oath never to remarry would receive permission to obey the overriding

---

[87] This is exactly the definition and function of fatwas in Islamic law: a mufti is approached by someone and asked a specific question concerning his or her situation that calls for legal advice. The mufti (preferably a mujtahid—the highest level of expertise in a Muslim setting) will then issue a fatwa.

[88] Cohen unearthed about 200 such medieval responsa, which had some connection to Gen. 1:28.

[89] Ibid., 167.

[90] Isaac b. Sheshet Perfet, *Shu''uT* (Vilna, 1879), 4b, cited in Cohen, *"Be Fertile and Increase,"* 171.

[91] The Karaites were a Jewish sect that arose in ninth-century Persia. They were purists who followed the Torah exclusively, while casting aside all previous oral and written traditions (hence the Talmud).

commandment of procreation. In some cases—though more rarely—bigamy was even permitted in order to encourage procreation.[92]

In the midst of these communal struggles to cope with their minority status, however, Jews found themselves defending their sacred text and its laws before critics of the majority religion, whether Christian or Muslim. Cohen correspondingly notes a distinctive revalorization of biblical exegesis among medieval Jewry. Still, a return to a more literal meaning of the texts called into question the Talmudic legacy with its rich spectrum of views. At the very least, it impelled many scholars to view the inherited cosmogony in a new light. In particular, rabbis started to focus their attention on the second part of Gen. 1:28. What is this "human dominion?"

For Cohen, it is Sa'adya Gaon (d. 942) who espouses the most rationalistic interpretation of the humanity's calling to rule over the created order. A towering scholar originally from Egypt (cf. n. 25), author of a systematic Hebrew grammar and lexicography and influential translator of the Hebrew Bible in Arabic, Sa'adya spent his last twenty years near Baghdad, as leader (Gaon) of the Jewish academy of Sura.[93] He is one of a handful of Jewish scholars mentioned in Ibn al-Nadim's (d. 995) *Fihrist*, a comprehensive index to Muslim authors and works. Levenson believes he might be the first scholar in Rabbinic Judaism clearly to "associate the image of God in Gen. 1:26-27 with humanity's God-like rule over creation."[94] In his *Tafsīr* of Genesis (in Arabic), Sa'adya completely passes over the blessing or command to "be fruitful and multiply" and comments at great length on the meaning of humankind's dominion over creation:

> "Ruling" includes the [use of] equipment by which man may gain dominion over the animals. Over some of them [he has dominion] with mines and hobbles and over others with cords and reins and yet others with pits and collar, hunting equipment and *adbaan*.[95] Others are with cages and towers and the like until God teaches [man] everything [about this].

---

[92] Ibid., 178.

[93] Haïm Zafrani and André Caquot, *La Version arabe de la Bible de Sa'adya Gaon: l'Écclésiaste et son commentaire 'Le Livre de l'ascèse,'* Collection Judaïsme en Terre d'Islam, 4 (Paris: G.-Maisonneuve & Larose, 1989), 14; see also "Sa'adya Ben Yosef," in *EI²*, vol. 8:661-62.

[94] Levenson, *Creation and the Persistence of Evil*, 112.

[95] Cohen translates from the Hebrew from an Arabic/Hebrew version, M. Zucker, ed., *Commentary on Genesis* (New York: Jewish Theological Seminary of America, 1984), 53-54, 258-59. His text is clearer in English. People's rule over animals is accomplished through a range of devices: "over some with fetters and bridles, over some with ropes and reins, over some with enclosures and chains, over some with

> [Ruling over] "Fish" includes [the use of] tactics in hunting fish from the bowels of the sea and rivers, preparing those permissible [for eating] with cooking utensils so that [one] can eat it, taking pearls from the shell, benefiting from the parts of the skin and bones that one prepares, and whatever applies to this.[96]

Sa'adya goes on to describe humanity's ability to devise machines to dig wells and dam rivers that can then power mills; to build all manner of crafts to sail the seas and bridges to carry people and equipment over large rivers; and to devise instruments to study the stars and calculate time. He refers to human ingenuity in using herbs both to feed and cure people; and in using animals to do their work and provide more food for them:

> "And in the entire Earth" alludes to God's giving him the wisdom [to know] how to build houses, fortresses, and castles and to plow the Earth with various seeds and plants. Also how to extract gold and so with silver and bronze and copper from mines; how to make vessels and ornaments in crafts, and also how to make equipment for farm work like ropes and the plow and making carpentry equipment like saws and axes; [making] equipment for sowing clothes like pants, like the weaver's beam, and [making] writing utensils like pens and inks and the like.[97]

I would also like to point out the apologetic aspect of Sa'adya's commentary and theological thought. Several Jewish academies were then functioning in Abbassid Mesopotamia, but they were facing dire challenges through internal strife and Muslim polemics on the outside. It was Sa'adya's intellectual skills that helped to turn the tide, allowing the Jewish communities to thrive once again. It was Moses Maimonides who later wrote, "If it had not been for Sa'adya, the divine religion might well have almost disappeared, for he made clear its mysteries and strengthened its weak points by spreading it and supporting it by his word and pen."[98]

Speaking of his commentary on the book of Ecclesiastes, Hraïm Zafrani and André Caquot note that by naming it "*Tafsīr*," Sa'adya intended this Arabic text for Jewish and non-Jewish consumption on a

weapons of the hunt, over some with cages and towers, and so on...", cited in Cohen, *"Be Fertile and Increase,"* 184.

[96] Linetsky, *Rabbi Saadiah Gaon's Commentary on the Book of Creation*, 115.

[97] Ibid. He then quotes from Exod. 31:3-4 in order to illustrate his point: (speaking of Bezalel, the multi-skilled craftsman who organized the building of the Tabernacle for Moses) "And I completed in him the knowledge from God with wisdom and understanding and with knowledge of all the crafts and the carving stone to arrange."

[98] Maimonides, *Epistle to the Yemen*, ed. and with an Introduction by A. Halkin (New York: American Academy for Jewish Research, 1952), 64, cited in Paul B. Fenton, "Sa'adya Ben Yosef," *EI²*, 661.

popular level: "One sees in this an effort to present a text in a rationally acceptable way and allowing a simple and easy reading."[99] Yet it was not only the style and format that he fashioned for this purpose, but his theological approach as well. At a time when Mu'tazilite theology was sanctioned by the Abbassid rulers, Sa'adya consciously (it would seem) borrows some of their tenets in his own theological approach. One of his best-known works is a philosophical treatise based on the five Mu'tazilite principles, *Kitāb al-amānāt wa-l-iʿtiqādāt* (*The Book of Beliefs and Opinions*).[100]

Fenton remarks that in each of the Arabic introductions to the books of his Arabic translation of the Hebrew Bible, Sa'adya studiously avoids anthropomorphisms (a distinct Mu'tazilite concern), supports a rational approach to the existence of God and uses a similar reasoning to refute the Muslim contention that the Hebrew scriptures had been abrogated.[101] Thus, in commenting on Gen. 1:29, which states that God gave to humanity plants and seed, trees and fruit for human consumption, Sa'dya writes, "Here too I have said…would they use wisdom. 'Behold I have given you' indicates that God implanted in the intellect of man the knowledge of all plants that are for sustenance whether grains, trees, or vegetables. Likewise knowledge of those [plants] by which he may be healed."[102]

Finally, while considering the statement in Gen. 2:15 "Then God took the man and placed him in the Garden of Eden," Sa'adya comments, perhaps defensively, "The statement 'And God took' does not imply coercion for [God] Exalted takes no direct effect on the actions of people rather [the placing] is [merely] an instruction."[103] Michael Linetsky offers a footnote, originally by M. Zucker in his posthumous 1984 Hebrew translation of this commentary: "Here the Gaon follows the Mu'tazila notion that no coercion on the part of God may be executed even if it is not in a matter of good and bad."[104] Within the wider context of the present book, it is clear that the project of theology can only be carried out in a particular context. Here we see Sa'adya articulating the truths of the Bible, while consciously appealing to the minds of the majority Muslim population and (perhaps) unconsciously being influenced by some of their own theological tenets. No doubt, Christians living in Muslim lands were acting likewise.

---

[99] Zafrani and Caquot, *L'Écclésiaste et son commentaire*, 14-15.

[100] This is the English title of the translation made by Samuel Rosenblatt (New Haven: Yale University Press, 1948).

[101] Fenton, "Sa'adya Ben Yosef," 662.

[102] *Rabbi Saadia Gaon's Commentary on Genesis*, 116.

[103] Ibid., 134.

[104] Ibid., 134 n. 57.

Christian Commentary on Genesis 1:28

Though late antiquity and medieval Christian literature is more voluminous that its rabbinic counterparts, it offers less commentary on this
particular verse. For one thing, Gen. 1:28 is not quoted in the New Testament (or even alluded to), and furthermore, it raised thorny theological
problems related to the Christian doctrine of original sin: how is such a
this-worldly concern with reproduction to be related to God's plan of
redemption (mostly other-worldly)? And to what extent did the Fall
incapacitate human dominion on earth? Also, the dominion theme of
Psalm 8 is picked up by two New Testament writers, but only in relation
to Christ, the "Second Adam" (1 Cor. 15:27; Eph. 1:22; Heb. 2:6-8).
Thus, in the Epistle of Barnabas, dated somewhere around 100 CE, one
can detect an anti-Jewish polemic. The command to multiply and fill the
earth is conflated both with the command to enter the land flowing with
milk and honey and with the new covenant mediated by Jesus with all of
his followers, be they Jewish or Gentile. The result is that this "primordial blessing, originally addressed to his son, Jesus, bespeaks of the ultimate triumph of the church."[105] At the same time, most Church Fathers
were eager to build on the doctrine of the image of God in Adam and
Eve in order to affirm with their pagan neighbors that human civilization
flows from God's good creation. Didymus the Blind, for instance, in
spite of the tendency of the Alexandrian school to allegorize and spiri-
tualize such scriptures (following Origen) wrote in terms that prefigure
the work of Sa'adya and others:

> "And master it" signifies an extensive power, since one cannot say of him
> who has limited power that he has dominion. God has made this gift to
> the human being…in order that land for growing and land for mining,
> rich in numerous, diverse materials, be under the rule of the human being.
> Actually, the human being receives bronze, iron, silver, gold, and many
> other metals from the ground; it is also rendered to him so that he can
> feed and clothe himself. So great is his dominion the human being has
> received over the land that he transforms it technologically—when he
> changes it into glass, pottery, and other similar things.[106]

While the Eastern patristic fathers often laid greater stress on the inner
side of dominion (mastery over one's sinful passions), whether they
belonged to the Antiochene, Cappadocian or Alexandrene school of
exegesis, they all affirmed to some degree human rule over creation as
well as its theological implication after the new covenant. But while the
issue of dominion "was rarely a controversial issue for them," it was the

---

[105] Cohen, *"Be Fertile and Increase,"* 225.
[106] Ibid., 227.

sexual reproduction part of the verse that grabbed their attention, as was the case for rabbinic scholars.[107] As celibacy became the ideal path to spiritual greatness in the third and fourth centuries, much discussion revolved around whether this verse concerned the state of people before or after the fall. This dilemma was only exacerbated by the various heresies, often spawned by the influence of popular schools of Greco-Roman thought—heresies which Church leaders did their best to refute. At stake here, then, is a healthy doctrine of creation that recognizes God's unequivocal affirmation of the goodness of his handiwork. Dualistic and Gnostic heresies sought to devaluate the physical side of human existence. For the adherents of such doctrines, spirituality can only be found in asceticism and a life-style of otherworldliness at the expense of any positive engagement with one's social or physical environment.

The situation was similar in the Western Church, but Augustine of Hippo (d. 430) set the tone for much Christian thought in the following centuries with his systematic grasp of theology and his assiduous attempt to relate creation to redemption in the warp and woof of a seamless narrative. For him, the blessing of "be fertile and multiply" was repeated for humans to show that, despite their fall from grace, human reproduction was not only desirable, but inherently good. Human beings rule on the earth by virtue of their bearing the divine image. However, though created with a midway status between angels and animals, human rebellion against God's commandments has lowered their ability to fulfill the original blessing of dominion and procreation. And while other writers before him had greatly emphasized the effect of sin on the primordial blessing, Augustine wanted his readers to consider "how many and what sorts of good things his [God's] providential goodness has infused into all that which he created." He explains:

> First, that blessing which he had conveyed before the sin, stating "Be fertile and increase and fill the earth," he did not wish to withhold even after the sin, and the fecundity thereby granted has remained in the condemned species. The guilt of sin could not remove the wonderful power of the seed—and even more wondrous, the power by which the seed is produced—instilled and somehow ingrained in human bodies... Despite his condemnation, he did not remove all that he had given; otherwise all would cease to exist. Nor did he remove that power from human capability, even when he inflicted a punishment of subjection to the devil, for he did not even exclude the devil himself from human domination.[108]

---

[107] Ibid., 229.
[108] *De Civitate Dei* (*The City of God*), vol. 22:24 (no publication details given), cited in Cohen, *"Be Fertile and Increase,"* 246-47.

For Augustine, the primordial blessing was only partially mitigated by human sin. As a result, the mixed blessing/command to procreate and have dominion still holds today. Nevertheless, it must be said that he inherited the theological consensus that "be fruitful and multiply," as it related to the state of paradise, could not literally mean sexual relations. Since Jesus characterized the post-resurrection state as angelic—no marrying in heaven (Mt. 22:30; Mk 12:25; Lk. 20:35)—scholars applied this retroactively to the original state in the Garden of Eden. Augustine's view on this evolved over time,[109] but even when he still held to the allegorical interpretation (the union of males and females was only spiritual) he wrestled with the plain meaning of the text—after all, the same blessing is given to the birds who obviously reproduce after their kind, as do all animals and plants. So the spiritual meaning is over and beyond the literal one, though without canceling the latter. With time, however, he came to believe that, had sin not entered the human race, people would have multiplied in the Garden until the fixed number of beings was reached and then a transformation would occur when God would lead them into their final spiritual state.[110]

Augustine, the bishop of Hippo (on the outskirts of today's Annaba, Algeria), was not just struggling against the dualistic current among Church Fathers who had long favored celibacy over marriage and union with God over the creation mandate for humanity to fill and master the earth. An even stronger incentive to embrace the physical side of God's trust to humans came from the fact that he had originally converted from being a follower of Manicheism (a strongly dualistic theology) to becoming one who embraced Christianity of the Church councils. As he progressed along the road that led from dualism "to the monistic theology of the Bible, Augustine accorded greater value to the historical sense of Scripture" and the necessity of first giving precedence to its literal meaning.[111] This kind of holistic appreciation of human nature as intended by its Creator is, in a sense, a return to its original Semitic worldview (whether he realized it or not). Throughout the Hebrew Bible, God is equally concerned about people's spiritual, social and physical welfare. Accordingly, the Hebrew word *shalom* (a cognate of the Arabic *salām*), typifies the ideal of divine deliverance that always harkens back to a solid theology of creation. The human person who is submitted to God, and who is thus a follower of his revealed ways, experiences peace within

---

[109] Cf. ibid., 247-59.

[110] Ibid., 250. This is a quote from *De Genesi ad litteram imperfectus liber*, his twelve-volume exegetical masterpiece written between 401 and 414.

[111] Cohen, *"Be Fertile and Increase,"* 257.

himself or herself, peace with other human beings, and harmony with the natural world. Hence, when the prophet Jeremiah (ch. 29) addresses the Israelites recently exiled to Babylon, he exhorts them with words that plainly refer to Gen. 1:28:

> [1] Thus says the Lord of hosts, the God of Israel, to all the exiles… [2] Build houses and live in them: plant gardens and eat their produce. [3] Take wives and have sons and daughters; take wives for your sons, and give your daughters in marriage, that they may bear sons and daughters; multiply there, and do not decrease. [4] But seek the welfare (*shalom*) of the city where I have sent you into exile, and pray to the Lord on its behalf, for in its welfare you will find your welfare. (RSV)

This exhortation to the Israelite refugees to pray for the "peace" (*shalom*) of their adopted city is God's way of saying to people devastated by war and destruction to rethink the presuppositions of their spirituality. In a revolutionary way, Jeremiah hears God telling the people of Israel that their faith is not dependent on a particular location—including the Temple of Jerusalem and its system of animal sacrifices.[112] Whatever God's particular purposes for the Jews, they must also reconnect with his designs for all peoples and remember that his grace and mercy beginning at creation encompass all the nations and cultures on the one earth he created. That interconnectedness of humankind and the corresponding holistic view of spirituality are akin to the Muslim concept of *tawḥīd*, the unity of God, which implies the unity of all that he fashioned in the world and the harmony humankind experiences when it carries out its deputyship of creation according to God's designs revealed in the Qur'an and Sunna.[113]

From the above discussion, it is obvious that for Jews and Christians over the centuries the idea of human dominion was much less controversial than the corresponding idea of human procreation. For the medieval rabbis, the blessing of Gen. 1:28 took the form of a command with clear covenantal ramifications.[114] It was integrated within the larger corpus of Mosaic law and was "limited in its application to free Jewish males, excluding slaves, women and Gentiles."[115] For churchmen of the same

---

[112] Cf., e.g., *The Interpreter's Bible*, vol. 5 (Nashville, TN: Abingdon Press, 1984), 1018.

[113] This thought is developed more fully in the next chapter.

[114] Recall that Sayyid Qutb had spelled out the covenantal connection between the human vocation as God's trustees, the primeval covenant (Q. 7:172), and God's specific covenant with Muslims, particularly after Jews had forfeited their own part in God's covenant with them (cf. Chapter 9).

[115] Cohen, *"Be Fertile and Increase,"* 311.

period, while Augustine's literal interpretation of "be fruitful and multiply" was now taken for granted, it could easily be combined with the earlier figurative or allegorical interpretation, which also laid claim to covenantal identity. In that vein, only followers of Christ, the Second Adam, may inherit the primordial blessing in its full sense. Our topic here, however, is dominion, or, as Cragg puts it, the human *imperium*— not the blessing/command of procreation. Actually, none of the qur'anic creation narratives allude to the command to be fruitful and multiply. Having said that, the command would have generated no controversy in the early Muslim community had it been added. If anything, the Qur'an chides Christians for their monastic practices (Q. 57:27). Marriage and procreation are held in high esteem in all the texts, while celibacy is castigated.

The most striking Muslim–Christian parallel appears when one connects political fortunes to a robust human dominion exegesis. Just as the initial Muslim conquests happened with lightning speed and dazzling success, so the width, breadth and longevity of Islamic civilization, including its art and sciences, contributed in the medieval period to a healthy sense of human mastery over creation. Certainly this is the feeling exuding from Razi's commentary on the human caliphate and his praise of human reason. Cohen sees the same kind of ethos rising from the experience of Christians who had inherited Rome's political power and cultural achievements. Thus it seemed natural that the philosophical focus of Greco-Roman philosophy on the human person would rub off on Christians as well:

> From the anthropocentric perspective of the pagan or in a Christian world view that focused on the deity's assumption of human existence, the limits and extent of human capability might sensibly rank high on a list of problems that occupied the intellectual. The fact that from the fourth century onward Christians actually enjoyed dominion in the Greco-Roman world—while Jews never did—invariably added to their interest in the subject. Experience stimulated intellectual inquiry.[116]

Interestingly, the Greco-Roman influence intensified in the twelfth century as Roman law was rediscovered, allowing the Western Church to develop canon law. For the first time, Gen. 1:28 was seen as expressing natural law—though a natural law revealed by God in his creation and thus distinct from Roman natural law.[117] In the next century, thanks to Europe's discovery of things Islamic in Spain and on the edges of their temporary foothold in the "Holy Land," Aristotle regained popularity and

---

[116] Ibid., 268.
[117] Ibid., 290-94; also Dalacoura, *Islam, Liberalism and Human Rights*, 6.

Christian scholasticism under Thomas Aquinas approached its zenith. For Aquinas, human reason is the connection between creation, natural law and the Bible. Human beings thus participate in the divine law and consult their ethical compass to create a more just human society—a theme also dear to many Muslim jurists of the period. The idea of natural law in turn paved the way for the growing attention to human rights in the following centuries, as I will now attempt to demonstrate.

As we continue to follow the Christian exegesis of Gen. 1:28, by the late Renaissance period, the Christian debates on the procreation side were waning—though simultaneously heating up in Catholic–Protestant polemics[118]—while the dominion side gained in importance. Yet the focus of the human mandate over creation had little to do with "creation" and almost everything to do with the "human." As in the preceding centuries, it was assumed that as science expanded, the technology people developed would grow more complex and sophisticated. Further, in theological terms, "[t]he progression from *imago Dei* to dominion is clear," yet, tellingly, the impact of human creativity in mastering the physical world was not even considered prior to the Industrial Revolution. Indeed, ecology did not become a science till the second half of the twentieth century. What mattered before that was anthropology, and in particular, the question of the specific place of human beings in the universe. But with the rapid social changes already discernible in the sixteenth century, Gen. 1:28 ceased to be the focal point of discussions altogether. As I have noted already, the discourse about humankind was increasingly non-religious and sought to answer questions about the new social realities of cities, states, and the role of the emerging bourgeois class. Having said that, the origin of the modern idea of human rights is more likely religious and not secular—though it has more to do with Gen. 1:26 and 27 (*imago Dei*) than with v. 28.

Our story begins with the sixteenth-century Reformation, which represented a watershed in at least three ways. First, the politico-religious monopoly of the Pope was shattered in Europe. And second, building on

---

[118] Cohen alludes to the polemical use of this verse by Martin Luther, Philip Melanchton and the Calvinist David Pareus. For them, the command "be fertile and multiply" was obviously being undermined by the Roman Catholic emphasis on celibacy. Melanchton in particular argued explicitly on the basis of natural law: "this creation or divine ordinance in man is a natural right... So it is ridiculous for our opponents to say that originally marriage was commanded but that it is no longer commanded... Natural right is really divine right, because it is an ordinance divinely stamped on nature" (*The Book of Concord: The Confessions of the Lutheran Church*, ed. and trans. Theodore G. Tappert [Philadelphia: Fortress Press, 1959], 241, cited in Cohen, *"Be Fertile and Increase,"* 308).

the Renaissance ideals of reason, scientific research, and the more subjective values of beauty, both in the arts and nature, the reformers and
their kin celebrated the individual right to read and interpret the sacred
texts—an emphasis that led to the invention of the printing press. More
sadly, however, the Reformation ushered Europe into a period of political
strife and social turmoil.

Yet it was in the cauldron of English Church rivalries and persecutions
that the concept of human rights was born. Glen Stassen, in his landmark
book *Just Peacemaking*, describes how the "free churches" (Mennonites,
Congregationalists and Baptists), or Puritans, were the driving force
behind the Levellers, "the first democratic political movement in modern
history,"[119] who provided the army that enabled Parliament (mostly
Presbyterian) to defeat the increasingly tyrannic King Charles in the
1640s and advocated among the masses for a genuine electoral democracy.[120] Remarkably, Glen Stassen, a Christian ethicist, devoted almost an
entire chapter in his *Just Peacemaking* to the original research he
conducted on the most articulate spokesmen for the Leveller movement,
Richard Overton.[121]

Originally part of the band of Puritans who broke off from the Church
of England in 1607 (a branch of which sailed to America on the
Mayflower), Overton, in a series of pamphlets he wrote during the 1640s,
helped to fuel the popular revolt that in 1649 led to the execution of King
Charles and to the English Bill of Rights in 1689. What is especially
striking about Overton's writing—apart from his biting wit and humor,[122]

---

[119] G. E. Aylmer, ed., *The Levellers in the English Revolution*, with 22 illustrations
(London: Thames & Hudson, 1975), 9. For other general documents of the movement, see Don Marion Wolfe, ed., *Leveller Manifestos of the Puritan Revolution*,
with a Foreword by Charles A. Beard (London/New York: T. Nelson & Sons, 1944).
[120] Glen H. Stassen, *Just Peacemaking: Transforming Initiatives for Justice and
Peace* (Louisville, KY: Westminster/John Knox Press, 1992), 140.
[121] John Lilburne was the leader of this movement that lasted only five or six years. It
was first allied with Cromwell, but later turned against him, as it was becoming clear
that the latter was no friend of the Puritans. Indeed, the Leveller movement, accounting for all the diversity within its ranks, was chiefly composed of various Christian
groups that had broken off from both the Church of England (Anglican) and the
Presbyterian Church (Scotland). If anything, writes Aylmer, Overton was more
"secular" than Lilburne—that is, though his arguments are based on overt Christian
principles, Overton was more systematic in spelling out his political theory.
[122] Stassen gives special attention to one booklet—"his masterpiece"—*The Arraign-
ment of Mr. Persecution*, in which he marshals ten accusers (among them "Mr.
Sovereignty Christ," and "Christian"), eight of which declare him guilty and charged
of exile "because he causes wars, insurrections, bloodshed, and hatred, and destroys
peace" (*Just Peacemaking*, 146).

is that he consistently appealed to the natural rights and freedom of all people, regardless of creed, including the Turk (Muslim), Jew or pagan. This was based on the creation truth that all people without distinction are fashioned in God's image and on the Gospel teaching that all are "the objects of God's love shown in Christ's sacrificial death on the cross."[123] Finally, Overton, in his triple appeal to reason, the Bible and the bitter experience of the oppressed and persecuted, called for the enactment of laws to protect not just the civil and political liberties of all, but also the economic rights of the poor and disenfranchised.

Stassen then compares this early seventeenth-century defense of human rights with another Christian appeal, this time by Vatican II in the early 1960s. Drawing on David Hollenbach's work,[124] he shows how the UDHR stimulated a fresh theological reflection in the Roman Catholic Church, enabling its leaders to move away "from a somewhat hierarchical understanding of natural law to a human rights understanding" more congruent with the pluralistic configuration of global society.[125] Tellingly for this project, Stassen demonstrates that the idea of human rights is ideally suited to become a universal ethic that fights injustice and inequalities in the name of human dignity and that this is possible because people from all faiths and cultures, drawing on their particular theological or philosophical resources, can agree on its "*intention and application*," though not on its "*source*."[126]

Admittedly, human rights in the seventeenth century were also grounded in a secular ideology, mostly in reaction to the sectarian bloodshed of the sixteenth and seventeenth centuries. While the horrific Thirty Years War was still raging, the Dutch Hugo Grotius (d. 1645) became the first to erect a consistent theory of natural rights for all human persons by virtue of their reason, independently of God.[127] Yet to describe his theory as "secular" is misleading—he was an earnest Protestant debating in a Protestant context. Protestants tended to integrate their biblical interpretation within the Renaissance humanist perspective more deliberately in this period than their Catholic colleagues were willing to do. Hence the lines of demarcation, as shown by Richard Tuck, are more

---

[123] Ibid., 148.

[124] David Hollenbach, *Justice, Peace and Human Rights: American Catholic Social Ethics in a Pluralistic World* (New York: Crossroad, 1988).

[125] Stassen, *Just Peacemaking*, 155.

[126] Ibid., 156.

[127] Dalacoura, *Islam, Liberalism and Human Rights*, 8. The topic is vastly more complex than I can explain here. See Richard Tuck's *Natural Rights Theories: Their Origin and Development* (Cambridge: Cambridge University Press, 1979), and especially his chapter devoted to Hugo Grotius (58-81).

between political conservatives and radicals. Grotius's theory tended to veer toward authoritarianism, yet his stress on "interpretative charity" (we assume, for instance, that slaves or their ancestors did not voluntarily renounce their natural right to freedom—hence all potentially have inalienable rights) gave inspiration to those who would question the authority of despotic rulers or unrepresentative parliaments.[128] What is clear, however, is that the Enlightenment notion of natural rights that was later developed by Thomas Hobbes (d. 1679) and John Locke (d. 1704) originated in a Renaissance culture deeply rooted in the Judeo-Christian conviction that humanity was created in God's image and thereby called to populate and master the earth in his stead. What Hobbes and Locke did, in the aftermath of over a century of intra-European violence in the name of Christianity, was to set aside the covenantal overlay of medieval theology (something Jewish exegesis shared with its Christian counterpart) and reclaim the universal dimension of the Genesis narrative, which is—by far—the simplest reading of Genesis 1.

No doubt human rights is a central concern today, and though I agree with Stassen and others that the concept is amenable to adoption and adaptation by a variety of religious and non-religious perspectives, the other side of dominion has also been forced into the forefront in the last fifty years or so: the negative human impact on the physical world blamed on Gen. 1:28 by a prominent ecologist in 1967. Muslims as well as Christians will have to deal with Lynn White's classic critique of Christianity as the root cause of the current ecological crisis,[129] precisely because it is grounded in a particular doctrine of creation, which sees humanity as having been endowed and charged by God to rule over the created order.

Much of this has been answered in the previous section and I will have to leave that for the moment.[130] What should be noted here, however, is

---

[128] Ibid., 143-55. Tuck argues that part of the reason the Levellers were not able to sustain their movement was due to the inconsistency of their theory. Yet his contention that they still believed people could renounce their rights and thus still did not believe in inalienable rights for all contradicts at least the writings of Overton as represented by Stassen. More work is needed in this area.

[129] Lynn White, "The Historical Roots of Our Ecological Crisis," *Science* 155 (1967): 1203-7. The literature consistently cites this article (and subsequent reprintings of it elsewhere) as the seminal statement of the "case against Christianity."

[130] I will simply outline some of Colin E. Gunton's arguments, cross-referencing others along the way:

1. Human dominion is taught in narrative form in the Gen. 1 and 2 passages: "Neither care of a garden nor the naming of the creatures represent a form of absolute domination" (*The Triune Creator: A Historical and Systematic Study* [Grand

that in the last few decades numerous books have come out by theologians from all stripes calling for a more holistic theology of creation.[131] What is more, whether in ecumenical or evangelical circles, more action is being taken to mobilize Christians to exercise their God-given calling to take care of the earth.[132] Indeed, Muslims and Christians must pool their resources in this field.[133]

Rapids, MI: Eerdmans, 1998], 197). To name is a way of establishing a relationship of symmetry, in this case, of asymmetrical reciprocity.

2. "Eden is a garden, and not a paradise. This is an important distinction. In paradise, the fruits simply fall off the trees on to our tables; in a garden, trees have to be tended" (ibid.).

3. Though the Hebrew for "having dominion" (*rada*) can mean the use of force, it is dramatically restricted here by the fact that "both humankind and animals were apparently initially vegetarian [which] suggests that dominion did not extend to killing animals" (Jewett, *Who We Are*, 353). The permission to kill animals for food comes only after the Flood. Meanwhile, the prophetic pictures of Paradise depict a world free of violence between people and animals (Isa. 11:6-9; 65:25). Clearly violence is not God's intention (Stanley Hauerwas, *In Good Company: The Church as Polis* [Notre Dame, IN: Notre Dame University Press, 1995], 113). For an excellent discussion of this, see also Paul Beauchamp's commentary on Ps. 8 (*Psaumes nuit et jour* [Paris: Éditions du Seuil, 1980], 159-60).

4. "[T]he doctrine of the image of God represents a relation, primarily to God the creator and secondarily to the other creatures, animate and inanimate alike" (Gunton, *The Triune Creator*, 198). This carries with it a built-in ethic. Larry Rasmussen is much more vocal about it: "All lives are integral to earth. All share in the integrity of creation. All belong to the Community of Life. All deserve names and the recognition of differences that bestow character... As a species, we're the newcomers... And while we cannot do without otherkind's membership in earth's Community of Life, otherkind can do without ours" (Jewett, *Who We Are*, 36-37).

5. As I will say later, from a Christian perspective creation cannot be understood in the Bible without christology. Hence Gunton's contention that "Jesus is the true image of God and the means of the restoration of its true form in others—those who are 'conformed' to his image (Romans 8:29)" (*The Triune Creator*, 198). This perspective naturally opens up the discussion to values of servant leadership, care for the poor, the oppressed, and marginalized of society, as well as the significance of the incarnation—God's choice to enter into his fallen creation to redeem it from within through his cross and resurrection. This then is the basis for the "New Heavens and the New Earth" and "Behold, the dwelling of God is with men" (Rev. 22:1, 3, RSV).

[131] See also Brennan R. Hill's excellent work on the subject from a Catholic perspective: *Christian Faith and the Environment: Making Vital Connections*, Ecology and Justice: An Orbis Series on Global Ecology (Maryknoll, NY: Orbis, 1998), especially Chapter 2.

[132] Increasingly, the lines between the two movements (and especially in this area) are fading. One telling sign is Maurice Strong's Foreword to Dayton Roberts's book,

Cohen's study too came out of a desire to respond to this kind of criticism. In fact, in the second page of his Introduction, he refers to White's seminal article, which attributed the attitude of "ruthlessness toward nature" so prevalent in Western civilization to the "Judeo-Christian teleology."[134] Here and there throughout the book he comments on this issue, mainly to point out that ancient and medieval theologians, whether Jewish or Christian, rarely if ever considered the Genesis mandate as a license to exploit creation in a destructive way. As he puts it in his Conclusion, "[t]he words of Gen. 1:28 surely addressed the status of human beings, but in a manner that was more insightful and complex than modern ecological advocates have recognized."[135] On the one hand, the blessing as command expressed the idea of covenant, which implies responsibility to the Creator. On the other, while the idea of humanity in God's image naturally led to the idea of human dominion over the rest of creation, it was the sexual dimension that largely captivated Jewish and Christian exegetes. Human fulfillment had to pass through both sides of the Gen. 1:28 equation, yet neither side ever sanctioned a ruthless exploitation of natural resources. How could the creature, who most intimately represented God on earth, in good conscience destroy his Creator's work? Surely, this would be a betrayal of humankind's high calling.

The present context in which Christians and Muslims are doing theology includes the global reality of today's human family—what I have called here postmodernity. Humanity will not survive if it does not find a way to work together, peacefully and constructively. Either it will destroy itself through an ecological crisis of apocalyptic proportions, or it will unleash against itself its newfound weapons of mass destruction. The two

---

Patching God's Garment (cf. Chapter 1, n. 109). In this book Roberts introduces his readers to the Au Sable Institute, an evangelical think-tank and retreat center for Christian ecologists in the woods of central Michigan. Finally, the Christian Environmental Network, an evangelical initiative, has been active for over two decades now (see http://www.creationcare.org).

[133] The aforementioned chapter by Roger Tim ("Divine Majesty, Human Vicegerency") is a significant Muslim attempt to grapple with the issue. He concludes that it is largely a hermeneutical issue for today: "In short, the potential effect of early Islam on the environment depends in large measure on how the *khilāfat Allāh* given to humans is interpreted. If it is interpreted in an anthropocentric way, seeing the purpose of creation as serving humans, then the result may be exploitation of the earth. If, however, the vicegerency of humans is seen as ultimately subordinate to Divine sovereignty and will, then human authority over the creation becomes responsibility to care gratefully for the environment that belongs to God and serves Allah's will" (ibid., 10). The same applies to the Bible.

[134] Cohen, *"Be Fertile and Increase,"* 2.

[135] Ibid., 309.

crises are related. As Larry Rasmussen so forcefully puts it, "Earth's distress is a crisis of culture":

> More precisely, the crisis is that a now-globabilizing culture *in* nature and wholly *of* nature runs full grain against it. A virile, comprehensive way of life is destructive of nature and human community together—this is the crisis. Soils, peoples, air, and water are being depleted and degraded together... It is the failure to submit human power to grace and humility, and to work "toward the habitation of the places in which we live" on terms that respect both human limits and the rest of nature's. Life-as-we-have-come-to-know-it is eating itself alive. Modernity devours its own children.[136]

In this light, I contend that it is all the more urgent to get the present Muslim–Christian theological project underway in order to tackle some of these problems together. But it will always be a project under construction, partly because theology always develops in a changing context, and partly because the dynamics of a more practical cooperation will in turn affect our theological discussions.

Pushing the Limits in Dialogue

In the same spirit of dialogue, building on the rather non-controversial idea of humanity's God-given caliphate, I offer the next three points, then, both as areas needing further study, and as distinctly Christian (or Jewish) doctrines, but which in the context of dialogue, could still enrich the Muslim perspective.

(1) After the sixth day of creation God rested from his work (Gen. 2:3).[137] As Barth has argued, the goal of creation is not the creation of

---

[136] Rasmussen, *Earth Community, Earth Ethics*, 7-8, emphasis original. The most imaginative and bold initiative that I am aware of on the Muslim side is the partnership of the UK-based Islamic Foundation for Ecology and Environmental Sciences with the UK branch of the youth-oriented Lifemaker movement. Together they produced a 23-page booklet guiding Muslim households through the practicalities of "reducing their carbon footprints." See http://ifees.org.uk/index.php?option =com_content&task=view&id=82&itemid=27. None of the qur'anic verses touch on the khilafa, however.

[137] Muslims have consistently argued that God, by definition, does not need to rest. Christians and Jews would answer that this is a figure of speech, or rather a symbol referring to the rhythm God has built into his creation. The sabbatical pattern starts with the weekly day of rest. It goes on to the seventh year in which the land must lay fallow and debts be remitted. Then, after seven cycles of these seven years, comes the year of jubilee, at which time all land goes back to the original owner, all debts are cancelled and the like. It would still be a useful topic to discuss between Muslims and Christians (and Jews of course) in relation to social justice.

humanity but rather the sabbath rest of God.[138] God surveyed the work of his hands and pronounces an overwhelmingly satisfying "it is very good!" The sabbath is the relational dimension that God built into his creation. "This speaks of mutuality in which, in this sabbath of creation, God waits for the response of creation."[139] For Moltmann, it is the "feast of creation," God's peace and joy stamped on his creative work, as well as its consummation.[140] It also closely connects worship to ecology and social justice in the practices which Yahweh commands of Israel: the release of debts and the land left fallow in the seventh years, as well as the amazing Year of Jubilee every seven cycles of seven years, in which all slaves are freed, all debts remitted and all land given back to its original owners (Lev. 25). As Levenson puts it, "We see, then, that Genesis 1:1–2:3, the priestly cosmogony, presents creation as an event toward the rest of God, with which it closes, a rest that signifies an act of redemption and social reform and an opportunity for human participation in the sublime quietude of the unopposed creator God." The Sabbath year, the Jubilee, or the "year of the Lord's favor" is possibly at the very center of Jesus' ministry (Lk. 4:19). Jubilee is also associated with the eschatological day of the Lord, which has now come (the beginning of the "countdown," as it were) in the person of Jesus the Messiah.

(2) For the Christian, the image of God in humanity has a dynamic character because it is primarily relational (though not exclusively).[141] This is so because God in his very being is a fellowship of loving persons. Contemporary Christian theology has rightly made this a central point: God is a triune God. Yet, for obvious reasons, this becomes a touchy issue in Muslim–Christian dialogue. From this perspective, the *imago dei* is not only the affirmation of the Trinity, it leads directly to Christology. In Moltmann's words, "This is to say that through the Son the divine Trinity throws itself open for human beings. The Son becomes human and the foundational image of God on earth. Through the Son,

---

[138] Dyrness, *The Earth is God's*, 136.

[139] Ibid., 14.

[140] "But according to the biblical traditions creation and the sabbath belong together. It is impossible to understand the world properly as creation without a proper discernment of the sabbath. In the sabbath stillness men and women no longer intervene in the environment through their labour. They let it be entirely God's creation. They recognize that as God's property creation is inviolable; and they sanctify the day through their joy in existence as Gods' creatures within the fellowship of creation" (Moltmann, *God in Creation*, 277).

[141] What I mean is that Clines's view of humanity as God's representatives called to embody his purposes in the world (like the king who places his image in conquered territory—the third view I presented) is primary and that the relational is an important dimension of that.

human beings as God's image on earth therefore acquire access to the Father."[142] Despite the difficulties this engenders in conversation with Muslims, however, this must remain the consistent Christian witness.

Finnish theologian Veli-Matti Kärkkäinen[143] is convinced that attention to the triune nature of the Godhead is a necessary touchstone for Christians in dialogue with other faiths. Precisely because the knottiest problem of interfaith dialogue is a meaningful encounter between the "I" and the Other, the Christian proposal to see the divine identity in terms of unity-in-diversity helps to bring a solution. Thus Kärkkäinen agrees with a growing theological consensus that "the Trinity as communion gives room for both genuine diversity (otherwise we could not talk about the Trinity) and unity (otherwise we could not talk about one God)."[144] He confesses that he has not yet developed a satisfactory theology of religions that stems from this Trinitarian insight (that will be left up to his next volume), but in the meantime he finds it best to conclude with a quote from Vanhoozer, who has also examined this question:

> the Trinity is the Christian answer to the identity of God. The one Creator God is Father, Son and Spirit. This is an identification that is at once exclusivistic and pluralistic.[145] And because this God who is three-in-one has covenanted with what is other than himself—the creature—the identity of God is also inclusivistic. The Trinity, far from being a skandalon [stumbling block], is rather the transcendental condition for interreligious dialogue, the ontological condition that permits us to take the other in all seriousness, without fear, and without violence.[146]

---

[142] Moltmann, *God in Creation*, 243. This is not in contradiction to the first point: though the redemptive sacrifice of Christ brings about our adoption into God's family (e.g. John 1:12; Eph. 1:5), the ontological distinction between Creator and creature has not been abolished, but rather the original fellowship that God and his human creatures enjoyed in the Garden is restored. In that sense, our "sonship" is both similar and different from that of the God-man, Jesus Christ. It is through his incarnation and faithful living out of the divine image as a human being (to the point of obedience unto death) that we humans are invited to be (and rule) with our Creator in the renewed creation forever.

[143] See Appendix D for more detail on his theology of religions.

[144] Kärkkäinen, *The Trinity and Pluralism*, 177.

[145] By "pluralistic" Vanhoozer actually means "inclusivistic," in the way I defined it in Chapter 5. I agree with Pinnock, Kärkkäinen, Vanhoozer and other evangelicals that while reconciliation with God comes only through the cross of Jesus, the Spirit of God is drawing many to himself outside the church and often people with no clear knowledge of who Jesus is and what he did for them.

[146] Kevin J. Vanhoozer *The Trinity in a Pluralistic Age: Theological Essays on Culture and Religion* (Grand Rapids, MI: Eerdmans, 1997), 70-71, cited in Kärkkäinen, *The Trinity and Pluralism*, 182.

Just as Christians must listen to Muslims on the issue of *tawḥīd*, I am urging my Muslim friends to listen to the theological implications of humanity's *khilāfa*. Though I have little space to argue this here, I see in Cragg's aforementioned article on "Islam and Incarnation" both a sensitive reading of Islam and an almost inescapable argument for the intimate connection between prophethood and "sentness," between Muhammad, the prophet of Islam and Allah who sends him. As recipient and speaker of the "word from heaven," the prophet's life unfolds in parallel to the unveiling of the Book. "The Qur'an, as divine word, is intensely a human phenomenon, and it takes its place vitally in human history."[147] The affirmation that the prophet is "only" human does not denigrate the revelation, but actually enhances its status through the human agency. What is prior (the sending source) is what counts.

Then Cragg proceeds to explain the Christian idea of incarnation as a different yet similar "association" between the human and the divine. I would continue his argument along the lines of the final communicative act of God in his perfect "*khalīfa*," the Only Son, or the first and last flawless human image of God. Interestingly, Watt entertained this idea as well:

> This discussion has been more concerned with divine sonship than with the concept of man as God's *khalifa*. There are similarities between the two conceptions, but also differences. So far Muslims have done little to work out the implications of being God's *khalifa*, though generally accepting this interpretation of the Qur'anic passage. Until Muslim thinking on this matter has developed further, it would be premature to say anything more about it here.[148]

Finally, I have found in Shabbir Akhtar a Muslim writer who seems to understand this central Christian concept better than others (though for him it remains "scandalously" complex). I quote him at length to show

---

[147] Cragg, "Islam and Incarnation," 131 (cf. Chapter 7, n. 117). See again Rahman's view of inspiration in Chapter 4 for a (liberal) Muslim exposition of this view. In his book, *Islam*, Rahman comes even closer to Cragg's point here. Commenting on the verse which says, "The troubled spirit has brought it down upon your heart that you may be a warner" (Q. 26:194), he argues that the revelation cannot be external to him, since it literally was placed by God in his heart. The Qur'an must have an "intimate connection with the word and the religious personality of the Prophet," besides its undeniable "otherness and verbal character." Hence, the Qur'an is entirely the Word of God and, in an ordinary sense, also entirely the word of Muhammad (*Islam*, 2d ed. [Chicago/London: University of Chicago Press, 1979], 31).

[148] W. Montgomery Watt, *Islam and Christianity Today: A Contribution to Dialogue* (London: Routledge & Kegan Paul, 1983), 137.

that we can, from both sides of the divide, make an effort to rethink our own theology in creative ways, as we truly listen to the Other:

> It seems to me that the doctrinal complexity of Christianity is intimately tied up with its concern to record and partly resolve a central perplexity, namely, the riddle of God's moral involvement with a human nature that is so strikingly recalcitrant to divine guidance. It is very much to the credit of the Christian faith that it recognises—feels and lives in creative tension with—this puzzle. The scandalously intricate collection of Christian doctrines is in fact the direct result of taking seriously the characteristically human threads of the theological fabric: sin, suffering, our moral sense, and the human demand for a divine accountability to the human.[149]

He goes on for another page and a half to unpack the Old/New Testament dynamic in a way that shows profound sensitivity. This is the kind of dialogue Christians, Muslims, Jews and others will experience at deepest levels of human consciousness, primarily because doctrine is meant to be lived out.[150] As we struggle together for human dignity in the midst of heart-wrenching tragedies, inevitably we risk injuring our own souls in the process. The hope, however, is that we will all uncover new and deeper truth.

In this vein, Muslim–Christian theological discussions today could fruitfully benefit from discussing afresh the issue of God's relationship to the world he created. As I see it, N. T. Wright has put this question in a helpful way.[151] There are basically three options. The first, the paradigm offered by pantheism, is by definition off-limits to monotheists: God's space is identical to that of the physical universe. God is in everything and everything is God. Developed in the first century by the Stoics, this view has recently gained in popularity, as the influence of Eastern religions in the West amply testify. Option Two, chosen by Lucretius and Epicurus a century before, posits two distinctly separates spheres: God's and ours. People should get used to the fact that they are alone in this world; the gods are distant and unconcerned with human affairs. In a Christian context, however, if taken to an extreme, this could lead to the idea that this world is essentially evil and that only by finding some secret spirituality (as in Gnosticism) or by death could one escape its

---

[149] Shabbir Akhtar, *A Faith for All Seasons: Islam and the Challenge of the Modern World* (Chicago, IL: Ivan R. Dee, 1990), 175.

[150] See also n. 34 above on Zaki.

[151] See Wright, *Simply Christian*, 60-66. The distinction between Option Two and Option Three becomes crucial at several points in the rest of this work. This is a popular book, yet deceivingly simple. It rests on the painstaking research and careful exposition of his previous technical works.

negative impact. This also led, with the advent of Enlightenment thinking to the position of Deism, according to which God—if he exists at all—is entirely remote from human existence.

Option Three is represented in the Hebrew worldview of the Old Testament and is more fully developed in the New Testament. It states that heaven and earth, though distinct, overlap and connect at many points. If this sounds confusing, suggests Wright, it is because reality itself from a human perspective is complex. Thus God's presence and deliberate interventions on earth form the warp and woof of the Hebrew scripture. Witness, for example, Jacob's dream: he sees a ladder spreading from heaven to earth. Moses sees a burning bush and hears God's voice. Later, God appears to Moses through the thunder, lightning and shaking of Mount Sinai, culminating with the divine delivery of stone tablets with God's law engraved on them by God's own hand. God also gives Moses the design for the Tabernacle, intriguingly "the Tent of Meeting," where he plans to converse with Moses on a regular basis. Finally, God's awesome presence in and around the "ark of the covenant" finds its fulfillment in the Jerusalem Temple, which for the Israelites was the unique location, where heaven touched the earth and God dwelt among his people.

Yet not only does incarnation make supreme sense under Option Three, but also the following: the sacrament of the Eucharist (bread and wine mediating in some sense the body and blood of Christ's redemptive sacrifice); prayer (wherein God and persons commune and requests, praises and thanksgiving are offered to God and mediated through his Spirit); finally too, the knotty paradox of revelation—how God's eternal word comes to be mediated in finite, fallible human language, in such a way as truly to communicate God's will and purposes to his human deputies.

(3) The last central yet thorny issue in the dialogue is the role of the Fall. Pakistani-born theologian Michael Nazir-Ali (now Bishop of Rochester, UK) notes that two movements developed in the Early Church. The Fathers from the eastern part of the Roman Empire, including Irenaeus and Clement of Alexandria, emphasized humanity's creation in God's image (using the Greek word *eikon*). The result was a positive and empowering view of human spirituality and moral awareness, along with the rational capacity to further the study of the material universe and the world of ideas. Here the continuity between God's nature and humanity's was underscored—an intentional Option Three move—with the work of the Holy Spirit enabling persons to fulfill their destiny in becoming more and more like their Creator. By contrast, the second movement, initiated

later in the west by Augustine, put more weight on the reality of human sin as "not only deprivation of the positive good intended by God but also the depravity which is its result and also its ground."[152] Individuals since Adam and Eve are born into a world of fragmentation and distortion, in which the very image of God in them has been tarnished. For this infectious evil has tainted not only individuals but also their social relationships and the institutions of human society. Repentance and faith will now have to precede people's effective use of their God-given gifts for the betterment of society.

As I noted above, creation is tied in the Scriptures to redemption, which in turns leads to re-creation in Christ and the promise of the new heaven and the new earth. The devastating reality of human rebellion and entrapment in evil is a central Christian doctrine, whether in Orthodox, Catholic or Protestant theologies. I also believe that as Christians and Muslims work together on issues related to ecology and social justice the fact of human sin is actually faced head-on by both sides. For the Christian this is not the full solution. As Akhtar recognized in the inner logic of Christianity, only in the cross of Jesus Christ does God lay the seeds of reconciliation between God and people, between people, and between people and their environment. Only in this definitive act of redemption is the promise guaranteed—that promise he made to those who trust in the person of Messiah that indeed one day all things will be made new (Rev. 21 and 22). Yet, as we await the final fulfillment of God's kingdom inaugurated in the person of Jesus of Nazareth, we join hands with all people of good will (especially fellow monotheists) and seek to bring a greater measure of justice, peace and compassion to human society. Indeed, we are convinced that these are signs of the kingdom to come.

Our work together, then, will increase in effectiveness as we improve at properly diagnosing the human condition, a diagnosis which includes a distressing propensity to wreak havoc and "spread corruption on the earth," as the Qur'an would have it—a thought more developed in the next chapter. At the same time, we proceed with a common hope of God's power to overcome evil, even now as we follow through, humbly and wholeheartedly, with our vocation to manage the earth in his stead.

---

[152] Michael Nazir-Ali, *Citizens and Exiles: Christian Faith in a Plural World* (Cleveland, OH: United Church Press, 1998), 112. In 2002 Nazir-Ali was rumored in the British press to be a contender for the position of Archbishop of Canterbury.

Part III

TOWARD AN APPLIED MUSLIM–CHRISTIAN THEOLOGY OF TRUSTEESHIP

Chapter 11

A COMMON THEOLOGY OF TRUSTEESHIP

We now come to the third part of this work, the part which seeks to connect the theory and data of the previous chapters in order to provide a viable theological foundation for Muslims and Christians to re-envision their joint God-given caliphate and carry it out more faithfully on this planet. I seek to go further than before, both in the on-going Muslim–Christian conversation about hermeneutics and a theology of humanity, and in the practical business of making this earth more of a community. This will involve a struggle for greater justice for the poor and marginalized and the promotion of a culture of peace and conflict resolution, often in direct confrontation with the seemingly ubiquitous tentacles of the global capitalist matrix.

Possible Steps Forward in Theology

First, I will look at the views of several Muslim authors on the concept of *tawḥīd* and *khilāfa*. Then, with Ted Peters, I will explore the interaction of hermeneutics and eschatology as a way of advancing the idea of the human caliphate; finally, with Catholic theologian Hans Küng and evangelical ethicist Glen Stassen, I will explore at greater depth the ethics of human rights in the globalized context of postmodernity. I argue, from both Muslim and Christian sources, that the way forward will require a greater integration of ethical theory and theology, which on the Muslim side will also include legal theory.

Humanity, Creation and Tawḥīd
As seen repeatedly throughout this study, the challenge of a careful human management under God of the earth and its creatures has become considerably more difficult in the last few decades. The growing gap between rich and poor, between the G-8, the G-20 and the G-77,[1] is inti-

---

[1] The Group of 77 (G-77) traces its roots back to 1964, starting as a lobbying block within the UN. Now it counts 133 member countries and its first official gathering

mately related to the looming ecological crisis of the planet and the seemingly unstoppable neoliberal global agenda imposed by the World Bank, the IMF and the largest transnational corporations. Our theology, therefore, has to speak directly to the issue of eco-justice. Recall Rasmussen's call to "earth community" as a total way of life and perspective, a paradigm that should govern all human activities. This means that as God's appointed caliphs we are to care for the earth—the gracious home he has granted us. And we are to care for one another, because this mandate to rule was entrusted to all of us equally.

This also implies that Christians and Muslims would do well to develop further their doctrine of creation as the starting point for their understanding of humanity. Gunton calls this a "theology of nature." He joins Murphy and the rationale behind much of this study as he gives "an account of what things naturally are, by virtue of their createdness":

> In the Middle Ages this was generally provided, as we have seen, by what we call the Platonic–Aristotelian synthesis. In the modern West, it was provided by what I now called foundationalism, the common foundations supposedly provided by reason and science. With the loss of foundationalism, we lose the common framework within which our culture was ordered and our moral difficulties approached. Like the medieval, the modern enterprise has collapsed, or is collapsing, under the weight of its own inadequacies. That is the truth in antifoundationalism. But its danger is in its loss of any framework for the ordering of culture—the fragmentation that is the concern of so much modern thought.[2]

In a world losing its modern foundations, we are naturally led back to the idea of the "wager," first mentioned by Pascal, and now brought back to the fore by Ricœur: "You must understand in order to believe, but you must believe in order to understand."[3] Gunton puts it differently. We run into an apparent circle, he says, when we look at the doctrines of revelation and creation. Yet, manifestly, the former depends on the latter. Then he describes in his own words the wager proposed by Naugle (cf. Chapter 4): we must choose to believe that the semiotic nature of the human person in society points to a Creator who designed the human mind to

was in Havana, Cuba in April 2000. Its resolutions have consistently called for the remaking of the world financial system, the forgiveness of debts to poor nations and measures toward the reduction of the growing disparities between haves and have-nots. It is also known as the Non-Aligned Movement.

[2] Gunton, *The Triune Creator*, 99.

[3] We also saw it in Naugle and Vanhoozer and will encounter it again in Peters. It is simply another way of describing the word "faith"—though only one aspect of faith. All trust placed in God's promises as grounded in the scriptures involves an element of risk, and hence, a kind of wager.

connect to reality through its cognitive powers (in a real but limited way). At the same time, we would not know about creation apart from the revelation provided by the scriptures. "Is this circle a vicious one?" he asks, going on the reply, "Not, it appears to me, if one important point is made. In making it, I share the concern of the intratextualists and post-moderns who claim that we are unable to gain an absolutely and objective transcendent perspective upon our world, but are in *certain respects* limited to our historical and conceptual situation."[4]

Gunton then shows that the two doctrines should be taken at two different levels. The doctrine of creation is primary, as it is a teaching about the material world, including our own species ("first-order"); the doctrine of revelation that shows us how we got the belief in creation in the first place is a "second-order" doctrine (*contra* Barth). For me, this is crucial, for it clears the way for Christians and Muslims to agree on common theological ground without touching the contradictory elements in our respective holy texts. Yet, as we continue to explore together how the Creator expects us to fulfill our mandate with regard to the earth and one another in the light of postmodernity, we may need to adjust our traditional approaches. In fact, the caliphate of humanity—our stewardship or deputyship—may be too anthropocentric.[5] Should we not be thinking in more holistic terms? Many feminist theologians would agree with Rasmussen that people must take a humbler role in the scheme of creation. The Korean feminist Chun Hyun Kyung speaks for other Asians and Africans on ecotheology when she writes,

> Many eco-feminists reject the spirituality of traditional Western Christianity, which is based on Greek and Hellenistic dualism, hierarchy of beings and an androcentric bias. Creation theology in this tradition puts human beings, especially man, at the centre of the universe. Man has

---

[4] Ibid., 100, emphasis original.

[5] To be fair, we would have to agree on the definition of "anthropocentric." I appreciate Rasmussen's emphasis, but if taken too far, as some environmental activists do, the protection of animal species might come before the preservation of human life. Bernhard Anderson rightly points out that we have a good deal of wisdom to learn from the rest of the Torah: "Some of Israel's laws place restrictions on the careless harming of and spoiling of nature, for instance the law prohibiting the taking of a mother bird (Deut. 22:6), or the command not to muzzle an ox while it treads the grain (Deut. 25:4)" (*From Creation to New Creation*, 130). He also cites Deut. 20:19-20, which forbids an attacking army to cut down trees. Then comes the haunting question from the holy text: "Are trees in the field human beings that they should come under siege from you?" My point is: human life does have priority, and for that very reason, we must do all we can to preserve the environment in view of generations to come.

"dominion over" all other beings in the cosmos, and God has increasingly become the transcendental Other who has power over the whole universe. This God has been used by men colonizers as an ideological weapon for domination, exploitation and oppression. When God becomes a white, rich European man, white European man becomes a god for all other people and beings in the universe. Therefore, eco-feminists are looking for an alternative spirituality which is able to respond to their need for affirming the sacredness of the cosmos.[6]

This conversation merits to be widened to include other ecotheologians represented in David G. Hallman's volume, *Ecotheology*, along with all monotheists wrestling with the twin crises of growing poverty and environmental destruction.[7] Yet it would also benefit from a wider interaction with the Islamic concept of *tawḥīd*. *Tawḥīd*—the unicity of God and of

---

[6] Chun Hyun Kyung, "Ecology, Feminism and African and Asian Spirituality: Towards a Spirituality of Eco-Feminism," in *Ecotheology: Voices from the North and South*, ed. David G. Hallman (Geneva: WCC; Maryknoll, NY: Orbis, 1994), 176. For a consistent argument using *tawḥīd* and *khalīfa* in the service of Islamic feminism, see Nimat Hafez Barazangi's chapter "Vicegerency and Gender Justice in Islam." The divine empowerment of all humanity (male and female) as his *khulafāʾ* on earth implies: (1) consensus of vision, that is, equality in *ijtihād* ("self-exertion to know, understand, and realize values in present conditions"); (2) consensus of power—all have a minimum of knowledge and power to participate in the affairs of the community; (3) consensus of action—equality in material and educational need. "Without this active role, the individual may not change history, which is part of the accountability involved in carrying out the message of *al Khilafah*. To assume that only males carry this message is an act of injustice not only to females but also to Allah's Minhaj" (in *Islamic Identity and the Struggle for Justice*, ed. Nimat Hafez Barazangi, M. Raquibus Zaman and Omar Afzal [Gainesville: University Press of Florida, 1996], 92).

[7] Hallman helped to organize a Canadian ethical think-tank on global issues on the tail of the Rio Summit. This group ("The Commons Group") "involves politicians, government bureaucrats, members of NGOs, scientists, ethicists, theologians and representatives of international agencies. Our meetings explore the ethical value dimensions of sustainable development." True, Christians are relative latecomers to this conversation but the present crisis evident to all represents "an evangelical opportunity." Therefore, it behooves people of faith to initiate cooperation with persons of other faiths as well as non-believers in these profoundly ethical questions ("Beyond 'North/South' Dialogue," in Hallman, ed., *Ecotheology*, 6). Further, Hallman helped to draft with the United Church of Canada the document One Earth Community. The first three principles read: "Human societies must bear a responsibility towards the Earth in its wholeness. To be both people-oriented and ecologically sound, all development strategies must be founded on a just international economic order, with priority for the world's poor. Life-styles of high material consumption must yield to the provision of greater sufficiency for all" ("Ethics and Sustainable Development," in Hallman, ed., *Ecotheology*, 266).

all things under his creative sovereignty—is the starting point of virtually all Islamic theologies. How this might be tied in with a more holistic Christian theology of creation can be seen from two brief synopses of Muslim views, those of writers Jamal Badawi and Isma'il al-Faruqi.

Badawi starts with this verse: "O mankind! We created you from a single (pair) of a male and a female, and made you into nations and tribes, that ye may know each other (not that ye may despise each other). Verily the most honoured of you in the sight of God is (he who is) the most righteous of you" (Q. 49:13). This oft quoted verse is the foundation for the respect of the rights and dignity of each and every human being. It stresses the unity of the human family because of their common origin and at the same time it affirms their diversity. God himself sanctions, indeed, rejoices in the peoples of the earth he has created as many different nations and tribes.

This is part of the "trust" (*amāna*), reasons Badawi, for if we have inherited the earth as a whole family then we are accountable to exploit its resources on an equal footing.[8] Our accountability is to our Creator God, whose essence, according to Badawi, can be summed up in the concept of *tawḥīd*, "the Oneness, Uniqueness and Incomparableness of Allah (God)."[9] He summarizes the conviction of Muslims scholars about *tawḥīd* in three points:

1. To believe in the One and Only True God (Allah) as the *sole Creator*, Sustainer and Cherisher of the universe.
2. To believe that Allah alone is worthy of *worship* and the unshared Divine authority.
3. To believe in the *unity of the essence and attributes* of Allah, which are all attributes of absolute perfection.[10]

---

[8] Badawi makes explicit what the conditions of this trust are: The method of acquisition of property should be legitimate—excluding theft, extortion, cheating or other Islamically illegitimate dealings; the enjoyment of one's property should not infringe on the similar rights of others; the owner should be mentally capable of looking after his/her property or else a guardian may act on his/her behalf (e.g. in the case of a minor); each must pay whatever is due on his or her property; all Muslims are to pay the *zakāt* (2.5 percent of one's annual revenues and net worth), which is neither a tax nor a tithe. "It is above all a highly rewardable act of worship and an application of *tawḥīd* as it relates to property" ("Relief and Development: An Islamic Approach," unpublished paper presented at the conference "Muslim and Christian NGOs: A Dialogue on Relief, Development and Cooperation," June 17–18 [1996]: 1-22, sponsored by the Canadian International Development Agency. Thanks to David A. Robinson [representing World Vision] who participated in this conference and was able to provide me with a copy of Badawi's paper).

[9] Ibid., 5.

[10] Ibid., emphasis original.

This description of monotheism is, of course, the hallmark of Islam, and it has some fascinating implications for community development. For one, since the One God is the originator of all there is, everything finds not only purpose but also order in him. According to Cragg,

> It is legitimate, with many recent Muslim writers, to see this technological competence as attaining, scientifically, what was latent in the religious theme of *Tawḥīd* or unity. For the sense of order upon which science rests and proceeds, which is indeed the core of its faith, derives from and grows with the sense of a single sovereignty such as Islam proclaimed.[11]

The late Temple University, Palestinian-American scholar Isma'il al-Faruqi, in a paper read in a Muslim–Christian conference, argued that the essence of Islam is *tawḥīd* and that this fact presents itself in six different principles. First, "there is no god but God" means that there is an ontological disparity between God and his creation. This distinction is muddled in Christian theology, in his view, because of the incarnation. Second, creation is related to God as its ultimate cause and its ultimate end.[12] Purpose, meaning and order, therefore, run all through the created world. Third, *tawḥīd* means that humankind can understand and manipulate creation. The purpose of religion is a transformed creation.[13] The fourth principle is that humankind is the only creature that has been granted free will. Only people are capable of moral decision and are therefore accountable for their choices. Al-Faruqi continues, "Fifthly, *tawḥīd* means the commitment of man to enter the nexus of nature and history, there to actualize the divine will. It understands that will as pro-world and pro-life and hence, it mobilizes all human energies in the service of culture and civilization." This, he contends, is in direct contrast to Christianity: "Instead of assuming him [man] to be religiously and ethically fallen, Islamic *Da'wah* acclaims him as the *khalifah* of Allah, perfect in form, and endowed with all that is necessary to fulfill the divine will indeed, even loaded with the grace of revelation."[14] His last

---

[11] Cragg, *The Mind of the Qur'an*, 134. *Tawḥīd* also becomes the starting point for liberation theology. In the words of Asghar Ali Engineer, "*Tawhid* in liberation theology implies not only unity of God but also unity of mankind in all respects. An Islamic society or *Jami'i Tawhid*, does not approve of any form of discrimination whether based on race, religion, caste or class. A truly *tawhidi* society is one which ensures complete unity among mankind and for that it is necessary to created a classless society" ("On Developing Liberation Theology in Islam," *Islam and the Modern Age* 13.2 [1982]: 118).

[12] Al-Faruqi, "On the Nature of Islamic Da'wah," 39 (cf. Chapter 4, n. 5).

[13] Ibid., 40.

[14] Ibid., 41. Al-Faruqi was not always so polemical in tone. His article "Meta-Religion Towards a Critical World Theology" was published posthumously in the

point is more of a summary: *tawḥīd* restores to humanity the dignity other religions have taken from him. "Salvation" is not an Islamic word. It is *falāḥ*, or "the positive achievement in space and time of the divine will."[15]

Though the differences in theology represented by the Christian view of sin should not ever be minimized (indeed "salvation" is central to the gospel), al-Faruqi may be reacting to a traditional Calvinist theology (including Karl Barth's "christomonism"), which has very little patience or room for creation. Badawi, writing twenty years later, is interacting with Christian NGO personnel in a context where the crying needs of the poor and a degraded environment call out for active and creative cooperation. Theology never lags far behind.

Hermeneutics, Khalīfa *and Eschatology*
The human caliph of God is born into his or her cultural worldview, but he or she is not condemned to being entrapped by it—in saying this I am subscribing to Benhabib's "weak" postmodernism. Human beings are

---

American Journal of Islamic Social Sciences (3.1 [1986]: 13-57). As I see it, he follows an inclusivist strategy, with some inspiration from John Hick perhaps (cf. Appendix D), which treats other monotheistic faiths as Revelation—up to a point: "Islamic meta-religion assumes that every religion is God-ordained, until it is historically proven beyond doubt that the constitutive elements of that religion are human made" (ibid., 56). This must be placed next to the following: "Islamic meta-religion honors human reason to the point of making it equivalent to revelation in the sense that neither can discard the other without imperiling itself" (ibid., 57). In the end, the Jewish or Christian readers will not recognize their own faiths as described by al-Faruqi. They feel in him condescension and not respect, despite his commendable attitudes. What is more, the whole enterprise of "Islamizing the Social Sciences" (of which he was a prime mover) is based on this modern, foundational epistemology. Arkoun and Moosa Ibrahim, and even Ahmed and Sardar, operate in a different world—a postmodern one.

[15] Ibid., 41. Badawi, for his part, sees four implications of this. First, there is the Islamic belief in the unity of all revelatory messages. The *tawrāt* (Torah), *zabūr* (Psalms) and *injīl* (Gospel), as well as the prophets God used to reveal them and those mentioned therein—all of these Scriptures are affirmed in Islam. Second, the human race is one and there is unity within its diversity. Third, Islam affirms the holistic dimension of human life—there are no distinctions between sacred and secular, spiritual and mundane. The human being is not a dichotomy in the Greek way of seeing things, but a harmonious blend of physical, mental, emotional and spiritual. Finally, Badawi sees that *tawḥīd* means unity between the present life and the life to come. "As such, individual and collective decision-making is guided by a time scale which is not limited by one's life span, the life of one or more generations, or even the life of all generations" ("Relief and Development," 6). The implications for ecology are plain: before God we have a responsibility to bequeath a clean and well functioning environment.

called to use their God-entrusted imagination and energies to rebuild the world so that it might more closely reflect the original purposes of their Creator. The reimaging part for the believer is theology. In this section I use Peters as my guide in reconstructing the world in a way that I hope will engage the support of Muslims. Also, because theology today is carried out in the context of an increasingly postmodern world, I offer Peters's framework as an illustration of what a theology "without foundations" might look like. All the diagrams that follow are mine, and in that sense, they represent my own interpretation of what Peters is advocating—through the filter of my purpose of advancing Christian–Muslim dialogue.

Hermeneutical concerns are at the heart of Peters's project. And to begin with, he contends, one has to go back to the core ideas of the Enlightenment. The very idea that there is a distance between knower and known and that this distance is problematic is a "modern" invention:

> In fact, it is only because of modern thinking that one could even suggest the possibility that the Bible's meaning is strange, anachronistic, or no longer valid. Augustine in the fifth century did not think the Bible was out-of-date. Nor did Thomas Aquinas in the thirteenth century or Martin Luther and John Calvin in the sixteenth century think of the Bible as old-fashioned. Only those who come after the rise of natural science and the Enlightenment pit what Scripture says against what we learn from other sources. The distancing of ourselves from what was said in the ancient world of the Bible is due to a fundamental shift in our way of thinking, a shift that marks the difference between the premodern and the modern eras.[16]

Modernity's worldview prizes three values above all: natural science and the scientific method; a secular understanding of the world and the self; and a worldwide cry for freedom. When the world came to be pictured as a great machine, the need to keep the divine presence as a necessary part of the explanation for things began to diminish, and, in the end, actually got in the way of "rational thinking." So too, this secular viewpoint carried over to society and politics. The French Revolution in its revolt against a monarchy by divine right embarked on a course of de-christianizing Western culture: "The spread of democracy is assumed to be a human achievement. We moderns no longer count on angels to help us. Secularity is a way of understanding life that simply accepts the natural world to be the only world."[17] Here Cragg would interject that modern secularity as a fact of life for much of the world represents a

---

[16] Peters, *God—The World's Future*, 8-9.
[17] Ibid., 9.

healthy challenge to all faiths, including Islam. In fact, in what I consider to be one of his best books (*Returning to Mount Hira'*), he argues that the original revelation at Mount Hira' is not only in harmony with much of what Christians believe, but provides a creative resource for Muslims to witness to the relevance of their faith in today's pluralistic context:

> The Quranic bid to bring all things through human *islam* into divine conformity has now to measure and fulfil that mandate in the dimensions that today sets for it. Secularity has to be religiously acknowledged as the sphere of faith's task, not the realm of its anathema. Islam has to do with things in trust which He has willed should be ours to bring, in as much as His purposes for them come about only when we make them our own.[18]

At the same time, continues Peters, secularity also drives a wedge between the human subject and the world of phenomena. The objective world operates "like a Toyota Camry—that is, according to a closed system of fixed mechanical laws." By contrast, the inner world of the human person is the locus of freedom. People are free to vote, choose what cars or cereal brands they want to buy, or decide what kind of occupation they will pursue. "We moderns have come to believe that this freedom to decide puts us (not God) into the driver's seat of the cosmic Camry."[19] The result comes as no surprise: as people celebrate their freedom from all the former shackles of tradition, the sacred dimension is sucked out of nature and society. Further, with the help of psychotherapy and various revolutionary ideologies and movements, the last half of the twentieth century has witnessed a fight for freedom on every front, from the liberation from all self-inhibitions to the struggle for freedom from poverty and oppression all over the world.

Peters calls this modern perspective "critical consciousness." The best word to describe it is distanciation. The drama of the real world unfolds as we human subjects observe, describe and analyze. Thus everything we look at becomes objectified, but as we affirm our non-participation, we become at the same time alienated from nature. But never mind: what is important is objective truth, which, by definition, must be impersonal. This so-called objectivity, sadly, carries with it a serious down side: "The

---

[18] Kenneth Cragg, *Returning to Mount Hira': Islam in Contemporary Terms* (London: Bellew Publishing, 1994), 139. Cragg's use of *"islam"* above refers to the human act of submitting one's will to the greater will of God. God's sovereignty through the mandate of human trusteeship deliberately (and mysteriously) sets limits on itself by risking human rejection and rebellion. But without freedom of choice, the trust has neither meaning nor possible accountability.

[19] Peters, *God—The World's Future*, 10.

flip side of this objectivity is unchained subjectivity." Ethical values, like individual tastes, are defined by individuals.[20]

And so it is that the contemporary struggle between objectivism and relativism is really a creation of the modern critical consciousness—a form of "sibling rivalry," as Peters puts it. Both are children of the modern mind. In the following quote, note the clash between the modern view of the self and the strong version of the postmodern I have been highlighting in this book:

> As the battle rages, the relativists accuse the objectivists of mistaking their own culturally determined perspective for what is universal or permanent, which means the objectivists are blindly purveying ethnocentrism. The objectivists counter by accusing the relativists of self-contradiction: if the relativists claim that their position is universally true, then the relativist position is said to transcend the limits of its own cultural conditionedness, and, hence, the position undermines itself.[21]

Missing here, of course, is the "weak" version of the postmodern self and its grasp of reality. In Chapter 4 we saw the importance of recognizing the situated nature of the self (in community and culture) and its critical realist approach to the knowledge of things outside of itself. Hence I opted for Naugle's view of rationality as context-dependent and Barbour's four criteria of truth (correspondence, coherence, scope and fertility). Along similar lines, Peters argues that we must adopt a more comprehensive vision of the truth than the dead-end clash of relativists vs. objectivists. Also, by leaving out our own human feelings, evaluations, convictions and beliefs, we are turning our back on an important part of reality. "Critical distance through objectification has made us forget that there is only one reality that includes both external objects and human subjects in relation to one another."[22] This also means that to think of human freedom in terms of individual autonomy is unrealistic and destructive in the end. As individuals we are tied to families and other social subgroups; we are also connected to the socioeconomic and political conditions of our society; and finally, we are dependent upon our physical environment. Now, as we approach the brink of ecological collapse (according to some scenarios), it becomes imperative that we humans think in more holistic ways.

Peters's description of postmodernism is brief, but very much in line with other accounts we have seen. I will only add his two categories: postmodern deconstructionism, which could also be seen as the radical

---

[20] Ibid., 12.
[21] Ibid., 13.
[22] Ibid.

critique from the above relativist position, or Benhabib's "strong" version
of postmodernism; and "the reconstruction of wholeness." Here he
quotes Murphy's vision of postmodernity as seeing the whole as greater
than the sum of the parts, "a complex mutual conditioning between part
and whole."[23] What is needed is a philosophy of holism that can overcome the destructive tendency of modernism to think in dichotomous
categories and thus fragment reality. The healing must both be epistemological and ontological. To this I would add, echoing Murphy,
"semantic holism," by relating critical realism to a viable hermeneutic of
sacred texts.

I attempt to summarize Peters's presentation here in figurative form.
His greatest insight, to my mind, is his relating the three periods of
thought (premodern, modern, and postmodern) to three stages in a
believer's faith.[24] Though the Christian's faith can be fully authentic at
each stage, it will not be healthy for a person who enters faith (and all
do) at the "naïve world-construction" stage to stay there—especially if
the motive is systematically to refuse all the criticism leveled at this
worldview by the second level, "critical deconstruction." This first level
he calls "atavistic fundamentalism, a defiant defense of biblical literalism."[25] With the second stage comes radical criticism—the modern
onslaught of rationalism—"critical deconstruction." Though this second
stage has meant the loss of faith for countless people (look at Europe
especially), "We can no longer live naïvely in the world of the Bible...
[O]ur naïve relationship to the symbolic world of the Bible is broken," he
contends.[26]

Peters cautions his readers that no stage is intrinsically better than
another and that the mission of the church is not to move people from
one end of the spectrum to the other. Yet he does not hide his preference
for the third stage, "postcritical reconstruction." The only way to counter

---

[23] Ibid., 17-8.

[24] Ibid., 22-23. I believe this applies also to Muslims. Most Jews in Western societies
have experienced these kinds of paradigm shifts, though not the "Ultra Orthodox,"
or *Haredim*, to be sure.

[25] Ibid., 29. He makes a helpful distinction here: "One aspect of the debate that makes
it unnecessarily bitter is the frequent failure on the part of liberal and neoorthodox
proponents to distinguish between the first naïveté and fundamentalism. These two
are not exactly the same. Naïve literalism is precritical. Fundamentalism is anticriti-
cal. There can still be considerable intellectual integrity at the level of compact
naïveté. The problem of bibliolatry [worship of the Bible instead of the God of the
Bible] arises only as a defiant response to the attack launched by critical consciousness. Bibliolatry does not belong to the first naiveté proper."

[26] Ibid., 25.

the devastating doubt of the previous mindset is to follow Ricœur's advice (like Pascal of old) and wager. Theology, in fact, is a hermeneutical circle. He quotes Ricœur: "We must understand in order to believe, but we must believe in order to understand." In essence, this is a post-Cartesian version of Saint Anselm's old formulation, perhaps best captured by the wager:

> A wager is a risk, a bet. In this case—and in this book—we are betting that a hermeneutic of belief in the Christian gospel will be more fruitful for living in the world than the skeptical conclusions produced by a hermeneutic of suspicion. We will not forget our doubts. But we will press on, trying to understand ourselves and the world around us in light of the symbols of divine revelation. The wager is a form of hypothetical belief, a self-entrustment to the world of meaning created by Christian language.[27]

As we have seen, this hermeneutical problem is not made easier in the postmodern global context, especially because it is in itself a response to the predicament of modernism. The three stages of thought are very much related. Nor is the postmodern shift a clean break from the modern mindset. As persons we go back and forth between these perspectives, depending on what the subject matter might be. The world of objective rational discourse still dominates most discussions in criminal or legal affairs, scientific discussions, or the everyday business of shopping and managing one's possessions. Also, the premodern perspective naturally colors much of religious life, at least in the confines of liturgy, prayer and sacred rites.

Now I present the three stages in graphic form, both as reflective of the history of Western thought, and as possible stages in a person's life of faith.

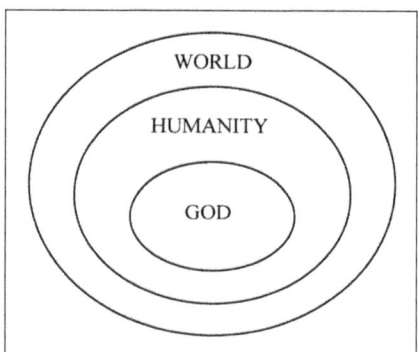

Figure 11. *Naïve World Construction*

27 Ibid., 30.

Figure 11 presents the integrated but precritical world of "naïve world-construction." This would equally apply to the three monotheistic faiths. The world–subject distance has not yet become a problem, but all flows ontologically and experientially from God in concentric circles.

Figure 12 represents Peters's "objective" side of the debate between both critical positions of modernism, that is, objectivism and relativism. Epistemologically, most scientists are critical realists. This is a recent turn. Figure 14 would work, however, for both critical and naïve realists (the modern mentality until, say, Einstein, or certainly Polanyi and Kuhn—though not all scientists are consistent in their epistemology).

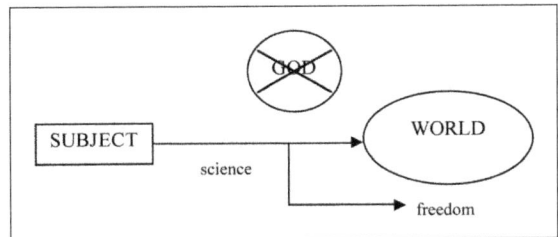

Figure 12. *The Objective Critical Consciousness*

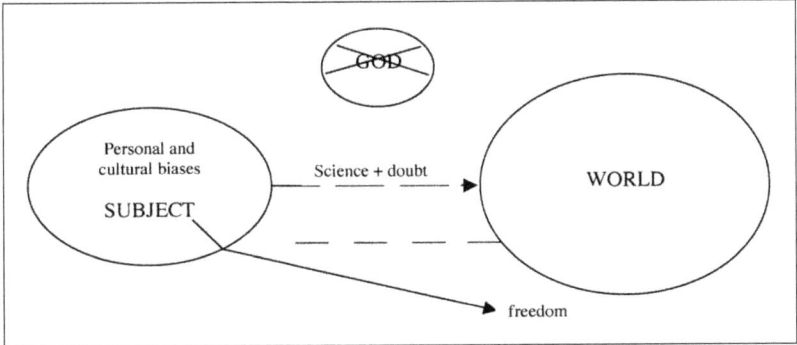

Figure 13. *Peters's Critical Deconstruction—From Modern Relativistic to Postmodern Deconstructionist*

Figure 13 attempts to illustrate the position held by most in the West today, even on a popular level. Though the relativist side of critical consciousness is mostly held with regard to values (and not about science, at least not for the average person), it can be more radically applied to all areas of human experience. This is why I am collapsing the two positions into one. Postmodern deconstructionism (or Benhabib's "strong" version), after all, is only the logical conclusion of Descartes's radical doubt—and doubt has been part of the Western tradition ever since. Also, freedom continues to be on the agenda, though in this last phase the basis for any universal ethic has collapsed and freedom is left to the individual or a particular ethnic group to work out.

In this study I am contending with Peters that a better model must be set up—one which could be embraced by all monotheists in particular. As humans in the twenty-first century we cannot return to a naïve realist perspective, but neither can we afford the radical deconstructionist version of postmodernism. Along with Murphy, Barbour, Naugle, Peters, and others, I would affirm a holistic vision based on a critical realist epistemology, coupled with a non-foundationalist theology.

In conclusion, I offer two sketches of possible directions from here, first from a monotheistic perspective on epistemology, then from a specifically Christian perspective. Whereas the first is synchronic (Fig. 14), the second is diachronic, sharply affected by time, space and history (Fig. 15).[28] I advocate with Peters that both Muslims and Christians adopt *vis-à-vis* their scripture a hermeneutic of trust beyond doubt—a perspective which takes seriously the critical methods of modernism and the doubts raised by the postmodern deconstructionists. It is at once a decision (1) to step inside the world created by the symbols of scripture and tradition; (2) to adopt them as one's own; (3) to test continually their compatibility with the world with which one interacts; (4) to listen daily to God speak through the text (the *hidāya* promised by the Qur'an, and the Holy Spirit promised by Jesus); and finally, (5) to keep on integrating those insights and putting them to use by reshaping the world as our Creator would have us do as his empowered deputies on earth. The largest circle represents the overall sovereignty of God as Creator and Sustainer of the universe. Whereas human beings as God's deputies have

---

[28] Whether in connection to Cragg or not (this is one of Cragg's arguments throughout), Rahman sees the interaction of theology and history as his key hermeneutic: "For modern thinkers such as Fazlur Rahman it was vital to make sense of revelation in historical terms. If history was to make any impact in understanding a transcendent revelation, then it was necessary to explore the interface of revelation with the world. An insistence on the complete 'otherness' of the Qur'an, as orthodoxy required in order to minimize the Prophet's involvement in the revelatory process, was not only historically inaccurate in his view, but also contrary to the Qur'an itself. Historically, it was difficult to ignore the fact that revelation itself commented on matters that affected the prophet's personal behavior and travails… Revelation was entirely from God and at the same time the locus of revelation was the 'heart' of the Prophet where it is vouchsafed in historical time" (Moosa, *Revival and Reform*, 13). This is very close (arguably identical) to a Christian view of revelation.

been granted a measure of free will in order to fulfill their mandate on earth, they do so within the scope of God's creative and determinative presence.

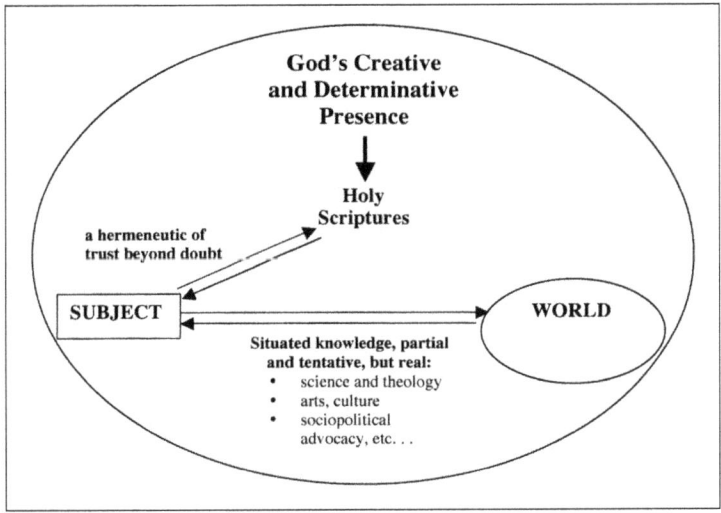

Figure 14. *A Monotheistic Epistemology of Holism*

Naturally this is not an individualistic exercise, neither for the Christian nor for the Muslim believer. The symbols are lived out in a local congregation and brought to life through liturgy, prayer, and the sacraments.[29] But as point 3 shows, a postmodern hermeneutic of trust

---

[29] Peters rightly questions the validity of the "translation model" modern theology has heavily favored: "Much as a translator interprets meaning from one language into another, modern theologians have been seeking new formulations for apparently out-of-date symbols and beliefs. They seem to assume that the meaning of the Christian faith could be cut loose from its original situation" (*God—The World's Future*, 15). With the coming of postmodernism this model needs modification for two reasons. First, the language and worldviews of the Bible are not so foreign to us, since modernity has generously planted its roots in ancient Greek thought. Second, Christianity offers some basic symbols, which cannot be translated—the cross, in particular. Their meaning must be reinterpreted in the light of new situations, but symbols must remain as they are. He concludes, "What permits the gospel to ride out the centuries, traveling from one age to another and one language to another, is not its translatability. It is rather the protean power of its symbols to emit new meanings in new contexts" (ibid., 16). I agree with him: the gospel is more than a verbal or written message. But this is precisely what makes it so "translatable" in other cultures. Lamin Sanneh argues that it is this quality of the gospel message (the translatability of the life and work of Jesus of Nazareth into forms and meanings of

must remain self-critical, open to new possibilities, ready to hear God's voice in unexpected settings and through unlikely instruments. Both the Christian and the Muslim must be ready to adjust or modify their theology (not their core beliefs) in the light of new insights gleaned from their conversation. For all of us too, obedience to exercise our God-shaped caliphate in the world will inevitably raise new questions. In the end, as Ricœur is fond of reminding us, we will be called to smash the idols of modernism and postmodernism again and again. The human heart is prone to create new idols and serve them with complacency. At each new turn we will need to be confronted with the uncompromising "Hear, O Israel: The Lord our God, the Lord is one. Love the Lord your God with all your heart and with all your soul and with all your strength" (Deut. 6:4-5, NIV).[30]

Figure 15 is my Christian interpretation of the human caliphate alluded to in these pages. It represents my best attempt so far to illustrate what I read in Peters and in the gospel. I find Peters's theology congenial, while noting his lack of sensitivity to the Western–non-Western dichotomy, as well as to the implications of the present north–south disparities and clashes. I have often mentioned liberation theology as a necessary partner in the course of this book. To that could be added biblical work attuned to postcolonial studies. An example of someone sensitive to issues of power is the recent work of Cuban-American biblical scholar Fernando Segovia. He uses the methods of the discipline of postcolonial studies (by definition multidisciplinary) to work on theology as a biblical scholar, a cultural critic and a constructive theologian. By drawing on the insights of postmodernist approaches to hermeneutics (particularly reader-response criticism), he crafts a theology that remains attuned to readers of the Bible living in impoverished settings both

---

indigenous cultures in Africa) that not only caused its rapid spread on that continent in the last two centuries but that also generously contributed to movements of indigenization, national aspirations and democracy. Here the contrast with Islam resides in this play on the word "translatable": the New Testament was always translated into the vernacular languages, which meant that the meaning and impact of the gospel took flesh more completely within the various cultures it penetrated (Sanneh, *Translating the Message*).

[30] Sparked originally by Pope Benedict XVI's controversial lecture at the University of Regensburg in September 2006, 138 leading Muslim scholars and leaders representing all regions of the world and all Muslim sects (Sunni, Shi'i and Sufi) published an open letter to the Pope and to all Christian leaders ("A Common Word"). Their central message was that what unites Muslims and Christians is the very core of their faith: love for God and love for their neighbor (see http://www.commonword .org).

socially and geographically, and this through a paradigm that focuses "on contextualization and perspective, social location and agenda, and thus on the political character of all compositions and texts, all readings and interpretations, all readers and interpreters."[31] Following Jesus, who lived out the prophetic vision of Jahweh delivering the poor, bringing justice to the oppressed and welcoming the outcasts, the theologian will necessarily engage in a mode of discourse that "may be described as profoundly ideological." At the same time, contends Segovia, one cannot replace the Western hegemony of modern theology by another imperialist anti-Western theology. What is needed is a plurality and diversity of voices, "a theology of engagement and dialogue, committed to critical conversation with other theological voices from both margins and center."[32]

The same could be said about theologian Georg Rieger, who builds his theology around God's primary concern for the outcast and excluded.[33] Though he is focused (intentionally) on the North American scene, his work lends direct support to the present thesis: "Theology is reconfigured in light of the challenges of the global market and a postmodern world in which difference and the loss of foundations are not only celebrated ideas but manifest in structures of exclusion and everyday marginalization, suffering, and oppression."[34] By drawing his theological method from French psychoanalyst Jacques Lacan, he opts for a postmodern hermeneutics that is both critical of four major theological currents and supportive of their strategies (or "discourses") when placed side by side one another: the "liberal" turn to the self (the modern paradigm), the "neoorthodox" turn to the Divine Other (Karl Barth), the "postliberal" turn to language and the texts of the church (deconstructionism), and the "liberation" turn to the Other.

Rieger recounts how his own theology experienced a paradigm shift while becoming acquainted with the poor in the slums of West Dallas. Notice in the following paragraph how he consciously opts for a "weaker" version of postmodernism:

> Truth has to do with openness and listening, with being shaped by rather than with being in control over the subject matter. The turn to others does not copy the classical liberal turn to the self. Marginalized people are not substitutes for the controlling position of the self. At the same time, truth

---

[31] Ibid., 119-20.

[32] Ibid., 123.

[33] Georg Rieger, *God and the Excluded: Visions and Blindspots in Contemporary Theology* (Minneapolis, MN: Fortress Press, 2001).

[34] Ibid., 15.

does not become a relativistic concept either. Not unlike modern foundationalism, postmodern relativism is the privilege of those who are in control. Once modern claims of universality and identity break down, those in charge have the luxury to withdraw into their gated communities and churches where they can make their own rules, and thus the pluralistic play of difference becomes fun.[35]

But still, postcolonial and liberational perspectives fall short of the breadth of vision as exemplified in the essays of Hallman's Ecotheology, or in Rasmussen's "earth community." As Mitzman reminded us, postmodernity is the fateful rush toward the double wall of global injustice and environmental collapse. We need the Segovias and the Riegers, but we also need a theology that is more holistic, one that weaves together creation and the end of creation, or eschatology.

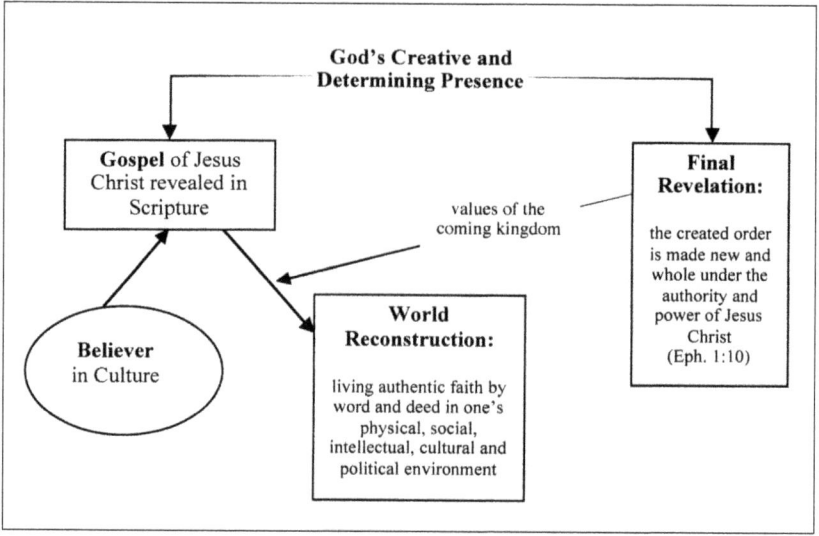

Figure 15. *A Christian Postmodern World Reconstruction Perspective*

As a Christian, then, I am arguing that creation is given to us by God to observe, analyze and reconstruct in the postmodern context, not just in terms of science, technology and the arts, but more importantly, as a means to fulfill our calling as God's representatives on earth with the dimensions of social justice, compassion and witness to his redeeming love in Jesus Christ. This is so, by virtue of our creation in God's image and further, by virtue of our re-creation in the image of him who is the

---

[35] Ibid., 186-87.

perfect human image of the Creator, our Lord Jesus Christ.[36] As a believer I move forward with other fellow believers, finding guidance by God's Spirit who brings the gospel to life for me—that Good News revealed in time and authoritatively enshrined in the Bible. But it is not only from the past that I receive direction. The gospel is the revelation of God's definitive work of redemption for people first, and for all the created order as well. God's work of re-creation began in Jesus of Nazareth; is continued in this age through the work of the Holy Spirit in and through his human agents; and will be consummated when he returns, judges all, and sets up his eternal kingdom in a creation made new. This vision of wholeness restored guides all our present efforts, as the next section indicates.

My main thesis, then, is that the vision offered in Figure 14 represents both Bible and Qur'an in a constructive theology of creation, with humanity as God's caliphs (*khulafā'*) on earth—a vision clear enough to empower and to guide Muslim and Christian efforts toward an authentic and embodied global ethic at this time in history (cf. the next section). Both the intellectual atmosphere—starting with its epistemology (post-modernism)—and the enfolding globalization dynamic of our twenty-first century world (postmodernity) do indeed warrant a new effort on the part of Muslims and People of the Book to reinterpret their faith accordingly. Indeed, there is hope beyond the destructive secular waters of modernism and postmodernism.

The central challenge for people of faith today is one of meaning. Peters frames the hermeneutical question in a way that is sensitive to the needs and concerns of our postmodern contemporaries, and particularly the anti-globalization activists, be they peasants, workers or intellectuals: "How can the Christian (Muslim or Jewish) faith be made intelligible amid an emerging postmodern consciousness that, although driven by a thirst for both individual and cosmic wholeness, still affirms and extends such modern themes as evolutionary progress, future consciousness, and individual freedom?"[37]

How then can the sacred text and its central symbols be made to come alive in the minds of readers today in a way that preserves the intent and integrity of the original message? This question cannot be resolved by

---

[36] Thus Gunton writes, "To be created in the image of God is not simply to be made and upheld by the triune God; it is, more specifically, to be upheld by him through the Son, that is, through Jesus Christ, who is the true image of God" (*The Triune Creator*, 207).

[37] Peters, *God—The World's Future*, 8.

theologians in isolation. It will be worked out in action, in the praxis of living out our God-given mandate to care for his creation—the earth itself and all it contains, starting with our suffering fellow human beings.

But there is one last point. One of the themes of postmodernity is "future consciousness." As People of the Book we all look ahead to the "hereafter," where we will stand before God's throne of judgment. Yet might there not be a more earthy, holistic, vision of the future, one which could inform and guide our efforts as God's caliphs now? As I have intimated above, from a Christian perspective there can be no doctrine of creation without eschatology at the same time. The work of creation is continued in Mary's womb when God takes the flesh of this Galilean Jewish peasant[38] and directly implicates himself in the work of re-creating a universe marred by humanity's rebellion. Christ's resurrection thus became the concrete pledge that not only our bodies will be resurrected on the last day, but that God has indeed begun to fulfill the prophet Isaiah's vision of a New Heaven and a New Earth (Isa. 65:17; 66:22; Rom. 8:18-22; Rev. 21:1-4).[39]

Certainly there are images of well-watered gardens in the qur'anic Paradise. But in the Qur'an the transition between this world and the next is more a case of discontinuity. So far, then, I have found no answers to my question. How could the Islamic vision of the "sacramental earth" (cf. Chapter 6) be carried into a vision of God's future dealing with the world that could inspire our efforts to preserve it today? I believe this is possible as we look together at the ethics of creation. For this reason, the rubric "values of the coming kingdom" in Figure 17 (see next chapter) is expanded in the next section. The fact remains that unless we Christians

---

[38] The parallel is made in the Qur'an, though not conceived of course in the category of incarnation: "The similitude of Jesus before God is that of Adam; He created him from dust, then said to him: 'Be': and he was" (Q. 3:59).

[39] Gunton adds this: "It is better to understand the incarnation of the eternal Son in the flesh as the beginning of an eschatological act of renewal, in which the true *telos*, direction, of the creation is restored from within… At the other end of Jesus' life, the resurrection is also to be understood as an eschatological act, new because it is the creation of the form of being which belongs to the end of time" (*The Triune God*, 223-24). Part of the difficulty from the Islamic side is that the first creation is not affected by human sin, thus the need for re-creation is absent. Yet the facts on the ground are the same for all: the earth today is sick, both its people and its air, soils, water and creatures, great and small. My sense is that more could be made of the use of God's "signs" (*ayāt*) in the Qur'an. God is certainly still active in his creation and, further, he desires to work through his *khulafā'* to cleanse and heal the earth. Yet the only eschatological dimension here is that of accountability. We humans will be asked to give account for the way we have squandered the gifts of nature.

and Muslims consciously develop a common ethics out of our theology of creation, we will miss a powerful tool for social and environmental change.

The Ethics of Trusteeship

Actively working on interreligious ethics since the 1980s, Hans Küng was also one of the prime movers behind the 1993 *Declaration of the Parliament of the World's Religions*. His third drafting of the pre-conference declaration remained virtually unchanged throughout the discussion by the 200 plus delegates from all the major religious groups of the globe in Chicago from August 28 to September 4, 1993. In this moving document signed by the great majority of participants, a basic recognition stands out—that there can be "no better global order without a global ethic."[40] In this document, the UN's Universal Declaration of Human Rights is reaffirmed, but "[w]hat it formally proclaimed on the level of rights we wish to confirm and deepen here from the perspective of an ethic: the full realization of the intrinsic dignity of the human person, the inalienable freedom and equality of all humans, and the necessary solidarity and interdependence of all humans with each other."[41]

In his own book devoted to this topic, Küng avers that the task is made easier with simply Islam, Judaism and Christianity in consideration: "it is the one unconditional in all that is conditioned that can provide a basis for the absoluteness and universality of ethical demands, that primal ground, primal support, primal goal of human beings and the world that

---

[40] Küng and Kushel, eds., *A Global Ethic*, 21. For a penetrating analysis of the complex issue of how to balance the "politics of difference" (e.g. self-determination for oppressed ethnic groups), distributive justice, and the ideals of democratic equality in a universal sense, see Michael Walzer's *Thick and Thin: Moral Argument at Home and Abroad* (Notre Dame, IN: Notre Dame University Press, 1994). Using Geetz's anthropological model, he uses "thick" to refer to local socio-political and cultural realities, and "thin" to refer to the commonality of human life across human cultures. See also the collection of essays in Max L. Stackhouse, Peter L. Berger, Dennis P. McCann and M. Douglas Meeks, eds., *Christian Social Ethics in a Global Era* (Nashville, TN: Abingdon Press, 1995). The differences of emphasis and perspective among them are instructive in themselves.

[41] Küng and Kushel, eds., *A Global Ethic*, 20. The first of four directives reads, "Commitment to a culture of non-violence and respect for life." This respect for all other human beings includes a respect for the environment: "As human beings we have a special responsibility—especially with a view to future generations—for Earth and the cosmos, for the air, water, and soil. We are all intertwined together in this cosmos and we are all dependent on each other" (ibid., 26).

we call God."[42] In this new opportune time of "postmodernity," he calls all believers in the one God to unmask and denounce all the false gods of modernity (science, technology, and industry, or capital).[43]

Here are the salient points of Küng's global ethic:

1. The beginning of an ecumenical quest for a global ethic is to recognize that truth does exist and that it includes right living as well as right belief. Each religion will therefore start out by taking "a critical look at one's own history of past failure and guilt."[44]

2. In addition to the specific criteria each religion has for itself, there are potentially universal ethical criteria. Long before the Enlightenment posited the autonomous human being, each tradition had its criteria for defining the truly human person against the horizon of the absolute. The result of an interreligious conference in Paris in 1989 was a positive affirmation of the universal applicability of the *humanum* (European humanism influenced by Christianity): "that would be morally good which allows human life to succeed and prosper in the long term in its individual and social dimension…(including their relationship to society and nature)."[45]

3. The Golden Rule is applicable to all. In Küng's own words,

> We must treat others as we wish others to treat us. We make a commit-
> ment to respect life and dignity, individuality and diversity, so that every
> person is treated humanely, without exception. We must have patience and
> acceptance. We must be able to forgive, learning from the past by never
> allowing ourselves to be enslaved by memories of hate. Opening our
> hearts to one another, we must sink our narrow differences for the cause of
> world community, practicing a culture of solidarity and relatedness.[46]

Beyond the Muslim participants in the 1993 conference, the ethic of "solidarity and relatedness" finds a growing echo in Muslim circles. We saw that this was an idea Muhammad Iqbal cherished. Recall also Anouar

---

[42] Hans Küng, *Global Responsibility: In Search of a New World Ethic* (New York/London: Continuum, 1996), 53.

[43] "Now in the postmodern period they all seem largely to have been demytholo-gized and deideologized, in other words, relativized. And in this new world constel-lation we should not replace them with a new idol, for example the 'world market,' to which all values would have to be subordinated, but with renewed faith in the one true God. True religion, which thus relates to the one and only absolute, again has a new opportunity in the postmodern period—no more no less" (ibid., 54). Note that his use of "postmodernity" (and Peters's as well) does not include my own socio-economic and political content.

[44] Ibid., 81.

[45] Ibid., 90.

[46] Küng, *The Declaration of the Parliament of the World's Religions*, 15.

Majid's *Unveiling Traditions: Postcolonial Islam in a Polycentric world* (cf. Introduction). Another veteran in this respect is Kenyan-born Ali A. Mazrui, the Andrew D. White Professor-at-Large Emeritus at Cornell University. A specialist on African Islam and postcolonial politics who has lectured on five continents, Mazrui has consistently advocated a global vision of peace and intercultural understanding—especially for his fellow Muslims. In a 1994 essay on the nuclear threat, he makes two important points.[47] The nuclear age has been difficult for Muslims in general, and, in desperation, countries with scientific know-how such as Egypt and Pakistan might join hands in the nuclear race with the oil-rich Arabian Gulf countries. On the one hand, this might actually force the world to denuclearize and help settle difficult conflicts, including the Israeli–Palestinian one. On the other hand, he warns, terrorism could also go nuclear: "[I]f the cultural imbalances between Israeli and Arab or white and black deepen this sense of desperation, we cannot rule out the possibility that the weak might acquire nuclear devices from radical friends elsewhere. Powerlessness also corrupts—and absolute powerlessness can corrupt absolutely."[48] His conclusion seems wise:

> [N]uclear disarmament is not enough. We need to reduce the risk of war... Islam and Africa will have to join forces in a search for a more viable world order. One day the warriors of Africa and the Mujahidin (fighters for Islamic justice) must put away their swords and spears and celebrate the liberation of Planet Earth from the specter of chemical weapons, nuclear war, and excesses of injustice in human affairs.

A global ethic for Mazrui, then, should include peacebuilding along with an ethic of nonviolence. Yet it should also look at the issue of race as a lens through which to observe and understand the grave injustices perpetrated on the weak and downtrodden. The same year his essay was published, Mazrui presented a paper at Chandra Muzaffar's Just World Trust (JUST) conference in Kuala Lumpur, Malaysia, on the theme "Rethinking Human Rights."[49] This time he highlighted the racial component of the growing oppression in the post-Cold War world. Using the dual society of Rwanda as a backdrop (Tutsis vs. Hutus), Mazrui pointed

[47] Ali Mazrui, "The Nuclear Option and International Justice: Islamic Perspectives," in Barazangi, Zaman and Afzal, eds., *Islamic Identity and the Struggle for Justice*, 95-116,

[48] Ibid., 112.

[49] Ali Mazrui, "Human Rights between Rwanda and Repatriations [*sic*; from the content of the article it is clear that it should read "Reparations"]: Global Power and the Racial Experience," in *Human Wrongs: Reflections on Western Global Dominance and its Impact upon Human Rights*, ed. Just World Trust (Penang: Just World Trust, 1996), 188-211.

to the dual nature of American society, fundamentally polarized between black and white, and to the growing reality of global apartheid since the white-led socialist countries ceased their advocacy for Third World causes. Perhaps the most graphic illustration of this is Africa's growing impoverishment: "The continent still produces what it does not consume, and consumes what is does not produce"; in other words, from an agricultural viewpoint, Africa lives on a "dessert and beverage economy."[50]

Between the structural causes of racism put into place since the Western conquests of the fifteenth century and the attitudes that continue to uphold racial prejudice in the West, the fall of the Warsaw Pact states has not been good news for the poor of the world—nearly all people of color. Though the 1990s witnessed a "micro-retribalization" in the Balkan conflicts (add to that Russians vs. Ukrainians and Czechs vs. Slovaks), is it not possible, asks Mazrui, to consider the great strides forward made by the European Union as a kind of racial "macro-retribalization"? He continues his query: "For both Eastern and Western Europe, is the White world closing ranks? Will we see a more united White world, and potentially more prosperous, presiding over the fate of fragmented and persistently indigent black world? Is this the prognosis of the 21st century?"[51]

Larry Rasmussen has recently examined this question of race as a Christian ethicist. His essay focuses on the environmental justice movement (shortened to EJ movement) that, despite its diversity, whether in the USA or in developing countries, is fighting for survival if only because more powerful human forces wreak havoc with the air they breathe or water they drink.[52] Environmentalist movements, on the other hand, advocate for the survival of all species "on the basis of *an assumed common good*"—a moral perspective that seeks to preserve and transmit to future generations the multifaceted wealth of the natural world as we know it today. "Yet justice and a race/class/gender/culture analysis, together with a concentration on urban conditions and those of the urban, rural, and reservation poor, have not been part of this 'common' good as normal fare."[53] By contrast, notes Rasmussen, the core question posed by

---

[50] Ibid., 197. By this he means that they produce "cocoa, coffee, tea and other incidentals for the Northern dining table."

[51] Ibid., 195. This was shockingly evidenced in the aftermath of hurricane Katrina (August 2005) in New Orleans, where the overwhelming majority of victims and refugees were poor and black.

[52] Larry Rasmussen, "Environmental Racism and Environmental Justice: Moral Theory in the Making?," *Journal of the Society of the Christian Ethics* 24.1 (2004): 3-28.

[53] Ibid., 6, emphasis original.

EJ advocates is the following: "What constitute healthy, livable, sustainable, and vital communities in the places we live, work, and play, as the outcome of interrelated natural, built, social, and cultural/spiritual environments?"[54] Here we come back to the embodied self as described by Benhabib, immersed in culture and collective traditions, though not without a voice of her own or his own, at least according to less strident communitarian versions.

Not surprisingly, the EJ movement traces its roots to Martin Luther King Jr, who, in active solidarity with the Memphis garbage workers, pointed to the connection between civil rights and environmental justice. Rasmussen notes that it was in this setting in 1982 that the term "environmental racism" first popped up and subsequently stuck. This was during an organized protest against a PCB landfill in Warren County, North Carolina, a predominantly African American county. Recall how Stewart Burns traced King's transition from "civil rights" to "human rights" as both a natural African American worldview adjustment to the modern autonomous self and a deliberate theological maneuver to preserve the holistic and communitarian doctrine of humanity found in the Hebrew scriptures and the gospel.[55] For Burns, King was returning to a "preindustrial meaning of rights"[56]—a Christian one as we saw it powerfully displayed in the pamphlets of Puritan Richard Overton. The Enlightenment view was sadly lacking: "This perspective diverged sharply from the classic liberal ideology of unbounded rights, owned by isolated, unencumbered selves devoid of community ties."[57]

Rasmussen's essay revolves around the same idea, albeit with a different twist, framed as it is by a remark made by noted African American writer James Baldwin: "In the church I come from—which is not at all the same church to which white Americans belong—we were counseled, from time to time, to do our first works over." He then went on to explain: "Go back to where you started, or as far back as you can, examine all of it, travel your road again and tell the truth about it. Sing or shout or testify or keep it to yourself: but *know whence you came.*"[58] What Rasmussen interprets these "first works" to be is nothing less than

---

[54] Rasmussen explains in a footnote that this is his summary of Charles Lee's discussion in "Environmental Justice: Building a Unified Vision of Health and the Environment," *Environmental Health Perspectives* 110.2 (April 2002): 141-44.

[55] Cf. the first few pages of Chapter 3.

[56] Burns, *To the Mountaintop*, 10.

[57] Ibid., 323. This was already quoted twice in Chapter 3.

[58] James Baldwin, *The Price of the Ticket: Collected Nonfiction, 1948–1985* (New York: St. Martin's Press, 1985), xix, quoted in Rasmussen, "Environmental Racism," 3, emphasis Rasmussen's.

the core of one's cultural understanding, as mediated and shaped by one's religious sensitivity:

> First works, those by which we expect to work out our salvation, are layered deeply in psyche and society. They generate the "normative gaze" that frames and guides feeling and thought.[59] They fund our personal habits and those of our institutions. They show up in our modes of production and reproduction, our cultural sensibilities, our basic aesthetic, intellectual, and moral values. They comprise, at day's end, nothing less than our way of life.[60]

Glen Stassen and David Gushee insert their moral theory precisely at this point: "it takes *community* to shape a person with integrity of character."[61] Along with character ethicists, they maintain that "character is formed not by self-made individuals, but by the shaping, encouraging and correcting influence of community."[62] The signs of a good character are specific virtues; and these virtues can only be cultivated through specific practices; and all of this is impossible to nourish outside of community. Enlightenment liberalism spawned "democracy, religious toleration and basic liberties," but it could not "nurture the kind of character needed for constitutional democracy to function in a healthy way."[63] Further, in an era when cultures are forced to cohabit with one another in such close proximity, any move away from ethics as rules and acts toward agents in particular contexts will have the best chances of promoting a common ethical discourse.[64] Finally, following Catholic ethicist Joseph Kotva, they find in character ethics a necessary antidote to "the modern individualistic and rationalistic ethical theories" that seek "to base behavior on universal reason and impersonal duty."[65] Rather like Polanyi, who emphasized the apprenticeship side of knowledge transmission in the sciences, Stassen and Gushee agree with Kotva that character ethics rightly prioritizes the master–disciple model of ethical learning.

---

[59] I pass on Rasmussen's footnote here: "I take this phrase from Cornel West and his discussion in *Prophesy and Deliverance! An Afro-American Revolutionary Christianity* (Philadelphia: Westminster Press, 1982), 53ff."

[60] Rasmussen, "Environmental Racism," 3.

[61] Glen H. Stassen and David P. Gushee, *Kingdom Ethics: Following Jesus in Contemporary Context* (Downers Grove, IL: InterVarsity Press, 2004), 56.

[62] Ibid., 57.

[63] Ibid.

[64] They quote from Joseph J. Kotva Jr's *The Christian Case for Virtue Ethics* (Washington, DC: Georgetown University Press, 1996), 8-9. The exact quote is: "Character Ethics moves the focus from rules and acts to agents and their contexts."

[65] Stassen and Gushee, *Kingdom Ethics*, 58.

Notice too how this flows from the metaphysics of an embodied self: "It is in being guided by, following after, and imitating masters or worthy examples that we learn to recognize and *embody* the emotional and intellectual dispositions, habits, and skills designated by the virtues."[66]

Muslims will no doubt think immediately of the crucial role played by the Sunna in their spiritual lives—following the prophet in his words and acts. Muhammad, as the prophet who received and transmitted the revelations later transcribed in the Qur'an, embodied those revelations in his life and is therefore eminently worthy of being emulated.[67] For Christians, Jesus repeatedly called on people to "follow" him, particularly his disciples who literally ate, slept and traveled with him for three years. In fact, his call was addressed to all:

> Come to me, all of you who are weary and carry heavy burdens, and I will give you rest. Take my yoke upon you. Let me teach you, because I am humble and gentle at heart, and you will find rest for your souls. For my yoke is easy to bear, and the burden I give you is light.[68]

Now back to James Baldwin. Character ethics and "returning to one's first works" are parallel moral tracks. Both emphasize personal character in a holistic way. Both praise actions that spring forth from deeply felt convictions that are also molded in a community's way of seeing the world and embody specific virtues. What counts is personal integrity, or truthtelling—a consistency between convictions, words and actions—widened to include harmony with one's community and its cherished ideals.

One might object that (1) this is too narrow for a global ethic that by definition is universal in scope; and that (2) this runs afoul of my use of Benhabib's modified version of Kant's deontological ethic in Chapter 2. In a fascinating recent study, Bishop Cragg takes on the question of religious diversity and the necessity to find among the religions a common ethical ground for human rights. I will begin to answer both objections by offering some of Cragg's remarks on the issue.

---

[66] Kotva, *The Christian Case*, 80-81, quoted in Stassen and Gushee, *Kingdom Ethics*, 58, emphasis added.

[67] S. H. Nasr no doubt speaks for all Muslims when he writes, "no Muslim can have any virtue that was not possessed in the most eminent degree by the Prophet. More specifically, the Prophet exemplifies the virtues of humility; nobility, magnanimity and charity; and truthfulness and sincerity. For Muslims, the Prophet is the perfect model..." (*The Heart of Islam*, 35).

[68] Mt. 11:28-30, NLT.

Cragg explores the tangled issues of identity in Christianity, Judaism, Islam, Hinduism and Buddhism through their use of the pronouns "I," "thou," "we" and "ours." To begin with, the Asian mind remains skeptical about the self-confident and self-assertive meaning of the "human" proclaimed in the Semitic religions. Perhaps best exemplified by the Arjuna of the Bhagavad Gita or in the practice of Theravada or Hinayana Buddhism, destiny's goal in the East is defined by a transcendental unity between human and universal being—all of this starting "from a first long brooding on 'being-in-itself' and that self frustrated by the pain of transience and caught up in puzzling mortality."[69] Judaism, by contrast, is the faith most attuned to its historical and ethnic roots. Cragg calls it "pronounal Jewry," the communal sense of being "God's own people." Tragically for Jews over the centuries, "chosen-ness" has meant harassment and persecution over and over again, with the horror of the Holocaust as its most poignant landmark. Yet the ideology of Zionism, now that the Jews have had a homeland for over half a century, may be posing more problems and queries than it initially proposed to solve.[70] The eternal tension between "pronounal Jewry" at home in its land and Jewish personalism (with the self-encounter of the "I" in its center)—Jews residing in "the land" and Jews negotiating various identities in the diaspora—may have become exacerbated.

While Christians since the fourth century have often resorted to state power to press their communal and political advantage, have persecuted minorities and waged war on Muslims during the Crusades, this behavior can only be seen as an aberration of the teachings of Jesus, who assured the Roman governor Pilate, "my kingdom is not of this world." The issue is more complex with the followers of the Meccan prophet, Muhammad, explains Cragg in two successive chapters: "The Muslim Personal Pronoun Singular" and "The Muslim Personal Pronoun Plural."[71] From the personal pronoun "I" in the *shehada* ("*I* testify that there is no god but God…") to the individual's standing before God on judgment day, the Qur'an makes it abundantly clear that each person will be held accountable for his or her own sins, and only for his or her own sins.[72]

---

[69] Kenneth Cragg, *Faiths in their Pronouns: Websites of Identity* (Brighton/Portland, OR: Sussex Academic Press, 2002), 4.

[70] See the prolific work of Jewish theologian Mark H. Ellis on this. Among his most recent works, see *Practicing Exile: The Religious Odyssey of an American Jew* (Minneapolis, MN: Fortress Press, 2002); *Toward a Jewish Theology of Liberation: The Challenge of the 21st Century* (Waco, TX: Baylor University Press, 2004).

[71] Ibid., 94-126.

[72] Ibid., 99. Two phrases in particular bear this out: "There is no burdening of a burdened that is not his own burden" (Cragg's literal translation, Q. 6:164; 17:15;

Perhaps the most stunning symbol of "Islamic personalism" is the Muslim prayer (*salat*): "There are few more intimate actions than washing one's own hands and feet, arms and face, and doing so in obligatory preface to *Salat*." Leaving one's shoes at the door is to follow in Moses' footsteps—"the enduring symbol of the will to hallow and be hallowed in the approach to Allah and the ordered sequences of the prayer-rite."[73] Finally, Islam is intensely personal in the first twelve years of Muhammad's preaching in Mecca. It was often at great cost that individuals braved the retribution of Mecca's powerful clans and surrendered body and soul to the call of the One God.

Yet the transition from persecuted minority in Mecca to political rule in Medina—divinely confirmed by the military victory at Badr, "the day of criterion" (*Yawm al-Furqān*)—was not just a turn of political fortunes.[74] The emigration to Medina in 622, the Hijra, is so central to the Islamic worldview that it marks the beginning of its calendar. Hence the founding narrative of the Islamic faith transformed it "into the most confidently political of world religions."[75] Indeed, the qur'anic injunctions to "obey God and his messenger" merged into the caliphal institution upon Muhammad's death and despite many twists and turns over time and in the wake of many empires, Muslims still feel instinctively that their faith requires the conjunction of *dīn* and *dawla*. As he critiques this position, Cragg notes that this view is far from unanimous among Muslims today. Yet the tension between religion and rule, while present for all religious groups, remains at the core of the Islamic identity.[76]

39:7; 53:38), and "God does not call any soul to account but for its own" (2:286; with parallels in 6:153; 2:223; 4:84).

[73] Ibid., 94. See also S. H. Nasr's more Sufi perspective: "The *ṣalāh*, however, is incumbent upon all Muslims, for it is the guarantee of our living in accordance with our theomorphic nature [he quoted elsewhere the *ḥadīth*, 'He created man upon His Form,' 15] as beings reflecting God's Names and Qualities and the means whereby we stand directly before God to address Him as His vicegerents on earth" (*The Heart of Islam*, 132).

[74] "In that they succeeded, whether at Badr or later, their faith was confirmed. Either way, the stern vicissitudes conjoined the field of war with the soul of faith. The physical and the religious were in the juncture and in the event... Fighting had its sanction from religion as crucial to religion's security" (ibid., 116).

[75] Ibid., 117.

[76] I do not have the space to comment here on recent attempts to forge an Islamic theology of peacebuilding and nonviolence. For a sampling, see Ralph Crow, Philip Grant and Saad Eddin Ibrahim, eds., *Arab Nonviolent Struggle in the Middle East* (Boulder, CO: Lynne Rienner, 1990); Osman Bakar, *Islam and Inter-Civilizational Dialogues: The Quest for a Truly Universal Civilization* (Kuala Lumpur: University

Now back to the first objection: how can a global ethic be fashioned out of a communitarian impulse? Without making any reference to the Sudanese martyr, Mahmoud Muhammad Taha, or to his articulate mouthpiece in the US, Abdullahi Ahmed An-Na'im,[77] Cragg suggests that Muslims dig deeper into the Meccan recesses of their faith, particularly since they now live as minorities in many more countries than before.[78] On this point however, I credit Cragg with a better knowledge of the Qur'an than An-Na'im: the universal message of the Qur'an is not confined to the Meccan suras.[79] At the same time, An-Na'im would agree with Cragg and Küng—and many Muslims scholars already cited here—that building a universal ethic of human rights with the specific contributions of religious people is an urgent task. The concept of "being human" must trump any other identity, be it racial, ethnic or religious. This may not always be easy—the forces of identity politics are notoriously tenacious, observes Cragg:

> However, it grows ever more clear that, in terms of a twenty-first century, there must needs develop a sense of religious claims and constraints in the socio-moral order as ready to acknowledge the global shape of human well-being and surrender the intent to legislate and dominate unilaterally inside the political nationhoods they have traditionally characterized. There are concepts of human rights across all nations that demand recognition and implementation through all national borders, with a writ mandated to require them—a writ not to be thwarted by warnings that what obtains within our borders is only the business of our rule and the arena of our *Shariʿah*. In the quest for a global ethic for a global ecology, the claims of a basic humanism deserve to override—and not to be overridden by—the exclusives of territorial religion.[80]

The second objection to my supporting Baldwin's motto to "do our first things over" or Stassen and Gushee's character ethics concerned its compatibility with Benhabib's version of Habermas's discourse ethics. Can an ethic stay in tune with local values and ideals and yet carry the weight of universal human concerns? Cragg believes that peoples of

of Malaya Press, 1997); and Mohamed Abu-Nimer, *Nonviolence and Peacebuilding in Islam: Theory and Practice* (Gainesville: University of Florida Press, 2003).

[77] An-Na'im, *Toward an Islamic Reformation.* Taha was executed in 1985 for being an "apostate" by the Numeiri government.

[78] In this sense, Cragg is narrowing his argument already developed in *Returning to Mount Hira'.*

[79] An-Nai'm has often been criticized for this reason. See in particular Abdulaziz Sachedina, *The Islamic Roots of Democratic Pluralism* (New York/London: Oxford University Press, 2001).

[80] Cragg, *Faiths in their Pronouns*, 184.

various cultures and faiths bring to the global table their own resources for "a basic humanism." Certainly this was the conclusion of the 1993 Parliament of the World's Religions. Furthermore, it is not incompatible with Benhabib's more secular (and postmodern) twin principles:

> (1) that we recognize the right of all beings capable of speech and action to be participants in the moral conversation—I will call this *the principle of universal moral respect*; (2) these conditions further stipulate that within such conversations each has the same symmetrical rights to various speech acts, to initiate new topics, to ask for reflection about the presuppositions of the conversation, etc. I call this *the principle of egalitarian reciprocity*. The very presuppositions of the argumentation situation then have a normative content that precedes the moral argument itself.[81]

Allow me to illustrate one way of doing this within an Islamic framework. Arkoun has recently revisited his doctoral thesis of the 1960s ("L'Humanisme Arabe au IVe–Xe Siècle"). With the advent of resurgent Islam in many forms, he is concerned that Muslims add their voices to the current discussion over the values to adopt in guiding humankind in their new globalized context. He begins by recalling the happy medieval synthesis of God–humankind–nature. This was a "theocentric humanism," he remarks.[82] The eighteenth-century European Enlightenment shattered that unity "by man proclaiming himself a transcendental and autonomous subject."[83] But the Islamic humanist synthesis (to which contributed both Jewish and Christian thinkers) of the ninth and tenth centuries CE was itself eroded by the imposition of Sunni orthodoxy, first under the Seljuk Turks, then under the Ottomans. The colonial period only reinforced this ossification of doctrine and the "closing in of groups on their traditions which functioned everywhere as an ethico-political security system for each of these social units thus constituted."[84] Nor did the postcolonial period after World War II help in any way. The pressures of the new nationalisms simply became additional barriers to

---

[81] Benhabib, *Situating the Self*, 29, emphasis original.

[82] Here is Arkoun's summary of the medieval worldview: "All things come from God and return to him: the world, history, animated beings; nature is inhabited by God, entrusted to humankind as a space and means [to attain] Salvation—by which we mean the earthly pilgrimage necessarily leading to the last Judgment, Resurrection, eternal Life in either Reward or Retribution" ("Peut-on parler d'humanisme en contexte islamique?," in *Compilation and Creation in Adab and Lugha: Studies in Memory of Naphtali Kingerg (1948–1997)*, ed. Albert Arazi, Joseph Sadan and David J. Wasserstein, Israel Oriental Studies, 19 [Tel Aviv: Eisenbrauns, 1999], 15).

[83] Ibid.

[84] Ibid., 20.

recovering the authentic humanist spirit of Islam in its classical period. Arkoun's assessment of Islam today is pessimistic:

> Contemporary Islam expresses the historical ruptures that have relegated to the *unthought* all the humanist gains, all the positive virtualities of evolution we have presented above. The urgency of the political battles, the care to safeguard the local organizations from the often brutal interferences of various central powers since the 5th/11th century, have reduced Islamic thinking to a narrow scholasticism, repetitive, preoccupied with conservation and the protection of dogmas, indispensable principles and "values" for the *formal* legitimization of these powers.[85]

Among the eight points he advances for the reinterpretation of Islamic thought today I will single out the last three for consideration.[86] The first is a decision for Muslims to engage in a critique of modernity—a task that has so far been monopolized by the West. As we have seen, Sardar, Esack, Soroush and others are moving in this direction. Still others are willing to probe cultural and sociological issues that affect relations between Muslims and traditional Western societies (recall Akbar Ahmed's project), but few are willing to wrestle with the Qur'an in the light of postmodern hermeneutics, epistemology and ethics.[87]

Shabir Akhtar is a good example of a scholar who realizes what needs to be done in this respect, but in the end stops short of actually taking the leap. In closing his impressive book, *A Faith for All Seasons*, Akhtar repeats his thesis "that Muslims have failed to interpret and appropriate Islam properly for the needs of the modern age." The main failure, he has argued, is "the shallowness of intellectual responses to the challenges of modernity and an increasing shallowness in the life of faith."[88] Yet for all of his soul-searching questions and commendable zeal to engage the

---

[85] Ibid., emphasis original. See my comments on his *Encyclopaedia of the Qur'an* article in the beginning of Chapter 4.

[86] The first five are: (1) authoritarian regimes must open up to democratic experiments; (2) there must be a strict respect for the freedom of thought, creation, diffusion, and communication; (3) a radical reforming of education is needed at all levels with new curricula, especially in the social sciences; (4) the religious spheres must be opened to scientific inquiries, leading to the kind of theological and philosophical discussions that promoted the classical humanism; (5) there must be freedom of expression for all the subcultures in the Muslim world (though he must be thinking about his own Kabyle-Berber culture, he only mentions Indonesia, India, Iran, the ex-Soviet Republics, and the like).

[87] I have written elsewhere about Ibrahim Moosa and Khaled Abou El Fadl, both of whom make important contributions to Islamic studies with a self-conscious postmodern stance. Both wrote chapters in Safi, ed., *Progressive Muslims*.

[88] Akhtar, *A Faith for All Seasons*, 213.

modern person with the claims of Islam, he tends to fall back on a traditional fideistic position. No doubt he is widely read and discusses with ease several Christian theologians, and thus is willing to wrestle, for instance, with the claims of historical criticism—but only on a surface level with regard to the Qur'an. After a long and lively discussion of revelation and inspiration, he finally comes down to the position (for him wrongly denigrated) of "fundamentalism": "the position that scripture contains a basic source of wholly correct guidance."[89] Revelation, when all is said and done, is "an all or nothing affair." Thus every fact in the Holy Book must be error free—historical, geographical, and factual in every sense. Yet Akhtar has been willing to reconsider the issue in two more recent articles.[90]

The second condition posited by Arkoun is "the active participation in the actualization of a geopolitical outlook freed from all the hegemonic strategies that have until now dominated 'national' policies."[91] Indirectly referring to Western hegemony, he is calling for a more equitable distribution of power and a wider network of solidarity among all the nations of the world. This comes out more clearly in the last condition, which is, more or less, Küng's global ethics."[92] Arkoun proposes "the development

---

[89] Ibid., 74.

[90] In the first he argues that the classic doctrine of the inimitability of the Qur'an leaves to be desired, but that "some kind of beauty attached to the truth is needed if one is to win allegiance for a vision" ("The Limits of Internal Hermeneutics: The Status of the Qur'an as Literary Miracle," in *Holy Scriptures and Judaism, Christianity and Islam*, 112). In the next article he is more candid about his sincere intellectual puzzles and questions about the Qur'an as a committed Muslim. In the context of modernity, "Muslims themselves should study their scripture with as much detachment and objectivity as possible" ("Critical Qur'anic Scholarship and Theological Puzzles," in *Holy Scriptures and Judaism, Christianity and Islam*, 123). After examining three central assertions of the Qur'an (including that humanity is *homo islamicus*), he decides that this cannot be backed up by empirical evidence. This must be explained by the fact that the Qur'an was "addressed to a group of people whose mood and temperament as well as dogma and perspective differed radically from that prevalent today in modern industrial societies." This contemporary belief—that no facts are value-free—is a "paradigmatic shift," which, he writes in conclusion, "we cannot ignore if we are to make the Qur'an relevant to modern humanity" (ibid., 127).

[91] Arkoun, "Peut-on parler d'humanisme en contexte islamique?," 21.

[92] This is in direct parallel with Esack's passionate conclusion, expressing the one certainty that he does embraces amid the swirling waters of hermeneutics: "The struggle for justice, gender equality and the re-interpretation of Islam so that it legitimates and inspires a comprehensive embrace of human dignity is one to which I am deeply committed. My own humanity is intrinsically wedded to this struggle in its various forms" (*Qur'an, Liberation & Pluralism*, 261).

of a modern ethical system of thought which combines critical philosophical values and an effort to awaken a moral understanding [or 'conscience'] freed from religious, nationalistic, political, and economic
constraints arbitrarily imposed by authoritarian regimes and the various
orthodoxies."[93] This is a forward-looking ethic, he explains, "aiming at
the widening of consensus on the rights and obligations of the human
person...a humanist 'order' demanded by all today."[94] Such a program,
he argues, is perfectly in line with the humanist vision of classical Islam.
In order for this vision to be realized in our modern age (he uses the term
"postmodern" with reluctance), it must willingly fuse its horizons with
the other "humanisms" which both the West and Muslims still ignore:
"the religions and cultures of Asia must, therefore, express themselves
and put forward their contributions to a humanism which, for the first
time, will actually deserve the qualifier of truly universal."[95]

In my estimation, this answers the second objection I raised. Arkoun's
ethical vision is plainly in line with the "basic humanism" pleaded by
Cragg, while both recognize the need for a global consensus on fundamental human rights, together with some latitude for local expressions of
those rights, taking into account cultural diversity. Recall that in my
discussion of indigenous peoples' rights (Chapter 2), both Paul Beal and
Seyla Benhabib were seen to advocate a similar strategy. Human rights
should trump the claims of nation-states in the case of Aboriginal
peoples, argues Beal; cultures are not reified entities, Benhabib reminds
us, but rather are constantly evolving customs, values and perspectives,
particularly as they are intensely shaped and reconfigured by the many
other worldviews that are increasingly knocking against each other in the
powerful river current of globalization.[96]

It is precisely the notion of discourse ethics tied to a robust ideology
of democratic governance that allows us to match the twin concepts of a
situated subject in community and culture and a global ethic. That they
are framed in secular terms need not frighten or deter the religious person.
Benhabib's joint principles of "universal moral respect" and "egalitarian
reciprocity" (the Golden Rule in effect) will in fact gain considerable
authority and legitimacy from a common Muslim–Christian theology of
human trusteeship as mandated by their sacred texts. The caveat here is
that such a theology be intentionally extended to include a moral theory

---

[93] Arkoun, "Peut-on parler d'humanisme en contexte islamique?," 21.
[94] Ibid.
[95] Ibid.
[96] Benhabib, *The Claims of Culture.*

that accommodates both a holistic vision of character ethics[97] and a deontological ethic capable of sustaining a universal vision of human dignity and inherent worth. By way of proposing one possible direction of Muslim–Christian dialogue on this issue of ethical theory, I present in Figure 16 Stassen and Gushee's four dimensions of holistic character ethics.

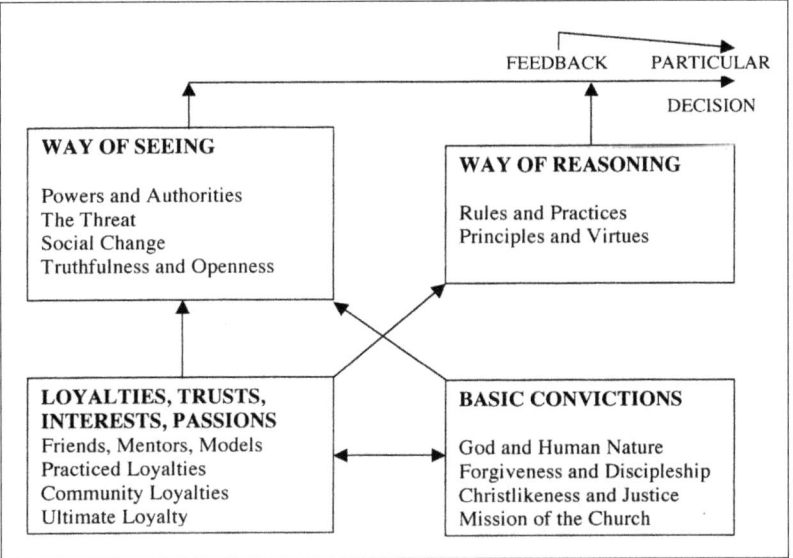

Figure 16. *Stassen and Gushee's Four Dimension of Character*[98]

[97] This is in opposition to a legalistic, rule-based ethics. Stassen and Gushee cite three other ethical models: (1) situationism (each case or situation demands a different ethical decision); (2) principlism (principles trump rules, particularly when two happen to conflict); (3) contextualism or narrative ethics ("rules and principles find their place within the context of core theological beliefs" and the founding narrative of a community's faith). On this score, both Islam and Judaism have historically opted for legalism with a touch of principlism, particularly in the modern period (think of Muhammad Abduh). Stassen and Gushee opt for a modification of the third view, with again a more dynamic view of moral norms as embodied in specific practices that mirror virtues (or principles) and connect to the narrative that sustains the community of faith. For Muslims, the revelation at Mount Hira, the Meccan and Medinan periods and the watershed experience of the Hijra represent some of the high points of the Islamic narrative. The Hajj is in fact an "embodied drama that gives continuity and coherence to life and its practices"—a phrase originally applied to Communion or the Eucharist for Christians (Stassen and Gushee, *Kingdom Ethics*, 124).
[98] Ibid., 59.

Space will not allow me to comment in detail about Stassen and Gushee's application of ethical theory to a specifically Christian perspective, except to note two things. First, a universal ethics of human rights will of necessity involve the "way of seeing" and "basic convictions" dimensions of this model—particularly the "human nature" aspect of the latter. This is what this book is claiming: as Muslims and Christians widen their common understanding of their God-given human calling on earth, they contribute to the moral strengthening of a universal concept of human dignity. Second, I offer some detail on the "way of seeing" dimension, as it bolsters another key argument I have put forward throughout this study: the more we (Muslims and Christians) sharpen our common vision of the world as it is, the more effective we will become in translating into reality our vision for a human caliphate based on mutual respect, compassion and justice—ethical norms that flow out of our texts and founding narratives.

By "power and authorities" Stassen and Gushee mean how people articulate a concept of political authority. In particular, to what extent are "the powers that be" accountable to God's standards of justice? One can point to a venerable tradition among Muslim jurists that called for the overthrow of despotic rulers, though it tended to be overshadowed for long periods by a staunchly conservative attitude to power.[99] Clearly, the theme of justice is central to the Qur'an.[100] Likewise, Stassen and Gushee point out that the word "justice" appears in the Bible 1060 times in its two Hebrew words and two Greek words.[101] In a chapter devoted to justice, they demonstrate that Jesus strongly identified with the justice tradition of the Hebrew prophets. The book of Isaiah in particular, which Jesus quotes more often than any other, makes a strong connection

[99] See Khaled Abou El Fadl, *Rebellion and Violence in Islamic Law* (New York: Cambridge University Press, 2001).

[100] Justice has always been a central concern in Muslim theologizing—recall Sayyid Qutb's *Social Justice in Islam*. See also Majid Khadduri, *The Islamic Conception of Justice* (Baltimore, MD/London: The Johns Hopkins University Press, 1984); *Islamic Identity and the Struggle for Justice*; Muhammad Hashim Kamali, *Equality and Justice in Islam* (Cambridge: Islamic Texts Society, 2002); and David L. Johnston, *Evolving Muslim Theologies of Justice: Jamal al-Banna, Muhammad Hashim Kamali and Khaled Abou El Fadl:* (Penang: CenPRIS and Universiti Sains Malaysia Press, 2009).

[101] Stassen and Gushee, *Kingdom Ethics*, 345. One of the reasons this is so little recognized is that the Hebrew *tsedaqah* is most often translated "righteousness" in English Bibles. Many versions also obscure the translation of *mishpat* (the KJV reads "judgment"). This is unfortunate, because *tsedaqah* means "delivering, community-restoring justice," while "*mishpat* means judgment according to right or rights, and thus judgment that vindicates the right especially of the poor or powerless" (ibid.).

between God's coming rule and the establishment of justice among all the nations.[102] As seen above, justice is also at the heart of the jubilee tradition in the Hebrew Bible and forms the framework behind Jesus' inaugural proclamation of his mission (Lk. 4:18-19). Further, recent scholarship has dramatically uncovered the economic and political power disparities that threatened to undo the fabric of first-century Palestinian Jewish society. Plainly, seen against this backdrop, an ardent concern for social justice undergirded and sustained Jesus' bold confrontation of the Jewish powers in Jerusalem. Specifically, his overturning of the money-changers' tables and his chasing them out of the Temple was a brazen act of civil disobedience that not only denounced the corrupt practices of the ruling religious elites, but exposed their grievous exploitation of the poor in the name of religion.[103]

Jesus' words and actions give great hope to Palestinian Christians in particular, writes Naim Ateek. In a world that links power so closely with justice, God in Jesus plainly affirmed that, as the Creator who upholds justice in the universe, he stands on the side of the poor and powerless. People, by contrast, left to their own devices resort to power to impose their version of justice. God thunders his disapproval in the words of the prophet Micah:

> Woe to those who devise wickedness and work evil upon their beds!
>
> When the morning dawns, they perform it, because it is in the power of their hand.
>
> They covet fields, and seize them; and houses, and take them away; they oppress a man and his inheritance.[104]

The achieving of peace between warring parties only comes through a measure of justice. "If you want peace," wrote Pope John XXIII, "work

---

[102] The chapter on justice is preceded by one on love. This forms a section on the two central Christian ethical norms and is followed by four chapters representing the four crucial arenas in which these norms are put to use and tested: truthtelling, race, economics and creation care.

[103] The seventy-member Jewish high council (Sanhedrin) was under the thumb of three or four of the wealthiest Jerusalem families. Hence there was an unmistakable collusion between the religious leaders and the rich—all of whom gladly collaborated with the Roman occupiers in order to assure their power. Jesus, of course, by claiming a higher authority and by promoting the prophetic message of justice and compassion for the poor and outcasts (by word and deed), represented a grave threat to their monopoly of power—and all the more so because the crowds were increasingly following him and not them.

[104] Micah 2:1-2, quoted in Ateek, *Justice and Only Justice*, 124.

for justice."[105] Having said that, justice alone rarely brings the end result sought by Jesus: reconciliation.[106] As Ateek puts it, "Absolute justice not only restores their rights [those who have been wronged] but also has a way of condemning and humiliating the wrongdoer." The result may be more brokenness and alienation, with the cycle of revenge and violence left intact rather than healed. "What we need in the Israel–Palestine conflict is a way in which justice can be exercised so that the ultimate result would be peace and reconciliation between and within each people and not the fragmentation and destruction of either or both."[107] Here again emerges the central hermeneutic of the Bible for Ateek: "Above all, peace means reconciliation… The cross expresses the costliness of peace. Christians understand this well when they experience reconciliation with God through Christ, who is 'our peace.'" This experience of reconciliation enables the believer "not just to live peaceably with one's neighbor, but also to be a peacemaker, like Jesus, to risk all to make peace."[108]

The second point under the "way of seeing" dimension in Stassen and Gushee's scheme is "threat." By this they mean how people perceive the cause of what is deemed wrong with the world. They borrow the social science term "threat perception" in this context. This is why I devoted a whole chapter to the description and analysis of the "double wall" currently standing before the rush toward globalization—part of the situation I have called postmodernity. Who and what is to blame for the ecological nightmare and growing disparities we face as a human race? The "what" is related to the modern conception of the autonomous self, married to an ideology of unregulated free markets (an idea going back to Hobbes), the commodification of the commons, and all of that ominously combined with the military and economic expansionism of the Western powers. Considering that the post-Cold War period is above all the tale of the remaining superpower's callous exploitation of laissez-faire capitalism in order to extend its imperial designs, the "who?" question is not so difficult either. As the Columbia historian and Middle East specialist Rashid Khalidi has claimed, the recent American adven-

---

[105] Pope John XXIII, *Pacem in Terris* (New York: Paulist Press, 1963), 76, cited in Ateek, *Justice and Only Justice*, 147.

[106] For example, in the Sermon on the Mount: "Therefore, if you are offering your gift at the altar and there remember that your brother has something against you, leave your gift there in front of the altar. First go and be reconciled to your brother; then come and offer your gift" (Mt. 5:23-24, NIV).

[107] Ibid., 139.

[108] Ibid., 149.

tures in Afghanistan and Iraq represent a "perilous path" trodden in the early twentieth century by the imperial predecessors, Britain and France.[109] Yet we know from David Harvey's analysis that the "new imperialism" of the United States is sapped at its base by the contradictory forces of state empire-building and the forces of capitalistic over-accumulation (which are global, especially since 1973).

Plainly, the United States, as the most powerful state to emerge from WWII, has consistently sought to expand its empire in the wake of Britain's crumbling one. Another indispensable perspective on this process—more damning even than Joseph Stiglitz's insider view of the World Bank—comes from John Perkins, a top-level economic forecaster and important player in the American race to global domination in the 1970s and '80s. His book is particularly *à propos* here, because evil must be named and those responsible must come clean of their misdeeds for redress and healing to take place. Appropriately, his book is entitled, *Confessions of an Economic Hit Man.*[110]

Perkins had been a Peace Corps worker in Ecuador in the late 1960s, and his abilities both in crosscultural communication and in political and economic analysis caught the attention of the National Security Agency (NSA, the largest American espionage agency). He was soon hired as an economist at MAIN, an international consulting firm that worked closely with the World Bank. Then one day he was approached in covert fashion by a woman named Claudine Martin, whose business card described her as "Special Consultant to MAIN." "I've been asked to help in your training," she informed him.[111] In the next few sessions held secretly in her apartment (even Perkins's wife was never to know about this), she proceeded to explain that his mission was to be an "economic hit man" (EHM) and that once he chose to follow this course, he would be in it for life—no turning back. He would be sent to various countries and for each one he would be expected to inflate the GNP (Gross National Product) growth forecast, which was to result from a huge investment in the country's infrastructure—railroads, electrical grid, telecommunication systems, water works, oil industry and the like. Further, in so doing, Perkins was to convince any skeptics who might object, so that each country would be persuaded to take out mammoth loans. This strategy

---

[109] Rashid Khalidi, *Resurrecting Empire: Western Footprints and America's Perilous Path in the Middle East* (Boston: Beacon Press, 2004; with new Introduction, 2005).
[110] John Perkins, *Confessions of an Economic Hit Man* (San Francisco: Berrett-Koehler, 2005).
[111] Ibid., 14.

was not only to extend the reach and profitability of US companies. The calling of the EHMs is unique. In Claudine's words,

> We're a small, exclusive club… We're paid—well paid—to cheat countries around the globe out of billions of dollars. A large part of your job is to encourage world leaders to become part of a vast network that promotes U.S. commercial interests. In the end, those leaders become ensnared in a web of debt that ensures their loyalty. We can draw on them whenever we desire—to satisfy our political, economic, or military needs. In turn, these leaders bolster their political positions by bringing industrial parks, power plants, and airports to their people. Meanwhile, the owners of U.S. engineering and construction companies become very wealthy.[112]

When and where was this profession given birth? The history of EHMs began in Iran in 1951, when the highly popular (and democratically elected) prime minister, Mohammad Mossadegh, stood up to the British oil company that had been given free reign in the country up till then. When Mossadegh nationalized all branches of the Iranian oil industry, Britain turned to the US for help; yet both realized that to intervene militarily would bring in the Soviet Union. A better plan was devised:

> Instead of sending in the Marines, therefore, Washington dispatched CIA agent Kermit Roosevelt (Theodore's grandson). He performed brilliantly, winning people over through payoffs and threats. He then enlisted them to organize a series of street riots and violent demonstrations, which created the impression that Mossadegh was both unpopular and inept. In the end, Mossadegh went down, and he spent the rest of his life under house arrest. The pro-American Mohammad Reza Shah became the unchallenged dictator. Kermit Roosevelt had set the stage for a new profession, the one whose ranks I was joining.[113]

Empire-building had always been on the American agenda. Now a new, more subtle and effective means had been discovered. International corporations would nicely carry out the agenda of the intelligence agencies and handsomely profit from it as well. The US and its European sister empire builders were developing a symbiotic relationship with the corporate world and international organizations like the World Bank and the WTO. This tri-partite alliance Perkins calls the "corporatocracy," or "the new elite who had made up their minds to attempt to rule the planet." This was "a close-knit fraternity of a few men with shared goals," who "moved easily and often between corporate boards and government

---

[112] Ibid., 17.
[113] Ibid., 18.

positions."[114] The archetypal global empire builder was Robert McNamara, who moved from president of Ford Motor Company to secretary of defense under presidents Kennedy and Johnson, and finally to president of the World Bank.

From Indonesia to Panama, from Saudi Arabia to Iran and Ecuador, Perkins turned out to be a ruthlessly successful EHM. Yet, from the beginning, he struggled with doubts and suffered pangs of conscience. His heart went out to the poor he had occasion to visit, knowing that he was personally responsible—at least in part—for their growing poverty. He also had many conversations with Panama's modern hero, Omar Torrijos, who had broken with a long line of US-beholden dictators and protectors of predatory firms such as the United Fruit Company, later bought by George H. W. Bush. Torrijos had managed to unite his country in a bid to sever its ties with the US, yet without selling out to the Soviet Union as Castro and Allende in Chile had done. He was simply a nationalist who believed in a people's right to self-determination. As a favor to a leader he personally admired, Perkins cut him a better deal he had cut anywhere else. At that time Carter was president—indeed, an exceptional time in US history when human rights topped the agenda. In 1977 Carter, to the dismay of the "corporatocracy," negotiated the handover of the Canal to Panama. Perkins too kept his promises to Torrijos— "I made sure our studies were honest and that our recommendations took into account the poor."[115]

What happened in this case only reinforced Perkins's anger and guilt. When the EHMs fail, the US sends the "jackals" to finish the job. This took one form in Guatemala. Jacobo Arbenz, Guatemala's democratically elected leader, boldly set out to confront the monopoly and ruthlessness of United Fruit in his country and initiate a comprehensive land reform program. In 1954, the CIA ousted him with a coup, replacing him by right-wing dictator Castillo Armas. United Fruit resumed its activities, the land reforms were cancelled and thousands of political opponents were jailed. In the case of Torrijos, the jackals attacked more directly. Reagan had now been elected and a new era of unabashed empire-expansion began. Just as Ecuador's popular leader, Jaime Roldos, was killed in

---

[114] Ibid., 26.

[115] Ibid., 102. An interesting sideline to this part of the story is that Perkins met the author Graham Greene, who was beginning to write—uncharacteristically—a nonfiction book on Torrijos. This was in Panama City and the book that came out several years later was titled *Getting to Know the General* (New York: Pocket Books, 1984). Greene was a personal friend and admirer of Torrijos and expressed his concern in the conversation with Perkins that the latter's life was in danger (pp. 106-7).

a helicopter crash,[116] a few months later, as Torrijos persisted in his adamant refusal to renegotiate the Canal Treaty with Reagan, he died in a plane crash.[117] The jackals had come after him.

I only bring up this information, which is available in many places,[118] to highlight a wider theological issue. The "who" behind all the evil of this world cannot be limited to a superpower, or the G8, or the tycoons of the corporatocracy. Reinvisioning the world as Muslims and Christians will not only lead us to take honest stock of how power is concentrated and deployed across the globe in military, economic and political ways, but also to look at root causes—drawing from the "basic convictions" dimension. On one level, Perkins's book is an excellent example of one person confessing his guilt for having chosen to be a part of an oppressive system, one that destroyed countless lives around the world. In his own words, the book is first and foremost a confession:

> It is the confession of a man who allowed himself to become a pawn, an economic hit man; a man who bought into a corrupt system because it offered so many perks, and because buying in was so easy to justify; a man who knew better but who could always find excuses for his own greed, for exploiting desperate people and pillaging the planet; a man who took full advantage of the fact that he was born into one the wealthiest societies history has ever known, and who could pity himself because his parents were not at the top of the pyramid; a man who listened to his teachers, read the textbooks on economic development, and then followed the example of other men and women who legitimatize every action that promotes global empire, even if that action results in murder, genocide, and environmental destruction; a man who trained others to follow in his footsteps. It is my confession.[119]

---

[116] Eyewitnesses claim that out of concern for his security (Roldos knew the CIA was after him), he used two helicopters. One of his security officers made a last-minute change and Roldos rode in the decoy aircraft. It blew up (ibid., 156-57).

[117] Ibid., 158.

[118] For example, see Howard Zinn, *A People's History of the United States* (New York: Harper & Row, 1980); Diane K. Stanley, *For the Record: The United Fruit Company's Sixty Years in Guatemala* (Guatemala City: Centro Impresor Piedra Santa, 1994); Gerard Colby and Charlotte Dennet, *Thy Will Be Done, The Conquest of the Amazon: Nelson Rockefeller and Evangelism in the Age of Oil* (New York: HarperCollins, 1995); P. W. Singer, *Corporate Warriors: The Rise of the Privatized Military Industry* (Ithaca, NY/London: Cornell University Press, 2003); Jim Garrison, *American Empire: Global Leader or Rogue Power?* (San Francisco: Berrett-Koehler, 2004).

[119] Perkins, *Confessions of an Economic Hit Man*, 223-24. The best recent book on a Christian wrestling with American empire-building is written by three theologians and one expert on international relations: David Ray Griffin, John B. Cobb Jr, Richard Falk and Catherine Keller, *The American Empire and the Commonwealth of*

And so evil at the personal level, or sin, must be faced and confessed for restoration and reconciliation to take place. Each and every one of us makes bad choices and commits evil, perhaps on a smaller scale; nevertheless, that sin merges with and undeniably contributes to the wider evil we encounter in the world. I agree with Cragg when he writes that original sin in Christianity has been badly misunderstood by many. There is no hapless heredity by which an individual is born guilty before committing anything wrong.[120] Instead, the Bible is saying that "birth ushered us into a world enmeshed in social tangles and historical legacies."[121] As the Qur'an testifies, Satan is granted power by God to ensnare and entrap humans at every corner, "except Thy Servants amongst them, sincere and purified (by Thy grace)."[122] The Prophet himself ardently struggled with some verses deceitfully inspired by the Devil and is told by God to ask for forgiveness on several occasions.[123] There is hope, however, and Cragg connects this to the human calling of deputyship—as expressed in a collective following of God's path: "O ye who believe! Guard your own souls: if ye follow (right) guidance, no hurt can come to you from those who stray."[124]

But there is another level of evil here. Perkins mentions the word "system." Sin undoubtedly adheres to sociopolitical and economic structures, as his narrative so graphically illustrates.[125] The Apostle Paul

God: A Political, Economic, Religious Statement (Louisville, KY/London: Westminster/John Knox Press, 2006).

[120] Cragg writes that Paul's expression "being in Adam" meant "a human situation into which birth introduced us, where factors were already at work from a long past and from which we could not expect to be exempt, since they were part of the ongoings of history. We 'inherited' them, not as a moral incrimination but as a circumstantial condition spelled out by history. In this sense we were all 'in Adam'" (*Faiths in Their Pronouns*, 100). See also his earlier statement on this issue in *Returning to Mount Hira'*, 78-79.

[121] Ibid., 214 n. 15.

[122] Q. 38:82-83 (Y. Ali).

[123] Q. 40:55; 41:6; 47:19. No specific sin is mentioned, but the emphasis is on the fact that Muhammad is a man like everyone else and therefore has to seek God's forgiveness.

[124] Q. 5:105 (Y. Ali).

[125] Perkins married a woman in the early 1980s whose father was chief architect at Bechtel, one of the world premier engineering firms. Two years after his resignation from MAIN, he founded an alternative energy company that did much better than all of its competitors (he suspected that large investments had come to his firm as a form of bribe in order to dissuade him from writing this book). Through his new professional activities and through his father-in-law's contacts, Perkins gained a privileged vantage point from which to view the rise of Enron, the demise of MAIN

denotes this dimension in these words: "For our struggle is not against flesh and blood, but against the rulers, against the authorities, against the powers of this dark world and against the spiritual forces of evil in the heavenly realms."[126] In the last two decades much ink has been spilled over these "principalities and powers" (KJV).[127] The reality of evil also emerges from the corporate workings of human institutions, ideologies that take on a life of their own,[128] as well as greed and racism, for example, at work in the sociopolitical structures of cities, nations and international organizations. In themselves the collective scaffoldings of society are good. We saw this in the semiotic nature of human personality, expressing itself in constant communication, building of social links and creating culture—all gifts from the Creator and part of humankind's divine caliphate. But these good structures can easily be infiltrated and empowered for evil by Satan and his hosts of fallen angels. We should not be naïve about the pervasive and corrupting leaven of racism that blinds European Americans to their "white privilege" and inflicts identities of self-hatred, rage, or—worse yet—the "model minority person" that "bends" herself or himself toward the dominant culture in idolatrous ways.[129] Also, think about the massacres committed in the 1990s in Rwanda, in the Balkans, or Algeria. I personally came away from viewing the film *Hotel Rwanda* with an eerie sense of horrifying darkness descending on a country from which there was no escape, except for the

(chiefly through the deregulation that allowed Bechtel, Halliburton and Enron to thrive) and the failed attempt of EHMs to secure Iraq. Again the jackals followed, this time in the form of an all-out invasion.

[126] Eph. 6:12, NIV.

[127] The most influential author here is Walter Wink, *Unmasking the Powers: The Invisible Forces that Determine Human Existence* (Philadelphia: Fortress Press, 1986); *Engaging the Powers: Discernment and Resistance in a World of Domination* (Philadelphia: Fortress Press, 1992); *When the Powers Fall: Reconciliation in the Healing of Nations* (Minneapolis: Fortress Press, 1998).

[128] Look at Hitler's Third Reich, or Stalin's embodiment of state communism, and also the corporate nature of neoliberal capitalism that is more than the sum of its parts (or people in leadership) and has been inflicting untold suffering around the world. Perkins talks about the second generation of EHMs, who have no idea of the suffering and injustice they are promoting. He observed that since Reagan came to power there has been a proliferation of EHMs, persons who for the most part were convinced they were helping the world's poor by giving them jobs and spreading "civilization." That has only increased the nefarious fallout of the ascending global empire led by the US.

[129] Brenda Salter McNeil and Rick Richardson, *The Heart of Racial Reconciliation*, with a Foreword by John M. Perkins (Downers Grove, IL: InterVarsity, 2004). See particularly their chapter on "Principalities and Powers."

grace of God. Close to a million corpses were found after three or four days of bloody rampage.

It is right that some of the Huttu leaders were prosecuted for war crimes and acts of genocide by the International Criminal Court in La Hague. But none of this explains the brutality and passion with which "ordinary" people massacred men, women and children on the basis of their identity cards (if "Tutsi" was inscribed on them). People are no doubt held responsible before God for their actions and Algerians, likely both islamist fighters and disguised government agents, who committed grisly crimes while wiping out entire villages in the mid-1990s, will answer for this on the Last Day. So too will both sides in the current conflict in Iraq—atrocities were committed by US troops in places like Fallujah and prison camps and there was daily indiscriminate killing of scores of civilians by the insurgency. The same can be said of Afghanistan, where no end to the conflict seems in sight.

Sadly, in September 2008, the world witnessed the greatest economic meltdown since the 1930s. Economist and Nobel laureate Paul Krugman wrote in March 2009 that the subprime loans crisis by then was only a small problem people were now looking back on with nostalgia. What unfolded, from the collapse of several pillars of the American and European financial system to a world economy grinding down to a halt, was nothing less than "a global debt crisis."[130] From a purely structural viewpoint, continues Krugman, the crisis was actually foreseen, ironically enough, by Federal Reserve Chairman Ben Bernanke in a speech that was meant to be reassuring. Its title was "The Global Saving Glut and the U.S. Current Account Deficit," an account of how US foreign debt has risen so dramatically in the first decade of the twenty-first century. Wary of borrowing too much capital after their own crash of 1997–98, the "Asian Tigers" decided to build up huge reserves of foreign assets and thus invest massively in other parts of the world.

Putting it in terms that would make David Harvey envious (recall his theory of capitalist "overaccumulation"), Krugman quips, "The result was a world awash in cheap money, looking for somewhere to go."[131] Some of this capital ended up in emerging European or Baltic economies like Iceland, Ireland and Estonia—nations now in deep trouble. But most of it ended up in the US, causing its foreign debt to mushroom. So "the revenge of the glut" took over:

---

[130] Krugman, "Revenge of the Glut."
[131] Ibid.

For a while, the inrush of capital created the illusion of wealth in these countries, just as it did for American homeowners: asset prices were rising, currencies were strong, and everything looked fine. But bubbles always burst sooner or later, and yesterday's miracle economies have become today's basket cases, nations whose assets have evaporated but whose debts remain all too real. And these debts are an especially heavy burden because most of the loans were denominated in other countries' currencies.[132]

In a sense, at this level no one is to blame. Asian economies crashed, and with solid evidence blamed the crisis on Western greed and their propensity to gamble with others' money. So they decided to build a hedge of protection. But why invest so blindly in US assets? Krugman notes the irony of this unbounded trust in the American banks:

Mr. Bernanke cited "the depth and sophistication of the country's financial markets (which, among other things, have allowed households easy access to housing wealth)." Depth, yes. But sophistication? Well, you could say that American bankers, empowered by a quarter-century of deregulatory zeal, led the world in finding sophisticated ways to enrich themselves by hiding risk and fooling investors.

Krugman then begins to dig beneath the surface of what went wrong in the 2000s. He sees another culprit, this time back in the early 1980s. Besides the usual complaints about the fallout of "Reaganomics" ("rising inequality and fiscal irresponsibility"), Krugman adds the signing of the 1982 Garn–St. Germain Depository Institutions Act.[133] In effect, this deregulatory legislation threw out the window all the careful safeguards put into place after the Great Depression. Now households could take out mortgages on new homes with hardly any money down. Other lending standards for consumers were also dramatically watered down, with the result that Americans began borrowing more and more:

in the 1970s Americans saved almost 10 percent of their income, slightly more than in the 1960s. It was only after the Reagan deregulation that thrift gradually disappeared from the American way of life, culminating in the near-zero savings rate that prevailed on the eve of the great crisis. Household debt was only 60 percent of income when Reagan took office, about the same as it was during the Kennedy administration. By 2007 it was up to 119 percent.[134]

[132] Ibid.
[133] Paul Krugman, "Reagan Did It," *New York Times* (May 31, 2009): http://www.nytimes.com/2009/06/01/opinion/01krugman.html?em.
[134] Ibid.

But what does all of this, you might object, have to do with the topic at hand, sin? Everything! Risking too much with too little capital is what characterized the behavior of people at all levels: from the lower income people who took out mortgages they could not in the least afford, to the real estate predators who sold them the mortgages, to the bank executives who bundled up these risky assets and divvied out these debts in various forms to banks elsewhere; from the credit rating agencies who turned a blind eye to these dodgy magic tricks (we all learned what a "credit default swap" was), to top government watchdogs who were also beholden to these people and "never saw it coming." From top to bottom human greed and cronyism was propelling this runaway juggernaut forward. In Krugman's words, it was "a broad national breakdown in personal responsibility, government regulation and financial ethics." So many people were involved:

> People who had no business buying a home, with nothing down and nothing to pay for two years; people who had no business pushing such mortgages, but made fortunes doing so; people who had no business bundling those loans into securities and selling them to third parties, as if they were AAA bonds, but made fortunes doing so; people who had no business rating those loans as AAA, but made a fortunes doing so; and people who had no business buying those bonds and putting them on their balance sheets so they could earn a little better yield, but made fortunes doing so.[135]

Perhaps there is no better illustration of human sin lurking in the motivation of individuals and lunging forward and amplified at each new level in the cogs of a financial system that wanton, selfish greed. And the ripple effects spread to the whole world, unfortunately (as it always turns out) coming around to bite the already impoverished and most vulnerable with starvation, disease and death.[136]

I am here urging Muslims and Christians to take stock seriously of the entrenched character of evil both in individuals and corporate entities. Herein is the purpose of adjusting our mutual perception of human nature

---

[135] Paul Krugman, "All Fall Down," *New York Times* (November 25, 2008): http://www.nytimes.com/2008/11/26/opinion/26friedman.html?_r=1&hp.

[136] *The Guardian* editorialized on the issue of hunger, now dramatically on the rise. Among some of the factors: "high food prices, the global recession, concurrent emergencies in Pakistan, Somalia, Sri Lanka, as well as slower burning conflicts such as Sudan." Over a billion people "are expected to go chronically hungry" in 2009. The World Bank, commenting on the current recession, "estimates that with every percentage point of decline in the growth rates of developing countries, another 20 million people will be pushed into poverty" (May 28, 2009): http://www.guardian.co.uk/commentisfree/2009/may/28/editorial-world-food-programme-aid.

and the world around us today: that we might witness against the oppressive structures that enslave and destroy the lives of people everywhere and that we might militate for the respect of each person by fostering more just and peaceful societies. But we will need to include spiritual remedies as well: combating the materialistic bent of our consumerist cultures and all other forces that dehumanize people and lead them into idolatry; returning to the basics of a personal relationship with God nurtured by prayer, worship in community, love for neighbor, care for the most disadvantaged, and a teaching of the sacred text which equips people to resolve conflicts peacefully and work for reconciliation.

That is what Stassen and Gushee mean by "social change," the third element of their perception dimension in ethical theory. I am here agreeing in part with social conservatives who see change as only possible through individual transformation. Indeed, for any society to move away from the self-centeredness of a callous life style that only aims to accumulate more goods, a radical re-centering must take place—a spiritual transformation that begins in each person.[137] Yet I also agree with those social liberals who advocate the changing of unjust laws and the reformation of an American system tragically beholden to the interests and dictates of the corporatocracy.[138] Social change comes about as we tackle the issues at both ends.

Finally, the fourth crucial variable to Stassen and Gushee's perception dimension is integrity and truthfulness—in a real sense, coming back to one's "first works." People submitted to God ("*islam*," by definition)

---

[137] Many Muslim authors state this as a priority—for example, see Chandra Muzaffar's recent collection of essays and articles, *Rights, Religion and Reform: Enhancing Human Dignity through Spiritual and Moral Transformation* (London/ New York: RoutledgeCurzon, 2002). He writes, "The prophets and sages of all religions knew what the modern world has largely forgotten: the only real solution for the challenges facing humanity is spiritual" ("A Spiritual Vision of the Human Being," 103). Drawing from Buddhism, Shintoism, Judaism, Christianity and Islam, he sees "the underlying metaphysical causes of human misery" as stemming largely from the human ego, the self-centeredness of each person (ibid., 105). Hence, while proposing solutions for the eradication of corruption in his country (Malaysia), he calls for a moral and spiritual transformation of each individual carried out at the family level that aims at combating the pervasive consumerism and materialism of today's society ("Establishing a Fully Moral and Ethical Society," 289-310).

[138] Jim Wallis's *God's Politics* is at its core a plea for a spirituality across religious boundaries that takes the best from both sides of the extremely polarized US political landscape and transforms society and the world on the basis of a common vision for justice and peace (*God's Politics: Why the Right Gets Its Wrong and the Left Doesn't Get it* [New York: HarperSanFrancisco, 2005]).

move forward with humility in a spirit of repentance. This is the profoundly religious attitude of a person awed by God's infinite power and holiness and keenly aware of one's own weakness and inclination to sin. Admittedly, for the Christian the cross of Jesus is absolutely central. Redemption is the key to spirituality itself and the beginning of faith leading to salvation. But Jesus also said, "You will know them by their fruit."[139] Here Muslims will agree that part of the needed vision or perception that they and Christians are called to embrace includes a sincere honesty about one's failings, repentance from the bad choices of the past, and a determination to act rightly. This is to produce good fruit. Growing in character is to take responsibility for our deeds, to notice where we have gone wrong and in repentance make a correction in our ethics.[140] It will also mean that we are truthful with one another, within and beyond our communities.

Chandra Muzaffar, the Malaysian scholar and activist, advocates a similar spiritual vision likely to catalyze united human efforts toward greater justice. Coming back to the idea of *tawḥīd*, he urges that "the human being, as God's trustee, must seek to nurture the oneness of humankind, guided by a vision inspired by his spiritual heritage."[141] The human caliphate, then, is at the same time a spiritual and a moral norm shared by many religions:

> In Christian and Islamic philosophy…the human beings position as God's trustee meant that she was free of domination and control by any earthly institution or individual… A conscious, liberated trustee of God will seek to determine his or her destiny, in communism with his or her fellow human beings, guided by those external, universal spiritual and moral values which lie at the core of God's message to humanity.[142]

Such a holistic ethics is not only desirable. It is also practical. Jim Wallis recounts how people of faith in the last decade brought about impressive changes globally, particularly in the area of debt relief for the most impoverished nations. Churches in Britain took a series of initiatives that fired up a larger movement to cancel these nations' debts. Appropriately, it was called the Jubilee 2000 movement, which soon brought on board a large coalition of religious and secular people across many nations. Later,

---

[139] Mt. 7:20.
[140] Stassen and Gushee, *Kingdom Ethics*, 68.
[141] Muzaffar, "A Spiritual Vision of the Human Being," in *Rights, Religion and Reform*, 117.
[142] Muzaffar, "Development and Democracy in Asia," in *Rights, Religion and Reform*, 22-23.

after the landmark UN Millennium Development Goals were signed by 147 nations,[143] the then United Kingdom Chancellor of the Exchequer, Gordon Brown, a committed Christian, pushed hard to see these goals realized in his own country. In a 2004 speech to a gathering of mostly religious NGOs, Brown, now UK Prime Minister, stated: "Let us not lose hope but have the courage in our shared resolve to find the will to act. And let us say to each other in the words of Isaiah 'though you were wearied by the length of your way, you did not say it was hopeless—you found new life in your strength.'" Then he explained, "The strength to fight poverty, remove destitution, end illiteracy, cure disease. The challenge for our time and for our generation. And let us achieve it together."[144]

A big step forward was taken under Brown and Tony Blair's leadership at the Gleneagles G8 summit in July 2005, with poverty reduction in Africa on the top of their agenda.[145] In a conversation with Wallis, Brown gave credit to the grassroots faith-based initiatives for this new window of opportunity: "The most important social movement in Britain since Wilberforce [the nineteenth-century parliamentarian who is widely credited for the abolition of slavery] was Jubilee 2000.[146] Without that campaign, led by your church people, our government simply could not have cancelled the debts of the poorest countries."[147] Throughout his

---

[143] This was an agreement to cut extreme world poverty in half by 2015, as well as resolutions pertaining to education and health.

[144] Wallis, *God's Politics*, 272.

[145] A month before the summit, the BBC reported on Gordon Brown's ambitious plan for this gathering. "He called for a doubling of European aid by 2010 and 100% debt relief, as well as an end to many trade subsidies" ("UK Pushes for Africa Debt plan," *BBC Online* [June 3, 2005]: http://news.bbc.co.uk/2/hi/business/4606197.stm).

[146] See also Nazir-Ali's chapter "Jubilee: A Theme for the Millennium," in *Citizens and Exiles*, 148-54.

[147] Wallis, *God's Politics*, 272. William Wilberforce (d. 1833) was an influential Member of Parliament who devoted his whole career to the abolition of slavery. This reference to the social impact of nineteenth-century evangelicals in Britain betrays Wallis's admiration for the achievements of that movement and forms the backbone of his appeal to American evangelicals to recapture this holistic vision today. In an email newsletter he wrote recently, "At heart, I am a 19th-century evangelical; I was just born in the wrong century. The evangelical Christians of the 19th century combined revivalism with social reform and helped lead campaigns for women's suffrage and child labor laws, and to abolish slavery. One of the most famous revivalists, Charles Finney, developed the idea of the 'altar call' in order to make sure he signed up all of his converts for the abolition movement. Today, poverty is the new slavery—imprisoning bodies, minds, and souls, destroying hope and ending the

book Wallis calls for people of all faiths to unite for the achievement of such goals.[148]

This leads me to a last remark about the perception dimension of a holistic ethics. I believe our ethical vision can be clouded by ideological factors. For instance, the reader might conclude from my use of critical theory and from my statements about the imperial designs of the United States that I oppose all forms of capitalism. This would be mistaken. Again, the virtues of humility and repentance serve us well—all of us. For one thing, I have been arguing for critical realism. The fact is that we only approximate truth on this side of heaven. While this is true in the sciences and in the pursuit of all forms of knowledge, it is all the more true in the pragmatics of human affairs. The polarization of US politics under the Bush administration was most regrettable for people of faith who wished to see sweeping changes implemented in their society in the name of justice and respect for the social, economic and cultural rights of all citizens.[149] The early months of the Obama administration have raised hopes worldwide for a more multilateral approach. President Obama's June 2009 speech in Cairo certainly seemed to inspire Muslims everywhere. Yet many others remained skeptical: how can one man turn a massive imperial structure around? Others, even less sympathetic, simply sighed, "Words are cheap."

Just like Barber's denunciation of McWorld, Perkins's attack on the corporatocracy is a plea for a restoration of the old American republic in lieu of the new global empire. That republic offered hope: "It was based on concepts of equality and justice for all. But it also could be pragmatic, not merely a utopian dream but also a living, breathing, magnanimous entity. It could open its arms to shelter the downtrodden."[150] In practice,

---

future for a generation" ("The G8 and Global Poverty: God is Acting," http://go .sojo.net/sojourners/notice-description.tcl?newsletter_id=33798808r=m7LrSbn1Em 3n).

[148] This is more urgent than ever. The 2005 UN Development Report reveals that 18 of the poorest countries of the world are now poorer than they were in 1990. The gap between rich and poor is widening, and unless some drastic measures are taken by the international community, the goal of halving extreme poverty by 2015 is unattainable ("Life 'Worse for World's Poorest,'" *BBC Online* [September 7, 2005]: http://news.bbc.co.uk/1/hi/world/americas/4222034.stm).

[149] The international conventions on human rights since the 1948 UDHR include the International Covenant on Economic, Social, and Cultural Rights (ICESCR) of 1966, and the International Covenant on Civil and Political Rights (ECCPR), also of 1966. While the US readily acknowledges the ECCPR, it continues to express great reservations about the ICESCR.

[150] Perkins, *Confessions of an Economic Hit Man*, 127-28.

however, it was also rooted in greed and prospered over the groans of slaves and the extermination of native populations.

For myself, the answer might be in some form of socialism, but not in a collectivism imposed by the state[151]—both Muslim and Christian theology has always valued private ownership, as can be seen in the (agrarian) principle of the biblical jubilee (Lev. 15) and the role of both Meccan entrepreneurs and Medinan farmers. Creative solutions to our world's complex problems will have to be hammered out in an ad-hoc fashion, in pragmatic ways that bear out the foundational values of justice and compassion. Further, we now know by experience that sustainable development takes place chiefly when people are enabled to start small businesses, are free to organize themselves in a variety of associations (whether for political, cultural, business or religious purposes), and when local economies can thrive by trading with neighboring districts. Compare two famous success stories in development literature, the Self-Employed Women's Association (SEWA) of India and the Grameen Bank of Bangladesh:

> Both are organizations of poor women microenterpreneurs organized to help one another with credit, training, support, and political activism. These organizations have small paid professional staffs, but the actual services provided to the membership are provided primarily by other members, acting as volunteers. These volunteer members are not paid anything, and all the program resources come from donations of time, money, or both. Even though some resources are used to pay the staff, very little of the money that SEWA and the Grameen Bank get comes from selling their services on the market, and very little of the revenue they get comes from government allocations. They are a perfect prototype of an NGO.[152]

People of faith should also listen to secular agents involved in the current flurry of "transnational social movements" (TSMOs),[153] the

---

[151] John Marsden makes a compelling case for a fruitful dialogue between Christian theology and the early humanist though of Karl Marx. We urgently need to retrieve a "Marxian utopia" that builds on Jürgen Moltmann's "theology of hope" yet goes beyond by joining forces with socialist and civil society movements all over the globe. Among other goals, we would aim for: "the reeducation of human values implied in the transition from universal self-interest to social solidarity, from the ruthless exploitation of nature to respect for the earth's resources, from nationalism and militarism to internationalism and a commitment to peace" (John J. Marsden, *Marxian and Christian Utopianism: Toward a Socialist Political Theology* [New York: Monthly Review Press, 1991], 175).

[152] Weaver, Rock and Kusterer, *Achieving Broad-Based Sustainable Development*, 210-11.

[153] Cf. the last section of Chapter 3.

activists of the "grassroots postmodernism" that resist the Western-led globalization steamroller and proclaim to one another on the Internet and in the street that "a better world is possible." I mentioned the ground-breaking work of the Zapatistas in the early 1990s, but there is much to be learned in Europe as well, including from people like French sheep-herder José Bové. In intellectual circles too, much can be gleaned. One of *Le Monde Diplomatique*'s bi-monthly magazines, entitled "Soulager La Planète" (Relieving the Planet), offers an article by Raymond Van Ermen (Belgian secretary-general of the European Environmental Bureau). He writes, "Western industrialization and its modalities of consumption and travel are largely responsible for the deterioration of the environment,"[154] condemning the concerted strategy of the Anglo-Saxon (US and Britain) style of capitalism that seeks to protect at all costs the competitiveness of its corporations from tougher environmental regulations and opting for a looser form of "self-regulation." He argues for a capitalism that would respect the environmental norms set up by the United Nations; that would cancel the debts of the LDCs (least developed countries) and would pressure governments to convert large portions of their military budgets for the benefit of developmental and environmental projects.

The next article in that issue of *Le Monde Diplomatique* is from Mohamed Larbi Bouguerra, professor at the University of Tunis, who notes that before the Rio Summit President Bush promised not to sign any convention that was "too costly for businessmen." Nonetheless, as the Rio Earth Summit demonstrated, "humanity must establish an unprec-edented cooperation."[155] He quotes M. Anil Agarwal from the Indian Center for Science and the Environment who proposes that every human being share the cost of the planet's capacity to absorb carbon dioxde. Thus each individual would be allotted 0.5 tons of CO_2 per year. Since Americans emit ten times that amount, they would pay the most (they estimate that $15 a ton is a reasonable figure). The Chinese and Indians, on the other hand would be owed large sums.[156]

---

[154] Raymond Van Ermen, "Intérêts capitalistes et responsabilité planètaire" ("Capi-talist Interests and Planetary Responsibility"), *Le Monde Diplomatique, Manière de Voir* 50 (March–April 2000): 10.

[155] Van Ermen, "Au service des peuples ou d'un impérialisme économique?" ("In the Service of Peoples or of an Economic Imperialism?"), *Le Monde Diplomatique, Manière de Voir* 50 (March–April 2000): 14.

[156] For an excellent treatment of this theme from a theological perspective, see John B. Cobb Jr's *Sustaining the Common Good: A Christian Perspective on the Global Economy* (Cleveland, OH: The Pilgrim Press, 1994). Arguing against free trade, he contends that a world government would be better than the current neoliberal model,

I end this section on ethics by indicating an urgent need to connect in innovative fashion ethical theory and Islamic law. As mentioned in several places of this work, the vast majority of Muslims worldwide are conservative—that is, they fall back on the traditional reflex of looking to their jurists as guides to socio-moral behavior in changing times. The weakness of Arkoun's analysis comes from his un-avowed disdain for *fuqahā'* (jurists, singular *faqīh*) and *'ulamā'*. So he searches for the "humanist" tradition of classical Islam almost exclusively in its ethical and literary texts. On my view, Abou El Fadl, as a trained jurist in both the Islamic and Western traditions, is more likely to impact mainstream Muslim ethico-legal thinking. Noting that "authoritarian hermeneutics have become rampant in contemporary Muslim societies," he explains "authoritarianism" as "a hermeneutic methodology that usurps and subjugates the mechanisms of producing meaning from a text to a highly subjective and selective reading."[157] At the same time, by operating within the accepted doctrinal framework of the Muslim faith (the Qur'an comes from God, who thereby authenticates Muhammad's prophethood) and by choosing to "accept the juristic tradition as part of the relevant community of meaning...and to work normatively from within that tradition,"[158] he seems to be seeking recognition as an authentic legal scholar (*faqīh*) and *mujtahid*.[159]

Having staked out his "progressive" take on traditional Islamic hermeneutics (not unlike my evangelical Christian position), Abou El

---

but that the better solution still would be a decentralized, regionally managed global system.

[157] Abou El Fadl, *Speaking in God's Name*, 5.

[158] Ibid., 31. There are two methodological alternatives, he argues: (1) authority could be based on normative rational principles, i.e., a rationalist approach; (2) a "hermeneutic approach"—go directly to the Qur'an as the sacred text and develop a theory of authority based on its divine origin. For Abou El Fadl this is the position of today's authoritarians—a position hopelessly naïve, in view of the fact that texts have to be interpreted by people in particular contexts. This is particularly problematic since language itself evolves rapidly over time. My own discussion of hermeneutics here is designed to interact in an affirming way with his writing, and the similar thinking of Soroush and Moosa.

[159] That is, someone who has met the stringent traditional qualifications as a jurist habilitating him to make independent judgments (*ijtihād*) in the light of new situations. In this sense, he is challenging the likes of the popular (and moderate) islamist Shaykh Yusuf al-Qaradawi, based in Qatar. I write "he seems," because he would probably disavow this statement. He is more involved in Western academia than in Muslim *fiqh* circles. Yet he has drawn considerable attention, and has received several death threats from radical groups.

Fadl defines three crucial issues related to authority in Islamic law: (1) competence or authenticity—the authority of the sacred text (here Qur'an and Sunna) to speak on God's behalf;[160] (2) determination—understanding and interpreting a command, but also "ascertaining the 'use' of a command";[161] (3) agency—who decides what is authentic, what a text means and how it should be applied, the individual believer or some kind of institutional hierarchy? As mentioned earlier, already in the first generation of Muslims, this issue led to the shedding of much blood. As the Kharajites maintained, we may live only according to what God has revealed. But who decides "what" God has revealed? Here is where Abou El Fadl inserts what he sees as a necessary ethical dimension. Indeterminacy is inevitable as people struggle to speak in God's name through the interpretation of the texts' "indicators" (*adilla*, sing. *dalīl*). Nobody can claim perfect knowledge of the Divine Will, but in order to sharpen the knowledge we already have we will have to go beyond the simple textual indicators on the basis of our sense of right and wrong.[162]

In practice, one "develops a knowledge of God...through a complex matrix of relationships that are collateral to the text." This includes a personal relationship with God nurtured through prayer and other devotional practices, as well as personal reflection on creation and the

---

[160] In this regard Abou El Fadl writes, "I made the faith-based assumption that the Qur'an is the immutable and uncorrupted World of God... As far as the Qur'an is concerned, the only pertinent issue is to determine its meaning. The Sunnah, and other historically relevant material, however, pose a very different challenge" (ibid., 87). For a good introduction to the crucial debates on the Sunna and its relative authority to the Qur'an in the last century, see Daniel Brown, *Rethinking Tradition in Modern Islamic Thought*, Cambridge Middle East Studies (Cambridge/New York/Melbourne: Cambridge University Press, 1996). With research both in Pakistan and Egypt, Brown shows that the early modern anti-*ḥadīth* movements lost steam, especially in the late twentieth century. Revivalists such as Qaradawi and ex-Muslim Brotherhood member Muhammad al-Ghazali are good examples of the new breed of popular theologians who forge a middle path between the *ahl-al-ḥadīth* (traditionists holding to the classical consensus on the Sunna—the *ḥadīth* interpret the Qur'an) and the likes of Fazlur Rahman for whom the Qur'an interprets the Sunna and who have moved to the fringes by going beyond scripturalism. On this, see Johnston, "*Maqāṣid al-Sharīʿa*," 149-87.

[161] Abou El Fadl, *Speaking in God's Name*, 25. This is where Abou El Fadl deals most directly with the issues I have raised concerning hermeneutics: (1) language is "semi-autonomous"—a culturally conditioned "system of bounded linguistic symbols" (ibid., 89); (2) readers can impose their meaning, but ultimately the meaning of sacred texts is determined by the parameters of what specific "communities of interpretation" consider "reasonable" (ibid., 90).

[162] Ibid., 93.

flow of history. Thus it can happen that friction develops between one's basic convictions and certain indications of the text. The individual instinctively exclaims, "This cannot be from God, the God I know!" In this case Abou El Fadl recommends the practice of what he calls a "conscientious pause," because "in the final analysis, Islamic theology requires that a person abide by the dictates of his or her conscience."[163] Sometimes, it would seem, ethical convictions must trump law, even laws derived from sacred texts and longstanding juridical traditions.

Heated debates between proponents of ethical objectivism (an act is good because it conforms to an objective goodness which God himself observes) and ethical voluntarism (an act is good because God commands us to perform it) raged from the third to the fifth centuries of Islam. The Ash'arite consensus in the end (opting for voluntarism) drew from the earlier positions of the *ahl-al-ḥadīth*, the "traditionists," while the Mu'tazilites embraced the more rationalist position of objectivism. This is a crucial epistemological issue on which one must now take a stand, contends Abou El Fadl. Does God act justly because he is bound by justice, or does he define what is just and good through his own decision-making?[164]

In his more recent contribution to *Progressive Muslims*, Abou El Fadl reinforces his ethical argument. Two aspects of the current influential paradigm of "Salafabism" (the *salafi* movement now co-opted by Wahhabism) are crucial here: (1) the sacred texts regulate all areas of human life; (2) the aesthetic or moral concerns of the interpreter have no bearing whatsoever on the reading of the text. Thus, "values like human dignity, love, mercy, and compassion are not subject to quantification, and therefore they cannot be integrated into legal determinations."[165] This has no doubt contributed to "a sense of rootlessness among modern Muslims." Though Salafabism is certainly behind much of the revivalism observed throughout the Muslim world and though it can lead to some ugly behavior,[166] the ideology of bin Laden and Ayman al-Zawahiri are only "extreme manifestations of the rather widespread theological orientation of Salafabism. In one sense they represent a generation that has been deeply wounded and crushed by the humiliation inflicted by Western postcolonial hegemony. Equally, their defensive use of the Islamic

---

[163] Ibid., 94.

[164] Ibid., 64.

[165] Abou El Fadl, "The Ugly Modern and the Modern Ugly," 59.

[166] His essay starts with the girls' school fire in Mecca in which several pupils died because the religious police (*mutawwa'un*) insisted they go back in to fetch their veils. He goes on "In recent times, Muslim societies have been plagued by many events that have struck the world as offensive and even shocking" (ibid., 34).

tradition is dissonant and dysfunctional. For Abou El Fadl, they are "orphans of modernity," and "their claim to an authentic lineage in the Islamic civilization is tenuous at best."[167]

We have all been saddened and outraged by bombings in London and Spain; yet they are almost daily occurrences in countries like Iraq, Afghanistan and Pakistan. Blind scripturalism will never be up to the task of forging a theology of peace, conflict resolution and human solidarity standing against all forms of economic exploitation and political oppression. As Christians and Muslims attempting to articulate a common theology of creation and of human trusteeship, we will not only have to raise the banner of ethical principles that unite humanity[168] but also weave them into a coherent ethical theory that finds its rightful place in our theology. Admittedly, the issue of human rights and global ethics has been hotly debated in Muslim circles for half a century now and its relation to Islamic jurisprudence is fraught with complexity.[169]

Tariq Ramadan, the Swiss professor of Philosophy and Islamic Studies and consultant to several European Unison committees, was interviewed by the British daily newspaper *The Independent* days after the London bombings. He had already issued a press release condemning the criminal acts.[170] At the very least, that young Muslims born and bred in the UK could choose to end their lives as suicide bombers points to a crisis in the Muslim community. Part of this phenomenon may be seen in the fact that the mainstream neglected to uproot a "radical and literalist discourse" in their midst, and part of it is a reflection of alienation and a severe identity crisis among the youth:

[167] Ibid., 61.

[168] No Muslim today, for instance, would disagree with Mohammad Hashim Kamali when he writes that "[t]he Qur'an's ubiquitous emphasis on ʿadl [justice] makes this clearly one of the cardinal objectives of Islam and an overriding theme of the Holy Book itself." This also includes all the prior revelations, so that we learn from the Qur'an that a sense of justice is both innate to humanity and therefore universal (*Freedom of Expression in Islam* [Kuala Lumpur: Berita, 1994], 267). Particularly relevant here are Q. 4:135 and 57:25.

[169] Cf. Johnston, "Rethinking Human Rights: A Common Challenge to Muslims and Christians," presented at the Conflict Transformation Project's Interfaith Dialogue in Rockville, MD, April 22–23, 2005; publication forthcoming in edited volume of other essays by Muslim and Christian scholars.

[170] Among other things, he stated, "The authors of such acts are criminals and we cannot accept or listen to their probable justifications in the name of an ideology, a religion or a political cause" ("Can this Erudite Swiss Lecturer Really Be the Man Branded by The Sun as 'the Acceptable Face of Terror'? Paul Vallely Meets Tariq Ramadan," *The Independent Online* [July 25, 2005]: http://news.independent.co.uk /people/profiles/article301486.ece).

> Young people are told: everything you do is wrong—you don't pray, you
> drink, you aren't modest, you don't behave. They are told that the only
> way to be a good Muslim is to live in an Islamic society. Since they can't
> do that, this magnifies their sense of inadequacy and creates an identity
> crisis. Such young people are easy prey for someone who comes along
> and says, 'there is a way to purify yourself'. Some of these figures even
> keep the young people drinking to increase their sense of guilt and make
> them easier to manipulate.[171]

At least four changes need to be implemented to remedy this, claims
Ramadan:

1. A critical mind should be instilled in the youth. While being
    encouraged to share in all the good aspects of European culture
    (from the non-essential aspects of food and dress to art, literature
    and values of democracy and pluralism), they should also learn to
    discriminate between the good and bad. Nonetheless, "[a]nything
    in Western culture that does not contradict the message of Islam
    can be accepted and integrated."

2. In the same spirit we must jettison the "us" vs. "them" mental-
    ity—the pronounal paradigm called into question by Cragg. This
    is the "binary vision of reality" that assumes that "everything
    Western is decadent and unIslamic."

3. The younger generation must reject their parents' strategy of
    seeking to remain invisible in the midst of Western society. To
    the contrary, they should speak out "both against those who are
    doing these things in the name of our religion and against those
    who say that being a loyal British citizen means blindly accepting
    all the decisions of the British Government. Ours must be a
    constructive and critically participative loyalty."

4. We must promote the ethical values common to all—human
    dignity, justice and compassion, so as to promote interreligious
    dialogue:

---

[171] Ibid. All other quotes below are taken from this article. See also Dan Murphy,
"Can Islam's Leaders Reach its Radicals?," *The Christian Science Monitor Online*
(July 14, 2005): http://www.csmonitor.com/2005/0714/p01s01-wome.html. Both the
Rector of Al-Azhar University, Sheikh Mohammed al-Tantawi, and Sheikh Yusuf
al-Qaradawi condemned the London bombings in the strongest of terms. Further,
most mosques in the US issued statements to that effect and devoted one or several
Friday sermons to a theological denunciation of those bombings. Having said that, I
have also heard that in many cases suicide bombings in both Israel/Palestine and Iraq
are not included in that condemnation. For many, the killing of civilians in those
cases is justified. My question is an ethical one: does the end justify the means?
Bishop Cragg has recently tackled this issue in *Faith at Suicide Lives Forfeit: Violent
Religion—Human Despair* (Brighton/Portland, OR: Sussex Academic Press, 2005).

> If you go back to the source of our religions you find common values. It's important to read the scriptures of the other faiths and see how the others interpret these common values. It's high time for Muslims to say that anti-Semitism is not acceptable. We have to ask questions of our own tradition and be self-critical about what is sectarian and racist. Only then can our society build a common future.[172]

A British person of color—one of the few to be a member of the House of Lords—sounds a similar clarion call. Bishop Michael Nazir-Ali, originally from Pakistan, celebrates the diversity of British society, yet reminds his readers that for all to contribute to the common good of society "there needs to be a framework in which we all lead our lives."[173] He suggests that the following Judeo-Christian value system—which, he observes, already has much in common with Islam—could serve as a starting point: (1) integrity of creation; (2) inherent dignity of all human beings (here he mentions the Qur'an's reference "to human dignity in terms of the stewardship [*khilafa*] of creation which has been entrusted to humankind [2.30, 6.165]"); (3) impartiality in justice; (4) importance of compassion, "especially for those who have been weakened by 'the slings and arrows of outrageous fortune.'"[174] In order for this framework to be adapted to a diverse society, it must be applied comprehensively, and thus include issues of social mores as well as political and economic policies. It will also have also have to eschew legalism, a temptation that has entrapped the followers of all religions. A third condition is that it renounce "Constantinian" Christianity, that is, the temptation to use political power to press its advantage. Finally, such an ethical framework must not be coercive, but rather seek to advocate for minority faiths so that everyone's voice might be heard and all be able to contribute to the well-being of society.[175]

This advocacy of interreligious dialogue in which one builds on human dignity-enhancing values common to all is also central to Seyyed Hossein Nasr's *The Heart of Islam*.[176] His own Sufi-oriented theology of humanity owes much to Ibn al-Arabi:

---

[172] Ibid.

[173] Nazir-Ali, *Citizens and Exiles*, 155.

[174] Ibid., 156.

[175] Ibid., 157-58.

[176] Cf. Chapter 9, n. 159. A scholar at the American University since 1984, Nasr is also a noted philosopher from the Perennial Philosophy school, which takes its inspiration from Aldous Huxley's *The Perennial Philosophy* (1945). Since then, other scholars have continued this search for the mystical common point of all religions, many of whom converted to Islam. Frithjof Schuon and René Guénon were particularly influential in Nasr's formative thinking. See Lewis Edwin Hahn, Randel E.

Islam believes that God breathed His Spirit into Adam and according to the famous *ḥadīth*, "God created Adam in His form," "form" meaning the reflection of God's Names and Qualities. Human beings therefore reflect the Divine Attributes like a mirror, which reflects the light of the Sun. By virtue of being created as this central being in the terrestrial realm, the human being was chosen by God as His vicegerent (*khalīfat Allāh*) and His servant (*ʿabd Allāh*). As servants human beings must remain in total obedience to God and in perfect receptivity before what their Creator wills for them. As vicegerents they must be active in the world to do God's Will here on earth.[177]

This double role of humankind as servant and trustee implies that any discussion of human rights must be balanced with a discussion about human responsibilities. Yet this is what all people of faith bring to the table. In the current discussion about freedom, Islam agrees with other faiths that true religion enables people "to overcome the clutches of the power of their lower souls," while the individualism of a consumerist society, "in the guise of freedom, only strengthens the bonds of slavery of our immortal soul to that powerful slave master within it that is the agent of rebellion, passion, concupiscence, and ultimately bondage."[178] For devotees of all religions, "to love God and obey His commandments is not considered a loss of freedom." "In these dark times," Nasr urges Muslims to be self-critical by denouncing "what goes on among themselves in the form of bigotry and fanaticism," but also to condemn the breakdown of the Western family and "the desecration and destruction of the natural environment."[179] Too often Muslims have been unfairly singled out as human rights violators. Here is what is needed, proposes Nasr:

> One of the roles of Muslims as members of a major world religion is to answer such questions in all honesty from the Islamic point of view. It is also to insist upon mutual respect between civilizations and the values they bear instead of accepting one-sided imposition. It is, moreover, to seek actively to cooperate with not only Westerners, but also with members of other cultures and civilizations to point to those values that we do all hold dear and that must be respected by everyone if we are going to live and function as human beings on a globe on which there seems now to be no other choice but to live in mutual respect with compassion and love for others or to perish together.[180]

Anxier and Lucian W. Stone Jr, eds., *The Philosophy of Seyyed Hossein Nasr*, Library of Living Philosophers (Chicago: Open Court, 2001).
[177] Nasr, *The Heart of Islam*, 276.
[178] Ibid., 292.
[179] Ibid., 304.
[180] Ibid.

In that spirit and in closing this section on the integration of ethics and theology, I raise an issue to which I have alluded indirectly several times. Ever since the Medinan battles of Badr and Uhud (624–25), Muhammad and his followers have proceeded with the unshakable conviction that (1) Mecca must be won over, even by force if necessary (Islam is the religion of Abraham who first built the Kaaba shrine); (2) in more general terms, when Muslims are outnumbered, they should fight, knowing that God will intervene miraculously on behalf of his people and grant them military victory, thus vindicating his revelation to the Prophet. As we read in the Qur'an, "Fighting is prescribed for you, and ye dislike it... Those who believed and those who suffered exile and fought (and strove and struggled) in the path of God [*jāhadū fī sabīli-l-llāh*], they have the hope of the Mercy of God; and God is oft-Forgiving, Most Merciful."[181] Though the jihad of arms is called by tradition the "lesser jihad" (the greater one is to control one's passions), it is no small matter in the Qur'an.

At the same time, one could point to the Bible, which records God calling his people to fight (Joshua's wars of conquest) and even to commit genocide (note, e.g., the treatment of the Amalekites at Exod. 17:14-16; 1 Sam. 15:1-9). And though Jesus clearly advocated nonviolence, his followers have left a violent trail from the Crusades, to the Reconquista and the fifteenth-century papal bulls sanctioning the plundering and exploitation of India and the Americas. Respected Indian Muslim journalist M. J. Akbar notes that Nicholas V's Bull of 1455 evoked a strategy by which the Indian Christians could become allies "against the Saracens and other such enemies of the faith."[182] This is hardly surprising—sadly, I confess. One might also point out that the long-used Christian cliché that "Islam was spread by the sword" is grossly exaggerated. T. W. Arnold's research of the 1890s has withstood the test of time: though forced conversions did occasionally occur, they were much more the exception than the rule.[183]

This is why Akbar's recent book, *The Shade of Swords*, is relevant here. His controversial thesis is that the current spate of jihadi violence draws directly from this central theme of the Muslim faith, sealed already in the Battle of Badr. The subsequent military campaigns of the Muslim armies are nothing short of phenomenal, and throughout history

---

[181] Q. 2:216, 218.
[182] M. J. Akbar, *The Shade of Swords: Jihad and the Conflict between Islam and Christianity* (London/New York: Routledge, 2002), 116.
[183] T. W. Arnold, *The Preaching of Islam: A History of the Propagations of the Muslim Faith*, 2d ed. (New York: Scribner's Sons, 1913).

this steely conviction that God ordains and orchestrates military prowess and political domination sustains the hopes and struggles of Muslim peoples. He explains,

> Islam, as the word itself implies, does not seek violence. Equally, Islam does not permit meek surrender either. There are circumstances in which all Muslims are commanded to fight to defend the faith. In such times war becomes a duty, and those who shirk it are condemned by the Quran. It may be a lesser jihad, but for some twelve hundred years only a comparative handful of infidel armies emerged from the lesser jihad either with their pride or their power intact.[184]

Akbar then retraces the history of Muslim expansion, with a special emphasis on the clash between Muslims and Christians—often more in the realm of ideas, economic rivalries and political influence than mere raw military clashing. In the modern period, he focuses particularly on the Subcontinent, first on India as the British gradually squeezed the life out of the Mughal Empire, then on Pakistan and neighboring Afghanistan. The common theme throughout, as the *ḥadīth* reflected in his book's title,[185] is the reward of Paradise for the martyr in God's cause and the call to submit to no one but him. "Think not of those who are slain in God's way as dead. Nay, they live, finding their sustenance in the presence of their Lord."[186] "They are not defeated by defeat. They wait, they keep the faith, and renew their jihad until they achieve the victory that Allah promised in His Bargain with the believer, specified so clearly in the Quran."[187]

My role as an outsider is not to argue how to negotiate the classical legal consensus that with respect to jihad had the violent verses abrogating (with disagreements on some details) the more peaceful ones,[188] but only to note that a lot of work remains to be done in the intersection of Islamic theology, ethics and law.[189] I will only venture two suggestions.

---

[184] Akbar, *The Shade of Swords*, 2

[185] "And know that Heaven lies in the shade of swords," from al-Bukhari's collection, *jihād* 112.

[186] Q. 3:169.

[187] Akbar, *The Shade of Swords*, xv. The "bargain verse" reads as follows: "God hath purchased of the Believers their persons and their goods; for theirs [in return] is the Garden of Paradise. They fight in His Cause and slay and are slain: a promise binding on Him in Truth, through the Law, the Gospel, and the Qur'an. And who is more faithful to His Covenant than God? Then rejoice in the bargain which ye have concluded: that is the achievement supreme" (Q. 9:111, Yusuf Ali).

[188] Again, see Peters, *Jihad in Modern and Classical Islam*.

[189] I have examined aspects of this issue, as mentioned above, in "A Turn in the Epistemology and Hermeneutics of Twentieth-Century *Uṣūl al-Fiqh*," and also in

On my view, the traditional scripturalist hermeneutic will not be up to the task—hence my presentation of a hermeneutics in this work that takes into account the contributions of postmodernism. Soroush, Abou El Fadl, Moosa and others, I feel, are on the right track. Second, in the spirit of Cragg's critique of the inherent tension between Mecca and Medina in Islamic theology, I am pleading with my Muslim brothers and sisters to reconsider the ethical implications of the early Muslim conquests. Just as I have forthrightly condemned the Crusades and Western colonialism as contrary to the spirit and letter of the gospel,[190] I would urge some soul-searching on the Muslim side. S. H. Nasr is typical of contemporary Muslim scholars, who quickly glosses the military conquests as "the rapid spread of the Arabs outside Arabia."[191] Granted, the early campaigns coincided with the significant weakening of the two empires to the north of Arabia (the Sassanian Empire crumbled with little resistance) and many Monophysite and Nestorian Christians initially welcomed the invading Muslims. Granted also, the Muslim empires centered in Baghdad and Cordova were greatly more tolerant than any counterpart

---

"*Maqāṣid al-Sharīʿa*". Also, I have analyzed the theologies of Moroccan scholar Allal al-Fasi (d. 1973) ("'Allāl al-Fāsī: *Sharīʿa* as Blueprint for a Righteous Global Citizenship?," in *Shari'a: Islamic Law in the Contemporary Context*, ed. Abbas Madanat and Frank Griffel [Stanford, CA: Stanford University Press], 83-103), Algerian thinker Malek Bennabi (d. 1973) and Tunisian activist Rachid Ghannouchi (b. 1941) ("Fuzzy Reformist-Islamist Borders").

[190] Note that evangelicals undertook an impressive campaign of apology for the Crusades (Reconciliation Walk) coinciding the 900th anniversary of the First Crusade (1995–1999), little mentioned in the press. Several of the routes taken were re-traveled by groups asking forgiveness from both civil and religious authorities (mosques and synagogues) along the way. They were interviewed on at least two occasions on Turkish television, for example. The culmination took place in Jerusalem in July 1999 with the participation of several historical churches in unity (see http://www.recwalk.net). I also was privileged to participate in the first two weeks of the Pilgrimage for Peace, a re-enactment of the ancient journey of the Magi, starting in Ctesiphon, near Baghdad in Iraq, October 2000. This was sponsored by an evangelical organization, the Holy Land Trust (http://www.holylandtrust.org) that sought to foster reconciliation between Muslims and Christians, protest the regime of sanctions that in the 1990s, according to the UN, was responsible for the death of tens of thousands of children, and honor Jesus two thousand years after the Mesopotamian wise men made the first journey to worship him after his birth. Traveling by foot and camel along the ancient caravan route, the participants in the trek finally arrived in Bethlehem on December 25, greeted by a crowd of cheering Palestinians, both Muslims and Christians (see the link to the pictures and commentary on that journey on the above website).

[191] Nasr, *The Heart of Islam*, 216.

in Europe at the time.[192] Nevertheless, the establishment and expansion of *dār al-Islām* in the first couple of centuries could not have taken place without bloodshed and political hegemony. Whatever justification was given in the past no longer holds today in a world community that is conscious more than ever of human solidarity and universal rights. Otherwise, "might makes right," and the European conquests beginning in the fifteenth century are also justifiable. As Cragg sees it, "In the quest for a global ethic for a global ecology, the claims of a basic humanism deserve to override—and not to be over-ridden by—the exclusives of territorial religion."[193]

The context of Muslim–Christian dialogue is changing. As Stassen and Gushee have reminded us, more than ever a joint claim to manage the earth in God's name will require us to lay hold of a common vision of the world we have inherited and will bequeath to our children—its environmental health and the measure of harmony and equity we can foster among its peoples. We are outrageously far from the ideal of environmental justice Rasmussen has pointed out and we desperately need to come back to our first works, back to "tending the Garden" and "being our brother's keeper."[194] But as I have been showing all along, much ethical and theological reflection is already under way on both sides.

---

[192] For a comprehensive study on this, see Bat Ye'or, *Decline of Eastern Christianity under Islam: From Jihad to Dhimmitude: Seventh Century*, trans. from the French by Miriam Kochan and David Littman (Madison, NJ: Farleigh Dickinson University Press, 1996); *Islam and Dhimmitude: Where Civilizations Collide*, trans. from the French Miriam Kochan and David Littman (Madison, NJ: Farleigh Dickinson University Press, 2002).

[193] Cragg, *Faiths in their Pronouns*, 134.

[194] Gen. 4:9. When God asks Cain where his brother is (he had just killed him), Cain retorts sarcastically, "I don't know. Am I my brother's keeper?"

Chapter 12

CONCLUSION

In this concluding chapter I wish to bring together the important strands of this study on a common theology of humanity and creation for Muslims and Christians. I begin with a short case study that illustrates the practical outworking of a number of principles emanating from this project. Then I list the main findings of the preceding chapters and some of the questions that will need to be kept in mind in the continuing conversation between members of both groups determined to act together as God's trustees for the common good, both locally and globally.

Peacebuilding in West Asia

By "West Asia" (we do speak of Central Asia, after all) I simply mean what is commonly called the "Middle East," except that it avoids the colonial connotation of a Eurocentric term. This is also Chandra Muzaffar's preferred term. This is the region where human civilization is thought to have started—even the biblical Garden of Eden is situated in present-day Iraq. I begin with a few remarks about the central dispute of the region and the one that squarely involves all three monotheistic religions, the Israeli–Palestinian conflict. The Christian Peacemaker Teams then serve as transition for a brief view of Muslim–Christian cooperation in American-occupied Iraq.

Hope in the Land of Abraham
The founding of the state of Israel, recognized within hours by both the USA and the Soviet Union, spelled for Palestinians the *nakba* ("catastrophe"): 420 villages razed in 1948 with 750,000 refugees dispersed in several countries; the military occupation of the West Bank and Gaza in 1967, with the establishment of Israeli settlements in the most strategic locations (in contravention of the 1947 Fourth Geneva Convention);[1] the

---

[1] Israeli officials have consistently argued that those conventions of 1949 (which they signed) cover the relations between two states. Here they do not apply, since the

most dramatic expansion of these settlements and parallel expropriation of Palestinian land since the beginning of the Oslo Peace Process; and now, after a second much bloodier uprising, the Israeli "separation wall" entrenches key Israeli settlements in the West Bank and makes it even more unthinkable that a viable Palestinian state will ever see the light of day.[2] Sara Roy, a senior research scholar at the Harvard Center for Middle Eastern Studies and a recognized expert on Gaza,[3] argues that the unilateral Israeli withdrawal from Gaza in 2005, far from providing a positive step toward peace, in fact succeeded in garnering international legitimacy for continued Israeli occupation.[4] With the Gaza Strip more effectively sealed off and isolated, one of the most densely populated areas of the world (three times that of Manhattan) and most impoverished (now on par with the poorest in Africa) now sees its prospects for economic development virtually obliterated—with ominous implications for an escalation of violence.

Looking back on the legacy of military occupation, the feminist theologian Rosemary Radford Ruether writes in the Preface to Naim Ateek's *Justice and Only Justice*, "Palestinians are the victims of a Zionist liberation theology and ideology. The Jewish exodus from oppression in Europe is the rationale for their conquest. The Jewish claim to the promised land is their dispossession. Jewish peoplehood

Palestinians do not form a state. Since the Oslo Agreements of 1993, this line of reasoning appears rather more disingenuous.

[2] Jamal Juma, coordinator of the grassroots Palestinian Anti-Apartheid Wall Campaign (http://www.stopthewall.org), expresses well the frustration of Palestinian civil society working hand in hand with scores of Israeli peace activists. President George W. Bush had just invited for the first time Prime Minister Mahmoud Abbas to the White house and promised a $2 million aid package to the Palestinian Authority. Yet for Juma this does nothing to alleviate the "devastating new realities" now being put into place: "The Apartheid Wall and accompanying infrastructure of Jewish-only bypass roads, military zones and settlements, are rapidly moving towards the permanent ghettoisation of the Palestinian people." The aid package will mostly serve to bolster the occupation project by reinforcing existing checkpoints and settlements, while doing nothing to remedy the above-mentioned "crimes." It just reinforces the "Bantustanisation of Palestine," just as the South African regime once controlled its African dependents in separate and semi-autonomous homelands (called "Bantustans") (Jamal Juma, "Israeli Apartheid," *Al-Ahram Weekly Online* [June 2–8, 2005]: http://weekly.ahram.org.eg/2005/745/re5.htm).

[3] Sara Roy, *The Gaza Strip: The Political Economy of De-development* (Washington, DC: Institute of Palestine Studies, 1995); see also her *Failing Peace: Gaza and the Palestinian-Israeli Conflict* (London: Pluto Press, 2006).

[4] Roy, "Gaza's Future: 'A Dubai on the Mediterranean,'" *London Review of Books* 27.21 (November 3, 2005). Online: http://www.lrb.co.uk/v27/n21/roy_01_.html.

excludes the existence of Palestinians as a people. Jewish redemption is Palestinian oppression."[5]

In 2001, two years before his death, Edward Said wrote from South Africa where he had been invited to participate in a conference on values in education. His old time friend, the South African minister of education Qader Asmal, had persuaded Mandela to address the conference on the first evening. Eloquent as usual, and gripping as well, Mandela spoke some words that led Said to think afresh about the Palestinian predicament. He reminded the audience that "our struggle is not over," and that the campaign against Apartheid "was one of the great moral struggles" that "captured the world's imagination."[6] But he went on to describe the anti-Apartheid campaign as more than just a struggle against racial discrimination. In fact, its goal had been broader. It was designed as a means "for all of us to assert our common humanity" and aimed to bring about coexistence, tolerance and "the realization of human values."

Said then proceeds to explain his puzzlement. Why have the Palestinians failed after all these years to capture the world's imagination in the same way the anti-Apartheid campaign did from the start? The Jews also, "a people with a tragic history of persecution and genocide," were considered heroic and justified in taking the land that colonial Britain had promised them (according to the common interpretation of the 1917 Balfour Declaration). Not so the Palestinians:

> Yet, for years and years, few paid attention to the conquest of Palestine by Jewish forces, or to the Arab people already there who endured its exorbitant cost in the destruction of their society, the expulsion of the majority, and the hideous system of laws—a virtual Apartheid—that still discriminates against them inside Israel and in the occupied territories. Palestinians were the silent victims of a gross injustice, quickly shuffled offstage by a triumphalist chorus of how amazing Israel was.[7]

Years later, in spite of a Palestinian liberation movement active since the late 1960s, except for a few accolades and checks written by mostly Arab countries, the strategic balance today leans heavily in Israel's favor. Who is to blame? The impeccable Jewish propaganda machine? No, of course, though it will continue to use all the means at its disposal to justify their hold on the Palestinian people. But it is precisely because of "useless acts of terrorism" on the part of Palestinians over time that the Jewish lobby has been able to claim the moral high ground. As a result,

---

[5] Ateek, *Justice and Only Justice*, xii.
[6] Edward Said, "The Only Alternative," *Al-Ahram Weekly Online* 523 (March 1–7, 2001): http://www.ahram.org.eg/weekly/2001/523/op2.htm.
[7] Ibid.

oppression continues unabated. As a black South African reporter wrote in one of the local newspapers while on a visit to Gaza, "Apartheid was never as vicious and as inhumane as Zionism: ethnic cleansing, daily humiliations, collective punishment on a vast scale, land appropriation, etc., etc."[8] The fault, concluded Said, lies clearly with the Palestinian leadership and their inability to set consistent and compelling goals to their national struggle.

"The Only Alternative," as his title runs, is to reimagine the conflict, change paradigms and shift from ethnic conflict to common humanity. For some time Said had been advocating a one-state solution—a bi-national state for both Jews and Palestinians.[9] But now, with the horrific violence now spiraling out of control, his words seem more prophetic. Could this be the setting for a bold application of common human trusteeship? I believe so. Both sides must accept that "the other is here to stay." Where the Palestinians have been wrong is holding on to "the preposterous hope that a volatile American president would give us a state," instead of focusing "on ending the military occupation as a moral imperative or on providing a form for their security and self-determinism that did not abrogate ours."[10] The only solution is, "Two people in one land. Or, equality for all. Or, one person one vote. Or, a common humanity asserted in a bi-national state."[11]

Said was not optimistic that the Palestinian leadership would listen to him.[12] The last six years since his untimely death in 2003 have sadly substantiated this hunch. Gaza since 2007 is in the hands of Hamas, whose leadership came out even stronger from the brutal Israeli attacks

---

[8] Ibid.

[9] Cf. Said's landmark article "The Poverty of Nationalism," *The Progressive* 62.3 (March 18, 1998): 27-29. Though he is harsh in his criticism of Jewish nationalism (Zionism), he is even harsher with regard to Palestinian incompetence, corruption and backwardness: "the fundamental challenge that Israel poses is to ourselves—our inability to organize, our inability to dedicate ourselves to a basic set of principles from which we do not deviate, our inability to marshal our resources single-mindedly, our inability to devote all our efforts to education and competence, and finally, our inability to choose a leadership that is capable of the task" (ibid., 29). Yet his article holds out the hope that, once Israelis acknowledge the wrong they have done to Palestinians, coexistence must be the goal, "free of ethnocentrism and religious intolerance."

[10] Said, "The Only Alternative."

[11] Ibid.

[12] He ends the article with the question: "Is the current Palestinian leadership listening? Can it suggest anything better than this, given its abysmal record in a 'peace process' that has led to the present horrors?" (ibid.).

of December 2008; and the West Bank is ruled by Fatah, with its president Mahmoud Abbas praised by the West and only tolerated by his own.

In spite of this depressing reality, this section is about the hope of an emerging civil society on both sides, actively promoting peace and reconciliation through nonviolent means. And in terms of moral theory, the dimension of perception, or vision, becomes crucial. Indeed, it is possible to imagine that a minority of Israelis and Palestinians who refuse the "us vs. them" rhetoric could turn the tide. It is true that as both peoples are grieving and reeling under the impact of the bloodshed, a majority on both sides is crying for revenge. Yet other voices are calling for dialogue, and, what is more, they are risking their lives to demonstrate solidarity with the other side.

Tanya Reinhart, professor of linguistics at the University of Tel Aviv, wrote at the end of March 2001 in the Israeli daily *Yediot Aharonot* that among Palestinians a new identity was being forged in the crucible of the second intifada:[13]

> Khan Yunis, isolated between the fences of Gaza prison, has thus been freed also of the control of the Palestinian Authority and started to form popular committees that lead the communal struggle, like in the previous Intifada.
> Now Ramallah awakens and prepares itself for a popular democratic uprising and civil disobedience. At the same time, the voice of Bir Zeit University and many others is heard, calling to strive for cooperation with the Israeli opponents of the occupation, like in the previous Intifada.[14]

The landscape is also beginning to change on the other side:

> On the Israeli side, 140 academics have published on the 20th of March an ad in three Palestinian newspapers. "We extend our arms to you in solidarity with your just cause" they open, and express their wish "to cooperate with you in opposing the IDF's brutal policy of siege, closure and curfews." In the spirit of Mandela and Said, they too believe that this cooperation "may serve as a precedent-setting example for future relations between the two communities in this country, our shared country."[15]

People were beginning to reimagine the contours of past Israeli–Palestinian realities, and during the 2000s, in spite of—or even because of—worsening conditions, peace activists on both sides have only

---

[13] The Arabic word intifada refers to a popular uprising. Literally it means "a shaking off [of a yoke, a condition of bondage]."

[14] I recently accessed this article (8/9/05) from her University of Tel Aviv website: "Right for both People" (March 27, 2001): http://tau.ac.il/~reinhart/political /RightForPeoples.html.

[15] Ibid.

intensified their efforts. Israeli activists have continued to ride buses into the "Territories" and defiantly to remove the ditches and mounds of dirt the Israeli Defense Forces (IDF) have built to isolate Palestinian towns and villages, to join with Palestinian villagers to protest the construction of the partition wall and to help with the harvest of olives in villages threatened by the expansion of Israeli settlements.

As the wall was snaking its way through the northeast district across from the Israeli settlement, Ariel (June 2004), Reinhart began an article thus, "Along the route of the separation barrier in the West Bank, a new culture is springing up: on one side, soldiers and bulldozers; on the other, Israelis and Palestinians embracing the land and the trees, trying to save them both"—a vision of eco-justice, purely and simply.[16] This new culture is breaking all the old taboos:

> The breathtaking scenery of the Ariel district has been sliced up by the new roads that the rulers have built for their own exclusive use. Beneath them lie the old roads of the vanquished. There, on the lower level, is where the other Israel–Palestine treads. Israeli youths arrive in settlement buses and then make their way on foot and in Palestinian taxis among the checkpoints. They trek between the villages in groups or alone. Some sleep in the villages. Others will travel the same route the next day to reach the demonstration. Everywhere they go they are greeted with blessings and beaming faces. "Tfaddalu," the children in the doorways say, as if they had never heard of stone-throwing. Like the inhabitants of other Palestinian villages along the route of the fence, those in the Ariel area have opened their hearts and their homes to the Israelis who come to support their non-violent resistance to the barrier that is robbing them of their land.

The Israelis who willingly face the tear gas, and at times even the rubber bullets of their army, do so, Reinhart tells us, because they know that international law stands above the law of their own country. They have breathed a new air and are alive to their own conscience. They have also tasted the joy of solidarity with people who, though they daily feel the weight of oppression, have welcomed them as friends. A new vision is emerging for both sides: "what brings them back, day after day, is the new covenant that has been struck between the peoples of this land, a pact of fraternity and friendship between Israelis and Palestinians who love life, the land, the evening breeze. They know that it is possible to live differently on this land."[17]

---

[16] Tanya Reinhart, "Standing Against the Claws of the Wall," *Yedionot Aharonot* and *Ynet*, translated from the Hebrew by Mark Marshall and Edeet Ravel (June 23, 2004): http://www.ynet.co.il/articles/0,734.L-2936546,00.html.
[17] Ibid.

A more recent article posted on the website of Israeli peace movement Gush Shalom tells a story with even wider ramifications. "From Both Sides of the Fence" narrates the common peaceful protest organized on July 7, 2009 in the village of Wadi Muhammad, near Bethlehem.[18] About 200 Israeli and international activists answered the call of the villagers and a Hebron association for the unity of Muslims, Christians and Jews ("Sons of Abraham") and met on either side of the six-meter wall that, among other evils, separated many of the villagers from their agricultural lands. The meeting place itself gave eloquent witness to the injustice created by the wall: at the bottom of a deep valley it follows the path of the main north–south settlement highway which passes over a bridge some 200 meters above.

The actual meeting of the two groups was, delightfully, much more diverse than originally planned. For one thing, several Israelis managed to come in from the West Bank side and a Protestant church in neighboring Beit Jala brought with them some 200 German peace activists. As Gush Shalom described the scene, "[t]he demonstration reached its climax when the band of drummers, which accompanied it, filled the wadi with the thunder of their drums, and the protesters from both sides put their hands on the fence in a symbolic act of protest and brotherhood."[19] Some of the posters brandished were: "Erase the shame! Dismantle the fence!" "People are stronger than Fences", "Bad fences create bad neighbors." And then this one, "Neighbors, YES! Apartheid, NO!"[20]

Yale professor of genetics and human rights activist Mazin Qumsiyeh points out that the positive side of globalization and the information revolution (with the Internet at its heart) is that "borders are dissolving; communication, intermarriage and relocation are creating a new world." While even "Liberal Zionists" support this vision, they still insist that the present state in which full citizenship is reserved for Jews is still necessary for the unforeseeable future. Yet, people "with a vision of plurality and democracy must strive to shorten the time to achieve this vision rather than fight it or waiting for it to materialize spontaneously."[21]

[18] Anonymous. Online: http://zope.gush-shalom.org/home/en/events/ 1247062614/.
[19] Ibid.
[20] President Jimmy Carter, Southern Baptist Sunday School teacher, has remained very involved in the Israeli–Palestinian peace process in the 2000s. His Carter Center monitored the 2006 legislative elections, declaring them to be as fair as any he had seen anywhere. He also wrote a book that created a storm of controversy at home, *Palestine: Peace not Apartheid* (New York: Simon & Schuster, 2007).
[21] Mazin B. Qumsiyeh, *Sharing the Land of Canaan: Human Rights and the Israeli–Palestinian Struggle* (London: Pluto, 2004), 202. Qumsiyeh comes from a Palestinian Christian family in Beit Sahour, next to Bethlehem.

Qumsiyeh here is stressing the critical dimension of vision when it comes to articulating a common plan of action for Muslims, Jews and Christians—a vision to foster human dignity in measurable fashion for the most dispossessed and disheartened.[22]

For Qumsiyeh, Said and many others, invoking the UDHR and international law as a tool for addressing thorny regional conflicts is only a starting place. Activists face major psychological hurdles on both sides: a majority of Palestinians and Israelis "still lives in the past," Qumsiyeh notes. Israelis hang on to the status quo—despite the growing resentment created among Palestinians and the international community by discriminatory laws inside Israel and the growing brutality of military occupation and colonial expansion in the West Bank. "Many Palestinians still believe that it is possible to reconstitute an Arab or even a wholly Muslim Palestine and reverse the wheels of history."[23] Indeed, people naturally "cling to their past, their emotions, and their tribalism." But the world is changing and people everywhere on the globe are beginning to embrace a new paradigm. Hence, Palestinians and Israelis, without ignoring history, are now called to emphasize some elements over others. Qumsiyeh points to some crucial choices:

> Should we emphasize the prosperity, peace, and unity that Islam and the Arabic civilization brought to the Middle East? Or should we mourn the loss of the diversity of languages and cultures that existed before that? Should we emphasize the tolerance and coexistence of Jews, Christians, and Muslims (e.g. in Spain or Al-Andalus)? Or should we emphasize the oppression of Assyrian Christians and the genocide of Romanian [*sic*] Christians by the Ottoman Turks?[24] Should we discuss the ethnic cleansing of some Canaanites by invading Hebrews? Or should we talk about the coexistence, trade, and neighborly relations in the Kingdom of Israel at the time of Solomon and David? Should we talk about the golden era of Arab sciences, mathematics, medicine, astronomy, and law? Or should we speak of the occasional problematical behavior of some Muslim rulers (e.g. in India with the suppression of Buddhism and Hinduism)? Should we celebrate the incredible ability of the monotheistic religions to make people work together for good deeds and as a team of devout people looking to better human life on earth? Or should we mourn the loss of individualism that ensues from the dogmatic practices of these religions?[25]

---

[22] Recall that this concept was reinforced by Stassen and Gushee's twin pillars of "way of seeing" and "basic convictions" among the four dimensions of ethical character.

[23] Ibid., 210.

[24] Presumably he is referring to the first great genocide of the twentieth century, that of the Armenians by the Ottoman Turks during World War I.

[25] Ibid., 212.

All religions, after all, have been used to commit war crimes and violate human rights. Thus, in the case of "Canaan"—Qumsiyeh's attempt to undermine all religious claims to that piece of land—the preferred solution in the long term is "the abolition of the privileges of the coloniz-ers and the creation of a democratic, egalitarian system anchored in a constitution guaranteeing equality, with the complete abolition of all forms of discrimination against the natives, along with the establishment of a framework capable of creating a pluralistic society."[26] While this is unattainable in the short term, he identifies a strategy of short-term and intermediate goals for achieving a joint Israeli–Palestinian pluralistic democracy. The means for achieving these goals are all nonviolent, as any military solution to the conflict has long been exhausted.

One of these instruments proved extremely powerful in bringing down the wall of apartheid in South Africa: divestment and economic boycotts. Qumsiyeh devotes a section of his website to this topic. Boycotts are simply a way to reduce the financial support of the occupation machinery in place. But considering the billions of dollars in aid the US government sends Israel every year and the near monopoly Israel enjoys in the Palestinian market for its own products, it may be more effective as simply a tool of international protest and political pressure.[27] Nonethe-less, boycotting Israeli products, including its tourist industry, could create some economic hardship with political consequences.

As for the divestment strategy, Sabeel, the Ecumenical Center for Palestinian Liberation Theology (founded by Ateek), issued a compre-hensive statement including both the theological and ethical issues sur-rounding it, and practical ways to use this peacebuilding tool.[28] Appendix A provides a short presentation of Sabeel's vision of peace built on justice and reconciliation, through the pen of long-time activist Nora Karmi. In a letter addressed to Pope Benedict XVI on the occasion of his visit to Jerusalem, she does not shy away from spelling out the suffering of Palestinian Christians, alongside their Muslim brethren. Sabeel in fact has been behind much of the movement by US mainline Protestant denominations on selective divestment from corporations that benefit from Israeli occupation. The Presbyterian Church U.S.A. was the first to issue a position statement in July 2004, the same month the International

---

[26] Ibid., 214.

[27] Several Israeli NGOs are calling for this as well. See in particular the group Matzpun (meaning "conscience" in Hebrew): http://www.matzpun.com.

[28] It is a link on the Sabeel website, dated April, 2005: http://www.sabeel.org /?page=article&id=89. "Sabeel" is an Arabic word meaning both a path and a public fountain.

Court of Justice (ICJ) in La Hague ruled 14-1 that the separation wall contravened international law.[29] Then in February 2005 the World Council of Churches urged all of its 347 member churches to consider moving in the same direction.[30] Several so far have responded to this appeal. Yet, as the New England Conference of the United Methodist Church puts it in their June 2005 Resolution, no one sees this initiative for divestiture as significantly affecting either the Israeli economy or the profit margin of the multinationals involved. Rather "[t]he goal is to make all United Methodists and other Americans aware of their relationship to companies that benefit from the Israeli occupation and give them an opportunity to withdraw from such relationships, so they are not participants in human rights violations that violate Christian principles and international law."[31] These are nonviolent actions people can engage in from a distance, whether secular or religious. I now turn to the direct involvement by people of faith seeking to make peace and establish common bonds with those of other faiths—Muslims and Christians, in particular.[32]

---

[29] It also ruled unanimously that the West Bank, Gaza Strip and East Jerusalem were occupied territories subject to the rules laid out by the 1949 Geneva Conventions. In August 2005, the Evangelical Lutheran Church in America issued a resolution condemning the separation barrier ("Peace not Wall," *Al-Jazeera Online* [August 14, 2005]: http://english.aljazeera.net/NR/exeres/A8B31654-754C-4343-A818-243857E86793.htm), but took no position on divestment. They were following the recommendation of their bishop in Israel and Jordan, Mounib Younan, "who lamented that the barrier was splitting his Jerusalem congregation into three sectors. He said the wall had made it almost impossible for the congregation to meet, for children to get to their schools and for adults to reach their places of work." He added in an emotional speech, "Our church believes in bridges, not walls; trust, not fear; dialogue for justice and peace, not more reason for division" (ibid.).

[30] Stephen Brown, "World Council of Churches Gives Nod to the Israeli Divestment Proposal," *Ecumenical News International* (February 21, 2005): http://www .episcopalchurch.org/3577 58769 ENG HTM.htm.

[31] Resolution on Divesting from Companies that Are Supporting in a Significant Way the Israeli Occupation of the Palestinian Territories, in Susanne Hoder, "Divestiture of Funds that Support Israeli Occupation of Palestinian Territories," posted on the New England Conference Website (June 20, 2005): http://www.neumc.org /news_detail.asp?TableName=oNews_PJAYMY&PKValue=60.

[32] As I review this manuscript for the last time, I find it difficult to be optimistic, however. An opinion piece by Walter Rodgers, who spent five and a half years as bureau chief for CNN in Jerusalem, sums up my main worry: the vast majority of Israelis want peace, but without sacrificing the settlements (at least the large settlement blocks): "Why Israel Will Thwart Obama on Settlements," *The Christian Science Monitor* (August 25, 2009): Online: http://www.csmonitor.com/2009/0825 /p09s01-coop.html. Rodgers also points to the moral weakness and consequent lacking of leverage of the American position on this issue. Whey prodded on the

My two research trips to the West Bank (1998 and 1999), and to Hebron in particular, allowed me the opportunity to spend time with the Christian Peacemaker Team (CPT), Christian peace activists who have maintained a presence in the heart of that city since 1995.[33] Their goal is to reduce violence by providing a buffer between the Israeli settlers (six housing units with about 400 people altogether) and the Palestinian population of nearly 150,000. Hebron was divided in two by the Hebron Agreement of 1997. "H1" designates the majority of the city, which, since then, is under Palestinian Authority control. The very center, which includes the Tomb of the Patriarchs (Abraham, his wife Sarah, Isaac and Rebecca, Jacob and Leah are buried there), is under the control of the IDF ("H2"). About 2000 Israeli soldiers are there to protect 400 settlers. Since the start of the current intifada (September 28, 2000) the 40,000 Palestinians living in that enclave have been mostly under 24-hour curfew. CPTs apartment is also there.

One of the projects CPT has initiated is a partnership with churches in the USA and Canada that are willing to sponsor Palestinian families whose homes have been destroyed by the Israeli Defense Forces (IDF). As the Israeli settlements continue to expand, more and more land is expropriated from the Palestinians, and, consequently, houses that stand in the way are bulldozed. Others are razed simply because building permits are not attainable and families were forced to build anyway. This "Campaign for Secure Dwellings," therefore, was meant to provide encouragement to families who had thus been victimized, and at the same time challenge North American Christians to stand in solidarity with Palestinian Muslims (the Hebron District has no Palestinian Christians). On two occasions I visited the family our church sponsored, Ghaleb Abu Rajab, his wife and seven children.

It was not unusual for CPT to hear from a family who had received the order to evacuate, and then to rush over and literally stand with the displaced family on their roof, defying army orders. Of course, within minutes, they would be bodily carried out by the soldiers. Yet a nonviolent act of protest and solidarity had been witnessed. Early on, CPT established links with Israeli peace activists and it was not long before

---

issue of settlements, Israelis often retorted to Rodgers, "Well, you did this to your indigenous populations too, didn't you?"

[33] I prayed with them several times during those research trips. My wife and I had made their acquaintance in 1995–96 when they regularly attended the services of St. George's Cathedral (East Jerusalem), mainly because of the excellent discussions on Canon Naim Ateek's sermons that would follow after a time of coffee and cakes, conducted in English and Arabic.

their Israeli friends witnessed this aspect of their work and decided to launch the Israeli Coalition Against Home Demolitions (ICAHD).[34] The founder, Jeff Halper, toured the USA with Salim Shawamreh (whose home had been demolished) in 1999, informing his American audiences about the destructive and unjust nature of Israeli policies in the Occupied Territories.[35] In a 2001 lecture I attended in Orange County, California, he predicted that the Oslo peace plan would soon collapse because of the strategic location (constantly reinforced) of the Israeli settlements and attendant bypass roads. Sadly, he was right.

Perhaps the most important work CPT activists do is to report what they see on a daily basis through emails. For someone who has never been to the Palestinian territories, the emails become a means of getting acquainted with a people and their plight in a personal and graphic way. Most of the members who come and go (the Israelis will not grant them residence papers, so they rotate in and out as tourists, with about six there at a time) are Mennonites, Quakers, or from other Protestant denominations. A Catholic priest who joined them, Bob Holmes, sent an email the day he left about the children of Hebron. One paragraph explains the curfew for H2 children:

> The curfew, 24 hour-a-day house arrest for the 40,000 Palestinian fami-
> lies in H2, has been imposed most days since the new intifada began at
> the end of September 2000. Its purpose, according to the Israeli military,
> is to protect the 400 Israeli settlers living in the city of Hebron. In fact, it
> is being used as collective punishment of the Palestinians of Hebron for

[34] Both CPT and ICAD work closely with Palestinian NGOs, such as the Land Defense Committee and LAW (Palestinian Society for the Protection of Human Rights and the Environment). The networking they do largely contributes to the growth of civil society on both sides of the Green Line.

[35] Salim Shawamreh's house had just been destroyed for the third time. In a quarterly Israeli peace movement electronic mailing, *The Other Israel* (started in the early 1980s by the Israeli Council for Israeli–Palestinian Peace, ICIPP), while a picketing of the Defense Ministry in Tel Aviv was going on, the participants received news "of the military involved in a wild spree of destroying 'illegal' Palestinian houses all over the West Bank. At least 19 families were left homeless in the space of a few hours. Rabbi Arik Asherman, of Rabbis for Human Rights, had been detained while trying to block the way of the bulldozers which destroyed the Shawamreh Family home at Anata Village, north-east of Jerusalem. On the following day, April 5, some thirty of us—including the recently released Asherman and his fellow Rabbi Yechiel Geitsman—were on the spot in Anata, starting work on rebuilding the Shawamreh home. It is a place and a name which had already become a symbol. Three consecu-tive times already the family's home was destroyed, only to have it rebuilt each time by the joint labor of Israelis and Palestinians" ("Briefing April 7, 2001," *The Other Israel*, available at http://otherisrael.home.igc.org/index.html).

acts of violence against settlers or soldiers. The children of Osama Bin Munqeth school on the top of a hill in H2 have learned some even harsher lessons. Their school is now a military base with a tank in the schoolyard meting out another form of collective punishment on the neighbourhoods below.[36]

Holmes had visited the Director of Education for the Hebron District in September and was told that Hebron urgently needed 34 new schools. Yet the Israeli authorities systematically refused to grant them permits. "Not one permit for a new school was granted by the military authority and schools were forced to double scheduling—morning and afternoon sessions." CPT works with whom they can to bring hope—and even offer modest solutions, if possible:

> In the village of Qilkis, one kilometre away from Hebron, 400 elementary students awaited the completion of their new school. The builders received a "stop work" order from the military authority. It was being built too close to a new bypass road to the Israeli settlement of Haggai. Four teachers conduct a "secret" school in a rented house for 100 students in Qilkis. Together with CPT, the teachers and parents had planned a tent school on the building site to draw attention to the injustice of the "stop learning" order. With the advent of the new intifada the principal, fearing bullets and tear gas from the soldiers, postponed the tent school.
>   The children are learning. The lessons are harsh.[37]

Yet these Christian activists also experience a lot of joy. Diane Roe describes a party that took place in the village of Idna—just inside the Green Line (1948 demarcation line between Israel and the West Bank, then under Jordanian control)—in May 1999. She was one of two CPT members who, with two partner families from Hebron (the Jabber and al-Attrash families), attended this joint Israeli–Palestinian barbecue sponsored by a Peace Now chapter from Tel Aviv. Roe describes the scene as that of a country carnival, with women baking bread and handing it to passers-by, with much dancing, feasting and even face-painting for the children. The dancing especially was memorable: "Two youths took the Israeli and Palestinian flags to the center of the circle so that from a distance it appeared that the flags themselves were dancing. Israeli and Palestinian men and women danced together forming a spiral."[38] Though parties of this type change little the harshness of day to day life in the

---

[36] Robert Holmes, "Lessons Learned…and Not," Email list (April 7, 2001) (Webster, NY: CPTnet.editor.guest.524947@Mennolink.org).
[37] Ibid.
[38] Diane Roe, "May Fourth Reflection," Email list (May 12, 1999) (Webster, NY: CPTnet.editor.guest.524947@Mennolink.org).

West Bank, they are powerful symbols pointing to another way of imaging the future and life together as two people:

> For this one afternoon, on one section of the Green Line there was a zone of peace. The Jabbers and the Al Atrashes still had to go back to the tents they have been living in since their houses were demolished by the Israeli military. But they went back feeling surrounded by the warmth and good will of their neighbors. And Atta [Jabber] seemed to still be dancing in his heart on Saturday when he recalled the festivities of the day before. That happiness was short-lived. On Sunday the bulldozers returned to the Bakaa, once again digging into the side of the hill where our friend Atta has lived most of his life.[39]

So Roe asks the crucial question, along with her friends from both sides, hoping, praying that the logic of power and ethnic separation will some day crumble:

> Which ideology will prevail? Will the Green Line be an open border where Israeli and Palestinian friends and neighbors can reach out to each other? Or will it become a military zone guarded by armed settler militia? Will Israelis and Palestinians be able to return to Atta's land to help him rebuild his demolished house and replant uprooted trees? Or will the hillsides of Jabber land be eaten up by the settlement expansion? Stay tuned.[40]

Peacemaking can also be a potent force for interreligious dialogue. Julie Hart of CPT expresses hope after an interfaith meeting of grassroots leaders in the Hebron area.[41] She singled out three participants who added their voices to the dialogue. First, a Palestinian (Muslim) journalist

> envisions Jews, Christians and Muslims living as neighbors in a representative democracy where human rights apply equally to all citizens, regardless of religion, ethnicity, race, or gender… Equality would be the foundation of this society in distribution of all resources: land, water, building permits, license plates, jobs, education, health care and freedom of speech and movement. He believes such a society should be the goal of all good Muslims, Christians and Jews, for "this is God's will for all people."

Second, a rabbi, originally from the US but now living with his family in Jerusalem, "hopes for a society where all persons, regardless of religion or ethnicity will be treated as equals. In the beginning, God created humans in God's image and thus people of God must treat all humans

---

[39] Ibid.

[40] Ibid.

[41] Julie Hart, "Vision of Hope," *CPTnet* (June 30, 2000). With a new system in place, one can go to their website and browse through the archives. From now on I will give the URL from there: http://www.cpt.org/archives/2005/jun2000/0021.html.

equally." The rabbi sees equal access to resources of the land in two representative democracies. Justice would flow down like mighty waters to all people. Finally, a Palestinian Christian woman...

> born during the formation of the Israeli state, aches with the daily oppression she and her fellow Palestinians experience in the occupied territories. But, she explains, "Our common God is a God of justice." She envisions a constitutional democracy for the Palestinians that protects and guarantees all people's rights, responsibilities, and duties without discrimination. "At every turn the principle of justice must be upheld. Unless justice is rendered and security is achieved, the solution must be rejected because it will not endure."

This kind of common envisioning matched with nonviolent direct action can truly impact a culture for peace.

Writing now eight years later, CPT has continued its work in Hebron and expanded at times to the Bethlehem area (they stood with people during the shelling back and forth from the mainly Christian village of Beit Jala) and now maintains a presence in the sheepherding village of al-Tuwani in the southern hills of Hebron where new illegal Israeli settlements have been sprouting like weeds (cf. Appendix B). Art Gish, a veteran Mennonite activist, wrote about his experiences in peacemaking and Muslim–Christian dialogue in Hebron, the only Palestinian city without a native Christian presence.[42]

Peacebuilding in Post-Saddam Iraq

CPT sent its members to Baghdad in the run-up to the 2003 coalition force invasion of Iraq, both as a protest against what seemed like an inevitable war and to act as human shields in protection of a water processing plant. Among the many projects they have tackled in Iraq, I present one example here that illustrates well how Muslims and Christians are cooperating for the sake of peacebuilding.[43]

The practice of nonviolent direct action is no stranger to Iraqis, though it has been overshadowed by the brutal wars of the last few decades. In November 2004 Ayatollah Ali al-Sistani called for a nonviolent march to

---

[42] Arthur G. Gish, *Hebron Journal: Stories of Nonviolent Peacemaking* (Scottdale, PA: Herald Press, 2001). See also Tricia Gates Brown, ed., *Getting in the Way: Stories from the Work of the Christian Peacemaker Teams* (Scottdale, PA: Herald Press, 2005).

[43] CPT in Iraq has held vigils, visited families of prisoners, interviewed released prisoners, started a campaign to inform US Christians about the plight of prisoners in Iraq, and advocated on behalf of Iraqi civilians with the Provisional Authority leaders (especially for the destruction of unexploded ordnance).

Najaf in order stave off a violent confrontation between the American military and the Sadr militia groups. CPT had the opportunity to participate in and witness the effectiveness of this bold initiative. Then, in the holy Shi'ite city of Karbala (where the fourth caliph of Islam, Ali, Muhammad's cousin and son-in-law, is buried), members of CPT developed a relationship with a local human rights organization, which then asked them to set up a five-day training seminar for nonviolent activism. The outcome of this session (January 22–26, 2005) was the founding of a Muslim Peacemaker Team in Karbala. I quote from the February 2 communiqué I received by email:

> Some of the topics covered in the training included stories of non-violent peacemaking, the power of non-violence, the spirituality of non-violence and planning for public action. On the last day, the trainers covered various smaller topics, including trauma and self-care, working with media and human rights documentation.
>
> In response to the stories and exploration of the power of non-violence, participants asked the questions, "How did that work?" and "Can we do that here?" The group also explored the roots of non-violence in the Muslim tradition and told the CPTers that Islam has a firm tradition of non-violence rooted in the teachings of the Qu'ran and in the teachings of the Prophet Mohammed. During each day's session, the trainees had opportunities to facilitate sessions, be the daily photographer, log keeper, time-keeper, convener and process observer. The concept of assigning roles for the day was new to the trainees and they greeted it with great enthusiasm.
>
> In the course of the training, participants shared stories of suffering and trauma they experienced under Saddam Hussein and during the wars in which Iraq has participated, including the most recent war with the United States and the subsequent occupation. The trainees said they feel compelled to use their suffering for peacemaking instead of avenging wrongs done to them.[44]

Peggy Gish reported subsequently that the trainees had officially formed a Muslim Peacemaker Team (MPT) with a steering committee that decided to establish "goals, bylaws, and plan for facilitating another non-violence training for students and staff at the Ahl Ul Beit University in Karbala."[45] In the course of their discussions, questions were raised about how MPT might spread to the rest of Iraq and beyond in the Muslim world. Another theme that dominated the conversations was how the

---

[44] "Muslim Peacemaker Team Training in Karbala," *CPTnet* (February 2, 2005): http://www.cpt.org/archives/2005/feb05/0004.html.

[45] Peggy Gish, "Overcoming the Divide," *CPTnet* (February 2, 2005): http://www.cpt.org/archives/2005/feb05/0013.html.

current tension between Shi'ites and Sunnis might be healed—a crucial topic for peacebuilding in Iraq is to move ahead. Here are some excerpts provided by Gish:

> - "In the 60's, Muslims, Christians, and Jews lived together peacefully in Karbala. It was after the 1991 uprisings against Saddam Hussein that his regime helped spread tension among the different groups. Now MPT can help remove the barriers."
> - "I see now that we are way behind in grasping the concepts of nonviolence. Nonviolence asks us to deal with the divisions in our own country."
> - "We must not make excuses. We as Iraqi, are complicit with the mass graves and killings in our past. We must begin with ourselves to build a new Iraqi humanity. The suggestion of going to Falluja helped me to understand more deeply what nonviolence calls us to. We must move on to overcome the divide."[46]

Talk about a "new Iraqi humanity" might sound idealistic. However, it was followed up by daring action. A June email by Joe Carr reports that MPT (now with twenty members) was able to raise $20,000 on the Internet in April 2005 for the purchase of needed medical supplies, which were then donated to the main hospital in Fallujah, the Sunni town that was devastated in November 2004 by occupation forces.[47] CPT accompanied the delegation to Fallujah. Then, in May 2005, MPT worked side by side with their Sunni compatriots to clear the rubble and help rebuild Fallujah. Carr comments that "[t]he Fallujans received their help gratefully and participants considered the action a transformative experience." One MPT member was overheard saying, "We proved, in a simple way, that peaceful living can exist," while another commented on the role his faith was playing in his service: "We started MPT because we believe that the real spirit of Islam is mercy and forgiveness." Moreover, they were convinced that nonviolence was more powerful and effective than military might:

> - "Occupations have all the weapons except the peace; we can use this weapon. In war, only the young men can resist. With nonviolence everyone can participate."
> - "Violence happens when democracy disappears," one said, "and the U.S. is using this violence to justify staying. We need to rebuild ourselves by ourselves. We need assistance, but it should be like the assistance CPT gave to MPT. They didn't control, they only gave inspiration and an example."[48]

---

[46] Ibid.
[47] Joe Carr, "The Weapons of the Muslim Peacemaker Teams," *CPTnet* (June 21, 2005): http://www.cpt.org/archives/2005/jun05/0027.html.
[48] Ibid.

This little window into "West Asia" has enabled us to witness women and men from many backgrounds "asserting their common humanity," in the words of Nelson Mandela. Edward Said and Mazin Qumsiyeh's secularized vision of a "pluralistic and democratic society" for Israel–Palestine is not all that different from the aspirations of the Israeli and international activists standing with Palestinian villagers against the separation wall. Paralleling the experiences of CPT activists both in Palestine and Iraq (Muslims, Jews and Christians), all long to affirm their mutual human dignity and reap the joy of celebrating life together—which they often experience in the midst of harsh and heartrending circumstances.

What all of these vignettes share is the ethical power of a common vision for peace. True, plurality of cultures, perspectives and experiences among people may just as well lead to a celebration of rich diversity in unity as it might lead to conflict and war. Yet part of the post-Cold War reality of a world condemned to find the means for its collective survival on a dangerously polluted planet is that through the Internet and other media people are talking to each other from across the globe like never before. This creation design by God is well expressed in both Bible and Qur'an. According to South Asian scholar Yusuf Ali's translation, "O mankind! We created you from a single (pair) of a male and a female, and made you into nations and tribes, that ye may know each other (not that ye may despise each other)."[49] My hope here is that common theological reflection on the implications of humanity's trusteeship would increase communication, understanding and concrete practices of peacemaking and social justice among Christians and Muslims.

Reasonable Findings

Throughout this study I have aimed to describe both the philosophical underpinnings of last century's turn away from the Cartesian modern paradigm (postmodernism) and the current global reality that I have called postmodernity—the sequence of US-led reconfiguration of the Bretton Woods international finance arrangements in 1973, the neoliberal turn of globalization initiated in the 1980s and its radicalization after the fall of the Iron Curtain. To be sure, I relinquish any claim that my description and analysis of these emerging realities is totally accurate or in any sense exhaustive. With regard to all human inquiries, I have argued for a metaphysics and epistemology of critical realism—a real world exists outside of the human mind and human beings can gain true

---

[49] Q. 49:13.

though only partial and provisional knowledge of that world. Equally, no one can predict with certainty how the current world order might develop in the coming decades. If Mitzman and Harvey are even partially right, then we do face a looming crisis— we are already weathering a monetary and financial crash that almost matches the 1930s depression, with dire consequences for the political stability of the present world configuration.[50] If anything—barring the possibility of a nuclear holocaust—our present economic recession could open up new possibilities of reinvisioning local economies and politics and their relationship to a reconfigured international order. Now I propose to summarize the salient points of my argument in this highly inter-disciplinary work (cf. Fig. 17).

Following a brief overview of postmodernity (Chapter 1), I sought to ask the basic philosophical questions that (a) brought on the modern paradigm of the lone thinker establishing reality through the rational processes of the mind and (b) those that led thinkers away from that paradigm. Philosophy is a good starting place. Recall E. J. Ashworth's definition, "Philosophy aims at intellectually responsible accounts of the most basic and general aspects of reality."[51] This is what I attempted to accomplish by examining the various possibilities of defining the triangle of self, language and world within an intellectual landscape colored by postmodernism. Using Seyla Benhabib as a guide in evaluating "strong" postmodern claims, I examined successively the theses of the death of the subject, the death of history and the death of metaphysics.

---

[50] Confirmation for this comes from many quarters. Among others, economist Lyndon H. LaRouche Jr addressed the delegates who had come to his June 28–29, 2005 seminar in Berlin with these words, "suddenly, very soon, the entirety of the present world monetary-financial system will collapse. It will come like a Summer [*sic*] thunderstorm, far more devastating than anything we have experienced during the recent two centuries" ("It Happened in Berlin Last Week," *Executive Intelligence Review* 32.28 [July 15, 2005]: 4). A Russian delegate, both an economist and a member of the Russian Duma, Sergei Glaznev, called for "a new architecture of the world financial system," as "the present dollar-based speculative financial system... is going to collapse anyway"("We Need a New World Financial Architecture," in ibid., 41). It is now shaped like a pyramid, "based on injustice, on fraud, on unequal and imbalanced exchange in the world." The American Federal Reserve controls only 40% of the global flow of dollars. Worse yet, the quantity of dollars in circulation today (including Treasury bonds) "is 25 times higher than the amount of the American gold and currency reserves. It means that there is nothing under the dollar, except the demand which is generated by growing speculative activity" (ibid., 42). Whatever new system emerges (some kind of new Bretton Woods agreement with fixed rates of exchange and an international reserve to back it up), China, India and the oil-rich Arab nations, he says, will have to be included as equal players.
[51] Cf. Chapter 2, n. 11.

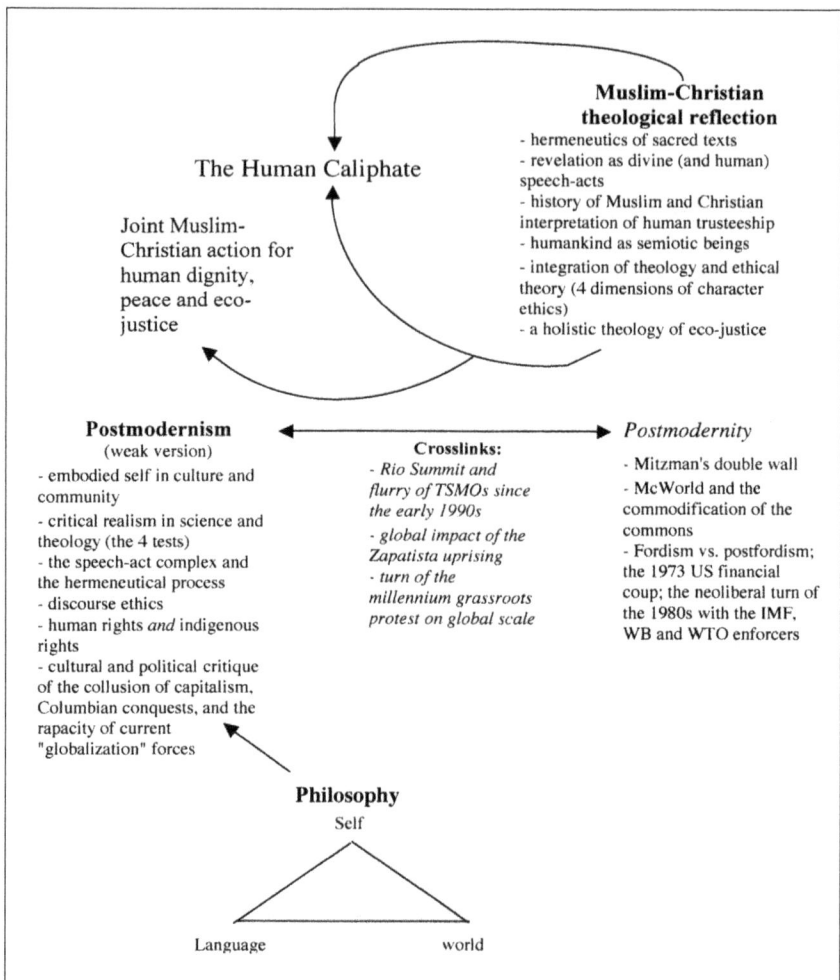

Figure 17. *Flowchart of this Project's Main Ideas*

With Nancey Murphy and Ian Barbour's help, I concluded from the groundbreaking work of Michael Polanyi in 1958 (followed by Thomas Kuhn, Imre Lakatos and others) that there was good reason to jettison the modern or Cartesian paradigm of the disembodied thinker who apprehends the world, either as a scientist or a theologian. At the same time, the simple fact that rationality in the modern mode was called into question did not spell the end of meaningful inquiries into the nature of the world, whether on the physical or metaphysical levels. Following David Naugle's thesis of the semiotic human person, we could make a "reasonable" wager (a leap of faith nonetheless) that what persons could partially and tentatively learn about their physical environment might also point them to the meaningfulness of ethical and theological discourse about a Creator who has revealed something about himself in his creation and the written texts we attribute in some way to him. This is about semiotics, and it nicely fits the qur'anic use of the word *ayāt* that refers both to revealed verse and God's signs in nature.[52]

Human knowledge, then, is inevitably related to meaning, which in turn represents the interpretation of "facts" persons advance in order to further their research. Barbour shows how the structure of scientific inquiry closely parallels the process of religious inquiry and experience. In both cases, knowledge is expanded through symbols, analogies and models. Most significantly, true knowledge is mediated through a four-step process: (1) agreement with data; (2) coherence; (3) scope; (4) fertility. This critical realist insight affects not only theology and science—as one would expect—but deeply impacts the hermeneutics of sacred texts. As Murphy demonstrated, the atomism and reductionism of the modern paradigm left only two options in a philosophy of language: language is either expressivist (conveys only emotions—the classical liberal Protestant version) or referential (the fundamentalist and early evangelical versions starting in the 1920s, or "plenary and verbal inspiration of scripture"). Both hold some truth; yet, in fact, with insights gained from philosophers J. L. Austin and John R. Searle, we know that successful communication through language is more holistic than that. A receiver (reader or hearer) understands the intended meaning (uptake) when three conditions are met: (1) the content is conveyed; (2) the illocutionary force of the statement comes through, (3) as well as its perlocutionary effect. Further, in the progression of how hermeneutics as a discipline progressed, the initial focus on the author soon became eclipsed, first by the text itself (semiotics, structuralism, poststructuralism) and then by the reader (reader-response theories). While all three play important roles in meaning production, Kevin Vanhoozer warns us not to lose sight of the author. This is especially crucial if the above-mentioned wager is to have any meaning. If human beings can communicate by means of various signs, the religious person reasonably posits the existence of a Creator who communicates with his human creatures through a variety of means, including a written text. Hence, we take several steps away from the reductionism and atomism of modern hermeneutics.

---

[52] The Bible often refers to the same idea (e.g. Ps. 19:1, "The heavens declare the glory of God; the skies proclaim the work of his hands. Day after day they pour forth speech; night after night they display knowledge").

I also pointed to the political implications of the movements of dissent from the modern epistemological paradigm of the lone, disembodied, and rational self. Philosopher of religion John Clayton uses a case study of Thomas Jefferson to show that the Enlightenment project sought above all to create a public space in which the voice of reason would command a general consensus of views. Yet, in order to achieve that, Jefferson had to displace completely historic Christianity (Calvin for him was an atheist and dangerous sectarian[53]) and impose the contours of "rational" faith that for him took the shape of Unitarianism. Kant's categorical imperative in ethics should be seen in the same light: an attempt to bypass the perceived contentiousness of religious faith in order to craft a secular ethical discourse for modern states. The irony of this enterprise is that Jefferson's "rational theology" turns out not to be as universal or tradition-free as he proclaimed it to be. In fact, it "ends in a paradox created by its own foundationalist pretensions to speak with a universal and neutral voice, when its tone is more nearly parochial and partisan."[54] For Clayton this is rather like John Hick today, who seeks to impose a similar kind of Jeffersonian solution to the cacophony of religious pluralism.

Habermas has attempted to inject some postmodern methodology into this vision of a Kantian public space, which would enable democracy and pluralism to flourish side by side. Benhabib, as we have seen, retrieves a good deal from Kant's work on discourse ethics;[55] only she adds the idea of "enlarged mentality" taken from Hanna Arendt and argues for a socially embedded self, strongly influenced (not determined) by culture and history, yet able to function as both author and character in the unfolding narrative of his or her life. Benhabib's plea for a "weak version" of postmodernism was designed to open the way for the liberationist vision of feminism and other causes committed to social justice. In fact, a good number of transnational movements in the 1990s rose up in the interstices of the neoliberal globalizing project (McWorld), which was recently invigorated by the fall of its socialist enemy. The "New World Order" proclaimed by George H. W. Bush at the conclusion of the First Gulf War of 1991 elicited many vocal opponents among the plethora of NGOs that converged on Rio de Janeiro the next year. Then it was the turn of Mexican peasants of Mayan descent (the Zaspatistas) to capture

---

[53] John Clayton, "Thomas Jefferson and the Study of Religion," in *The Future of Religion: Postmodern Perspectives*, ed. Christopher Lamb and Dan Cohn-Sherbok (London: Middlesex University Press, 1999), 94.

[54] Ibid., 99.

[55] Recall how she integrates her two substantive norms (universal respect and egalitarian reciprocity) to the concept of communicative action in six steps (cf. Chapter 2).

world attention as they protested—starting with violence but gradually giving way to nonviolent tactics—the painful fallout of the rapacious capitalist stranglehold on their country. The Utopian ideals of postmodernism were now adding fuel to the anti-globalization protests. These represent the "crosslinks" I portray in Figure 17.

Now back to hermeneutics. I began (in my Introduction) by stating that theology was simply the articulation of a people's understanding of their faith's sacred texts in response to the questions raised by their socio-economic and political context. In Chapters 6 through 9 I skimmed over thirteen centuries of Islamic interaction with the second-sura qur'anic account of God's announcement to the angels that he was placing Adam on earth as his *khalīfa*—his trustee, or vice-regent. We saw that from the beginning commentators stumbled on the second half of Q. 2:30, in which the angels protest God's establishing a human deputy on earth ("one who will make mischief and shed blood"). This led the earliest *mufassirūn* to posit Adam as the referent to the word *khalīfa*, and thus referring to the whole human race (how could a prophet wreak havoc on the earth?). The irony, however, is that because of the primary meaning of the root ("to follow after") and because of the political implications of the concept (the Umayyad caliphs advertising themselves as *khalīfat Allāh*), little was written in the classical period about humanity as ruling on earth in God's stead. Two factors, nevertheless, seem to have contributed to the fact that this meaning never entirely disappeared. First, the root *kh-l-f* contains within its semantic field the meanings of cultivating and ruling. Thus for instance Ibn Ishaq quotes Mujahid b. Jabr's commentary as favoring the meaning "cultivating" and Sufyan al-Thawri sees several qur'anic instances of this root with the connotation of "ruling."

Later theological disputes over free will and determinism (also with definite political overtones) tended to overwhelm initiatives to interpret a human caliphate as an active fulfilling of a divine mandate on earth. Additionally, the classical worldview, steeped as it was in Christian and Jewish lore (*al-isrā'īliyyāt*), was far more preoccupied with the prophetic role of Adam and whether this means prophets are immune from sin or to what extent they might be, or whether they rank above angels or not. They naturally reverted to the "succession" meaning as well, particularly since some early commentators had argued that jinn and/or angels had inhabited the earth before humans. In spite of this kind of speculation, we found that with the passage of time and with the blossoming of the arts and sciences and the social evolution of urban societies in the Muslim world even traditionalist authors like Ibn Kathir advanced rather bold statements: God's *khalīfa* refers to the human being who is uniquely gifted to inhabit and cultivate the earth, while producing civilization and prosperity for all. For others, including Zamakhshari and Razi, the methodology and concepts of Greek philosophy suddenly added greater significance to "the names" God teaches Adam. For them, these are the universals that sustain rational thinking and bring meaning to the particulars gathered through human observation. This is the process that sustains human knowledge and furthers scientific endeavors. No wonder, they reasoned, that the angels were commanded to bow to Adam. To this Razi added the weight of moral responsibility God puts on Adam's shoulders—a theme, as we saw, that becomes central in the modern period.

But while the traditionalists and rationalists who imagined an empowered role for humanity located it in science, civilization and ethical striving, they made no connection between that God-given capacity and God's own nature. The mystical commentators, by contrast, started with the *ḥadīth* that considers human nature analogous (in some way) to God's nature. Thus al-Ghazali developed the idea of an inner connection between Creator and human creature. So, the human caliphate for him, for Nisaburi, for Ibn al-Arabi and Jami, was a mission of mediation between Creator and creation. The human being as God's vicegerent is the *barzakh* that connects this world and the next and brings harmony to the universe, with a calling of both lordship and servanthood.

And yet, what all of these thinkers have in common is a worldview in which substantial mastery over the created order is more wishful thinking than reality—in spite of admirable advances in medicine, astronomy and other branches of science. For people all over the globe in medieval times, nature remained a fearful entity. Besides the brutal effect of natural disasters, wars and epidemics kept populations low with no expectation of significant growth in sight. It was not until the Industrial Revolution in Europe, with its attendant phenomena of massive industrialization, urbanization, inoculations and the spread of international capitalism, that a very different worldview began to develop. For the first time, a theology of human mastery over the created order came into focus—so much so that in the 1960s a growing realization of environmental degradation led secular scholars to blame the Judeo-Christian Genesis narrative for the debacle.

I have contended that the momentous pressures for social, economic and political change imposed on Muslims states in the colonial period account to a large extent in producing a new interpretation of the qur'anic notions of human trusteeship. From Shah Wali Allah in India to Muhammad Abduh in Egypt, a strong current of reformism (rightly dubbed "modernism," at least the nineteenth century) took hold of many Muslim intellectuals. I have further traced it in Ali Abd al-Raziq's drawing out the logical conclusion of Ibn Khaldun's distinction between *khilāfa* and *mulk*; in Bint al-Shati's humanistic distinction between *bashar* and *insān* (with the latter raising the torch of godly trusteeship through the fiery trials of moral decision-making); in Mawdudi's contextualization of a qur'anic human deputyship in order to fuel an Islamic reformist political platform in a modern state; in Shaykh Ahmad Yasin and Shaykh Nizar Ramadan's application of *khilāfa* to the Palestinian nationalist struggle for self-determination; and finally in the works of Muslim scholars seeking to do theology in a postmodern context, such as Ziauddin Sardar, Mohammed Arkoun, Ebrahim Moosa, S. P. Manzoor, Khaled Abou El Fadl and Addolkarim Soroush. It can even be argued that Rashid Rida's rejection of Ibn Khaldun's *khilāfa–mulk* distinction and advocacy for a renewed Caliphate is only the first instance of a twentieth-century islamist impulse to re-establish the Medinan *dīn wa-dawla* harmony in a nation-state context—also a work of theological contextualization.

There is no doubt that the holistic vision of the Medinan revelation derives its energy and scope from the oneness of God and from its parallels to the revelation God made to Moses on Mount Sinai. The banning of all idols and the worship of the One True God played out in the minute details of people's daily lives, from individual matters, to family relationships, to social, economic and political concerns. *Tawḥīd* is also at the heart of the Genesis 1 account of creation: "So God made man in his own image... And God blessed them... 'Be fruitful and multiply, and fill the earth and subdue it.'" Jeremy Cohen's summary is worth quoting again: "Of all God's creatures, humans alone bear the image of God. They alone engage in direct conversation with their Maker. And they alone receive authority over the natural world: a responsibility to populate it, to control it, and thereby to civilize it."[56] Perhaps the best example of such a holistic and ecumenical work of theologizing is that of fourth/ tenth-century Jewish scholar Sa'adya Gaon, whose academy of Sura was on the outskirts of Baghdad. Completely sidestepping past rabbinic controversies over the "be fruitful and multiply" command, the Gaon ("chief scholar") elaborated on the human dominion of the earth and its wonderful civilizing effects in his Arabic *Tafsīr* of Genesis.

Allow me to draw a parallel based on the phenomenology of religion between the medieval rabbis who closely tied the command to multiply with the covenant from Abraham to Moses for the Jewish people (even though the text plainly refers to all humanity) and some contemporary islamist writers who narrow the *khilāfat al-insān* to the Muslim commu-

[56] Cohen, *"Be Fertile and Increase,"* 1.

nity, starting with Sayyid Qutb. Recall that the Syrian traditionalist Abd al-Majid al-Najjar also connects the human caliphate to one's faithful observance of *shariʿa*. Might not both cases be more revealing of a fear of losing one's community identity and power in the face of outside pressures than of clear textual indications? This is why I have consistently pleaded for a move away from the literalism of a modern atomistic hermeneutic and toward an embracing of a critical realist epistemology and hermeneutics, which goes beyond Peters's "critical deconstruction" mandated by a strong version of postmodernism. I attempted to suggest a holistic theology of monotheism in Figure 14, in which the subject/self gains real yet imperfect knowledge of the world. In that scheme, much as we have gleaned from Naugle (humanity as semiotic persons), human knowledge is mediated by "God's creative and determinative presence," with the consequence that even our language, which is plainly inadequate for expressing the numinous, can nevertheless convey a relationship of love between God and people and a reverent attitude of submission for persons before their Creator. This is the wager of revelation, and indeed, of all human knowledge, mediated as it is through language.

Then finally, I turned to ethics. In the last chapter, from Larry Rasmussen's discussion of environmental justice, Ali Mazrui's "global apartheid," Kenneth Cragg's "pronounal Jewry" and Medinan communitarianism, to Stassen and Gushee's four dimensions of character ethics, I argued for an integration of ethics and theology at the point of creation. This was also the direction of Chandra Muzaffars "spiritual vision of the human being" and Mohammed Arkoun's "Islamic humanism," along with the theologies of Bint al-Shati, Sayyed Hossein Nasr and many others. But, as I noted too, proof-texting a universalist, peaceful and harmonious vision of humanity in the twenty-first century will not do. A comprehensive theology that makes hermeneutical decisions about texts that go in more violent and communitarian directions is also needed.

Understandably, hermeneutics took a lion's share of this project. The philosophical explications of how a subject relates to the world and how language mediates both knowledge and communication tightly knits epistemology, metaphysics and hermeneutics to one another. Then the fact that as Muslims and Christians we are attempting to understand this afresh in an intellectual climate that has questioned the philosophical (and political) legacy of the Enlightenment leads us to a fork in the road. We are faced with the double challenge of postmodernity and postmodernism; yet this challenge is also the chance to open the windows and let in some fresh air to revive and inspire new theological thinking. At least three centuries starting with the third/ninth in both Andalusia and Abbasid Baghdad were privileged moments of Muslim–Christian and Jewish intellectual and cultural cooperation. The twenty-first century could offer much more, particularly with the democratic stirrings of grassroots movements everywhere and current efforts to reform the United Nations. As I have reiterated, however, this will entail a renewed focus on hermeneutics.

In summary form, then, a better knowledge of the process of inter-pretation will help to:

1. Trace the career of texts over the centuries—either as Heath's "metahermeneutics," or Cohen's historical research into the use of the *dominium terrae* passage in both Jewish and Christian ancient and medieval writings.[57]
2. Discern the relationship between politico-military power and textual exegesis (e.g. the interplay between the political institu-tion of the Caliphate and the interpretation of Q. 2:30; US Protestant theology and frontier "Manifest Destiny"; Bible and Qur'an, colonialism, neo-colonialism, and the issue of apartheid as seen by Esack and his Christian counterparts).
3. Delineate a gradual shifting of epistemological thought and method from premodern times, to modern, and now to postmod-ern (the modern view of knowledge as tradition and context-independent must be discarded, though without adopting a "strong version" of postmodernism, which may undercut any attempt to embark on an ethical conversation on a global scale).
4. Understand the urgent need for an interreligious theology of global ethics in a world increasingly reeling under the brutal impact of neoliberal globalization and the growing destruction of our environment.
5. Appreciate the contribution of a postmodern, critical realist, non-foundationalist theology, while still holding on to Revelation as normative. Stability of meaning on essentials does not preclude new and sometimes jarring insights from critical textual and historical research, and fresh theological insights and paradigms as new questions are asked of the text in changed settings.
6. Undergird a holistic doctrine of creation through the semiotic nature of the human person—always seeking to communicate, analyze signs in various forms and media, and understand. Hence human culture in its great diversity and awesome creativ-ity is part of the matrix in which humans were created.

---

[57] Respectively, Heath, "Creative Hermeneutics," and Cohen, *"Be Fertile and Increase."*

7. Unleash the power of the text from semiotics (an arbitrary sign-system), to semantics (sentences as bearers of meaning), and finally to speech-acts, in which meaning includes the embodiment of the author's intention. We are no longer stuck with the modern either/or dead-end of referential versus expressivist hermeneutics, but now can embrace a holistic and dynamic view of language and meaning.[58]

8. Establish the metaphysical dimension of communication in general. Hermeneutics for people of faith point to God as Creator, who communicated verbally with Adam (and Eve), holds him accountable for his human words, and gives him the power to name things. Hence the general stability of meaning is assured, and a sacred text, in particular, becomes a communicative act par excellence.[59] Further, just as God communicates with people through revelation, he also guides and speaks to them through the inner stirrings of the soul, thus giving weight to the mystical experience common to all religions.

A big part of the challenge we face today as Muslims and Christians, therefore, is to rethink how God would have us live out our earthly caliphate in light of the texts we call sacred. Starting with creation texts and then putting them in the context of the whole canon, we must urgently design a common theology of humanity that would provide an inspiring and useful vision for how we might contribute to bringing more harmony and justice to our global human community. Having surveyed the ominous double wall described by Mitzman and others, I reached the conclusion that our theology must end up addressing both social justice (a more just access to the world's resources for all people everywhere) and ecological wisdom—in brief, a vision of eco-justice. We heard

---

[58] As a reminder, a text has three components which interact with each other and back and forth between author and reader: matter (prepositional content); illocutionary force (world to word, or word to world; nuances of irony, questioning, and various shades of emotion); and perlocutionary effect (cf. Fig. 9). In the end, meaning is a combination of what the author seeks to do through his or her words and what the reader/receiver understands.

[59] Recall Ricceur's concept of the text opening up a new world to the reader. Christians tend to emphasize the personal dimension—God initiating a relationship of love (a Father to his children) through the pages of the Bible—whereas Muslims have tended to emphasize the commands, or the Straight Path to follow. As I noted above, God's announcement to the angels that he will place his *khalīfa* on earth carries the force of a promise, a world-to-word fit. Thus he implicates himself in the success of his creature's mission, a point often argued by Cragg over the years.

Martin Luther King Jr address us in his last speech, using the imagery of the prophet Moses: "I've been to the mountaintop…I've looked over… and I've seen the promised land." For him, utopia was an American society in which black and white would equally share its bounty and together contribute to its wealth and creativity.

This is the utopian vision of hope that animates and inspires people, particularly those, whether peasants, Indians, intellectuals or Western youths, who at the turn of the millennium filled the streets to protest the policies of the IMF, WTO and WB in Seattle, Washington or Genoa. Others, from the Rio Earth Summit to the yearly World Social Forum gatherings (starting in 2000), insisted that an alternative exists to the current steamroller of neoliberal design, which we glibly call "globalization." "A better world is possible," they continue to cry out, where abject poverty is banished and where all nations, ethnic groups and civil society agencies contribute to a more prosperous, peaceful and healthy planet. Not surprisingly, a record number of people in cities the world over turned out on January 15, 2003 to voice their angry opposition to the looming American and British invasion of Iraq.

These aspirations for a better world that also found expressions in a flurry of NGOs—whether Jackie Smith's TSMOs or the grassroots postmodernism of Esteva and Prakash's analysis—represent the crossover between postmodernism and postmodernity as I have defined them. The other bridge between this philosophical movement and the harsh reality of a world globalized according to a free-market fundamentalist blueprint and under the thumb of a single dominant superpower is the intellectual critique of the modern paradigm of the lone and tradition-free self. We saw this expressed in a variety of movements and disciplines, and it corresponds nicely to emotional energy and vision expressed by such people as French peasant union leader José Bové. As of yet, the proceedings of the yearly conventions of the World Social Forum have yielded little in terms of concrete directions to follow. But the conversation goes on, and, if nothing else, the general disenchantment with the world as it is leaves room for other voices to join in and perhaps give some clearer direction—particularly religious voices. My purpose here is to point out the potential impact a united Muslim–Christian (add Jewish, Buddhist, Hindu, etc.) front could have in civil society and political arenas.

Widening the Circle of Dialogue

As we saw, the Palestinian question involves the related issues of sustainable development (Israelis control water resources and continue to destroy traditional Palestinian farmlands), colonialism, social justice and peacebuilding.[60] Already groups of Israeli Jews, Palestinian Muslims and Christians, as well as Western Christians are actively involved in a dialogue of common acts of resistance to the destructive forces of military, economic and political occupation. Significantly, the first "trialogue" of the three Abrahamic faiths in modern times met in Glion, Switzerland, in 1993, on the theme of "Jerusalem." Though it struggled to produce a common statement,[61] it did point to the fact that any lasting conversation between Jews, Christians and Muslims today must begin, or at least address the issue of Jerusalem and the Israeli–Palestinian conflict.

Dialogue must also be increased and widened, if only because of the deep anxieties that life at such a rapid rate of change creates today against a "backdrop of high risk" in a world where all the old certainties have been eroded.[62] Anxiety over humanity's ability to decimate its own species through weapons of mass destruction, or even disappear in a catastrophic ecological meltdown, is one thing. But people of faith must worry first and foremost about the ethics of their theology and practice, as Ali Mazrui and Chandra Muzaffar have rightly reminded us. We must move away from the individualism fostered by the modern Cartesian epistemological paradigm and reinforced by the Hobbesian ethic of self-interest. We must renounce the demons of empire and Mammon that drove a civilization of white Europeans to colonize the globe and continue to wreak havoc particularly in the form of eco-injustice. Finally, we must listen again to James Baldwin calling us to come back to our first works; to Malcolm X, who through his spiritual experience in Mecca transitioned from civil rights to human rights; and to Martin Luther King, who challenged his nation to change its laws to reflect the dignity of each and every person. He too proclaimed the notion of human rights by boldly speaking out against the Vietnam War.

Indeed, we human beings are embodied selves, people who live, think and speak as persons embedded in culture and tied to specific historical markings. Yet the postmodern context—what many call the "information

---

[60] See also Jeremy Milgrom, "'Let Your Love for Me Vanquish Your Hatred for Him': Nonviolence and Modern Judaism," in *Subverting Hatred: The Challenge a/Nonviolence in Religious Traditions*, ed. Daniel L. Smith-Christopher, Faith Meets Faith, an Orbis Series in Interreligious Dialogue (Boston, MA: Boston Research Center for the 21st Century; Maryknoll, NY: Orbis, 1998). Among other biblical principles, he sees that of restoration of stolen property as a basis for the compensation of Palestinians who lost their homes and land in 1948 (ibid., 134).

[61] Jutta Sperber, *Christians and Muslims: The Dialogue Activities of the World Council of Churches and their Theological Foundation,* Theologische Bibliothek Topelmann, 107 (Berlin: W. de Gruyter, 2000), 80.

[62] Giddens, *The Consequences of Modernity.*

age," with all of its built-in inequalities and injurious affronts to human communities, has prompted people of many backgrounds (including many non-religious people) to fill the streets and flood cyberspace with their indignant protests at the current international order. A movement of solidarity is afoot. Will this civil society movement grow stronger and impact governments? Human nature, as we have seen, is fickle and vulnerable to greater forces of evil on the prowl. This is why I have been calling for Christians and Muslims to look back again at their creation stories. What kind of theology of creation and humanity could sustain a movement that will effectively challenge the forces of division, racism, militarism and materialism, and turn the world's attention to the God who is coming to fulfill his promised kingdom of justice, peace and harmony?

Doing theology in the twenty-first century, as Joerg Rieger fittingly reminds us, is to listen to all the voices of the excluded. We cannot go on as if the death by starvation of over 30,000 children in our world each day is of no concern. The poorest of the poor have glued themselves to the urban giants of both "developed" and "developing" worlds. But who is listening to their pain?

> While the powers to be tend to assume that the suffering and pain of people on the margins is something that will eventually be cured by the system, we need to realize that the suffering and pain of those who are repressed first points to the truth about ourselves... [W]e cannot be fully human without others, without taking a deep look at what has been repressed and pushed into the recesses of our social unconscious all this time. The extreme forms of pain at the margins affect us all. The interest of the repressed is common interest.[63]

I leave the last word to Native American theologian George E. Tinker, whose vision of "reimaging creation" serves as a needed hermeneutical tool borrowed from the indigenous peoples of the earth.[64] As I intimated in Chapter 2, Fourth World peoples like those of Chiapas hold a key that can smooth the transition between modern and postmodern, human rights for individuals and human rights for communities, while at the same time subverting to some extent the current nation-state division of the world.

For Tinker, liberation theology has been a useful model to point out and attempt to redress the oppression of the poor and marginalized. Yet their analysis is drawn "from the modes of discourse of the Western academy." Tinker notes that any idea of development implies a particular view of humanity. Latin American liberation theology "identifies the preferential option for the poor with socialist and even Marxist solutions

---

[63] Rieger, *God and the Excluded*, 189.
[64] Tinker is a Native American from the Osage-Cherokee tribe.

that analyze the poor in terms of class structure. This overlooks the crucial point that indigenous peoples experience their very personhood in terms of their relationship to the land."[65] The "non-person" of their theology is thus reduced to a social class. In addition to their being up-rooted from their land, this developmental strategy puts the means of production into the hands of the poor, which in turn leads them to become exploiters of the native peoples and their natural resources.[66] Hence, the historic double oppression of the indigenous peoples—by the First (and Second—until 1989 at least) and Third Worlds:

> We share with our Third World relatives the hunger, poverty and repres-
> sion that have been the continuing common experience of those overpow-
> ered by the expansionism of European adventurers and their missionaries
> five hundred years ago. What distinguishes us from them are deeper,
> more hidden, but no less deadly effects of colonialism, which impact our
> distinct cultures in dramatically different ways. These effects are espe-
> cially felt in the indigenous spiritual experience, and our struggle for
> liberation is within the context of distinctive spirituality.[67]

[65] George E. Tinker, "The Full Circle of Liberation," in Hallman, ed., *Ecotheology: Voices from South and North*, 219.

[66] He uses strong language: "Many liberation theology and socialist movements promise indigenous peoples nothing better than continued cultural genocide. From an American Indian perspective, the problem with modern liberation theology, as with Marxist political movements, is that class analysis gets in the way of recogniz-ing cultural discreteness and even personhood. Small but culturally unique commu-nities stand to be swallowed up by the vision of a classless society, an international workers' movement or a burgeoning majority of Third World urban poor. This too is cultural genocide and signifies that indigenous people are yet non-persons, even in the light of the gospel of liberation" (ibid., 220).

[67] Ibid., 218. Tinker's book, *Missionary Conquest: The Gospel and Native American Cultural Genocide* (Minneapolis, MN: Fortress Press, 1993), is a moving, well-researched documentary on the American missionary movement among their own native peoples. In it, the American church (both Catholic and Protestant) is indicted for an intentional cultural genocide, along with an unintentional literal genocide of American Indians. The Euro–American colonial enterprise continues today, albeit in different forms. He hints at one solution at the very end: theology was a necessary ingredient for the European conquest of the Indian peoples. Presumably, the way to healing would also be through theology. Here the hermeneutical insight is capital: Christians (and I would add "Muslims") must be allowed the freedom to articulate their faith within their own cultural context, through their own worldview lens. The second (and last) explicit piece of advice for well-meaning white Christians is the following: "American Indian peoples need their white friends today more than ever. What we need, however, are genuine friends, not self-proclaimed friends who know what is best for us. We do not need so-called friends who would invite themselves in to pillage the remaining treasures of Indian spirituality, or well-meaning liberals who

The challenge for Christian theology, Tinker argues, is to reclaim a doctrine of creation that was lost in the modern era. This theology will also integrate the image of the world shared by all native peoples:

> Respect for creation and the recognition of the sacredness of all in creation is a deeply rooted spiritual base for American Indians, rooted in the soil of the tribal cultures of North America... It is a matter of related-ness and interdependence that finally results in a necessary relationship of interdependence with all nature.[68]

As a Lutheran, Tinker deplores the "christomonism" of his and other Protestant churches, whose first article of faith is almost always something like, "God's reconciling act in Jesus Christ." Obviously this is central to any affirmation of the Christian faith. The question is whether this is an appropriate starting place. "To make fall/redemption the beginning point in theological proclamation generates traumatic experiences of spiritual and emotional dislocation for American Indians which some people survive and many do not." On the other hand, when the proclamation of the gospel is made on the common ground of creation as sacred and good, "it can generate genuine healing and life-giving response." Moreover, Tinker continues, affirming the necessary balance and harmony of all living creatures has implications for justice and peace among all peoples as well:

> On the one hand a proper prioritizing of First Article/Creation concerns will enable the churches to appreciate and value the inherent spiritual gifts that many cultures, especially indigenous, tribal, fourth-world cultures, bring with them to Christianity... Secondly, ... We will discover that respect for creation can become the spiritual and theological basis for justice and peace just as it is the spiritual and theological basis for God's reconciling act in Christ Jesus and the ongoing sanctification in the Holy Spirit.[69]

---

would try to show us how to make the system work for Indians. Rather, we need friends who will join in the struggle against the continuing imperialism of Western, European-American culture. Genuine friends do not invade one another, physically or spiritually. Genuine friends do not prescribe for one another. But genuine friends do stand beside one another, supporting one another in times of need and crisis" (ibid., 120-21). This is also a good piece of advice for Western Christians relating to Muslims.

[68] Tinker, "The Integrity of Creation: Restoring Trinitarian Balance," in *Constructive Christian Theology in the Worldwide Church*, ed. William R. Barr (Grand Rapids. MI: Eerdmans, 1997), 209.

[69] Ibid., 207. CPT has also maintained a presence with the Christian indigenous group Las Abejas (the Bees) in Acteal (Chiapas), Mexico, which in the midst of military and paramilitary fighting has staunchly clung to its non-violent principles

Tinker throws down the gauntlet to white Christians of the West. The challenge, I might add, equally applies to Muslim countries in their treatment of minorities, religious or ethnic. Beyond that, he points us to the spiritual vision of humanity's trusteeship that encompasses all people, and particularly in the specificity of their ethno-cultural identity and religious convictions. How can this deep respect for the human person possibly harmonize with neoliberal development schemes (usually benefiting the elites and dominant tribal or ethnic groups), which are mostly imposed from the outside? How then can it survive the aggressive designs of the corporatocracy, the maws of McWorld as people's natural resources are pillaged, their people impoverished and indebted to foreign masters? Indeed, governments cannot solve all these problems alone. They will need to be supported, and at times confronted by citizen groups advocating for marginalized, dispossessed and despised persons and communities.

Ruling as God's representatives on earth will involve this kind of intervention on behalf of the "Fourth World," however these people might be defined locally. It will also involve adopting and developing a holistic theology of eco-justice, built on God's good creation and waiting in

(45 were massacred in a church in December 1997 while on their third day of a fast for peace). CPT has also sought to relieve the plight of tens of thousands of indigenous coffee farmers who have been driven from their lands by the violence and are now refugees in the highland county of Chenalho. Lynn Stoltzfus and Scott Kerr were invited by Las Abejas to attend the third National Indigenous Congress at the end of the Zapatista March, which reached the capital city on March 11, 2001. Forty out of the 56 indigenous peoples of Mexico attended the congress. Stoltzfus shows that the challenge to bring about peace is intimately tied to cultural and economic issues: "In the economic sphere, indigenous people have traditionally had communal ways of owning and managing land and natural resources. In Mexico, the government has dismantled communal landholding structures as a part of neoliberal economic reforms. Indigenous peoples have a cultural tradition of respect for the natural world and are working to preserve, protect and manage natural resources for the benefit of their communities. Without ways of protecting their communities and resources from the economic pressures of the market, the traditional indigenous ways of relating to the land will not be able to continue. For the Zapatistas, the Abejas and the other indigenous people represented at the National Indigenous Congress, peace is something that cannot come without maintaining their cultural and economic ways of life. In many ways, the dominant economic and cultural systems have been at war against the indigenous culture for 500 years, so any peace that does not deal with this violence will not be a true and lasting peace" (Lynn Stoltzfus, "Chiapas: Peace and Indigenous Rights," *CPTnet* [March 17, 2001]: httn://www.cnt. ors/archives/ 2001/mar2001/0014.html).

expectant hope for its fulfillment—in a partial way now, as we are called to spread his Kingdom on this earth and fully at the end of the age. This is where Christianity and Islam also converge in surprising ways. First, I offer an excerpt from Jesus' Parable of the Sheep and Goats (Mt. 25:31-40, NIV):

> When the Son of Man comes in his glory, and all the angels with him, he will sit on his throne in heavenly glory. All the nations will be gathered before him, and he will separate the people one from another as a shepherd separates the sheep from the goats. He will put the sheep on his right and the goats on his left. Then the King will say to those on his right, "Come, you who are blessed by my Father; take your inheritance, the kingdom prepared for you since the creation of the world. [70] For I was hungry and you gave me something to eat, I was thirsty and you gave me something to drink, I was a stranger and you invited me in, I needed clothes and you clothed me, I was sick and you looked after me, I was in prison and you came to visit me."
>
> Then the righteous will answer him, "Lord, when did we see you hungry and feed you, or thirsty and give you something to drink? When did we see you a stranger and invite you in, or needing clothes and clothe you? When did we see you sick or in prison and go to visit you?"
>
> The King will reply, "I tell you the truth, whatever you did for one of the least of these brothers of mine, you did for me."

Then the King (obviously God in the story) turns to those who had not cared for those in need and says, "I tell you the truth, whatever you did not do for one of the least of these, you did not do for me" (v. 45). And they were sent to an eternity away from his presence.

In the book of *aḥādīth* collected by Sahih Muslim, we read in a similar manner that God feels directly implicated by the way Adam's progeny treat one another—a direct consequence of our common creation-caliphate:

Abu Huraira reported Allah's Messenger (may peace be on him) as saying:

> Verily, Allah, the Exalted and Glorious, would say on the Day of Resurrection: O son of Adam, I was sick but you did not visit Me. He would say: O my Lord; how could I visit Thee whereas Thou art the Lord of the worlds? Thereupon He would say: Didn't you know that such and such servant of Mine was sick but you did not visit him and were you not aware of this that if you had visited him, you would have found Me by him? O son of Adam, I asked food from you but you did feed Me. He would say: My Lord, how could I feed Thee whereas Thou art the Lord of the worlds?

[70] Notice again the interlinking of the themes of creation, kingdom of God, and re-creation.

He said: Didn't you know that such and such servant of Mine asked food from you but you did not feed him, and were you not aware that if you had fed him you would have found him by My side? (The Lord would again say:) O Son of Adam, I asked drink from you but you did not provide Me. He would say: My Lord, how could I provide Thee whereas Thou art the Lord of the worlds? Thereupon He would say: Such and such servant of Mine asked you for a drink but you did not provide him, and had you provided him drink you would have found him near Me.[71]

As Stassen and Gushee emphasize, holistic character ethics, especially the ethics of Jesus, lead to *practices* that cement God's reign in one's life, family and community. Thus prayer and politics cannot be separated.[72] We are back to the Muslim concept of *tawḥīd*. As Jesus taught, "Not everyone who says to me, 'Lord, Lord' will enter the kingdom of heaven, but only the one who does the will of my father in heaven."[73] Solemnly and resolutely donning the mantle of our God-given trusteeship, we seek with God's grace active in our hearts to align our own will with God's will for ourselves and our world—with a special concern for the poor and needy. Seyyed Hossein Nasr rightly observed, "Every practicing Muslim, which includes the vast majority of the population of the Islamic world, could not but agree that his or her highest wish is none other than the prayer uttered by Christ, 'Thy Will be done on earth as it is in Heaven.'"[74]

Parting Words

Monotheistic convictions about creation, global ethics and human solidarity must translate into action. Perhaps because of the almost intractable problems of our planet, it is even more likely today than before the attacks of September 11, 2001 that Christians and Muslims are ready to sit together and map out a common theology of humanity to serve as a value-base for common projects in development, peacebuilding, social justice and environmental concerns. At the same time, I would have to agree with the spirit in which the longtime Saudi Oil Minister Shaykh Ahmed Zaki Yamani commented on Muslim–Christian dialogue. In a Foreword to Montgomery Watt's only book on the subject, he commends

[71] *Al-Jāmiʿ al-Saḥīḥ by Imam Muslim*, trans. Abdul Hamid Saddiqi, "Merit of Visiting the Sick," no. 6232, section MLXIII, Book 30, Vol. 4 (Lahore: Sh. Muhammad Ashraf, 1971).

[72] Their last section is entitled, "A Passion for God's Reign," with three chapters: "Prayer," "Politics," and "Practices" (*Kingdom Ethics*, 447-91).

[73] Mt. 7:21, NLT.

[74] Nasr, *The Heart of Islam*, 305-6.

the author for his eminent scholarship and irenic spirit and then notes, "it may be impossible in the present state of awareness of the Christian and Muslim worlds to reconcile their respective understandings of the most controversial points at issue between them." He hopes, nevertheless, that this will be achieved someday through God's infinite wisdom. Then he quotes the old Arabic proverb, "What cannot be achieved in its entirety, must not be abandoned in its entirety."[75]

Whether it be Toulmin's plea to regain a more Renaissance-like humility, or Benhabib's advice to engage cross-culturally in a practice of embodied discourse ethics, or the simple realization that truth will never be more than just approximated on this earth, we can leave the modern hegemonic certainties behind and begin truly to listen to one another. Again, "Now we see things imperfectly as in a poor mirror." Yet we live in the hope of the second phrase, "but then we will see everything with perfect clarity.[76] In the meantime, we remember the biblical proverb, "Without a vision, the people perish."[77]

As we have seen, the trusteeship of humanity offers us such a vision. Earth community must be our common paradigm as Christians and Muslims, and our doctrine of humanity must be rooted in creation. The God who called all his human creatures to be his trustees on earth promised to be with them as they cared for one another and for their groaning planet. Therefore, leaving behind all the politics of exclusion—including the modern "commodification of the commons"—we proclaim that the bounty of the earth is for all to share equally. And when it comes to the shameful reality of hunger today, we will insist that there is room at the table for all.[78] In doing so we will be following the well-trodden path of the Latin American base-communities: praxis (doing the Word), reflection on the Word, and more praxis.

As always, theology is constructed in context, and what better context could there be for Christians and Muslims doing theology together than in the trenches of human poverty, oppression, and despair? At the same time, since the vast majority of Muslims and Christians worldwide are conservative, I have offered here a way to take the sacred texts seriously,

---

[75] "Foreword," in W. Montgomery Watt, *Islam and Christianity Today: A Contribution to Dialogue* (London: Routledge & Kegan Paul, 1983), x. The Arabic reads, *Mā lā yudraku kulluhu lā yutrak kulluh.*

[76] 1 Cor. 13:12, NLT.

[77] Prov. 29:18, KJV.

[78] This is the slogan of an ecumenical agency I have always admired, Bread for the World. See David Beckman and Arthur Simon, *Grace at the Table: Ending Hunger in God's World* (Downers Grove, IL: InterVarsity Press, 1999).

without being bound to a naïve literalism. The resulting theology—applying the "canon" to real life situations—will be incarnated in Muslim, Christian and Jewish solidarity and active lobbying for a land (Israel/ Palestine) equally shared by two peoples; or as living out their faith together in rebuilding the broken nation of Iraq. It may be as northern advocates joining with local Muslim and Christian NGOs in Sarawak, Malaysia to combat the crippling effect of deforestation.[79] Then turning their attention to neighboring Indonesia, these activists might establish contact with groups there who are struggling to find ways to reconcile "settlers" and indigenous peoples, Muslims, Christians and animists. Earnest believers in this common trusteeship might also set their sights on the inner cities of the United States and blaze new trails for the redressing of injustice and the empowerment of minorities and recent immigrants, legal or not. In each situation, too, theology will be enriched; Muslims and Christians will uncover new facets of truth pointing to the love and mercy of the Creator; in turn, this will shed new light on the sacred texts and the hermeneutical process.

To be sure, in this uncertain journey forward in our postmodern, globalized world, all humans, from the First to the Fourth Worlds and from every culture, need to seize the hope that is from beyond themselves. My prayer is that all of us who believe in the God who created humanity to rule together in harmony on the earth would increasingly join hands and witness by word and deed to his love and care for his creatures—protecting the life of every human being while respecting the integrity of the ecosystems from which we all draw our sustenance. Whatever small steps we take will not only sow hope, but also shed more light as we continue to converse with one another and those of other faiths.

---

[79] Cf. Margaret E. Keck and Kathryn Sikkink, "Environmental Advocacy Networks," in *The Globalization Reader*, ed. Frank J. Lechner and John Boli (Malden, MA: Blackwell, 2000), 392-99; originally (in a longer version) in *Activists Beyond Borders: Advocacy Networks in International Politics* (Ithaca, NY: Cornell University Press, 1998). My doctoral dissertation devoted seven pages to this case study.

Appendix A

SEVENTH INTERNATIONAL SABEEL CONFERENCE, NAZARETH AND JERUSALEM, NOVEMBER 12–19, 2008: "THE NAKBA: MEMORY, REALITY AND BEYOND"

We are more than 200 Christians from five continents who have come together to commemorate the tragic events that occurred 60 years ago in the lives of the people of Palestine. While we have come to hear from and to offer our solidarity and support to the indigenous Palestinian community in both Palestine and Israel, we heard several testimonies and presentations by Christian, Muslim and Jewish speakers who bore witness to the injustices visited upon the Palestinian population of this land. They have seen more than 531 villages depopulated and destroyed, and the creation of more than 750,000 refugees who have not been allowed to return to their homes since 1948.

We recognize the irony in the coincidence that this year also marks the 60th anniversary of the Universal Declaration of Human Rights. The establishment of peace with justice requires that the full truth be told about the events of 1948 and the subsequent displacement of hundreds of thousands more Palestinian citizens in 1967, a process which has continued to the present day. The human rights of the Palestinian people continue to be crushed under a military occupation that dehumanizes both oppressed and oppressor. We share our conviction that it is only an acknowledgement of the full truth behind and within this current state of oppression that will lead to true freedom for all parties in the conflict.

Truth is essential for peacemaking. We acknowledge the truth that our silence about the status of the Palestinian people equals complicity in this ongoing tragedy. The status quo is a crime against humanity. As Christians, we can no longer be silent. Things worsen as each day passes. The so-called peace process is rather a consistent and persistent process of death and destruction, both physically and spiritually. The Nakba—the catastrophe that has been imposed and is still being imposed on the people of Palestine—continues unabated and unrestrained. The truth of it is silenced or ignored both in our churches and in our media. This must change if we are to be true to Jesus' call to be peacemakers.

We have been encouraged by the thousands of Palestinians and Israelis who have practiced methods of nonviolent resistance in seeking to bring an end to the current conflict. We lift up the practice of nonviolence as the most practical means of achieving peace in this situation where the balance of military power is so overwhelmingly one-sided and where the reliance upon violence only continues to make matters worse. We are concerned by the use of the Bible as an instrument of colonialism and exploitation by those who would enlarge the conflict. We reject the exclusivism presupposed in such an interpretive approach to biblical truth. We seek the reconciliation of all peoples throughout the world, and therefore call on our brothers and sisters in the worldwide church to speak out and act out the ministry of reconciliation.

We have been touched by the faces of children wherever we have gone. We have come to realize that an entire generation of children is being crippled because they have no access to the nutrition needed for normal growth and development, and thus endure spiritual and social alienation, violence and lack of opportunities which none of us would tolerate even for a day in our own communities. We remember the call of the Nobel peace laureates that the first decade of this new century be devoted to nonviolence. We hear anew the call of Jesus to "let the little children come unto me," to let them be placed in the center of the current picture of marginalization, thus challenging the international community with their vulnerability and their need for protection.

Therefore, we call upon all our churches and governments:

- to work with renewed energy for an end to this endlessly spreading military occupation;
- to insist on full implementation of all United Nations resolutions and all human rights requirements in international law which pertain to Israel's withdrawal from the occupied Palestinian territories and the right of return for Palestinian refugees;
- to insist on greater freedom of movement and more humane conditions in the occupied territories;
- to insist that Israel accord equal rights to all its citizens, Jewish and Palestinian alike;
- to divest themselves from investments in companies that enable the occupation;
- to insist that Israel lift its ongoing siege and collective punishments which prevent the free movement of people, goods and humanitarian aid in and out of Gaza; and finally

- to support the work of Sabeel in its efforts to build bridges of nonviolence between people in all the monotheistic religions represented in the region.

We have heard the call of urgency from our fellow Christians in this holy land. As in Jesus' own day, so Bethlehem lies under military occupation today surrounded by a prison wall. Our memories of the birth of The Child of Bethlehem 2000 years ago are contrasted and challenged by the reality of the children and the parents and the grandparents of Bethlehem today. As followers of that holy child, may our spirits meet in Bethlehem's streets as we join in prayers and actions for light and life! May we seek creatively to disturb the status quo with acts born of the Spirit of courage, love and truth.

Appendix B

CPT IN THE HEBRON DISTRICT

CPTnet
September 29, 2004
http://www.cpt.org/archives/2004/sep04/0034.html

HEBRON DISTRICT: CPTers Kim Lamberty and Chris Brown badly injured by settlers in the south Hebron hills

At about 7:15am on the morning of Wednesday September 29, 2004 settlers attacked Christian Peacemaker Team members Chris Brown and Kim Lamberty as they accompanied children to school. The children, from the village of Tuba, have experienced harassment from settlers in the past as they go to school in the village of al-Tuwani.

The five settlers, dressed in black and wearing masks, came from an outpost of the nearby Ma'on settlement and attacked Brown and Lamberty with a chain and bat. All of the children escaped injury by running back to their homes.

The settlers pushed Brown to the ground, whipped him with a chain and kicked him in the chest, which punctured his lung. They kicked and beat Lamberty's legs. She is not able to walk because of an injury to her knee and has a broken arm. The settlers also stole Lamberty's waistpack, which held her passport, money and cellular phone.

Lamberty and Brown were taken by ambulance to Soroka hospital in Beer Sheva for treatment. Hebron Team Support person, Rich Meyer, reports that the two CPTers told him they are receiving excellent care from Israeli doctors.

Children from four small Palestinian villages walk to a central school in the village of al-Tuwani. Because settlers have harassed the children since school began in September, and the Israeli police would not intervene to prevent the attacks, the villagers have sought the protection of international accompaniment. A coalition comprising Christian Peacemaker Teams, the Israeli group Tayush and members of Operation Dove, (an Italian Christian organization that undertakes accompaniment work similar to CPT's work), set up a presence in the village of al-Tuwani beginning on September 12, 2004. The three groups initially committed themselves to six weeks of accompaniment after members of these organizations witnessed settler attacks on children each time they made exploratory visits to the area.

Christian Peacemaker Teams, Operation Dove and Tayush plan to continue accompanying children to school in al-Tuwani.

CPTnet
November 13, 2004
http://www.cpt.org/archives/2004/nov04/0020.html

HEBRON: Nonviolent direct agriculture
By Nicholas Klassen

Now is the season for picking olives. And while the olive harvest should be an innocuous exercise, in Palestine it is often an act of resistance. Harvesting olives may result in a beating, an arrest, or general harassment. And so whether they intend to or not, farmers become political actors engaged in nonviolent direct action.

On November 1, Dianne Janzen and I picked olives with Palestinians and other internationals in the shadow of Otni'el settlement. We were immediately stopped by soldiers who insisted the group was too close to a military installation and did not have permission to be there. The Palestinians required authorization from a foreign government to glean olives on land they have cultivated for generations.

When questioned about their inflexibility, the soldiers offered a response so familiar to CPTers that it comes across like a recording. It goes something like this: "This area is closed for security reasons. We must protect against terrorists. Even if we recognize that not all Arabs are terrorists, we can't know which ones are. So we have to assume they all are." Regarding the specifics of harvesting olives outside the camp, the soldiers claimed that the pickers could look into the camp and plot an attack against it. (Later, we found that we could not look into the camp even when we were picking from trees adjacent to the fence that surrounds the camp.) I could not gauge whether the soldiers believed what they were saying, or if they found reciting the party line from prepared notes is easier than looking critically at their role in subjugating an entire people.

While the soldiers stood fast, Musa Muhamry of the Hebron Land Defense Committee negotiated with Israeli officials. After an hour, they gave the green light, and the soldiers stepped back. The olive pickers flooded the grove, hindered only by the need to avoid the empty beer bottles and other detritus soldiers lob into the olive grove.

On this day, the Palestinian farmers were able to gather their olives, but the incident provides a glimpse into some of the hurdles they face in harvesting all manner of agricultural goods. In other circumstances, Israeli settlers harass and attack farmers, and burn and tear up trees and crops. Meanwhile, settlements, army posts, and the separation wall consume more and more farmland throughout the West Bank. And the military restrictions that prevent or delay delivery of produce diminish the value of the Palestinian harvest.

The olive harvest continues until the middle of November.

CPTnet
March 1, 2005
http://www.cpt.org/archives/2005/mar05/0000.html

AT-TUWANI REFLECTION: A different vision of the future
by Art Gish

At a time when the policy of the Israeli government involves keeping Israelis and Palestinians separate (apartheid), Israeli peace activists are living according to a different vision for the future. Instead of bowing to fear, they reach out in love to their Palestinian cousins. Against the wishes of the Israeli government, these Israelis are traveling to the West Bank and maintaining relationships with Palestinians. Instead of accepting the repressive nature of the Israeli occupation, these activists are able to see the failings of their own government and their own people.

Palestinians trust these Israelis and have their phone numbers. Every time Israeli settlers attack the Palestinians here, Israeli activists will be there with their Palestinian friends in the next few days, if not in the next few hours. After the violent attack by settlers on 16 February in At-Tuwani, Israeli activists were out in the mountains with At-Tuwani shepherds the next day. They have come in the night when villagers were attacked by soldiers. They have slept in the village to help prevent attacks.

Each October and November, Israelis come to help protect Palestinians from settler attacks as the Palestinians pick their olives. Within two days after Israeli settlers destroyed 200 olive trees in the village of

Ma'im, Israelis were there replanting trees. In April and May, the Israelis come to protect the Palestinian farmers and their barley harvest, some of which settlers burn each year.

Most Palestinians never have any positive contact with Jews. Their only contacts with Jews involve settlers and soldiers—relationships that are oppressive and degrading. The people of At-Tuwani have a different perspective. Repeatedly I have sat with Israelis and Palestinians, Jews, Muslims, and Christians breaking bread together. These relationships are visible demonstrations of a future that could be.

CPTnet
March 4, 2005
http://www.cpt.org/archives/2005/mar05/0004.html

AT-TUWANI REFLECTION: Hungry sheep and hungry people
by Art Gish

The grass was tall and lush, not like the short grass on which the sheep had been grazing before we snuck around the mountain behind the illegal Israeli settler outpost. Since it had been a number of years since the shepherds have dared to risk getting so close to the outpost, the grass had had time to recover from overgrazing.

Members of Christian Peacemaker Teams and Operation Dove have been accompanying shepherds and their flocks as they risk settler attacks on both themselves and their sheep. Our accompaniment of shepherds is both to protect them from attacks, and to raise the broader issue of whether Palestinian shepherds should be allowed to graze their sheep on their own land.

Because illegal settler outposts lie all over the South Hebron Hills, and because the settlers, supported by the Israeli military, have declared huge areas around those outposts to be off-limits for Palestinian flocks, the land still available for grazing is limited and inadequate to support the local population.

Israeli military and police often deny Palestinian shepherds access to their own land. The Israeli government wants the South Hebron Hills for a military training area, and are attempting to force the Palestinians out. One method for achieving this goal is to keep the Palestinians from feeding themselves.

With decreased available land for grazing comes increased pressure on the fragile mountain sides in this area near the Judean desert. Overgrazing both decreases the amount of grass available to the sheep, and causes increased erosion of the precious topsoil, resulting in even less grass in coming years. With increased need to purchase feed for the animals and the necessity of having smaller flocks, comes decreased income for the shepherds. The result is increased poverty and hunger not only for the shepherds, but also less food for the larger Palestinian population, resulting in increased dependency on international food aid.

The response of the international community to this problem caused by the occupation has been to ship huge amounts of food to feed the Palestinian people. Although praiseworthy, a better solution would be to end the occupation and allow the Palestinians to feed themselves.

Appendix C

Ninian Smart's Typology
of Religious Experience

Rudolf Otto (1869–1937) contended that at the heart of religion for human beings lies the notion of the numinous—the quality of awe, majesty, and fascination inspired by the experience of the Other.[1] Ninian Smart contends that much anthropological research has been done since Otto's time and that the picture is rather more complex than that. Some of that complexity is revealed in the six-fold dimension of religious experience I cited from Barbour in Chapter 3. Three other distinctions should be made here, according to Smart. The first is between the personal model of the Beyond (the theist or monotheist) and the impersonal Indian model of a divine being behind the impermanence of the cosmos but which turns out to be *maya*—a mere illusion, as the privileged believer is led to experience through meditation.[2] The second distinction is somewhat parallel to the first: mystical vs. numinous kinds of experience. In the numinous mode, "the eternal lies, so to speak, beyond the cosmos and outside the human being."[3] In the Upanishads, for instance— writings that appeared around the time of the Buddha on the Subcontinent—one encounters a series of "identity statements": "I am the divine Being," and "That art Thou." This is usually interpreted as meaning: "That divine Being which lies behind the whole cosmos, which creates it and sustains it and constitutes its inner nature, is the same as what you will discover in the depths of your own Self, if you will voyage inward through self-control and the methods of meditation and purification of your consciousness."[4]

---

[1] From the Latin word *numen*, spirit in the animistic sense. Yet he took it further to indicate that quality of mystery and awe-inspiring experience of the Wholly Other.
[2] Smart, *Worldviews*, 48-54.
[3] Ibid., 64.
[4] Ibid., 60.

This statement begins with the numinous, however (the key word in Sanskrit here is *Brahman*, the power or the divine being),[5] but combines its insights with the second mode of religious experience, the mystical. In the latter mode, common to many religious traditions of the world, the devotee relies on the practices of prayer and/or meditation in order to climb up a ladder of spiritual experience. Near the top of that ladder lies the experience of union with the divine—whether conceived of in personal or impersonal terms.[6] If impersonal, the mystical experience will be interpreted as non-dual—the subject–object distinction disappears altogether, as with Buddhism in general, in which the experience of personal liberation becomes, paradoxically, the freedom of discovering the emptiness that is at the heart of the world. Thus one becomes just as much liberated from the belief in gods or God as from the fleetingness of the cosmos. In Advaita Vedanta Hinduism too, the saintly yogis who practice self-control and meditation finally accede to the highest level of knowledge, in which the self and the divine Being become one and the saints realize that the notion of "Creator" was simply an illusion.[7]

Naturally, the mystical traditions of Christians, Jews and Muslims take a distinctly different path. I quoted Akbar Ahmed in the Introduction as asserting that the Sufi saying "I am God" is not "so far-fetched." For him that kind of experience naturally flows out of the realization that we are meant to be God's deputies on earth, and that, consequently, God seeks to clothe us with his own qualities. The result is simply a theological emphasis on "the unity and integration of creation itself."[8] The mystical experience for monotheists, therefore, unfolds in the numinous mode, in which unity does not obliterate individuality, but rather enhances it through union with God. It could never be a non-dual experience.

A third distinction is also needed, Smart argues: "right wing" and "left wing" shamanism. Shamanism is a phenomenon common to all traditional religion in small-scale societies. The shaman is the villager or tribesman (or tribeswoman in many societies as well) who forges a bridge between the natural and supernatural world. Shamans either ascend to

---

[5] The numinous side, according to Smart, was emphasized in later Indian thought in the dramatized portrayal of the great gods, Shiva and Vishnu, along with the goddess Kali, all "replete with power, terror and love" (ibid.).

[6] Consider the Islamic mystical tradition that found a ready precedent in Muhammad's ascension to heaven from Jerusalem all the way up to the seventh heaven. Starting with al-Bistami (d. 873), many Sufis used the prophet's ascension as a model of mystical journeying upward along specific stages (Lapidus, *A History of Islamic Societies*, 175-76).

[7] Smart, *Worldviews*, 52.

[8] Ahmed, *Islam Under Siege*, 4.

heavenly realms or descend to the world of the dead, and for some cultures there would be no distinction between the two. In doing so, however, they acquire the ability to communicate with spirits who grant them powers of divination or healing. In some cases also, they are considered to have returned from the realm of the dead and therefore "can reenact dramatically the death and restoration to life of the sick person" by restoring their health. The long-time leader of the Chicago School of Religion, Mircea Eliade (1907–1986), did much to advance research on shamanism. He thought that the common experience of shamans who became "possessed" by a god or spirit was a precursor to the prophetic traditions of monotheistic faiths, in which a prophet becomes similarly "possessed" by God and is able to speak on his behalf.[9]

In this vein, we might consider two wings of shamanism, proposes Smart:

> The right wing focuses on the numinous experience of the Other, and the experience of the prophet is a special form of this. Institutionally, the successor to the prophet is the preacher, who tries to recapture something of the spirit of prophecy. The left wing focuses on the mystic or yogi, the one who practices the art of contemplation; institutionally, the successor of the mystical teachers of the past is the monk or nun.[10]

Considering the importance of the mystical tradition in Islam, both as the main means of missionary propagation of the faith in Africa and Asia and the Sufi brotherhoods as the warp and woof of many Islamic societies from the eleventh century onwards, it behooves us to take it into account with regard to the present project. First, the commonality of the mystical experience of the numinous in Islam and Christianity argues for the inclusion of a wider spectrum of views than the current mood of Islamic revivalism is prepared to allow. To think theologically together as Muslims and Christians today in the area of creation is to tap into a wealth of material from the past, on both sides. Second, the mystical experience of the numinous in both traditions reveals the limits of human language in describing what is actually taking place. The idea that God is "Wholly Other" and at the same time the Divine Lover who longs for us to be united with Him (a common thought in Sufism) presents a challenge to both Muslims and Christians—hence, the powerfully symbolic value of religious language, in the sacred texts, in commentaries about them, and for those who articulate the various aspects involved in the human–divine relationship. Finally, the realization of these paradoxes,

---

[9] Smart, *Worldviews*, 61.
[10] Ibid., 62.

which by definition will always be somewhat resistant to rational analysis, will induce us to proceed very humbly.

I offer Figure 18 as a way to imagine the various distinctions Smart proposed, and as a way to plot various religious experiences in a wider theological and anthropological perspective. It should be a reminder that the task of articulating theology is always done not only in particular socio-cultural contexts, but also within the horizon of a religious community whose beliefs are constantly reenacted in rituals, which include both worship and more mystical elements as well. For Muslims this entails a long and rich Sufi heritage, and for Protestants in particular, it would require a greater appreciation for the crucial role played by monasteries in the Catholic and Orthodox traditions. Protestants would come to appreciate as well their own past movements of revivalism (e.g. the Great Awakening of the eighteenth century) and, more recently, the Pentecostal and charismatic movements.

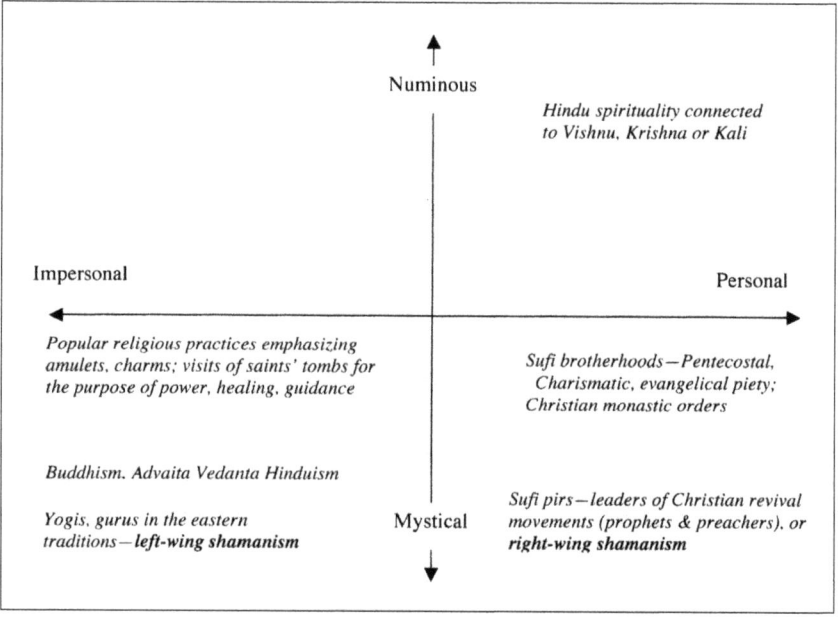

Figure 18. *Smart's Typology of Religious Experience*

The above chart is admittedly oversimplified, particularly as it draws an artificial line between worship and mystical experience, which occur on a continuum that is difficult to chart in practice. Many Muslims—but not all—would feel comfortable with the distinction between Sufi and traditional[11] Islamic practice. Yet many Christians would deny that there is any distinction between worship and mystical experience in the first place. This is particularly true in my own evangelical heritage, which has internalized much of the pietistic movement that arose in Germany in the seventeenth century, and that resurfaced in the "awakenings" in Britain and the United States. We see ourselves—at least in theory—as living in a personal relationship with God and seeking to be led on a daily basis by the inner guidance of the Holy Spirit. Even the Pentecostal movement, with its roots in the Azusa Street Revival of 1906, and the subsequent charismatic movement, which deeply affected both the Roman Catholic Church and mainline Protestant denominations in the 1960s and '70s, are characterized by a piety that is identical to that of the indigenous African and Asian churches that have exploded in the last two or three decades.[12] There is some justification, however, for separating traditions that emphasize ritual and theological norms from those that have arisen very recently and emphasize spiritual experience to a much greater extent.

[11] Here I find it impossible to find an appropriate word. If I use "orthodox," that would imply that Sufis are heretical, which is plainly false according to Muslim practice over the centuries. If I use "mainline," it would reflect the fact that with the movement of Islamic revivalism in the last three decades the number of Muslims who consider themselves Sufi has shrunk considerably. The term "traditional" is also inadequate, because Sufism was a dominant force in institutionalized Islam from about the eleventh until the twentieth century.

[12] Besides Lamin Sanneh's *Whose Religion Is Christianity?* (cf. Chapter 4, n. 25), see also Philip Jenkins' *The Next Christendom: The Rise of Global Christianity* (Oxford and New York: Oxford University Press, 2002).

Appendix D

TOWARD AN AUTHENTIC
MUSLIM–CHRISTIAN DIALOGUE

I begin with an ongoing conversation about these issues in Europe, intro-
ducing a thinker about whom I will have more to say in a later chapter,
the grandson of Hasan al-Banna himself, the Swiss Tariq Ramadan (b.
1962).[1] As a Muslim intellectual born and raised in the West, Ramadan
has painstakingly studied the foundation of Western civilization. His
Ph.D. thesis in philosophy was on Nietzsche and he now teaches philosophy at the College of Geneva and Islamic studies at the University of
Friburg (he holds a second doctorate in that field). From 1988 to 1992 he
served as principal of a Muslim school he helped found,[2] and the next
year (1992–93) he spent in intensive study of the Islamic disciplines in
the city of his grandfather, Cairo. An imam and a sought-after preacher
in mosques all over Europe, he is an active member of two large Muslim
organizations: Présence Musulmane (Islamic Presence) and Jeunes
Musulmans de France (Young Muslims of France). Author of over a
dozen books and a gifted public speaker and debater, Ramadan is frequently seen in the European media, and is regularly consulted by
various commissions in the European Parliament in Brussels.[3]

---

[1] I mentioned in the Introduction that al-Banna founded in 1928 the Muslim Brotherhood in Egypt.

[2] The school in Geneva is called "College Tariq Ramadan."

[3] Most of this information is from his website (http://www.tariq-ramadan.org). The
English version of the online Italian weekly, *Chiesa*, provides some intriguing
additional details about his views and goals. His book, *To Be a European Muslim*
(1999), was translated into fourteen languages. His brother Hani runs the more
conservative publishing house Tawhid and an Islamic center in Geneva. The author
indicates that Tariq denies having any connections with his brother's work as the
Islamic center is "accused of contact with the terrorist network of al-Qaeda" ("Tariq
Ramadan's Two-faced Islam: The West is the Land of Conquest," http://
www.chiesa.com [Rome, January 19, 2004]: http://213.92.16.98/ESW_articolo/0
,2393,42025,00.html).

Ramadan, for all of his progressive views, has his share of detractors as well. Catholic Islamicist Olivier Clément, for instance, calls attention to the dangers represented by Ramadan's approach to the Christian community of Europe. Clément sees Ramadan's desire to articulate Islamic tenets in the language of the European intelligentsia as a way to bring Islam into the vacuum produced by the eclipse of Christianity and Judaism from this society. Laying hold of the postcolonial model of center-periphery, Clément sees Ramadan's strategy as twofold: (1) train a generation of Muslim intellectuals to join the opinion makers in the global nucleus of power, the Europe–North America axis and influence its values, and (2) woo Europeans into the fold of a progressive Islam.[4]

It seems, therefore, that Clément deplores Ramadan's exclusivism and proselytizing zeal, which posits that Islam supersedes and corrects all previous revelations of the People of the Book (Jews and Christians) and that the Christian advocates for the poor and dispossessed (Mother Theresa, Abbé Pierre, Bishop Helder Camara) are actually crypto-Muslims, since Islam is the original religion of humanity at creation.[5] He quotes another French expert, Jacques Jomier, who opined that Ramadan's concern is "not the modernization of Islam, but the islamification [*sic*] of modernity." For Clément, Ramadan's fundamentalism is not far below the surface when he (Ramadan) writes, "Today the Muslims who live in the West must unite themselves to the revolution of the anti-establishment groups from the moment when the neoliberal capitalist system becomes, for Islam, a theater of war... The revelation of the Koran is explicit: whoever engages in speculation or cultivates financial interests enters into war against the transcendent."[6]

Is Ramadan in fact reviving the classical shariʿa bipolar division of the world between *dār al-ḥarb* ("abode of war"—i.e. non-Muslim countries) and *dār al-islām*? I believe this suggestion advanced by Clément ("a

---

[4] This is from the same online *Chiesa* article mentioned in the previous note. Ramadan, who was hired by Notre Dame University (at the Croc Institute for International Peace Studies) in 2004 was never given an entry visa by the US authorities. Despite a letter of protest issued by the American Academy of Religion and many other prominent individuals, Ramadan was never allowed into the country (no reasons given) and therefore had to resign his position.

[5] This is by virtue of the original covenant (Q. 7:172)—certainly the standard Muslim interpretation over time.

[6] Clément quotes from an article Ramadan wrote in the French journal *Pouvoirs*, number 163, but omits the page and the date. Also problematic for Clément was his public support for Hasan Turabi, the ideologue of the Sudanese Islamic revolution. Ramadan has since distanced himself from that position.

theater of war") is unjustified. As a figure of speech, I could also say that Christians ought to wage war in the name of Christ on all systems that dehumanize people, exploit and enslave them in any way. Behind Clément's suspicion of Ramadan's motives and worldview I sense a theological pluralism with regard to truth: he finds it difficult to tolerate that others could claim that their faith is "true," thus implying that all other faiths are "wrong."

Here I propose a brief outline of the categories Catholic theologian Paul J. Griffiths offers on these issues of religious diversity. My own position will then make more sense to the reader. Griffiths is concerned with a straightforward demarcation of the philosophical issues involved. Philosophers of religion, he argues, are concerned with the examination of religious claims and ask the following questions: (1) What does it mean? (2) In what sense is it true or false? (3) What does it mean to believe it or not to believe it?[7] The first question raises the issue of the meaning of a religious claim. Here is Griffiths' definition: "A religious claim…is a claim about the way things are, acceptance of or assent to which is required or strongly suggested by the fact of belonging to a particular form of religious life."[8] Religious claims are truth claims— concerned with truth and falsehood in the area of metaphysics: Does this or that assertion actually correspond to a state of affairs that can be verified to be the case? The central claims of any religion typically cover the nature of the world inhabited by humans, the nature of humans themselves, and recommendations or commands as to how human life should be conducted. One can easily observe, in this respect, a great variety of truth claims across the religious spectrum. A commonsensical reaction to this state of affairs is to say that some claims must be true and some must be false.[9]

The second question relates to the issue of "epistemic confidence," or to the degree of confidence that "an act of assent to a claim does indeed bring you knowledge, or ought to be made, or has been formed in the

---

[7] Paul J. Griffiths, *Problems of Religious Diversity*, Exploring the Philosophy of Religion Series (Oxford/Malden, MA: Blackwell, 2001), 16.

[8] Ibid., 21. His definition of religion helps to explain the above definition more fully: "a form of life that seems to those who inhabit it to be comprehensive, incapable of abandonment, and of central importance" (p. 7).

[9] Griffiths distinguishes between incompatibilities that are (a) *contradictory* ("if and only if each makes a claim to truth, both cannot be true, and yet one must be false"); (b) *contrary* ("both cannot be true, yet it is possible that neither is"); (c) and *non-compossible* ("if each prescribes a course of action and it is impossible for a single person to perform both") (ibid., 31-37).

right way, the way that such assents ought to be formed."[10] These questions are tied to a theory of knowledge, or epistemology—an issue that was addressed in relation to the interpretation of texts. Finally, the question about belief relates to people of other faiths, the "religious alien," as Griffiths put it (all those who inhabit your "religious home" are your "religious kin"). Here the question Griffiths explores is: What attitudes toward the religious alien are possible? This in turn brings up the central issue religions are concerned with: "salvation," or "What is the proper end of human beings?"[11]

Recall that it is precisely the truth question that occasions Clément's objections to Ramadan's views. He objects to Ramadan's claim that Islam as a system holds more truth than its competitors. Apparently he believes either that all religions make similar truth claims, or that, if they make different truth claims, there is no way to vindicate any of these claims one way or another. Either way, Clément seems to feel that there is little distinction between one person's religious home and that of others. Is this actually the case? The two issues of metaphysics and epistemology are intertwined, but I follow Griffiths in beginning with the question of metaphysics—What about the truth claims of various religious worldviews concerning the nature of reality? To answer this, Griffiths builds on the threefold typology that was originally proposed by Alan Race: pluralism, inclusivism and exclusivism.[12]

The philosopher of religion who has most forcefully argued the position of pluralism—that is, "placing all genuinely religious claims on a par with respect to truth"[13]—is John Hick. It is not that Hick minimizes the differences between religions, it is rather that he dismisses two categories of incompatibilities as only "apparent incompatibilities": (1) questions that can be settled by historical fact, but since it is not possible to verify such questions in a definitive way, they are not directly relevant to the question of truth; (2) disagreements about metaphysical questions related to the after-life, which are even less verifiable. The only category of difference that is "religiously important" (excluding all questions about what happened or what will happen, the nature of human beings and their environment): how one conceives, experiences and responds to the

---

[10] Ibid., 17.
[11] For Griffiths, the word "end" presents a useful pun: end as cessation of the process, or end as the culmination of a process that has achieved its purpose (ibid., 138).
[12] Alan Race, *Christians and Religious Pluralism: Patterns in the Christian Theology of Religions* (Maryknoll, NY: Orbis, 1982).
[13] Griffiths, *Problems of Religious Diversity*, 42.

"divine Reality."[14] Sure enough, contends Hick, religions make seemingly incompatible claims about the divine and the nature of the "proper end," yet in fact their incompatibility is only apparent. This is so, asserts Hick, because (following Immanuel Kant)[15] the differences are only at the phenomenological level (what we experience as humans) and not on the level of the "ultimately religiously Real." Hick is able to affirm this only because he has already defined "true" religion as "the transformation of human existence from self-centredness to Reality-centredness."[16]

Griffiths offers two main objections to this pluralistic view. First, by reducing all religious claims to a fundamental one, he dismisses the doctrines and teachings of particular religions and is therefore telling them that their understanding of their own religion is wrong. Second, he argues "that, in aggregate, the claims made by every actual religious community (the claims of Christians, of Buddhists, of Jews, of Muslims, and so forth) bear the same amount of truth—or at least this ought to be a working assumption."[17] Yet this can easily be contradicted by the facts: would he equally assent to the teachings of religious groups that have advocated murder, or mass suicide? No, Hick would reply, only those great "post-axial" religions such as Buddhism, Hinduism, Jainism, Judaism, Christianity and Islam count as "true" religions. For Griffiths, then, Hick is no longer fully pluralistic.

When it comes to truth, beyond the pluralistic view just examined, one might opt for exclusivism or inclusivism. Exclusivism is a view few—if any—religious groups have adopted: one would have to claim that nothing that is true of the teachings and doctrines of the home religion have ever been taught in other religions. In any case, this would not be a tenable position for either Islam or Christianity, which both trace their lineage back to Abraham and thus, more or less directly, through Judaism.[18]

The inclusive view with respect to truth admits the possibility that other religions may teach truth. Some would further conceive it possible that a particular alien religion may teach truths not found explicitly in the home religion (open inclusivism), and others would assert that "all alien

---

[14] See Griffiths's discussion on Hick in ibid., 40-44.

[15] Griffiths discusses Hick as a proponent of Kant's view of "parity with respect to truth" (pp. 37-44). He then looks at Wittgenstein's understanding of religious claims (pp. 45-50). Both sets of views support the pluralist hypothesis.

[16] John Hick, "On Conflicting Religious Truth-Claims," in *Problems of Religious Pluralism* (New York: St. Martins' Press, 1985), 89, quoted in Griffiths, *Problems of Religious Diversity*, 41.

[17] Griffiths, *Problems of Religious Diversity*, 44.

[18] Ibid., 53-56.

religious truths (should there be any) are already known to and explicitly taught by the home religion in some form" (closed inclusivism).[19] In this project I have adopted the position of inclusivism. In fact, the central argument of the present book is built on the explicit formulation of a qur'anic concept, the caliphate of humanity. I am not sure whether to call this "new truth" (thus making me an open inclusivist), because, as I will argue later, it so closely resembles the Genesis account of humanity as God's caretakers of creation. The biblical and qur'anic perspectives are more than just compatible; they are parallel and complementary.

More fundamentally, I begin with the premise that ultimate truth exists but that on this side of eternity we humans at best approximate our grasp of it. First, this is a statement about the nature and structure of being, or reality (ontology). We are back to Ian Barbour's statement of critical realism. Twentieth-century philosophy of science confirmed what Kant was groping for two centuries before: knowledge of this world is provisional, tentative, and approximate. We believe there is a physical world out there—let's call it Reality with an uppercase "r"—but our picture of it will never be an exact replica. This picture we may call "reality."[20]

Now we have transitioned from ontology to epistemology—a theory of knowledge, the second question that Griffiths proposed is involved in such a discussion. Reality transcends the physical world of our five senses; it must also include the values we humans cherish—justice, love, compassion, friendship, honesty and, above all, Truth. These ethical values are part of a universe that is more than just material. At the same time, we would go beyond Plato, Aristotle and the Greek philosophers, who simply posited a Supreme Principle that sustains all that is and provides the ground for its being. As monotheists, we would specify that this Supreme Principle is also a personal God. He is the one who guarantees that we, his creatures, can use our minds, explore the world he has graciously made available to us, and make statements about our discoveries that are true. Our minds do connect with Reality, albeit in a limited way. Moreover, we strive in our personal and social lives to fulfill an ethical ideal of peace, harmony and justice, and though our human systems can never achieve more than a pale reflection of that ideal, we nevertheless continue to forge ahead, believing that progress can be made.

[19] Ibid., 59. See Griffiths's discussion of inclusivism on the question of truth (pp. 56-64).
[20] This distinction was made by my anthropology/theology professor at Fuller Seminary, Charles H. Kraft. Much of his previous work in this area is summarized in his later work, *Anthropology for Christian Witness* (Maryknoll, NY: Orbis, 1996).

The idea of transcendent truth was, as we have seen, at the heart of the Enlightenment project (certainly at the Renaissance end of it). But it gradually became detached from the Creator who, for Jews, Christians and Muslims, was the original guarantor of truth. Now postmodernism (and those who foresaw it like Kant, Nietzsche, Sartre and others) seeks to draw out the conclusions of the Enlightenment in a consistent way. If humans are the measure of their knowledge, then indeed knowledge is itself very limited, because values are construed by persons in particular contexts. There are no metanarratives, no grand schemes of inevitable progress and development. My proposal here is that people of faith, and monotheists in particular, can learn from these insights, but without losing their connection to a transcendent God who simultaneously upholds Reality and provides guidance for those who wish to live out their calling as his deputies over his creation. Human reason, then, stands in need of revelation. Muslims look to the Qur'an, and Jews to the Torah, while Christians see God's revelation in the New Testament in addition to the Jewish Bible.

As stated above, my argument in this book rests on the assumption that all three faiths refer to the same Creator God who placed Adam and Eve on this planet he fashioned to be their home. As my discussion of hermeneutics sought to demonstrate, any reading of texts involves interpretation, which in turn is colored by a multiplicity of factors—from the personal to the influence of culture and specific sociopolitical issues. Since the topic of concern here is common to all—the nature of our human environment, physical and social—much can be learned from one another as we read our respective texts together and listen to our various perspectives. Openness is the necessary prerequisite for a mutual enriching, though at times it may feel unsettling. It takes courage, after all, to listen and entertain a thought that calls into question one of two of one's cherished formulas.

One of Griffiths's fine contributions, it seems to me, is his distinction between the question of religious truth claims and the question of salvation. It is in this second arena, however, that most discussions of exclusivism/inclusion/pluralism take place. In this context I am partly an exclusivist—along with the great majority of other religious people—and partly an inclusivist. Succinctly put, exclusivism here teaches that only one religion is true and that the others are ultimately false, thus tightly connecting truth, revelation, and salvation. As Griffiths puts it, "belonging to the home religion is necessary for salvation."[21] This point

---

[21] Griffiths, *Problems of Religious Diversity*, 150. Griffiths adds that "necessary" here is quite different from saying that belonging to the home religion is "sufficient"

likely represents the greatest source of discomfort for Clément in Ramadan's discourse. Exclusivism in Western postmodern society rings as bigoted intolerance. It is a position that will meet with resistance "for many late-modern inhabitants of forms of political and economic life that understand choice, preference, and tolerance as unsurpassable values."[22]

Griffiths tells us that among exclusivists there are two possible positions. Most conservative Christians, for example (evangelical and fundamentalists), maintain that God's emphasis on human freedom necessarily implies that most people will choose not to accept his gift of redemption offered in his son Jesus (restrictivists), while others, relying more on God's gracious design to save all, believe God will find a way to do so in the end (universalists—his position). But this is where I part with Griffiths. Apparently unbeknown to him, there is a middle way between restrictivists and universalists, which a growing number of evangelicals, including myself, take. As Canadian evangelical theologian Clark Pinnock shows, the sociological reality of religious pluralism over the last two or three decades in the West has forced more conservative theologians (and ordinary believers) to articulate a more intentional "theology of religions." In this regard, his argument for an "optimism of salvation" and a "hermeneutic of hopefulness" is worth considering.

As Christians stand before the biblical data in order to map out a response to the world's religions and the nations whence they have come, they do so through the filter of two different paradigms (in Barbour's

to salvation. That would be a statement on how one attains salvation. For those more familiar with Islamic history, it was the position of the Murji'ites ("those who postpone judgment") that led to the eventual Sunni "orthodox" position in the ninth and tenth centuries. They opposed three groups (though the first had virtually disappeared by then): the Kharijites who claimed that committing a "great" sin was enough to exclude a person from the Muslim fold (or lose one's salvation), the Shi'ites who maintained that salvation came only through submission to the charismatic leadership of the Prophet's family, and finally the Qadarites for whom free will meant that good works had to count a great deal in the balance of salvation. The general consensus since that time has been that to pronounce sincerely the *shahada* ("I testify that there is only One God and that Muhammad is his apostle") with the intention of following the other four pillars of Islam (prayer, Ramadan fast, zakat, and hajj) is sufficient for salvation.

[22] Ibid., 151. Griffiths cites two kinds of objections that are usually raised against exclusivism: (1) an epistemological one to the effect that such a claim could not be "verified" (i.e. by human reason or experience)—but that is circular reasoning, objects Griffiths, for what criterion could one use to support a "verifiable" refutation of that position in the first place?; (2) An ethical objection: this means that "vast numbers of people will not attain their proper ends" (p. 156).

sense). Pinnock calls these "control beliefs," by which he means "a large-scale conviction that affects many smaller issues."[23] But in fact they are paradigms (he also calls it a "megashift")—two very different set of assumptions with which one may come to the texts. Consider first of all the ruling paradigm since Augustine.[24] With his focus on God's election (predestination), Augustine set the course for a mindset of "fewness": God's salvific plan concerns only one nation originally (Israel) and then widens to include individuals in nations that embraced Christianity. With regard to the other nations, and especially other religions, the expectation is that only a few might come to a saving knowledge of Jesus Christ. With Cyprian also, the other famous North-African bishop, the slogan "outside the church, no salvation" begins to take hold. "Now it is not so much the passion to include as many as possible, but keep the standards of entry into salvation high."[25] The harsh attitude began to prevail that God, in his sovereignty, is perfectly justified in saving only a few, while sending the majority of humankind to hell. Jesus' statement that only a small number find the narrow gate (but that the road to hell is very wide indeed, Mt. 7:14) is interpreted as a paradigm for all times, instead of seeing that it might just be an observation about his own three-year ministry.[26]

The Greek Orthodox Church has historically held on to the optimism of salvation paradigm, while it is only in the last half century or so that it has made a return in the Western churches. There the "megashift" came first in the mainline Protestant churches (as seen in various declarations of the World Council of Churches), and then most dramatically in the Roman Catholic Church at the Vatican II Council (1962–65). The first of several pivotal documents issued during the Council was *Lumen Gentium* (The Light of the Nations). It acknowledged that Muslims worshiped God as Creator, "profess to hold the faith of Abraham, and together with us they adore the one, merciful God, mankind's judge on the last day."[27] And then, speaking more generally of Jews and Muslims:

---

[23] Clark H. Pinnock, *A Wideness in God's Mercy: The Finality of Jesus Christ in a World of Religions* (Grand Rapids, MI: Zondervan, 1992), 18.

[24] The first three centuries saw the dominance of the hermeneutic of hopefulness in the theology of the Greek fathers such as Justin Martyr, Clement of Alexandria, Origen, Theophilus of Antioch and others (ibid., 36-37).

[25] Ibid., 37.

[26] On this, see ibid., 154.

[27] Pinnock, as quoted Ataullah Saddiqi, *Christian–Muslim Dialogue in the Twentieth Century* (London: Macmillan Press; New York: St. Martin's Press, 1997), 36.

> Those who, through no fault of their own, do not know the Gospel of Christ or his Church, but who nevertheless seek God with a sincere heart, and, moved by Grace, try in their actions to do His will as they know it through the dictates of their conscience—these too may achieve eternal salvation.[28]

Notice that Islam and Judaism in this formulation are not as such "vehicles" of salvation. The Catholic position remains officially silent on this issue. Yet, through the Spirit of God, working to illumine people through many channels in culture, religion and their personal experiences and thoughts, people are led to follow the light they have received and find salvation.

As an evangelical, I can only make the paradigm shift through weighing what the scriptures say. I have long been amazed at what Pinnock calls the "pagan saints" of the Bible—people outside the covenant with the people of Israel whose faith is affirmed by God, like Abel, Enoch, Noah, Job, Melchisedek (a pagan priest to whom Abraham gives his tithe), Abimelech (a righteous Canaanite king who obviously fears the one God), Jethro (Moses' father-in-law and a Midianite priest), Balaam the soothsayer (who clearly hears God's words and passes them on), the Queen of Sheba, Naaman the Syrian general (who, once healed of his leprosy, is given permission to continue to worship at his pagan shrine) and the population of Nineveh that in its entirety repents at the preaching of Jonah.[29] Then in the New Testament we meet Mesopotamian astrologers who come to worship the baby Jesus and the Roman centurion Cornelius, who receives a vision that the apostle Peter is soon to visit (Acts 10). When Peter does pay him a visit (on the basis of another vision), he discovers him to be "a devout and God-fearing" man, who gives alms to the poor and prays regularly (v. 2). When Peter addresses the large crowd of Gentiles assembled in this man's home, he begins

---

[28] Quoted in ibid.

[29] Jesus specifically compares his prophetic preaching to that of Jonah in the polytheistic city of Nineveh, except that those people repented, whereas his contemporaries did not. He then declares to the crowd that "the Queen of Sheba will rise up against this generation on the day of judgment, because she came from a distant land to hear the wisdom of Solomon. And now someone greater than Solomon is here—and you refuse to listen to him." He goes on, "The people of Nineveh will also stand up against this generation on judgment day and condemn it, for they repented of their sins at the preaching of Jonah. Now someone greater than Jonah is here—but you refuse to repent" (Lk. 11:31-32, NLT; cf. Mt. 12:38-42). Clearly, both the Queen of Sheba and the Ninevites are people who, for Jesus, will enter paradise.

with the startling phrase, "I now realize how true it is that God does not show favoritism but accepts men from every nation who fear him and do what is right" (vv. 34-35).

I have no space to develop Pinnock's theology of religions here, but only hope to have shown that one's paradigm is crucial to one's position on people of other religions. So, following Pinnock and others, I am an exclusivist in my conviction that salvation comes only through the death and resurrection of Jesus, the Messiah and Son of God incarnate. But I am also an inclusivist in that I believe God's Spirit is actively involved in the lives of people from many religious backgrounds who have little or no knowledge of Jesus; and if they continue to respond in faith (trust) to God's tugging at their hearts and actively obey what they do know of God's will through their conscience and religious teaching, then they are on the path to God.

Furthermore, for those who, having been exposed to the gospel of Jesus, continue strongly to oppose it, who can possibly see into the future of anyone's life trajectory? As an evangelical theologian I take seriously the Reformation rediscovery of God's gracious and free gift of salvation, coupled with the invitation to make an explicit confession of faith (followed by baptism if at all possible), yet at the same time I must do justice to many biblical passages indicating that people will also be judged by their works (Mt. 25:31-46; Acts 10:35; Rom. 2:10-12; 1 Pet. 1:17; Rev. 20:12-13 and others). Add to that the gospel affirmation of God's justice (he is "no respecter of persons," that is, impartial), and I must confess to a strong hope that God's mercy and grace will in the end bring into his kingdom a vast and bounteous harvest, "a great multitude that no one can count from every nation, tribe, people and language" (Rev. 7:9)—especially so when thinking about individuals I know, whatever their religious belonging. God alone is Judge. What I do know, however, is that God's mercy draws from the infinite reservoir of his grace owing to Jesus' self-sacrifice on the cross—basically the official post-Vatican II Catholic position.

This particular combination of exclusivism and inclusivism promoted by Pinnock is refined and developed in a more Trinitarian way by Finnish theologian Veli-Matti Kärkkäinen, who came to Fuller Seminary in 2000.[30] To date, he has offered the most comprehensive analysis of Christian thinking on other religions. And he does so with the conviction that the parallel revival of interest in the doctrine of the Trinity (starting

---

[30] Pinnock cited other evangelicals who had pioneered the hermeneutic of hopefulness: C. S. Lewis, J. N. D. Anderson, Colin Chapman, Charles Kraft (also at Fuller), Stuart Hackett and John Sanders.

with Karl Barth in the 1940s) offers the most promise in this task.[31] Though he devotes a chapter to Pinnock in both of his recent books (*An Introduction to the Theology of Religions*, and *Trinity and Religious Pluralism*), he finds most affinity with Catholic thinker Gavin D'Costa, whose inclusivism is tied to his doctrine of the Trinity. In particular, the work of the Holy Spirit everywhere in the world "not only implies the presence of the triune God, but also the presence of the church."[32] He chooses thus to strongly endorse the recommendation of Vatican II documents for the churches to engage in dialogue with people of other faiths. Thus D'Costa enunciates two conclusions Kärkkäinen finds greatly useful: (1) though other religions are not salvific in themselves, they can play a key role in helping the church "penetrate more deeply into the divine mystery";[33] and further, (2) "Christianity may be 'fulfilled' by listening to what the Spirit is saying among other religions even if the Other does not bear the gift of God self-consciously."[34] Though Kärkkäinen may not agree with every detail of D'Costa's argumentation, he finds himself in sympathy with his general orientation. Note also that the present book reflects this kind of intentional engagement with Muslim thinking—interacting not only with the general tenets of the Islamic faith, but intentionally engaging with contemporary currents of Muslim interpretation. Indeed, the Spirit of God is at work in these circles, and I pray for the ability to discern and learn from these divine stirrings, through reading and personal conversations.

Finally, it is with regard to pluralism that I (along with Kärkkäinen—though I discovered him later) find D'Costa most helpful. On both the question of religious ends and epistemology, pluralism means that all the major religions are partly true, that no one in particular can claim final truth, and that salvation is equally possible, whichever route to God (however the divine is defined) one chooses to take. D'Costa, in the course of examining the position of the two most prominent advocates of this position, the Protestant John Cobb and Catholic Paul Knitter, avers

[31] Kärkkäinen has been prolific recently: *Pneumatology: The Holy Spirit in Ecumenical, International, and Contextual Perspectives* (Grand Rapids, MI: Baker Academic Books, 2002); *An Introduction to the Theology of Religions: Biblical, Historical, and Contemporary Perspectives* (Grand Rapids, MI: Baker Academic Books, 2003); *Trinity and Religious Pluralism: The Doctrine of the Trinity in Christian Theology of Religions* (Aldershot/Burlington, VT: Ashgate, 2004). Having surveyed the fields of Trinitarian studies and the theology of religions, he is now working on his own formulation of the integration of these two critical fields of Christian theology.

[32] Kärkkäinen, *Trinity and Religious Pluralism*, 69-70.

[33] Ibid., 72.

[34] Ibid., 73.

that pluralism is a distinctly "modern" theological construction of the concept of "god," which does violence to central beliefs of all the religions it seeks to pull together under one umbrella.[35]

This had also been the late Bishop Newbigin's contention in *The Gospel in a Pluralist Society*. For instance, he quotes Hick as defining "salvation" as "the transformation of human experience from self-centredness to God—or Reality-centredness."[36] What is reality, then? "Reality 'has no form except our knowledge of it.' Reality is unknowable," comments Newbigin. This neo-Kantian, postmodern epistemology now assumed by the pluralists is really a conversion of theology into anthropology, he argues—"a move about which perhaps the final word was spoken by Feuerbach who saw that the 'God' so conceived was simply the blown-up image of the self thrown up against the sky. It is the triumph of the self over reality... It is the authentic product of a consumer society."[37]

Many Catholics involved in Muslim–Christian dialogue (as in ecumenical circles)[38] assume that a necessary prerequisite for dialogue is to recognize in the other's faith a way of salvation (a step beyond Vatican II's inclusivist view). Ovey N. Mohammed, S.J., expresses this view in strong terms. Muslims see no point in dialoguing with Christians, he asserts, unless they recognize Islam as a way of salvation. Whereas Protestants have not been willing to say this, Catholics therefore will go further in dialogue: "The Roman Catholic understanding of mission as an invitation to faith through witness and proclamation follows from the church's teaching that the many religions can be possible ways of salvation and is in harmony with the Islamic understanding of mission."[39] He

---

[35] Gavin D'Costa, *The Meeting of Religions and the Trinity* (Maryknoll, NY: Orbis, 2000).

[36] John Hick and Paul K. Knitter, eds., *The Myth of Christian Uniqueness* (Maryknoll, NY: Orbis, 1987), 23, cited in Newbigin, *The Gospel in a Pluralist Society*, 78.

[37] Ibid., 169.

[38] I mean the dialogues organized by the World Council of Churches (WCC), the largest ecumenical body globally, but to which Roman Catholics and most evangelicals (as denominations) do not officially belong.

[39] Ovey N. Mohammed, S.J., *Muslim–Christian Relations: Past, Present, Future* (Maryknoll, NY: Orbis, 1999), 78. A similar view from the Muslim side comes from Temple University scholar Mahmoud Ayoub, who recently asserted that the greatest obstacle to Muslim–Christian dialogue is the reticence from both sides to admit that "both Christianity and Islam have in themselves the moral and spiritual resources to guide their followers to the way of salvation" ("Christian–Muslim Dialogue: Goals and Obstacles," in *The Muslim World* 94.3 [2004]: 315). For a comprehensive treatment of Muslim and Christian writings about each other, see Kate Zebiri, *Muslims*

then quotes approvingly Maurice Borrmans on a strategy of mission through dialogue, a strategy in which Muslims and Christians participate on an equal footing.

Here I would have to disagree. Of course, no dialogue can take place without all sides entering the discussion on the basis of equality. But this equal plane is not predicated on a particular theological position, like a judgment on the relative salvific value of other religions. Following Newbigin, Griffiths and D'Costa, I enter the dialogue as I am, as a Christian, with integrity and respect for the other at the same time. And I expect my Muslim brother or sister to do the same. I support Harold Vogelaar, from the Lutheran School of Theology in Chicago, when he writes that "very conservative roots, Christian or Muslim, can form a good basis for honest and fruitful interfaith dialogue."[40] Being conservative, however, does not mean one is not ready to learn from a person of another religion. Vogelaar bases his article on a quip he once heard from Bishop Kenneth Cragg in this context: "have, he said, an open door, an open hand, heart, mind, and, yes, an open creed."[41]

So far we have been dealing with scholars entering into predetermined—artificial, in some sense—venues for conversation. As Yale Church historian Lamin Sanneh rightly points out, Muslim and Christian communities have centuries of rich conversations and dialogue behind them in many parts of the world, and especially, Africa.[42] But this fact has been neglected in the recent Western literature on dialogue. "It has left dialogue vulnerable to popular mistrust, something specialists do, the fringe activity of a small minority that is itself on the fringes of its own religious tradition."[43] Yet dialogue should be about the shared life of communities living side by side. In this sense, dialogue is exclusive, inclusive and pluralistic all at the same time:

> Religions are exclusive in the sense of their unique, particular claims and their historical expression; they are also inclusive in the sense of those religious elements that find echoes in other traditions, and pluralist in terms of their own internal diversity and dynamism, but above all in meeting complex human needs and situations.[44]

and Christians Face to Face (Oxford: Oneworld, 1997). Zebiri cites Ayoub as one of several Muslim writers who adopt an irenic stance on Christianity (ibid., 162).
[40] Harold Vogelaar, "Open Doors to Dialogue," *The Muslim World* 94.3 (2004): 397.
[41] Ibid., 398.
[42] Cf. Chapter 4, n. 25.
[43] Lamin Sanneh, *Piety and Power: Muslims and Christians in West Africa* (Maryknoll, NY: Orbis, 1996), 6.
[44] Ibid. Sanneh continues his thought, demonstrating the necessity of widening the scope of dialogue: "Interfaith engagement proceeds by what is common to being

Having said that, the issue of proselytism, particularly in view of the history of Muslim–Christian relations in the modern period, must be faced head on. Both Christianity and Islam are faiths (unlike Judaism), which, at their core, impel their adherents to spread their beliefs to those around them with the hope of winning them over to their fold. In my years of residence in the Arab world I was often singled out as a promising convert to Islam—speaking fluent Arabic brought me half way there, I was told. And so I would patiently listen and do my best to answer, explaining why I found my Christian faith both reasonable and satisfying. Naturally, there were those whose tone became more aggressive. I remember at times feeling attacked in my own person, my own identity. I sympathize with Muslims who have felt this at the hand of Christians as well. John Borelli, a Catholic scholar in the forefront of Catholic–Muslim dialogue in the US, notes that in his experience, "the word 'mission' functions in the same way among Muslims as the word 'jihad' does among Christians."[45] But not only are polemical arguments that are carried on by both sides angry and divisive; they prevent Christians and Muslims from learning from one another.[46]

I am arguing that missionizing on both sides is normal, but one has to look at the conditions in which it takes place. Both Muslims and Christians have been guilty of exercising unethical pressure—the Muslims in the first centuries *vis-à-vis* the *ahl al-dhimma* (second-class protection status of Jews and Christians in Muslim-ruled lands), and Christians during the Crusades and the colonial period. Conversely, while the

religious as well as by the exigencies of complex human needs and situations, but most indisputably by what is unique and determinative of faith tradition. On those three grounds of common faith, common experience of life, and historical particularity, we have necessarily been involved with each other, whether by mutual foreclosure or by active solicitation. However we view religion, we cannot be content with one side talking just to itself about the other" (ibid.).

[45] John Borelli, "Christian–Muslim Relations in the United States: Reflections for the Future After Two Decades of Experience," *The Muslim World* 94.3 (2004): 327.

[46] Sanneh (*Piety and Power*) is a helpful source on Muslim–Christian relations in Africa. His research shows that Islam spread into West Africa mainly through trade, kings, and clerics (mostly of the Sufi type). The supremacy of the code in Islam prompted Ibn Khaldun to write that Christianity was not a true "missionary religion" because it did not believe in starting at the top and politically enforcing God's law among the people: "If the power of wrathfulness were no longer to exist in (man), he would lose the ability to help the truth become victorious. There would no longer be holy war or glorification of the word of God" (Ibn Khaldun as quoted in Sanneh, ibid., 10). *Jihad* was used in Africa, but much less than "the quietist way, which places emphasis on the structural aspects of establishing Islam, a reform process to which the name *tajdid* is given" (ibid., 17).

activities of Christian missions are well known, the activities of Muslim *da'wa* organizations are less so. Heribert Busse documents some of these efforts, naming such initiators as the Muslim World League, the Pakistani Society for the Islamic World Mission, the al-Azhar University, and the High Council for Islamic Affairs.[47]

A both timely and rich contribution to this discussion comes to us from Muhammad Hussein Fadlallah, the veteran Shi'ite leader of Lebanon—a land in which Muslims and Christians have lived alongside mostly in peace over the centuries. Fadlallah recently devoted a book to Muslim–Christian dialogue, *On the Horizons of Muslim–Christian Dialogue* (*Fī āfāq al-ḥiwār al-Islāmī al-masīḥī*)[48] and touches on the topic of proselyzing in several places. While "Christianity is a religion of evangelism (*tabshīr*)" and its adherents are called upon to "widen the circle of faith in person's life, acting in such a way as to bring into its fold the greatest number of people, whether they be pagans, Jews or Muslims," "Islam is the religion of *da'wa*," considering that "God's guiding any person to Islam represents the highest good in the love of God."[49] This, however, should not be a problem in itself, but in light of recent complaints from both sides in Lebanon and in particular due to the Muslim feeling that some evangelistic efforts on the Christian side may have political connections to the West (and its hegemonic designs in the world at large), these concerns must be discussed openly among the two communities. Yet Fadlallah remains optimistic. Muslims and Christians hold so much theological truth in common through the messengers of God from Abraham on, which can contribute a wealth of ideas and energy to the promotion of human dignity in our world.[50] Hence, the need to clear the air of misunderstandings related to proselytism on both sides so that these efforts can be pursued with high ethical standards and in harmony with common values. The most urgent condition, then, is an open exchange of views: "This issue must be one of objective dialogue in the arena of ideas—a process of competing, not of conflict."[51]

This issue was squarely faced three decades ago in a Muslim–Christian conference sponsored by the World Council of Churches (WCC) in 1976 in Chambésy, Switzerland. The document produced at the end stated the following:

[47] Heribert Busse, *Islam, Judaism and Christianity*, trans. Allison Brown (Princeton, NJ: Markus Wiener, 1998), 155-57.
[48] Muhammad Hussein Fadlallah, *On the Horizons of Muslim–Christian Dialogue* (*Fī āfāq al-ḥiwār al-Islāmī al-masīḥī*), 2d ed. (Beirut: Dār al-Malāk, 1998).
[49] Ibid., 110.
[50] See especially ibid., 456-57.
[51] Ibid., 111.

> The conference upholds the principle of religious freedom recognizing
> that the Muslims as well as the Christians must enjoy the full liberty to
> convince and be convinced... The conference recognizes that mission and
> *da'wah* are essential religious duties of both Christianity and Islam.[52]

The statement also explains why Muslims are still reluctant to cooperate with Christians. The general Muslim feeling is that Christians continue to offer missionary services of relief and development with shameful ulterior motives:

> Taking advantage of Muslim ignorance, of Muslim need for educational,
> health, cultural and social services, of Muslim political stresses and
> crises, of their economic dependence, political division and general weak-
> ness and vulnerability, these missionary services have served purposes
> other than holy—proselytism, that is, adding members to the Christian
> community for reasons other than spiritual.[53]

Indeed, if Christian mission leads Christians to rob Muslims of their human dignity then its actions are morally reprehensible. That this has been the case is certainly is the perception of the Muslim community. While this judgment may not apply to all Christian agencies in this kind of setting, I feel that with the backdrop of age-old grievances, we as Christians must go the extra mile to dispel this deep-seated suspicion.

That was precisely the step taken by the Chambésy conference. In view of the past abuse of service, it strongly urged "Christian churches and religious organizations to suspend their *diakonia* activities in the world of Islam."[54] This was necessary to clear the air, at least till the time came when mission could be re-established in a way that would be acceptable to both parties.[55] This study, in essence, is a proposal for a

---

[52] International Review of Mission, "Statement of the Conference on 'Christian Mission and Islamic *Da'wah*,' Chambésy, June 1976," *International Review of Mission* 65.260 (1976): 458-59.

[53] Ibid., 459.

[54] Ibid.

[55] I agree with Tunisian scholar and statesman Mohammad Talbi when he writes, "The surest means of making it impossible for it ever to renew the immense damage done in the past and the sins committed against reason, is to renounce any idea of using dialogue, either openly or in one's mind, as a means of converting the person we are talking to. If, in fact, dialogue is conceived as a new form of proselytism, a means of undermining convictions and bringing about defeat or surrender, sooner or later we shall find ourselves back in the same old situation as in the Middle Ages. It will merely have been a change in tactics" ("Islam and Dialogue: Some Reflections on a Current Topic," in *Christianity and Islam: The Struggling Dialogue*, ed. Richard W. Rousseau, S.J., Modern Theological Themes: Selections from the Literature, 4 [Scranton, PA: Ridge Row, 1985], 59).

different basis of mutual witnessing—a chance to put our faith into practice side by side and to proclaim to an increasingly secularized global society by word and deed that human life is infinitely precious in the eyes of our Creator.[56]

One of the participants at the Chambésy conference was Bishop Cragg. Though he is still alive and writing, it certainly is not too soon to speak of Cragg's legacy.[57] I mention him, not because of the sheer volume of his works over the years, nor because of the wide acclaim his solid erudition has earned for him, but because of the irenic and respectful spirit he has consistently maintained, with the pen and in person, in his discussion of Islam.[58]

His opening paper at Chambésy exemplifies the approach and spirit of what I am advocating and trying to accomplish here. Its title is "Contemporary *Takbir*: Muslim and Christian." In it, he attempts to develop the praise of God's greatness, or the *takbīr* (saying, "God is greater," *allāhu akbar*),[59] both in the Muslim and the Christian context. I will single out six principles worthy of emulation:

---

[56] Here I would agree with Maurice Borrmans, Catholic veteran of Muslim–Christian dialogue, when he writes, "Partisans of 'modernity' have henceforth the possibility of continuing to evolve or, instead, of degenerating into sterile uniformity. Should not believers all together proclaim to them, in relevant terms, the marvels of the cosmos, the dignity of humankind and the greatness of God? Many see here, with good reason, the threefold perspective of any true dialogue" (*Guidelines for Dialogue between Christians and Muslims*, trans. R. Marston Speight [Mahwah, NJ: Paulist Press, 1990], 26).

[57] This point is well made by Christopher Lamb in his book *The Call to Retrieval: Kenneth Cragg's Christian Vocation to Islam* (London: Grey Seal, 1997).

[58] Farid Esack, in his article, "Qur'anic Hermeneutics: Problems and Prospects," pays tribute to Cragg. Commenting on the difficulty of elucidating the relationship between the Qur'an and history, he brings Cragg into the discussion: "The most disturbing of those who have consistently challenged Muslim reluctance to confront these connections in our time is undoubtedly Kenneth Cragg. I say 'disturbingly' because he writes with a profundity and compassion that makes it difficult to dismiss him as 'just another orientalist'. Cragg's *The Event of the Qur'an*, to my mind, remains the most profound and moving account of the Qur'an's engagement with a living and dynamic context" (p. 119). Another liberal Muslim who has expressed appreciation for Cragg is Fazlur Rahman. In his introduction to *Major Themes of the Qur'an*, Rahman had noted about Cragg's *The Event of the Qur'an* and *The Mind of the Qur'an*, that his was "an extraordinarily sensitive response to Islamic scripture by a Christian" (p. xv).

[59] This is the initial phrase in the Muslim call to prayer, proclaimed five times daily from the minaret of every mosque in the world.

1. Cragg selects common themes: here the *takbīr*, which is obviously central to Muslim theology and practice, but which is very present as a driving force in Christian thought and spirituality as well.

2. He presents a sympathetic and respectful view of the Muslim position: without going into detail he puts forth a penetrating interpretation of the "God is greater" for both religions in three spheres (the witness of faith, the criterion of religion, and the crisis of society). Cragg has devoted his energy and heart to listening to Muslims speak of their faith.

3. He emphasizes the common ground: for instance, the stress on the necessity of the *takbīr* both in light of humanity's temptation to pride itself in its mastery of the natural order and its proneness to fear and doubt. On both sides theology underpins the human experience at a deep level.

4. Cragg does not shrink from a lucid analysis of the differences; using Mary's *Magnificat* as the Christian paradigm most akin to the Muslim *takbīr*, Cragg has this to say:

> There is a sense in which we cannot realize our common ground unless we discover the ground we do not share. If our old estrangements were mutually insensitive, our new alignments may be insensitively facile. The crucial question has to do with the nature of the "greatness" we affirm. *Takbir* and *Magnificat* may be verbally one, within their Arabic and Latin roots, and in their instinct complementary. But the criteria differ deeply by which we comprehend the concept they proclaim. The Christian must conclude that while our *Takbir* may, and must, be made in common, it must also be make apart.[60]

5. He recognizes the past faults and failures of the Christian movement: speaking of the temptation of people of faith to entrench themselves into the power and security of their institutions, he quotes T. S. Elliot in *Murder of the Cathedral*, a work which clearly indicts clergymen who in playing with political power loose their spiritual calling.[61]

---

[60] Kenneth Cragg, "'Greater is God': Contemporary *Takbir*: Muslim and Christian," *The Muslim World* 71 (1981): 38.

[61] Ibid., 33. In this article I do not believe Cragg is going far enough in this direction. There is a discernible imbalance in the challenges presented: they are mostly for the Muslims. Yet on this point past examples of Christian error abound, from the absolutism of the Byzantine Empire to the unholy alliance of Church and State in the colonial period.

6. He sends a challenge to Muslims: one of several challenges in this article is for Muslims to face squarely the interrogations of the contemporary world. By doing so, he aligns himself with a number of Muslim intellectuals who in the last three centuries, at least, have been calling their coreligionists to the same task. In the following passage he decries a "fortress" mentality, along with a host of "progressive" Muslims:

> There can be few attitudes more blasphemous than crying *Allahu akbar* with a clenched fist. For then we clearly mean: Down with our enemies. Religious faith, even piety in practice, may be a sadly inverted assertion of ourselves. Religious loyalty then becomes corporate enmity, and acrimony the stuff of our devotion. The rule of God has then become the rule of pride.[62]

As a more recent example of Muslim–Christian dialogue, the Inter-Religious Federation for World Peace (IRFWP), the Council of the World's Religions (CWR) and the International Religious Foundation (IRF) (and two colleges, the University of Karachi and Renison College in Canada) came together for the Renison Conference in June, 1995. In one of the essays presented, "Islamic–Christian Dialogue: A Muslim View," 'Izz al-Din Ibrahim offers the following as an important function and goal of dialogue—a view closely related to the present work and in line with Fadlallah above:

> A major issue in Islamic–Christian dialogue, and in fact in Islamic–Christian cooperation, is that adherents of the two religions should help one another in resisting the evils, crimes and wickedness which manifest in all places—unbelief, social injustice, unbridled materialism, moral corruption…the disregard of human rights, and the abuse of the environment on earth and space. These and their like are the great evils which confront humanity and which must be resisted by those who hear God's word in the Torah, the Gospel and the Qur'an, working together in mutual support."[63]

In the same volume, Abdullah Nooredeen Durkee, after showing that in his estimation no amount of discussion could reduce the gap between the "non-negociables" in Christianity and Islam, centers his presentation on the following qur'anic exhortation, "Vie together in good works" (Q. 3:114). He explains:

---

[62] Ibid., 36.
[63] 'Izz al-Din Ibrahim, "Islamic–Christian Dialogue: A Muslim View," in *Muslim–Christian Dialogue: Promise and Problems*, ed. M. Darrol Bryant and S. A. Ali (St. Paul, MN: Paragon House, 1998), 23. This volume assembled all the conference papers.

> Surely if Muslims, Christians and Jews worked out their various differ-
> ences by way of competing with one another in the carrying out of good
> works, everyone would benefit. Surely there can be no doubt that such a
> way of relating to one another would go a long way to alleviating, if only
> in a small way, so much of the oppression and suffering evident every-
> where around us."[64]

Competition in the field of good works rings as a familiar note to the Christian. Jesus exhorted his disciples in these words, "You are the light of the world... In the same way, let your light shine before men, that they may see your good deeds and glorify your Father in heaven."[65] However, it would be disingenuous to divorce good works from the call to witness to the truth of one's faith. Jesus declared that he was "the light of the world" (John 8:12), and one cannot be light—"glorifying the Father"—without lifting up the Son in one's words *and* deeds. In a similar manner, the Qur'an calls those who receive God's message through His apostle Muhammad to invite others to the truth: "Invite [from the root of the word *da'wa*] (all) to the way of thy Lord with wisdom and beautiful preaching; and argue with them in ways that are best and most gracious: for thy Lord knoweth best, who have strayed from His Path, and who receive guidance" (Q. 16:125). Witness—whether one labels it *da'wa* or mission—is a non-negotiable, I would argue, for Muslims and Christians. Thus, to come back to the beginning of this section, I find nothing offensive in Ramadan's exclusivism. Furthermore, I firmly believe (as he does, I am sure) that it is only God who converts. Our task is to sow the seed (to use our previous symbolism), and to make God's message known to others in a way that reflects the dignity of each person encountered. What people do with our message is between them and God. At the same time, we expect a bountiful harvest—God's mercy and grace are beyond our scrutiny.

---

[64] Abdullah Nooredeen Durkee, "Personal Thoughts about Muslim–Christian Dia-logue," in Bryant and Ali, eds., *Muslim–Christian Dialogue*, 118. Over twenty years ago, a World Conference on Religion and Peace (1970) took place in Kyoto, which welcomed delegates from very diverse faiths: Bahai, Buddhist, Confucian, Christian, Hindu, Jain, Jew, Muslim Shintoist, Sikh, Zorostrian. One of the findings of the conference was that the issues that united them were by far more important than the ones that divided them. The greatest common value was "a conviction of the funda-mental unity of the human family and the equality and dignity of all human beings" ("The Findings of the World Conference on Religion and World Peace, Held in Kyoto, Japan, from October 16 to 21, 1970," in *Islam and the Modern Age* 2.1 [1971]: 1).

[65] Mt. 5:14, 16 (NIV).

Ted Peters calls this view the "confessional universalist position," quite similar, in fact, to Griffiths's "exclusivist universalist position." It is a clear stance that articulates for anyone who questions the claims of the Christian faith, yet who at the same time seeks to listen and learn from the faith of others. He explains:

> It is confessional because it takes a stand regarding the gospel that has been borne through history by only one religious tradition, namely, Christianity. It is universal because its claims are ultimate—that is, they are thought to be valid for all people of all times and all places. In short, Christians have something to say. But in dialogue, they also listen.[66]

Perhaps this appendix on pluralism is best summarized by Peters's four conditions (theoretical and at the same time very practical), which he insists must be present, if interreligious dialogue is to be fruitful:

1. Each party has a position to put forth, in the true spirit of pluralism—"normative pluralism," as in the writing of sociologist Said Amir Arjomand,[67] or Habermas's discursive action. Commitment to one position is a necessary first step to dialogue.

2. An attitude of openness which seriously considers that the other traditions do have valid points to make, and that one still has truth to discover—a corollary of critical realism. The human mind does connect with reality and does apprehend truth, but

---

[66] Ted Peters, *God—The World's Future: Systematic Theology for a New Era*, 2d ed. (Minneapolis, MN: Augsburg Press, 2000), 353. This is a classic work on systematic theology with a twist: having covered the traditional foci of Christian theology in a trinitarian fashion (Father: creation; Son: redemption; Spirit: Church and eschatology), he adds "Ecumenic Pluralism" (Chapter 11, his theology of world religions) and "Proleptic Ethics" (Chapter 12).

[67] Arjomand writes that "it would be a grave mistake to assume, as is commonly done, that the pluralism of normative orders is the exclusive characteristic of the modern West." All the great civilizations of the "Axial Age" (500 BCE and on)— including the Islamic civilization of medieval times—had to deal in a practical way with competing heterodoxies and rival visions of salvation in their midst. In all these settings "the public representation of transcendent truth itself is a matter of contention within each religiocultural tradition." He goes on to apply this to our current context: "The very fact of this normative pluralism itself, I argue, has normative consequences. Modernity, I contend, should consist in the explicit acknowledgment of the diversity of moral orders and its elevation to a normative principle" (S. A. Arjomand, ed., *The Political Dimensions of Religion* [Albany, NY: SUNY Press, 1993], 4). To sum up my own position, I advocate normative pluralism (the whole point of Habermas's communicative action and the democratic project in general); but I reject theological pluralism.

always in an approximate and tentative fashion.[68] Each party
enters the conversation fully expecting to be enriched and to
enrich.

3. Dialogue, if it is genuine, requires the virtue of love. This means
   a will to trust the other's best intentions and to act and speak
   with the purpose of enriching all the parties involved in an
   atmosphere of enjoyment and growing unity.

Also needed is a commitment to invest the necessary time and effort in
order to do justice to the complex issues involved. There is no magic
wand to wave here—genuine dialogue will require hard work and the
will to struggle through misunderstandings, hurt feelings at times, and
just plain complexity.[69]

---

[68] Peters indicates that attitude is all-important. Dialogue cannot be conducted on the
model of a labor-management negotiation in which two adversaries assume that a
finite pie of wealth is going to be cut up. Inasmuch as one party wins, the other loses.
"Dialogue, in contrast, is not adversarial. Here, ironically enough, losing could be
winning. The spiritual pie is infinite in the wealth it offers the human soul. To lose—
which consists in giving up some aspect of one's position because a new and better
insight has come to replace it—results in a net gain of knowledge and understanding
and perhaps even a strengthening of faith" (*God—The World's Future*, 353-54).

[69] Peters notes that the prefix "*dia*" in dialogue ("through" or "throughout") is not the
"*di*" that is often assumed ("two"). He asks, "Could we think of a dialogue as a
conversation in which we talk a subject through, in which we exhaust its details and
nuances and implications and draw out its full significance?" (ibid., 354).

BIBLIOGRAPHY

Abbott, Freeland. "The Decline of the Mugul Empire and Shah Waliullah." *The Muslim World* 52 (1962): 115-23.

Abderraziq, Ali. *al-Islām wa-uṣūl al-ḥukm: baḥth fī al-khilāfa wa-l-ḥukūma fī al-Islām* (Islam and the Foundations of Political Rule: Research on the Caliphate and Government in Islam). Edited with comments by Mamdouh Haqqi. Beirut: Dār Maktabat al-Hayāh, 1966.

_____. *L'islam et les fondements du pouvoir*. New translation and Introduction by Abdou Filali-Ansary. Paris: Éditions La Découverte, 1994.

Abou El Fadl, Khaled. *Rebellion and Violence in Islamic Law*. New York: Cambridge University Press, 2001.

_____. *Speaking in God's Name: Islamic Law, Authority and Women* Oxford: Oneworld, 2001.

_____. "The Ugly Modern and the Modern Ugly: Reclaiming the Beautiful in Islam." In Safi, ed., *Progressive Muslims: On Justice, Gender and Pluralism*, 33-77.

Abu-Nimer, Mohamed. *Nonviolence and Peacebuilding in Islam: Theory and Practice*. Gainesville: University of Florida Press, 2003.

Abu-Nimer, Mohamed, and David Augsburger, eds., *Peace by, between, and beyond Muslims and Evangelical Christians*. Lanham, MD: Lexington Books, 2009.

AbuSulayman, AbdulHamid. "Islamization of Knowledge with Special Reference to Political Science." *American Journal of Islamic Social Sciences* 2.2 (1985): 263-89.

Adams, A. K. M. "Postmodern Biblical Interpretation." In Hayes, ed., *Methods of Biblical Interpretation*, 305-309.

Adams, Charles J. "Mawdudi and the Islamic State." In Esposito, ed., *Voices of Resurgent Islam*, 99-133.

Adul-Ghafur, Saleemah, ed. *Living Islam Out Loud: American Muslim Women Speak*. Boston: Beacon Press, 2005.

Ahmad, Aziz. "Political and Religious Ideas of Shah Wali-Ullah of Delhi." *The Muslim World* 52 (1962): 22-30.

Ahmed, Akbar S. "Ibn Khaldun's Understanding of Civilizations and the Dilemmas of Islam and the West Today." *Middle East Journal* 56.1 (Winter 2002): 20-45.

_____. *Postmodernism and Islam: Predicament and Promise*. London: Routledge, 1992.

_____. *Islam Today: A Short Introduction to the Muslim World*. London: I.B. Tauris, 1999.

_____. *Islam Under Siege: Living Dangerously in a Post-Honor World*. Cambridge: Polity Press, 2003.

Akbar, M. J. *The Shade of Swords: Jihad and the Conflict between Islam and Christianity*. London/New York: Routledge, 2002.

Akhtar, Shabbir. "Critical Qur'anic Scholarship and Theological Puzzles." In Vroom and Gort, eds., *Holy Scriptures and Judaism*, 122-27.

_____. *A Faith for All Seasons: Islam and the Challenge of the Modern World*. Chicago: Ivan R. Dee, 1990.

_____. "The Limits of Internal Hermeneutics: The Status of the Qur'an as Literary Miracle." In Vroom and Gort, eds., *Holy Scriptures and Judaism*, 107-12.

Allouche, Adel. "God's Caliph." (review article) *Muslim World* 79.1 (1989): 71-74.

Alverson, Keith, et al. *Environmental Variability and Climate Change*. International Geosphere-Biosphere Program Science Series, 3, 2001.

Amin, Samir. *Eurocentrism*. Translated by Russell Moore. New York: Monthly Review Press, 1989.

Anawati, G. C. "Fakh al-Dīn al-Rāzī." In van Donzell et al., eds., *The Encyclopaedia of Islam*, vol. 8:751-55.

Anderson, Bernhard W. *From Creation to New Creation: Old Testament Perspectives*. Ouvertures to Biblical Theology. Minneapolis, MN: Fortress Press, 1994.

Anderson, Ray S. *On Being Human: Essays on Biblical Anthropology*. Pasadena, CA: Fuller Seminary Press, 1982.

Appleby, R. Scott. *The Ambivalence of the Sacred: Religion, Violence and Reconciliation*. Lanham, MD: Rowman & Littlefield, 2000.

Arendt, Hanna. *Imperialism*. New York: Harcourt Brace Yanovich, 1968.

Arjomand, Said Amir, ed. *The Political Dimensions of Religion*. Albany, NY: SUNY Press, 1993.

Arkoun, Mohammed. "Contemporary Critical Practices and the Qur'an." In McAuliffe, ed., *Encyclopaedia of the Qur'an*, vol. 1:412-31.

―――. *Penser L'Islam Aujourd'hui*. Algiers: Éditions Laphonic ENAL, 1993.

_____. "Peut-on parler d'humanisme en contexte islamique?" In *Compilation and Creation in Adab and Lugha: Studies in Memory of Naphtali Kingerg (1948-1997)*, 11-22. Israel Oriental Studies. Tel Aviv: Eisenbrauns, 1999.

_____. "Religion and Society: The Example of Islam." In Cohn-Sherbok, ed., *Islam in a World of Diverse Faiths*, 134-77.

Arnaldez, Roger. *Fakhr al-Dīn al-Rāzī: Commentateur du Coran et philosophe*, Paris: Librairie Philosophique J. Vrin, 2002.

Arnold, T. W. "*Khalīfa*." *E. J. Brill's First Encyclopaedia of Islam*. Edited by M. T. Houtsma et al., 881-5. Leiden: E. J. Brill, 1987 [1954].

_____. *The Preaching of Islam: A History of the Propagations of the Muslim Faith*. 2d ed. New York: Scribner's Sons, 1913.

Asad, Talal. *Genealogies of Religion: Discipline and Reasons of Power in Christianity and Islam*. Baltimore: The Johns Hopkins University Press, 1993.

Ashworth, E. J. "Philosophy of Language." In Craig et al., eds., *Routledge Encyclopedia of Philosophy*, vol. 6:408-15.

Askari, Hasan. "Religion and State." In Cohn-Sherbok, ed., *Islam in a World of Diverse Faiths*.

Ateek, Naim S. *Justice and Only Justice: A Palestinian Theology of Liberation*. Maryknoll, NY: Orbis, 1989.

Austin, J. L. *How to Do Things with Words*. Cambridge, MA: Harvard University Press, 1962.

Aylmer, G. E., ed. *The Levellers in the English Revolution*. London: Thames and Hudson, 1975.

Ayoub, Mahmoud M. *The Qur'an and Its Interpreters*. Albany, NY: State University of New York Press, 1984.

Azad, Jagan Nath. "Iqbal, Islam and the Modern Age." *Islam and the Modern Age* 9.1 (1978): 35-66.

Badawi, Abdurrahman. "L'humanisme dans la pensée arabe." *Studia Islamica* 6 (1956): 67-100.

Badawi, Jamal. "The Earth and Humanity: A Muslim View." In *Three Faiths—One God: A Jewish, Christian, Muslim Encounter*. Edited by John Hick and Edmund S. Meltzer. Albany: State University of New York Press, 1989.

_____. "Relief and Development: An Islamic Approach." Paper presented at the conference "Muslim and Christian NGOs: A Dialogue on Relief, Development and Cooperation." Ottawa, Canada, June 17–18, 1996.

Al-Baghdadi, Abd al-Qahir b. Tahir. *Kitāb uṣūl al-dīn*. Istambul: Madrasat al-Ilāhiyyāt bi Dār al-Funūn al-Turkiyya, 1928.

Bahya b. Asher, "*Kad ha-Qemaḥ*." In *Kitve Rabbenu Baḥya*. Edited by C. B. Chavel, 3 vols. Jerusalem: Mosad Ha-Rav Kuk, 1969.

al-Baidawi, Nasir al-Din. *Anwār al-tanzīl wa-asrār al-taʾwīl* (The Lights of Revelation and the Mysteries of Interpretation). Edited by Mahmoud Abd al-Qadir al-Arna'ut, 2 vol. Beirut: Dār Sādir, 2001.

Bakker, Dirk. *Man in the Qur'an*. Amsterdam: Drukkerij Holland N. V., 1965.

Baldwin, James. *The Price of the Ticket: Collected Nonfiction, 1948–1985*. New York: St. Martin's Press, 1985.

Barazangi, Nimat Hafez. "Vicegerency and Gender Justice in Islam." In Barazangi, Zaman and Afzal, eds., *Islamic Identity and the Struggle for Justice*, 77-94.

Barazangi, Nimat Hafez, Raquibus Zaman and Omar Afzal, eds. *Islamic Identity and the Struggle for Justice*. Gainesville: University Press of Florida, 1996.

Barber, Benjamin. *Jihad Vs. McWorld: Terrorism's Challenge to Democracy*. New York: Ballantine Books, 1995, with a new introduction, 2001.

Barbour, Ian G. *Myths, Models and Paradigms: The Nature of Scientific and Religious Language*. London: SMC Press Ltd, 1974.

_____. *Religion and Science: Historical and Contemporary Issues*, A Revised and Expanded Edition of *Religion in an Age of Science* [1990]. New York: HarperSanFrancisco, 1997.

Barboza, David and Andrew Ross Sorkin, "Chinese Company Drops Bid to Buy U.S. Oil Concern," *New York Times* (August 3, 2005): A1 and C4.

Barraclough, Steven. "Al-Azhar: Between the Government and the Islamists." *Middle East Journal* 52.2 (Spring 1998): 236-49.

Barton, John and John Muddiman, eds. *The Oxford Bible Commentary*. Oxford/New York: Oxford University Press, 2001.

Bat Ye'or, *Decline of Eastern Christianity under Islam: From Jihad to Dhimmitude: Seventh Century*. Trans. from the French by Miriam Kochan and David Littman. Madison, NJ: Farleigh Dickinson University Press, 1996.

———. *Islam and Dhimmitude: Where Civilizations Collide*. Trans. from the French Miriam Kochan and David Littman. Madison, NJ: Farleigh Dickinson University Press, 2002.

Baudrillard, Jean. *Le miroir de la production*. Paris: Casterman, Tournail, 1973.

_____. *La société de consommation*. Paris: Gallimard, 1970.

_____. *Le système des objets*. Paris: Gallimard, 1968.

BBC News, World Edition Online. "Life 'Worse for World's Poorest.'" (September 7,2005): http://news.bbc.co.uk/1/hi/world/americas/4222034.stm (last accessed September 7, 2005).

Beal, T. K., K. A. Keefer, and T. Linafelt. "Literary Theory, Literary Criticism, and the Bible." In Hayes, ed., *Methods of Biblical Interpretation*, 159-67.

Beauchamp, Paul. *Psaumes nuit et jour*. Paris: Éditions du Seuil, 1980.

Beck, Ulrich. *Risk Society: Toward a New Modernity*. London: Sage, 1992.

Beckman, David, and Arthur Simon, *Grace at the Table: Ending Hunger in God's World*. Downers Grove, IL: InterVarsity Press, 1999.

Benhabib, Seyla. *The Claims of Culture: Equality and Diversity in the Global Era*. Princeton, NJ and Oxford: Princeton University Press, 2002.

_____. *Situating the Self: Gender, Community and Postmodernism in Contemporary Ethics*, New York: Routledge, 1992.

Bennett, Clinton. *In Search of Muhammad*. London/New York: Cassell, 1998.

Bettelheim, Bruno. *The Uses of Enchantment: The Meaning and Importance of Fairy Tales*. New York: Random House, Vintage Books, 1977.

Bint al-Shati (Aisha Abd al-Rahman). *Al-Qurʾān wa-qaḍāyā al-insān* (The Qur'an and Human Affairs). Beirut: Dār al-ʿIlm li-l-Malāyīn, 1982.

_____. *Al-Tafsīr al-bayānī li-l-Qurʾān al-karīm* (The Explanatory Commentary on the Wise Qur'an). Maktabāt al-dirāsāt al-adabiyya. 2 vols. Cairo: Dār al-Maʿārif, 1982.

Bohman, James. "Two Versions of the Linguistic Turn: Habermas and Poststructuralism." In d'Entrèves and Benhabib, eds., *Habermas and the Unfinished Project of Modernity*, 197-220.

The Book of Concord: The Confessions of the Lutheran Church. Edited and translated by Theodore G. Tappert. Philadelphia: Fortress Press, 1959.

Borelli, John. "Christian–Muslim Relations in the United States: Reflections for the Future After Two Decades of Experience." *The Muslim World* 94.3 (2004): 321-33.

Borrmans, Maurice. *Guidelines for Dialogue between Christians and Muslims*. Translated by R. Marston Speight. Mahwah, NJ: Paulist Press, 1990.

Bouguerra, Mohamed Larbi. "Au service des peuples ou d'un impérialisme économique?" ("In the Service of Peoples or of an Economic Imperialism?"). *Le Monde Diplomatique*, Manière de Voir 50 (March–April 2000): 12-14.

Boullata, Issa J. "Modern Qur'an Exegesis: A Study of Bint Al-Shati's Method." *The Muslim World* 64 (1974): 103-13.

Bové, José, and Francois Dufour. *The World Is not for Sale: Farmers Against Junk Food*. London/New York: Verso, 2001.

Böwering, Gerhard. "Chronology and the Qur'an." In McAuliffe, ed., *Encyclopaedia of the Qur'an*, vol. 1:316-35.

_____. *The Mystical Vision of Existence in Classical Islam: The Qur'anic Hermeneutics of the Sufi Sahl At-Tustari (d. 283/896)*. Studien zur Sprache, Geschichte und Kultur des Islamischen Orients. Berlin/New York: W. de Gruyter, 1980.

Brockopp, Jonathan. "Islam." In Neusner, ed., *Sacred Texts and Authority*, 31-59.

Brown, Chris. *International Relations Theory: New Normative Approaches*. New York: Columbia University Press, 1992.

Brown, Daniel. *Rethinking Tradition in Modern Islamic Thought*. Cambridge Middle East Studies. Cambridge: Cambridge University Press, 1996.

Brown, Stephen. "World Council of Churches Gives Nod to the Israeli Divestment Proposal." *Ecumenical News International* (February 21, 2005): http://www.episcopalchurch.org/3577 58769 ENG HTM.htm (last accessed July 15, 2005).

Brown, Tricia Gates. *Getting in the Way: Stories from the Work of the Christian Peacemaker Teams*. Scottdale, PA: Herald Press, 2005.

Brueggemann, Walter. *Texts Under Negotiation: The Bible and Postmodern Imagination.* Minneapolis, MN: Fortress Press, 1993.

Bryant, M. Darrol, and S. A. Ali, eds. *Muslim–Christian Dialogue: Promise and Problems.* St. Paul, MN: Paragon House, 1998.

Bulliet, Richard W. *The Case for Islamo-Christian Civilization.* New York: Columbia University Press, 2004.

Burbach, Roger. *Globalization and Postmodern Politics: From Zapatistas to High Tech Robber Barons.* London/Sterling, VA: Pluto; Kingston: Arawak, 2001.

Burns, Stewart. *To the Mountaintop: Martin Luther King Jr.'s Mission to Save America: 1955–1968.* New York: HarperSanFrancisco, 2004.

Busse, Heribert. *Islam, Judaism and Christianity.* Translated by Allison Brown. Princeton, NJ: Markus Wiener, 1998.

al-Bustani, Abdallah. *Al-Bustān: Muʿajam Lughawī Muṭawwal.* Beirut: Maktabat al-Lubnān, 1992.

Butler, Judith. *Gender Trouble: Feminism and the Subversion of Identity.* New York: Routledge, 1990.

Caputo, John D. "Circum., Circumfession: Fifty-Periods and Periphrases." In *Jacques Derrida.* Edited by Geoffrey Bennington and Jacques Derrida. Translated by Geoffrey Bennington. Chicago: University of Chicago Press, 1993.

_____. "Toward a Postmodern Theology of the Cross." In *Postmodern Philosophy and Christian Thought.* Edited by Merold Westphal, 202-25. Bloomington, IN: Indiana University Press, 1999.

Carr, Joe. "The Weapons of the Muslim Peacemaker Teams." *CPTnet* (June 21, 2005): http://www.cpt.org/archives/2005/jun05/0027.html.

Carré, Olivier. *Mystique et politique: lecture révolutionaire du Coran par Sayyid Qutb, frère musulman radical* (Mysticism and Politics: A Revolutionary Reading of the Qur'an by Sayyid Qutb, Radical Muslim Brother). Paris: Les Éditions du Cerf, 1984.

Carter, Jimmy. *Palestine: Peace not Apartheid.* New York: Simon & Schuster, 2007.

Chambers, Robert. *Whose Reality Counts? Putting the Last First.* London: Intermediate Technology Publications, 1997.

Chiesa Online. "Tariq Ramadan's Two-faced Islam: The West Is the Land of Conquest" (January 19, 2004): http://213.92.16.98/ESW_articolo/ 0,2393,42025,00.html (last accessed June 12, 2005).

Chittick, William C. "The Perfect Man in the Sufism of Jami." *Studia Islamica* 49 (1979): 135-47.

_____. *Principles of Ibn Al-'Arabi's Cosmology: The Self-Disclosure of God.* Albany, NY: SUNY, 1998.

Chopp, Rebecca S. *The Praxis of Suffering.* Maryknoll, NY: Orbis, 1989.

Christian Peacemaker Teams Electronic Briefings (CPTnet). "Muslim Peacemaker Team Training in Karbala" (February 2, 2005): http://www.cpt.org/archives.php (last accessed August 22, 2005).

_____. "The Weapons of the Muslim Peacemaker Teams" (June 21, 2005): http://www.cpt.org/archives.php (last accessed August 22, 2005).

Chua, Amy. "Our Most Dangerous Export: Imposing Free-Market Democracy on Iraq Has Unleashed Ethnic Hatred." *The Guardian Online* (February 28, 2004): http://www.guardian.co.uk/Iraq/Story/0,2763,1158215,00.html (last accessed February 28, 2004).

_____. *World on Fire: How Exporting Free-Market Democracy Breeds Ethnic Hatred and Global Instability*. New York: Anchor Books, 2004.

Clayton, John. "Thomas Jefferson and the Study of Religion." In *The Future of Religion: Postmodern Perspectives*. Edited by Christopher Lamb and Dan Cohn-Sherbok, 88-111. London: Middlesex University Press, 1999.

Clines, D. J. A. "The Image of God in Man." *Tyndale Bulletin* 19 (1968): 53-103.

Cobb, John B. Jr. *Sustaining the Common Good: A Christian Perspective on the Global Economy*. Cleveland, OH: The Pilgrim Press, 1994.

Cobb, John B., and Paul Knitter, eds. *The Myth of Christian Uniqueness: Toward a Pluralistic Theology of Religions*. Maryknoll, NY: Orbis, 1987.

Cohen, Jeremy. *"Be Fertile and Increase, Fill the Earth and Master It": The Ancient and Medieval Career of a Biblical Text*. Ithaca, NY: Cornell University Press, 1989.

Cohn-Sherbok, Dan, ed. *Islam in a World of Diverse Faiths*. Library of Philosophy and Religion. New York: St. Martin, 1999 [1991].

Colby, Gerard, and Charlotte Dennet. *Thy Will Be Done, The Conquest of the Amazon: Nelson Rockefeller and Evangelism in the Age of Oil*. New York: HarperCollins, 1995.

Collingwood, R. G. *Essay on Metaphysics*. Oxford: Clarendon, 1940.

Connor, Steve. "Melting Greenland Glacier May Hasten Rise in Sea Level." *The Independent Online* (July 25, 2005): http://news.independent.co.uk/world /environment /article301493.ece (last accessed July 25, 2005).

Cooper, David. "Genes for Sustainable Development: Overcoming the Obstacles to a Global Agreement on Conservation and Sustainable Use of Biodiversity." In *Biodiversity: Social & Ecological Perspectives*. Edited by Vandana Shiva et al. London/New Jersey: Zed; Penang: World Rainforest Movement, 1991.

Cornell, Vincent J. "Where is Scriptural Truth in Islam?" In Vroom and Gort, eds., *Holy Scriptures in Judaism*, 69-76.

Cowan, J. Milton. *The Hans Wehr Dictionary of Modern Written Arabic*. Ithaca, NY: Spoken Language Services, Inc., 1994.

Cox, Harvey. "The Market as God: Living in the New Dispensation." *The Atlantic Online* (March 1999): http://www.theatlantic.com/issues/99mar/marketgod.htm (last accessed October 18, 2004).

Cragg, Kenneth. *Am I Not Your Lord? Human Meaning in Divine Question*. London: Melisende, 2002.

_____. *The Event of the Qur'an: Islam in Its Scripture*. London: George Allen & Unwin, 1971.

_____. *Faiths in their Pronouns: Websites of Identity*. Brighton/Portland, OR: Sussex Academic Press, 2002.

_____. *Faith at Suicide—Lives Forfeit: Violent Religion—Human Despair*. Brighton/Portland, OR: Sussex Academic Press, 2005.

_____. "'Greater is God': Contemporary *Takbir*: Muslim and Christian." *The Muslim World* 71 (1981): 27-39.

_____. "Islam and Incarnation." In *Truth and Dialogue in World Religions: Conflicting Truth-Claims*. Edited by John Hick, 126-39. Philadelphia: Westminster, 1974.

_____. *The Mind of the Qur'an: Chapters in Reflection*. London: George Allen & Unwin, 1971.

_____. *The Privilege of Man: A Theme in Judaism, Islam and Christianity*. London: The Athlone Press, 1968.

_____. *The Pen and the Faith*. London: George Allen & Unwin, 1985.

_____. *The Qur'an and the West*. Washington, DC: Georgetown University Press, 2006.

_____. *Readings from the Qur'an*. Brighton: Sussex Academic Press, 1999.

_____. *Returning to Mount Hira': Islam in Contemporary Terms*. London: Bellew Publishing, 1994.

Craig, Edward et al., eds. *The Routledge Encyclopedia of Philosophy*. London/New York: Routledge, 1998.

Crone, Patricia, and Martin Hinds. *God's Caliph: Religious Authority in the First Centuries of Islam*. Cambridge: Cambridge University Press, 1986.

Crow, Ralph, Philip Grant and Saad Eddin Ibrahim. *Arab Nonviolent Struggle in the Middle East*. Boulder, CO: Lynne Rienner, 1990.

Curtis, Edward M. "Image of God (OT)." In *The Anchor Bible*, vol. 3. Edited by David Noel Freedman. New York: Doubleday, 1992.

Cutrofello, Andrew. "Jacques Derrida." In Craig et al., eds., *Routledge Encyclopedia of Philosophy*, vol. 2:307-23.

Dalacoura, Katerina. *Islam, Liberalism and Human Rights: Implications for International Relations*. Rev. ed. London & New York: I. B. Tauris, 2003.

Davis, Joyce M. *Between Jihad and Salaam: Profiles in Islam*. New York: St. Martin's Press, 1997.

D'Costa, Gavin. *The Meeting of Religions and the Trinity*. Maryknoll, NY: Orbis, 2000.

Déaut, R. Le. *Targum du Pentateuque*. Vol. 1. *Genèse*. Sources Chrétiennes, 245. Paris, 1978.

Déclais, Jean Louis. "La tenue d'Adam." *Arabica* 46 (1998): 111-18.

Delacoura, Katarina. *Liberalism and Human Rights: Implications for International Relations*. Rev. ed. London/New York: I. B. Tauris, 2003.

Denholm, Euan. "Congo's Marginalized Pygmies See Hope in Polls." *Reuters Online* (August 21, 2006): http://today.reuters.com/news/articlenews.aspx?type=worldNews&storyID=2006-08-22T010531Z_01_L15576369_RTRUKOC_0_US-CONGO-DEMOCRATIC-PYGMIES.xml& archived=False.

D'Entrèves, Maurizio Passerin, and Seyla Benhabib, eds. *Habermas and the Unfinished Project of Modernity: Critical Essays on The Philosophical Discourse of Modernity*. Cambridge: Polity Press; Oxford: in association with Blackwell, 1996.

Derrida, Jacques. *Margins of Philosophy*. Translated, with additional notes by Alan Bass. Chicago: University of Chicago Press, 1982.

Descartes, René. *Discourse on Method and Meditations*. Translated by Laurence J. Lafleur. Indianapolis: Bobbs-Merrill, 1960 [1637].

De Villiers, Marq. *Water: The Fate of Our Most Precious Resource*. New York: Houghton Mifflin, 2000.

Dews, Peter. "Communicative Rationality." In Craig et al., eds., *Routledge Encyclopedia of Philosophy*, vol. 2:459-62.

Dilthey, Wilhelm. *Gesammelte Schriften*. 12 vols. Leipzig: Teubner, 1914–58.

Docherty, Thomas, ed. *Postmodernism: A Reader*. New York: Columbia University Press, 1993.

_____. "Postmodernism: An Introduction." In Docherty, ed., *Postmodernism*, 1-31.

Doi, A. R. "The Islamic View of Freedom." *Islam and the Modern Age* 6.2 (1975): 41-60.

Donzel, E. van, et al., eds. *The Encyclopaedia of Islam*. New ed. Leiden: Brill, 1995 [1954].

Dozy, R. *Supplément aux Dictionnaires Arabes*. 2 vols. Paris: Maisonneuve Frères, 1927.

Dreyfus, Hubert L., and Harrison Hall. *Heidegger: A Critical Reader.* Cambridge, MA: Basil Blackwell, 1992.

Dunne, Timothy. *Inventing International Society: A History of the English School.* London: Macmillan, 1998.

Durkee, Abdullah Nooredeen. "Personal Thoughts about Muslim-Christian Dialogue." In Bryant and Ali, eds., *Muslim–Christian Dialogue*, 111-24.

Dyrness, William A. *The Earth Is God's: A Theology of American Culture.* Maryknoll, NY: Orbis, 1997.

Eaton, Charles Le Gai. "Man." In *Islamic Spirituality: Foundations.* Edited by Sayyed H. Nasr. New York: Crossroads, 1987.

Eco, Umberto. *A Theory of Semiotics.* Bloomington, IN: Indiana University Press, 1976.

Eickelman, Dale F. "Inside the Islamic Reformation." In Rubin, ed., *Revolutionaries and Reformers*, 203-6.

Eilberg-Schwartz, Howard. "Creation and Classification in Judaism: From Priestly to Rabbinic Conceptions." *History of Religions* 26 (1987): 357-81.

Embree, Lester. "Phenomenological Movement." In Craig et al., eds., *Routledge Encyclopedia of Philosophy*, vol. 6:333-43.

Ellis, Mark H. *Practicing Exile: The Religious Odyssey of an American Jew.* Minneapolis, MN: Fortress Press, 2002.

_____. *Toward a Jewish Theology of Liberation: The Challenge of the 21st Century.* Waco, TX: Baylor University Press, 2004.

Engineer, Asghar Ali. "On Developing Liberation Theology in Islam." *Islam and the Modern Age* 13.2 (1982): 101-25.

Ermath, Michael. *Wilhelm Dilthey: The Critique of Historical Reason.* Chicago: University of Chicago Press, 1978.

Esack, Farid. *Qur'an, Liberation & Pluralism: An Islamic Perspective on Interreligious Solidarity against Oppression.* Oxford: Oneworld, 1997.

_____. "Qur'anic Hermeneutics: Problems and Prospects." *The Muslim World* 83.2 (1993): 118-41.

Ess, Josef van. *Theologie und Gesellschaft im 2. und 3. Jahrhundert Hidschra: eine Geschichte des religiösen Denkens im früher Islam.* 6 vols. Berlin/New York: W. de Gruyter, 1991–97.

Esposito, John L., ed. *Political Islam: Revolution, Radicalism or Reform?* Boulder, CO/London: Lynne Rienner, 1997.

———. *Voices of Resurgent Islam.* Oxford/New York: Oxford University Press, 1983.

Esposito, John L., and John O. Voll, eds. *Democracy in Islam.* Oxford/New York: Oxford University Press, 1996.

_____. *Makers of Contemporary Islam.* Oxford/New York: Oxford University Press, 2001.

Esteva, Gustavo, and Madhu Suri Prakash. *Grassroots Post-Modernism: Remaking theSoil of Cultures.* London/New York: Zed, 1998.

Fadlallah, Muhammad Hussein. *On the Horizons of Muslim–Christian Dialogue (Fī āfāq al-ḥiwār al-Islāmī al-masīḥī).* 2d ed. Beirut: Dār al-Malāk, 1998.

Fakhry, Majid. *Ethical Theories in Islam.* Leiden: Brill, 1991.

Fanon, Frantz. *The Wretched of the Earth.* Translated by Constance Farrington. New York: Grove Weidenfield, 1968.

Farooqi, Muhammad Zubair. *Islam, The Muslim World-Community and Challenges of the Modern Age.* Islamabad: Isti'ara, 1997.

Farooqi, Waheed Ali. "A Qur'anic Solution to Man's Modern Predicament." In Said, ed., *Essays on Islam*, 201-10.

al-Faruqi, Isma'il R. "Meta-Religion Towards a Critical World Theology." *American Journal of Islamic Social Sciences* 3.1 (1986): 13-57.

_____. "On the Nature of Islamic *Da'wah*." *International Review of Mission* 65.260 (1976): 33-42.

Fenton, Paul B. "Sa'adya Ben Yosef." In van Donzell et al., eds., *The Encyclopaedia of Islam*, vol. 8:661-62.

Fickling, David. "World Bank Condemns Defense Spending." *The Guardian Online*, (February 14, 2004): http://www.guardian.co.uk/globalisation/story/0,7369,1147888 ,00.html (last accessed February 14, 05).

Fiorenza, Francis Schüssler. "The Crisis of Scriptural Authority." *Interpretation* 44 (October 1990): 353-68.

_____. *Foundational Theology: Jesus and the Church*. New York: Crossroad, 1984.

_____. "Systematic Theology: Task and Methods." In *Systematic Theology: Roman Catholic Perspectives*. Edited by Francis Schüssler Fiorenza and John P. Galvin, vol. 1:3-87. Minneapolis, MN: Fortress Press, 1991.

Fine, Arthur. "Scientific Realism and Antirealism." In Craig et al., eds., *Routledge Encyclopedia of Philosophy*, vol. 8:581-84.

Fish, Stanley. *Is There a Text in This Class? The Authority of Interpretive Communities*. Cambridge, MA: Harvard University Press, 1980.

Fisher, Ian. "Pope Urges Muslims to Confront Terrorism." *The New York Times* (August 21, 2005): A 16.

Fisher, Wolfdietrich. "Das Geshichliche Selbstvertändnis Muhammads and seiner Gemeinde zur Interpretation von Vers 55 der 24. Sure des Koran." *Oriens* 36 (*Festchrift* for Franz Rosenthal) (2001): 145-59.

Flax, Jane. *Psychoanalysis, Feminism, and Postmodernism*. Berkeley: University of California Press, 1990.

Fodor, James. *Biblical Narrative in the Philosophy of Paul Ricœur: A Study in Hermeneutics and Theology*. Cambridge: Cambridge University Press, 1990.

Foucault, Michel. *The Archeology of Knowledge*. Translated by A. M. Sheridan Smith. London: Tavistock, 1974.

Freeman, Michael. *Human Rights: An Interdisciplinary Approach*. Cambridge: Polity Press, 2002.

Friedman, Thomas. *The Lexus and the Olive Tree*. New York: Farrar, Straus & Giroux, 1999.

Fuller, Graham. *The Future of Political Islam*. New York: Palgrave Macmillan, 2003.

Gadamer, Hans-Georg. *Truth and Method*. 2d rev. ed. Translation revised by Joel Weinsheimer and Donald G. Marshall. New York: Continuum, 1993.

Ganai, Ghulam Nabi. "Muslim Thinkers and their Concept of Khilāfah." *Hamdard Islamicus* 24.1 (January–March 2001): 59-72.

Garaudy, Roger. "The Balance Sheet of Western Philosophy in this Century." *American Journal of Islamic Social Sciences* 2.2 (1985): 169-78.

Gardet, Louis. *La Cité Musulmane: Vie Sociale et Politique*. 4th ed. Paris: Librairie Philosophique J. Vrin, 1981 [1954].

Gätje, Helmut. *The Qur'an and its Exegesis: Selected Texts with Classical and Modern Interpretation*. Berkeley: University of California Press, 1976.

Gerholm, Tomas. "Two Muslim Intellectuals in the Postmodern West: Akbar Ahmed and Ziauddin Sardar." In *Islam, Globalization and Postmodernity*. Edited by Akbar S. Ahmed and Hastings Donnan, 190-212. London/New York: Routledge, 1994.

Ghadbian, Najib. *Democratization and the Islamist Challenge in the Arab World*. Boulder, CO: Westview, 1997.

Al-Ghazali, Muhammad. "Holistic Trend in Islamic Thought: Pioneering Contribution of Shah Wali Allah." *Hamdard Islamicus* 18.4 (1995): 41-55.

Giddens, Anthony. *The Consequences of Modernity*. Stanford, CA: Stanford University Press, 1990.

Gilliot, Claude. "Exegesis of the Qur'an: Classical and Medieval." In the *Encyclopaedia of the Qur'an*, vol. 2. Edited by Jane Dammen McAuliffe, 99-124. Leiden: Brill, 2002.

Gish, Arthur G. *Hebron Journal: Stories of Nonviolent Peacemaking*. Scottdale, PA: Herald Press, 2001.

Gish, Peggy. "Overcoming the Divide." *CPTnet* (February 2, 2005): http://www.cpt .org/archives.php (last accessed August 22, 2005).

Glaznev, Sergei. "We Need a New World Financial Architecture." *Executive Intelligence Review* 32.28 (July 15, 2005): 41-43.

Goldingay, John. *Models for Interpretation of Scripture*. Grand Rapids, MI: Eerdmans; Carlisle, PA: Paternoster, 1995.

Goulet, Denis. *The Cruel Choice: An New Concept in the Theory of Development*. Lanham, MD: University Press of America, 1985.

Gowan, Donald E. *From Eden to Babel: A Commentary of the Book of Genesis 1–11*. The International Theological Commentary. Grand Rapids, MI: Eerdmans, 1988.

Gramsci, Antonio. *Prison Notebooks/Antonio Gramsci*. Edited and with Introduction by Joseph A. Buttigieg. Translated by Joseph A. Buttigieg and Valentino Gerratana. New York: Columbia University Press, 1991.

Grandberg-Michaelson, Wesley. *Redeeming the Creation: The Rio Summit: Challenge for the Churches*. Geneva: WCC, 1992.

Grant, Jacquelyn. *White Women's Christ and Black Woman's Jesus: Feminist Christology and Womanist Response*. Atlanta, GA: Scholars Press, 1989.

Green, William Scott. "Introduction." In Neusner, ed., *Sacred Texts and Authority*.

Greene, Graham. *Getting to Know the General*. New York: Pocket Books, 1984.

Greider, William. *The Soul of Capitalism*. New York: Simon & Schuster, 2003.

Griffin, David Ray, John B. Cobb Jr, Richard Falk and Catherine Keller. *The American Empire and the Commonwealth of God: A Political, Economic, Religious Statement*. Louisville, KY/London: Westminster/John Knox Press, 2006.

Griffiths, Paul J. *Problems of Religious Diversity*. Exploring the Philosophy of Religion Series. Oxford/Malden, MA: Blackwell, 2001.

Garrison, Jim. *American Empire: Global Leader or Rogue Power?* San Francisco: Berrett-Koehler, 2004.

Gulet, Denis. *The Cruel Choice: A New Concept in the Theory of Development*. Lanham, MD: University Press of America, 1985.

Gunton, Colin E. *The Triune Creator: A Historical and Systematic Study*. Grand Rapids, MI: Eerdmans, 1998.

Habermas, Jürgen. *Between Facts and Norms: Contributions to a Discourse Theory of Law and Democracy*. Translated by William Rehg. Cambridge: Polity Press, 1996.

_____. *The Philosophical Discourse of Modernity: Twelve Lectures*. Translated by Frederick G. Lawrence. Studies in Contemporary German Social Thought. Cambridge, MA: MIT Press, 1987.

_____. *The Theory of Communicative Action*. Translated by Thomas McCarthy. Boston: Beacon Press, 1984.

Haddad, Yvonne Y. "Sayyid Qutb: Ideologue of Islamic Revival." In Esposito, ed., *Voices of Resurgent Islam*, 67-98.

Haeri, Shaykh Fadhlallah. *The Cow: A Commentary on Chapter 2: Surat Al-Baqarah* Reading: Garnet, 1993.

Hahn, Lewis Edwin, Randel E. Anxier and Lucian W. Stone Jr. *The Philosophy of Seyyed Hossein Nasr*. Library of Living Philosophers. Chicago: Open Court, 2001.

Hall, Anthony J. *The American Empire and the Fourth World: The Bowl with One Spoon*. Montreal/Kingston: McGill-Queen's University Press, 2003.

Hallman, David G. "Beyond 'North/South' Dialogue." In Hallman, ed., *Ecotheology*, 3-9.

_____. "Ethics and Sustainable Development." In Hallman, ed., *Ecotheology*, 264-83.

Hallman, David G., ed. *Ecotheology: Voices from the North and South*. Geneva: WCC/Maryknoll, NY: Orbis, 1994

Hamilton, Victor P. *The Book of Genesis: Chapters 1–17* The New International Commentary on the Old Testament. Grand Rapids, MI: Eerdmans, 1990.

Haney, Marsha Snulligan. "The Practice of Theological Engagement in Interreligious Dialogue: The Need for a Clarification." *The Muslim World* 94.3 (2004): 357-71.

Hart, Julie. "Vision of Hope." *CPTnet* (June 30, 2000): http://www.cpt.org/archives.php (last accessed October 19, 2005).

Harvey, David. *The Condition of Postmodernity: An Inquiry into the Origins of Social Change*. Malden, MA: Blackwell, 1989.

_____. *The New Imperialism*. Oxford/New York: Oxford University Press, 2003.

Hassan, Nik Mustapha Nik, and Mazilan Musa, eds. *The Economic and Financial Imperatives of Globalization: An Islamic Response*. Kuala Lumpur: Institute of Islamic Understanding Malaysia, 2000.

Hauerwas, Stanley. *In Good Company: The Church as Polis*. Notre Dame, IN: Notre Dame University Press, 1995.

Hawting, G. R., and Abdul-Kader A. Shareef, eds. *Approaches to the Qur'an*. London: Routledge, 1993.

Hayden, Patrick, ed. *The Philosophy of Human Rights*. Paragon Issues in Philosophy. St Paul, MN: Paragon House, 2001.

Hayes, John H., ed. *Methods of Biblical Interpretation*, with a Foreword by Douglas A. Knight, excerpted from the *Dictionary of Biblical Interpretation*. Nashville, TN: Abingdon Press, 2004.

Heath, Peter. "Creative Hermeneutics: A Comparative Analysis of Three Islamic Approaches." *Arabica* 36 (1989): 173-210.

Heidegger, Martin. *Basic Writings*. Edited by David Farrell Krell. New York: Harper & Row, 1977 [1947].

_____. *The Question concerning Technology and Other Essays*. Translated and with an Introduced by William Lovitt. New York: Harper/Row, Harper Torchbooks, 1977.

Hick, John. "On Conflicting Religious Truth-Claims." In *Problems of Religious Pluralism*. Edited by John Hick. New York: St. Martin's Press, 1985.

Hiebert, Paul G. *Anthropological Reflections on Missiological Issues*. Grand Rapids, MI: Baker, 1994.

_____. "The Flaw of the Excluded Middle." *Missiology* 10.1 (1982): 35-47.

Hill, Brennan R. *Christian Faith and the Environment: Making Vital Connections.* Ecology and Justice: An Orbis Series on Global Ecology. Maryknoll, NY: Orbis, 1998.

Hillmer, Mark. "The Book of Genesis in the Qur'an." *Word and the World* 14 (1994): 195-203.

Hinds, Martin, Jere L. Bacharach, Lawrence I. Conrad and Patricia Crone, eds. *Studies in Late Antiquity and Early Islam*. Princeton, NJ: Darwin Press, 1995.

Hoder, Susanne. "Divestiture of Funds that Support Israeli Occupation of Palestinian Territories." Posted on the New England Conference Website, June 20, 2005, http://www.neumc.org/news_detail.asp?TableName=oNews_PJAYMY&PKVa lue=60 (last accessed July 15, 2005).

Hodgson, Marshall G. S. *The Venture of Islam: Conscience and History in a World Civilization*. 3 vols. Chicago: University of Chicago Press, 1974 [paperback ed. 1977].

Hoebink, Michel. "Thinking about Renewal in Islam: Towards a History of Islamic Ideas on Modernization and Secularization." *Arabica* 46 (1998): 29-62.

Hollenbach, David. *Justice, Peace and Human Rights: American Catholic Social Ethics in a Pluralistic World*. New York: Crossroad, 1988.

Hollingham, Richard. "Icy Greenland Turns Green." *BBC News Online* (August 14, 2005): http://news.bbc.co.uk/1/hi/programmes/from_our_own_correspondent/ 4145034.stm (last accessed August 14, 2005).

Holmes, Robert. "Lessons Learned...and Not." *CPTnet* (April 7, 2001): http://www .cpt.org/archives.php (last accessed August 22, 2005).

Horowitz, David. *The First Frontier: The Indian Wars and America's Origins, 1607–1776*. New York: Simon & Schuster, 1978.

Hourani, Albert. *Arabic Thought in the Liberal Age, 1798–1939*. 2d ed. Cambridge: Cambridge University Press, 1983.

Hoyningen-Huene, Paul. "Thomas Samuel Kuhn (1922–1996)." In Craig et al., eds., *Routledge Encyclopedia of Philosophy*, vol. 4:315-18.

Hroub, Khaled. *Hamas: Political Thought and Practice*. Beirut: Institute of Palestine Studies, 2000.

Humphreys, R. Stephen. *Between Memory and Desire: The Middle East in a Troubled Age*. Berkeley: University of California Press, 1999.

Huntington, Samuel. *The Clash of Civilizations and the Remaking of the World Order*. New York: Simon & Schuster, 1996.

Ibn Kathir, Isma'il Imad al-Din Abu al-Fida. *Tafsīr al-Qurʾān al-ʿaẓīm* (Commentary of the Great Qur'an). Edited by Mustafa al-Sayyid Muhammad et al., 4 vols. Giza, Egypt: Muʾassasa Qurṭuba, 2000.

Ibn Khaldun, Abd al-Rahman b. Muhammad. *Muqqadimat Ibn Khaldūn*. Edited, annotated and introduced by Ali Abd al-Wahad Wafi. 2 vols. Cairo: Lajnat al-Bayjān al-ʿArabī, 1965.

Ibn Manẓūr, Muḥammad b. Mukarram. *Lisān al-ʿArab* (The Tongue of the Arabs). 6 vols. Cairo: Dār al-Maʿārif, 1981.

Ibrahim, 'Izz al-Din. "Islamic-Christian Dialogue: A Muslim View." In Bryant and Ali, eds., *Muslim–Christian Dialogue*, 15-27.

Ignatieff, Michael. "American Empire: Get Used to It." *New York Times Sunday Magazine* (January 5, 2003): 22-54.

_____. *Human Rights as Politics and Idolatry*. Edited and with an Introduction by Amy Gutmann. Princeton, NJ: Princeton University Press, 2001.

_____. "Human Rights, Sovereignty and Intervention." In Owen, ed., *Human Rights, Human Wrongs*, 52-87.

Izutsu, Toshihiko. *Ethico-Religious Concepts in the Qur'an*. Originally *The Structure of Ethical Terms in the Qur'an*, 1959. Montreal/Kingston: McGill-Queen's University Press, 2002.

The Independent Online. "Can this Erudite Swiss Lecturer Really Be the Man Branded by The Sun as 'the Acceptable Face of Terror'? Paul Vallely Meets Tariq Ramadan" (July 25, 2005): http://news.independent.co.uk/people/profiles/article301486.ece (last accessed August 18, 2005).

International Review of Mission. "Statement of the Conference on 'Christian Mission and Islamic *Da'wah*,' Chambésy, June 1976." *International Review of Mission* 65.260 (1976): 457-60.

Internet Encyclopedia of Philosophy. Edited by James Fieser. "Paul Ricœur" (2004): http://www.iep.utm.edu (last accessed November 22, 2005).

The Interpreter's Bible. Edited by George Arthur Buttrick et al. 12 vols. Nashville: Abingdon Press, 1951–57.

Iqbal, Muhammad. *The Reconstruction of Religious Thought in Islam*. Lahore: Shaikh Muhammad Ashraf, 1960.

Ismail, Salwa. *Rethinking Islamist Politics: Culture, the State and Islamism*. London/New York: I.B. Tauris, 2003.

Jameson, Frederic. *The Cultural Turn: Selected Writings on the Postmodern 1983–1998*. London/New York: Verso, 1998.

_____. *Postmodernism, or, The Cultural Logic of Late Capitalism*. Durham, NC: Duke University Press, 1991.

Jansen, Johannes J. G. *The Dual Nature of Islamic Fundamentalism*. Ithaca, NY: Cornell University Press, 1997.

_____. *The Interpretation of the Koran in Modern Egypt*. Leiden: Brill, 1974.

Jefferey, Arthur. *Foreign Vocabulary of the Qur'an*. Baroda: Oriental Institute, 1938.

_____. "Ibn Al-Arabi's *Shajarat Al-Kawn*: Introduction." *Studia Islamica* 10 (1959): 43-77.

Jenkins, Philip. *The Next Christendom: The Rise of Global Christianity*. Oxford/New York: Oxford University Press, 2002.

Jewett, Paul K. *Man as Male and Female: A Study in Sexual Relationships from a Theological Point of View*. Grand Rapids, MI: Eerdmans, 1975.

_____. *Who We Are: Our Dignity as Human: A Neo-Evangelical Theology*. Grand Rapids, MI: Eerdmans, 1996.

John XXIII (Pope). *Pacem in Terris*. New York: Paulist Press, 1963.

Johnston, David L. "Allāl al-Fāsī: *Sharīʿa* as Blueprint for a Righteous Global Citizenship?" In *Shari'a: Islamic Law in the Contemporary Context*. Edited by Abbas Amanat and Frank Griffel, 83-103. Stanford: Stanford University Press, 2006.

_____. "An Epistemological and Hermeneutical Turn in Twentieth-Century *Uṣūl al-Fiqh*." *Islamic Law and Society* 11.2 (2004): 233-82.

_____. "Fuzzy Reformist-Islamist Borders: Malek Bennabi and Rachid Ghannouchi on Civilization." *The Maghreb Review* 29.1-2 (2004): 123-52.

_____. "Hassan al-Hudaybi and the Muslim Brotherhood: Can Islamic Fundamentalism Eschew the Islamic State?" *Comparative Islamic Studies* 3.1 (2007): 39-56.

_____. "The Human *Khilāfa*: A Growing Overlap of Reformism and Islamism on Human Rights Discourse?" *Islamochristiana* 28 (2002): 35-53.

_____. *Evolving Muslim Theologies of Justice: Jamal al-Banna, Mohammad Hashim Kamali and Khaled Abou El Fadl.* Penang, Malaysia: The Centre for Policy Research and International Studies (CenPRIS) and Universiti Sains Malaysia Press, 2009.

———. "*Khalīfa*, Culture Change, and Muslim–Christian Cooperation in Hebron District Community Development." Unpublished paper, 1999.

_____. "Loving Neighbors in a Globalized World: US Christians and Muslims in the Mideast." In *Anxious about Empire: Theological Essays on the New Global Realities.* Edited by Wes D. Avram. Grand Rapids, MI: Brazos, 2004.

_____. "*Maqāṣid al-Sharīʿa*: Epistemology and Hermeneutics of Muslim Theologies of Human Rights." *Die Welt des Islams* 47.2 (2007): 149-87.

_____. "Rethinking Human Rights: A Common Challenge to Muslims and Christians." Paper presented at the Conflict Transformation Project's Interfaith Dialogue in Rockville, MD, April 22–23, 2005.

Jomier, Jacques. *Le commentaire coranique du Manar: tendances modernes de l'exégèse coranique en Égypte.* Paris: Éditions G. P. Maisonneuve, 1954.

Juma, Jamal. "Israeli Apartheid." *Al-Ahram Weekly Online* 745 (June 2–8, 2005): http://weekly.ahram.org.eg/2005/745/re5.htm.

Khadduri, Majid. *The Islamic Conception of Justice.* Baltimore/London: The Johns Hopkins University Press, 1984.

Khalidi, Rashid. *Resurrecting Empire: Western Footprints and America's Perilous Path in the Middle East.* Boston: Beacon Press, 2004, with new Introduction, 2005.

Kamali, Mohammad Hashim. *Equality and Justice in Islam.* Cambridge: Islamic Texts Society, 2002.

_____. *Freedom of Expression in Islam* (Kuala Lumpur: Berita, 1994).

_____. "Issues in the Understanding of Jihad and Ijtihad." In Hawting and Shareef, eds., *Approaches to the Qur'an,* 617-34.

Kant, Immanuel. *Groundwork of the Metaphysics of Morals.* Translated with notes by H. J. Paton. 4 vols. London: Hutchinson, 1948. Repr. New York: Harper & Row, 1964 [1785].

_____. *Perpetual Peace and Other Essays.* Translated by T. Humphrey. Indianapolis: Bobbs-Merrill, 1983.

Kaplan, Jeffrey. "Consent of the Governed: The Corporate Usurpation of Democracy and the Valiant Struggle to Win it Back." *Orion* (November–December 2003): 54-61.

Kärkkäinen, Veli-Matti. *An Introduction to the Theology of Religions: Biblical, Historical, and Contemporary Perspectives.* Grand Rapids, MI: Baker Academic Books, 2003.

———. *Pneumatology: The Holy Spirit in Ecumenical, International, and Contextual Perspectives.* Grand Rapids, MI: Baker Academic Books, 2002.

_____. *Trinity and Religious Pluralism: The Doctrine of the Trinity in Christian Theology of Religions.* Aldershot/Burlington, VT: Ashgate, 2004.

Kateregga, Badru D., and David W. Shenk. *A Muslim and Christian in Dialogue.* Scottdale, PA: Herald Press, 1997.

Kates, Robert W. "The Nexus and the Neem Tree: Globalization and a Transition Toward Sustainability." In *Worlds Apart: Globalization and the Environment.* Edited by James Gustave Speth. Washington, DC: Island Press, 2003.

Keal, Paul. *European Conquest and the Rights of Indigenous Peoples: The Moral Backwardness of International Society.* Cambridge Studies in International Relations. Cambridge: Cambridge University Press, 2003.

Keck, Margaret E., and Kathryn Sikkink. *Activists Beyond Borders: Advocacy Networks in International Politics.* Ithaca, NY: Cornell University Press, 1998.

_____. "Environmental Advocacy Networks." In Lechner and Boli, eds., *The Globalization Reader*, 392-99.

Kennedy, Paul. *The Rise and Fall of the Great Powers: Economic Change and Military Conflict from 1500 to 2000.* New York: Fontana Press, 1990.

Kenny, Martin. *Biotechnology: The University-Industrial Complex.* New Haven, CT: Yale University Press, 1986.

Kepel, Gilles. *Jihad: The Trail of Political Islam.* Cambridge, MA: The Belknap Press of Harvard University Press, 2002.

Kincaid, Harold. "Positivism in the Social Sciences." In Craig et al., eds., *Routledge Encyclopedia of Philosophy*, vol. 7:558-61.

Khor, Martin. "Globalization and Sustainable Development: The Choices Before Rio + 10." *International Review for Environmental Strategies* 2.2 (2001).

Klemm, D. E. "Hermeneutics." In Hayes, ed., *Methods of Biblical Interpretation*, 147.

Kohlberg, E. "Al-Tabrisi." In Donzel et al., eds., *The Encyclopaedia of Islam*, vol. 10:40-41.

The Bounteous Koran: A Translation of Meaning and Commentary by Dr. M. M. Khatib. London: Macmillan, 1986.

The Glorious Koran: A Bilingual Edition with English Translation, Introduction and Notes by M. Pickthall. London: George Allen & Unwin, 1976.

The Koran Interpreted. Translated by A. J. Arberry. London: George Allen & Unwin, 1955.

Korten, David C. *Getting to the 21st Century: Voluntary Action and the Corporate Agenda.* West Hartford, CT: Kumerian, 1990.

_____. *The Post-Corporate World: Life After Capitalism.* San Francisco: Berret-Koehler; West Hartford, CT: Kumerian, 1999.

_____. *When Corporations Rule the World.* West Hartford, CT: Kumerian, 1995.

Kotva, Joseph J. Jr. *The Christian Case for Virtue Ethics.* Washington, DC: Georgetown University Press, 1996.

Kraft, Charles H. *Anthropology for Christian Witness.* Maryknoll, NY: Orbis, 1996.

Krämer, Gudrun. "Cross-Links and Double Talk? Islamist Movements in the Political Process." In *The Islamist Dilemma: The Political Role of Islamist Movements in the Contemporary Arab World.* Edited by Laura Guazzone. Reading, PA: Ithaca, 1995, 39-67.

_____. "Visions of an Islamic Republic: Good Governance According to the Islamists." In *The Islamic World and the West: An Introduction to the Political Cultures and International Relations.* Edited by Kai Hafez. Translated from the German by Mary Ann Kenny, 33-45. Leiden: Brill, 2000.

Krugman, Paul. "All Fall Down." *New York Times* (November 25, 2008): http://www.nytimes.com/2008/11/26/opinion/26friedman.html?_r=1&hp.

_____. "The Chinese Connection." *New York Times Online* (May 20, 2005): http://www.nytimes.com/2005/05/20/opinion/20krugman.html?hp (last accessed may 20, 2005).

_____. "Reagan Did It." *New York Times* (May 31, 2009): http://www.nytimes.com/2009/06/01/opinion/01krugman.html?em.

_____. "Revenge of the Glut." *New York Times* (March 1, 2009): http://www
.nytimes.com/2009/03/02/opinion/02krugman.html.

Kuhn, Thomas. *The Structure of Scientific Revolutions*. Chicago: University of Chicago
Press, 1970 [1962].

Küng, Hans. *Global Responsibility: In Search of a New World Ethic*. New York/London:
Continuum, 1996.

_____. "Paradigm Change in Theology." In *Paradigm Change in Theology*. Edited by
Hans Küng and David Tracy. Edinburgh: T. & T. Clark, 1989.

Küng, Hans, and Karl-Josef Kushel, eds. *A Global Ethic: The Declaration of the
Parliament of the World's Religions*. New York: Continuum, 1993.

Kurtzman, Charles, ed. *Liberal Islam: A Sourcebook*. Oxford/New York: Oxford
University Press, 1998.

_____. *Modernist Islam, 1840–1940: A Sourcebook*. Oxford/New York: Oxford
University Press, 2002.

Kyung, Chun Hyung. "Ecology, Feminism and African and Asian Spirituality: Towards a
Spirituality of Eco-Feminism." In Hallman, ed., *Ecotheology*, 175-78.

Lamb, Christopher. *The Call to Retrieval: Kenneth Cragg's Christian Vocation to Islam*.
London: Grey Seal, 1997.

Lambton, A. K. S. "*Khalīfa*: (2) In Political Theory." In van Donzell et al., eds., *The
Encyclopaedia of Islam*, vol. 4:947-50.

Lane, Edward William. *Arabic–English Lexicon*. 8 vols. New York: Frederick Ungar,
1955.

Lapidus, Ira M. *A History of Islamic Societies*. 2d ed. Cambridge: Cambridge University
Press, 2002.

LaRouche, Lyndon H. Jr. "It Happened in Berlin Last Week." *Executive Intelligence
Review* 32.28 (July 15, 2005): 4-18.

Latouche, Serge. *In the Wake of the Affluent Society: An Exploration of Post-Development*. Introduced and translated by Martin O'Connor and Rosemary Arnoux.
London/New York: Zed, 1993.

Lawrence, Bruce B. *Shattering the Myth: Islam Beyond Violence*. Princeton, NJ:
Princeton University Press, 1998.

Lechner, Frank J., and John Boli, eds. *The Globalization Reader*. Malden, MA: Blackwell, 2000.

Lee, Charles. "Environmental Justice: Building a Unified Vision of Health and the
Environment." *Environmental Health Perspectives* 110, supplement 2 (April 2002):
141-44.

Lee, Robert D. *Overcoming Tradition and Modernity: The Search for Islamic
Authenticity*. Boulder, CO: Westview, 1997.

Levenson, John D. *Creation and the Persistence of Evil: The Jewish Drama of Divine
Omnipotence*. Princeton, NJ: Princeton University Press, 1988.

Li, Xiaorong. "'Asian Values' and the Universality of Human Rights." Report from the
Institute for Philosophy and Public Policy 16.2 (1996): 18-23.

Lia, Brynjar. *The Society of the Muslim Brothers in Egypt: The Rise of an Islamic Mass
Movement 1928–1942*. With a Foreword by Jamal al-Banna. Reading: Ithaca, 1998.

Lindbeck, George. *The Nature of Doctrine: Religion and Theology in a Postliberal Age*.
Philadelphia: Westminster Press, 1984.

Linetsky, Michael. *Rabbi Saadiah Gaon's Commentary on the Book of Creation*.
Northvale, NJ/Jerusalem: Jason Aronson, 2002.

Linklater, Andrew. "Citizenship and Sovereignty in the Post-Westphalian State." *European Journal of International Relations* 2.1 (1996): 77-103.

Linton, Ralph. *The Study of Man.* New York: D. Appleton–Century, 1937.

Loux, Michael J. "Nominalism." In Craig et al., eds., *Routledge Encyclopedia of Philosophy*, vol. 7:17-23.

Lovibond, Sabina. "Feminism and Postmodernism." In Docherty, ed., *Postmodernism: A Reader*, 390-414.

_____. "Feminism and Postmodernism." In *Postmodernism and Society*. Edited by R. Boyne and A. Rattansi, 154-86. Basingstoke: Macmillan Foundation/New York: St Martin's Press, 1990.

Lyotard, Jean-François. *L'Economie Libidinale.* Paris: Minuit, 1974.

_____. *The Postmodern Condition: A Report on Knowledge.* Translated by G. Bennington and B. Massumi. Manchester: Manchester University Press, 1984.

MacIntyre, Alasdair. *After Virtue: A Study in Moral Theory.* 2d ed. Notre Dame, IN: University of Notre Dame Press, 1984.

_____. *Whose Justice? Which Rationality?* Notre Dame, IN: University of Notre Dame Press, 1988.

Macpherson, C. B. *The Political Theory of Possessive Individualism: Hobbes to Locke.* Oxford: Oxford University Press, 1979.

Mahmoud, Muhammad. "The Creation Story in *Sūrat Al-Baqara,* With Special Reference to Al-Tabari's Material: An Analysis." *Journal of Arabic Literature* 26 (1995): 201-14.

Maimonides. *Epistle to the Yemen.* Edited and with an Introduction by A. Halkin. New York: American Academy for Jewish Research, 1952.

Majid, Anouar. *Unveiling Traditions: Postcolonial Islam in a Polycentric World.* Durham, NC/London: Duke University Press, 2000.

Makdisi, George. *Ibn 'Aqil: Religion and Culture in Classical Islam.* Edinburgh: Edinburgh University Press, 1997.

Malcolm X. *Malcolm X Speaks.* New York: Grove Press, 1965.

_____. *Malcolm X Speaks: Selected Speeches and Statements.* Edited by George Breitman. 5th print. New York: Pathfinder, 2002.

Malti-Douglas, Fedwa. "Postmoderning the Traditional in the Autobiography of Shaykh Kishk." In *Tradition, Modernity, and Postmodernity in Arabic Literature.* Edited by Kamal Abdel-Malek and Wael Hallaq, 389-410. Leiden: Brill, 2000.

Manzoor, S. Pervez. "Faith and Law: At the Crossroads of Transcendence and Temporality." *Muslim World Book Review* 18.3 (1998): 3-11.

_____. "Human Ecology and the Quest for a Universal Science." *MAAS Journal of Islamic Science* 7.2 (1991): 81-107.

Mao, Norbert. "Unevenly Yoked: Has Globalization Dealt Africa a Bad Hand?" *YaleGlobal Online* 3 (November 2003): http://yaleglobal.yale.edu/display. article?id=2721 (last accessed February 28, 2004).

Marsden, George. *Understanding Fundamentalism and Evangelicalism.* Grand Rapids, MI: Eerdmans, 1991.

Marsden, John J. *Marxian and Christian Utopianism: Toward a Socialist Political Theology.* New York: Monthly Review Press, 1991.

Martin, Richard C., and Mark R. Woodward (with Dwi S. Atmaja). *Defenders of Reason in Islam: Mu'tazilism from Medieval School to Modern Symbol.* Oxford: Oneworld, 1997.

Mawdudi, Abul A'la. "Human Rights in Islam." *al-Tawhid* 4.3 (1987): 59-103.

May, Rollo. *The Cry for Myth*. New York: Bantam Doubleday Dell, Delta, 1991.

Mayer, Ann Elizabeth. *Islam and Human Rights: Tradition and Politics*. 3d ed. Boulder, CO: Westview, 1999.

Mazrui, Ali. "Human Rights between Rwanda and Repatriations [*sic*]: Global Power and the Racial Experience." In *Human Wrongs: Reflections on Western Global Dominance and its Impact upon Human Rights*. Edited by Just World Trust, 188-211. Penang: Just World Trust, 1996.

_____. "The Nuclear Option and International Justice: Islamic Perspectives." In Barazangi, Zaman and Afzal, eds., *Islamic Identity and the Struggle for Justice*, 95-116.

McAuliffe, Jane Dammen, ed. *Encyclopaedia of the Qur'an*. Leiden: Brill, 2001.

McClendon, James W., and James M. Smith, *Understanding Religious Convictions*. Notre Dame, IN: University of Notre Dame Press, 1975.

McCloud, Aminah. "Reflections on Dialogue." *The Muslim World* 94.3 (2004): 335-41.

McMullin, Ernan. "A Case for Scientific Realism." In *Scientific Realism*. Edited by Jarret Leplin. Berkeley/Los Angeles: University of California Press, 1984.

McNeil, Brenda Salter, and Rick Richardson. *The Heart of Racial Reconciliation*. With a Foreword by John M. Perkins. Downers Grove, IL: InterVarsity, 2004.

McPheron, William, ed. "Fredric Jameson." *Stanford University Libraries* (1999): http://prelectur.stanford.edu/lecturers/jameson/ (last accessed December 10, 2004).

McPherson, C. B. *The Political Theory of Possessive Individualism: Hobbes to Locke*. Oxford: Oxford University Press, 1979.

Midrash Bereshit Rabba: Critical Edition with Notes and Commentary (Hebrew). Edited by J. Theodor and C. Albek. 3 vols. Jerusalem: Sifre Vahrman, 1903–36. Repr. 1965.

Miguez-Bonimo, José. *Doing Theology in a Revolutionary Situation*. Philadelphia: Westminster Press, 1975.

Milbank, John. "The Gospel of Affinity." In *The Future of Hope: Christian Tradition amid Modernity and Postmodernity*. Edited by Miroslav Volf and William Katerberg, 149-69. Grand Rapids, MI/Cambridge: Eerdmans, 2004.

Milgrom, Jeremy. "'Let Your Love for Me Vanquish Your Hatred for Him': Nonviolence and Modern Judaism." In *Subverting Hatred: The Challenge a/Nonviolence in Religious Traditions*. Edited by Daniel L. Smith-Christopher. Faith Meets Faith, an Orbis series in Interreligious Dialogue. Boston, MA: Boston research Center for the 21st Century; Maryknoll, NY: Orbis, 1998.

Miller, Sara B., and Amanda Paulson. "Despite More Jobs, US Poverty Rises." *Christian Science Monitor Online* (August 31, 2005): http://www.csmonitor.com/2005 /0831/p02s01-usec.html?s=itm (last accessed August 31, 2005).

Mitzman, Arthur. *The Iron Cage: An Historical Interpretation of Max Weber*. New York: The Universal Library, 1971.

_____. *Prometheus Revisited: The Quest for Global Justice in the Twenty-First Century*. Amherst/Boston: University of Massachusetts Press, 2003.

Mohammed, Ovey N., S.J. *Muslim–Christian Relations: Past, Present, Future*. Maryknoll, NY: Orbis, 1999.

Moltmann, Jürgen. *God in Creation: A New Theology of Creation and the Spirit of God*. San Francisco: Harper & Row, 1985.

Moosa, Ebrahim. "The Debts and Burdens of Critical Islam." In Safi, ed., *Progressive Muslims*, 111-27.

Moreland, J. P., and William Lane Craig. *Philosophical Foundations for a Christian Worldview*. Downers Grove, IL: InterVarsity Press, 2003.

Mkapa, Benjamin. "Giving Everyone a Place at Global Dining Table." *The Guardian Online* (February 16, 2004): www.guardian.co.uk/business/story/0,,1148792,00 .html (last accessed February 16, 2004).

Murphy, Dan. "Can Islam's Leaders Reach its Radicals?" *The Christian Science Monitor Online* (July 14, 2005): http://www.csmonitor.com/2005/0714/p01s01-wome.html (last accessed July 14, 2005).

Murphy, Nancey. *Anglo-American Postmodernity: Philosophical Perspectives on Science, Religion, and Ethics*. Boulder, CO: Westview, 1997.

_____. *Beyond Liberalism and Fundamentalism: How Modern and Postmodern Philosophy Set the Philosophical Agenda*. Valley Forge, PA: Trinity Press International, 1996.

Muslim, Ibn al-Hajj al-Qushayri. *Ṣaḥīḥ Muslim; Being Traditions of the sayings of the Prophet Muhammad as Narrated by his Companions and Compiled under the title Al-Jāmiʿ al-Saḥīḥ by Imam Muslim, with Explanatory Notes and Brief Biographical Sketches of Major Narrators*. Translated by Abdul Hamid Saddiqi. 4 vols. Lahore: Sh. Muhammad Ashraf, 1971.

Muslim World Book Review, eds. "Changing Stories." *Muslim World Book Review* 19.1 (1998): 41-42.

Muzaffar, Chandra. *Human Rights and the New World Order*. Penang: Just World Trust, 1003.

_____. *Rights, Religion and Reform: Enhancing Human Dignity through Spiritual and Moral Transformation*. London/New York: RoutledgeCurzon, 2002.

An-Na'im, Abdullahi Ahmed. *Islam and the Secular State: Negotiating the Future of Shari'a* (Cambridge, MA: Harvard University Press, 2008).

—*Toward an Islamic Reformation: Civil Liberties, Human Rights, and International Law*. Syracuse, NY: Syracuse University Press, 1990.

al-Najjar, Abd al-Majid. *Khilāfat al-insān bayna al-waḥī wa-l-ʿaql* (The Vicegerency of Man between Revelation and Reason). Beirut: Dār al-Gharb al-Islāmī, 1987.

_____. *Muqārabāt fī qirāʾat al-turāth* (Approximations on the Reading of Cultural Legacy). Beirut: Dār al-Badāʾil, 2001.

_____. *The Vicegerency of Man: Between Revelation and Reason: A Critique of the Dialectic of the Text, Reason, and Reality*. Translated by Aref T. Atari. Herndon, VA: International Institute of Islamic Thought, 1999.

Naqvi, Syed Nawab Haider. "Exogenous Shocks and Islamic Economic Response." In Hassan and Musa, eds., *The Economic and Financial Imperatives of Globalization*.

Nasr, Seyyed Vali Reza. *Mawdudi and the Making of Islamic Revivalism*. Oxford/New York: Oxford University Press, 1996.

Nasr, Seyyed Hossein. *The Heart of Islam: Enduring Values for Humanity*. New York: HarperSanFrancisco, 2002.

_____. "Sufism and the Integration of Man." In *God and Man in Contemporary Islamic Thought: Proceedings of the Philosophy Symposium Held at the American University of Beirut, February 6–10, 1967*. Edited and with an Introduction by Charles Malik, 144-51. Beirut: American University of Beirut Centennial Publications, 1972.

National Geographic Magazine. "Global Warming: Bulletins from a Warmer World." September 2004: 2-75.

National Research Council. *Abrupt Climate Change: Inevitable Surprises*. Washington, DC: National Academy Press, 2002.

Naugle, David K. *Worldview: The History of a Concept*. Grand Rapids, MI/Cambridge: Eerdmans, 2002.

Nazir-Ali, Michael. *Citizens and Exiles: Christian Faith in a Plural World*. Cleveland, OH: United Church Press, 1998.

Neusner, Jacob, ed. *Sacred Texts and Authority*. The Pilgrim Library of World Religions. Cleveland, OH: The Pilgrim Press, 1998.

Neuwirth, Angelika. "Cosmology." In McAuliffe, ed., *Encyclopaedia of the Qur'an*, vol. 1:440-58.

Newbigin, Lesslie. *The Gospel in a Pluralist Society*. Grand Rapids, MI: Eerdmans; Geneva: WCC, 1989.

Newby, Gordon D. *A History of the Jews of Arabia*. Columbia, SC: University of South Carolina Press, 1988.

Nietzsche, Friedrich. *The Birth of Tragedy and the Case of Wagner*. Translated with Commentary by Walter Kaufmann. New York: Random House, Vintage Books, 1967.

Nöldeke, Theodor. *Geschichte des Qorans*. 2d ed. Leipzig: Dietrich'sche Verlags-buchhandlung, 1926.

Nüsse, Andrea. *Muslim Palestine: The Ideology of Hamas*. Amsterdam: Harwood, 1998.

O'Neill, Onora. "Kantian Ethics." In Craig et al., eds., *Routledge Encyclopedia of Philosophy*, vol. 5:200-204.

Osborne, Grant R. *The Hermeneutical Spiral: A Comprehensive Introduction to Biblical Interpretation*. Downers Grove, IL: InterVarsity, 1991.

O'Shaughnessy, Thomas J., S.J. *Creation and the Teaching of the Qur'an*. Rome: Biblical Institute Press, 1985.

_____. *The Development of the Meaning of Spirit in the Koran*. Rome: Pontifical Oriental Institute, 1953

Owen, Nicholas, ed. *Human Rights, Human Wrongs*. Oxford/New York: Oxford University Press, 2003.

The Oxford Bible Commentary. Edited by John Barton and John Muddiman. Oxford/New York: Oxford University Press, 2001.

The Other Israel. "Briefing April 7, 2001." Electronic newsletter. http://otherisrael .home.igc.org/index.html (last accessed October 19, 2005).

Palast, Greg. *The Best Democracy Money Can Buy: An Investigative Reporter Exposes the Truth about Globalization, Corporate Cons and High Finance Fraudsters*. London/Sterling, VA: Pluto, 2002.

Paret, Rudi. "Signification coranique de *khalīfa* et d'autres derives de la racine khalafa." *Studia Islamica* 31 (1970): 211-17.

The Penguin Dictionary of Critical Theory. "Frankfurt School." Edited by David Macey, 139-40. London/New York: Penguin Books, 2000.

Perfet, Isaac b. Sheshet. *Shu''uT*. Vilna: beo-hesta'at Yehuda Leyb Eli'ezer Lipman Matz, 1879.

Perkins, John. *Confessions of an Economic Hit Man*. San Francisco: Berrett-Koehler, 2005.

Peters, Rudolph. *Jihad in Classical and Modern Islam*. Princeton, NJ: Marcus Wiener, 1996.

Peters, Ted. *God—The World's Future: Systematic Theology for a New Era*. 2d ed. Minneapolis, MN: Augsburg Press, 2000.

Pilger, John. *The New Rulers of the World*. London: Verso, 2002.

Pinnock, Clark H. *A Wideness in God's Mercy: The Finality of Jesus Christ in a World of Religions*. Grand Rapids, MI: Zondervan, 1992.

Pocock, J. G. A. "The Ideal of Citizenship Since Classical Times." *Queen's Quarterly* (Spring 1992): 55.

Polanyi, Michael. *Personal Knowledge: Towards a Post-Critical Philosophy*. Chicago: University of Chicago Press, 1958.

Al-Qadi, Wadad. "Caliph." In McAuliffe, ed., *Encyclopaedia of the Qur'an*, vol. 1:276-78.

_____. "The Term '*Khalīfa*' in the Early Exegetical Literature." *Die Welt des Islams* 28 (1988): 392-411.

Quayson, Ato. "Postcolonialism." In Craig et al., eds., *Routledge Encyclopedia of Philosophy*, vol. 7:578-83.

Qumsiyeh, Mazin B. *Sharing the Land of Canaan: Human Rights and the Israeli–Palestinian Struggle*. London: Pluto, 2004.

The Quran. Translated by Zafrulla Khan. London: Curzon Press, 1981.

The Qur'an. Translated by T. B. Irving (Al-Hajj Ta'lim Ali). Brattleboro, VT: Amana Books, 1991.

Al-Qur'an: A Contemporary Translation by Ahmed Ali. Princeton, NJ: Princeton University Press, 1984.

Quran: The Final Testament. Translated by Rashad Khalifa. Tucson, AZ: Islamic Productions, 1989.

The Holy Qur'an: Text, Translation and Commentary by Abdullah Yusuf Ali. Elmhurst, NY: Tahrike Tarsile Qur'an, Inc., 1987 [1934].

Qutb, Sayyid. *Fī ẓilāl al-Qurʾān*. 6 vols. Beirut: *Dār al-Shurūq*, 1973.

Race, Alan. *Christians and Religious Pluralism: Patterns in the Christian Theology of Religions*. Maryknoll, NY: Orbis, 1982.

Rahman, Fazlur. *Islam*. 2d ed. Chicago/London: University of Chicago Press, 1979.

_____. *Islam and Modernity: Transformation of an Intellectual Tradition*. Chicago: University of Chicago Press, 1982.

———. *Major Themes in the Qur'an*. 2d ed. Minneapolis: Bibliotheca Islamica, 1994, [1980].

_____. *Revival and Reform in Islam*. Edited and with an Introduction by Ebrahim Moosa. Oxford: Oneworld, 2000.

Ramadan, Tariq. *Western Muslims and the Future of Islam*. Oxford/New York: Oxford University Press, 2004.

Rasmussen, Larry. *Earth Community, Earth Ethics*. Maryknoll, NY: Orbis, 1996.

_____. "Environmental Racism and Environmental Justice: Moral Theory in the Making?" *Journal of the Society of the Christian Ethics* 24.1 (2004): 3-28.

_____. "Union Theological Seminary (New York) Faculty Panel of 9/20/01 on the Twin Towers Disaster." The Missionary Society of Connecticut (UCC): http://www.ctconfucc.org/resources/fromgroundzero/rasmussen.html (last accessed November 2, 2004).

Rawls, John. *A Theory of Justice*. Cambridge, MA: Harvard University Press, 1972.

Al-Razi, Fakhr al-Din. *Al-Tafsīr al-kabīr* (The Large Commentary). 32 vols. Cairo: al-Maṭbaʿa al-Bahiyya al-Miṣriyya, 1934–62.

Reid, Walter V. "Biodiversity, Ecosystem Change, and International Development." *Environment* 43.3 (2001).

Reinders, Johannes S. "*Imago Dei* as a Basic Concept in Christian Ethics." In Vroom and Gort, eds., *Holy Scriptures in Judaism*, 187-204.

Reinhart, Tanya. "Right for both People." Posted on her website March 27, 2001. Http://tau.ac.il/~reinhart/political/RightForPeoples.html (last accessed August 9, 2005).

Ricœur, Paul. *Fallible Man*. Translated by Charles Kelbley. Chicago: Regnery, 1965.

_____. *Symbolism of Evil*. Translated by Emerson Buchanan. New York: Harper & Row, 1967.

Ricœur, Paul, and André Lacocque. *Thinking Biblically: Exegetical and Hermeneutical Studies*. Translated by David Pellauer. Chicago: University of Chicago Press, 1998.

Rida, Rashid. *Tafsīr al-Qurʾān al-ḥakīm al-shahīr bi-Tafsīr al-Manār* (Commentary on the Wise Qur'an, Known at the Lighthouse Commentary). 12 vols. Beirut: Dār al-Maʿrifa, 1990.

Rieger, Georg. *God and the Excluded: Visions and Blindspots in Contemporary Theology*. Minneapolis: Fortress Press, 2001.

Rippin, Andrew. "*Tafsīr*," In van Donzell et al., eds., *The Encyclopaedia of Islam*, vol. 10:83-88.

Rizq, Yunan Labib. "Rural Start." *Al-Ahram Weekly Online*: *A Diwan of Contemporary Life* (553).698 (July 8–14, 2004): http://weekly.ahram.org.eg/ 2004/698/chrncls. htm (last accessed October 18, 2005).

Roberson, B. A. "The Shaping of the Current Islamic Reformation." In *Shaping the Current Islamic Reformation*. Edited by B. A. Roberson, 1-19. London/Portland, OR: Frank Cass, 2003.

Roberts, Dayton W. *Patching God's Garment: Environment and Mission in the 21st Century*. Monrovia, CA: MARC, 1994.

Robertson, Roland. *Globalization, Social Theory and Global Culture*. London: SAGE, 1992.

Robinson, Neal. *Discovering the Qur'an: A Contemporary Approach to a Veiled Text*. London: SCM Press, 1996.

Rodgers, Walter. "Why Israel Will Thwart Obama on Settlements." *The Christian Science Monitor* (August 25, 2009): Online: http://www.csmonitor.com/2009/ 0825/p09s01-coop.html.

Roe, Diane. "May Fourth Reflection." *CPTnet* (May 12, 1999): http://www.cpt.org /archives.php (last accessed October 22, 2005).

Rosenthal, Erwin I. J. *Political Thought in Medieval Islam: An Introductory Outline*. Cambridge: Cambridge University Press, 1958.

Roy, Olivier. *The Failure of Political Islam*. Translated by Carol Volk. Cambridge, MA: Harvard University Press, 1994.

Roy, Sara M. *The Gaza Strip: The Political Economy of De-development*. Berkeley, CA: Institute for Palestine Studies, 1995.

_____. "Gaza's Future: 'A Dubai on the Mediterranean.'" *London Review of Books* 27.21 (November 3, 2005): http://www.lrb.co.uk/v27/n21/ roy_01_.html.

Rubin, Barry. "Islamist Movements in the Middle East." In Rubin, ed., *Revolutionaries and Reformers*, 207-17.

Rubin, Barry, ed. *Revolutionaries and Reformers: Contemporary Islamist Movements in the Middle East*. Albany, NY: State University of New York Press, 2003.

Sa'adya, b. Yosef Gaon. *The Book of Beliefs and Opinions*. Translated from the Arabic and Hebrew by Samuel Rosenblatt. New Haven: Yale University Press, 1948.

Sabeel: The Ecumenical Center for Palestinian Liberation Theology. "The Fifth International Conference Statement" (April 24, 2004): http:// www.sabeel.org (last accessed August 10, 2005).

Sachedina, Abdulaziz. *The Islamic Roots of Democratic Pluralism*. New York/London: Oxford University Press, 2001.

Saddiqi, Ataullah. *Christian–Muslim Dialogue in the Twentieth Century*. London: Macmillan Press; New York: St. Martin's Press, 1997.

Safi, Omid, "Introduction: 'The Times they Are a-Changin'." In Safi, ed., *Progressive Muslims*, 1-15.

Safi, Omid, ed. *Progressive Muslims: On Justice, Gender and Pluralism*. Oxford: Oneworld, 2003.

Said, Edward. "The Only Alternative." *Al-Ahram Weekly Online* 523 (March 1–7, 2001): http://www.ahram.org.eg/weekly/2001/523/op2.htm.

_____. *Orientalism*. New York: Pantheon, 1978.

_____. "The Poverty of Nationalism," *The Progressive* 62.3 (March 18, 1998): 27-29.

Said, H. M., ed. *Essays on Islam: Felicitation Volume in Honour of Dr. Muhammad Hamidullah*. Karachi: Hamdard Foundation, 1991.

Sanford, Charles B. *The Religious Life of Thomas Jefferson*. Charlottesville: University Press of Virginia, 1984.

Sanneh, Lamin. *Piety and Power: Muslims and Christians in West Africa*. Maryknoll, NY: Orbis, 1996.

_____. *Translating the Message: The Missionary Impact on Culture*. Maryknoll, NY: Orbis, 1989.

_____. *Whose Religion is Christianity? The Gospel beyond the West*. Grand Rapids, MI: Eerdmans, 2003.

Sardar, Ziauddin, ed. *The Future of Muslim Civilization: Islamic Futures and Policy Studies*. London/New York: Mansell, 1987.

Sarij, Tan Sri Dato Seri Ahmad bin Abdul Hamid. "Preface." In Hassan and Musa, eds., *The Economic and Financial Imperatives of Globalization*.

Schimmel, Annemarie. "Creation and Judgment in the Koran and in Mystico-Poetical Interpretation." In *We Believe in God: The Experience of God in Christianity and Islam*. Edited by Annemarie Schimmel and Abdoldjavad Falaturi. New York: Seabury, 1979.

Schleiermacher, F. D. E. *Hermeneutics: The Handwritten Manuscripts*. Edited by H. Kimmerle. Translated by J. Duke and J. Forstman. A.A.R. Text and Translation Series 1. Missoula, MT: Scholars Press, 1977.

Schöck, Cornelia. "Adam and Eve." In McAuliffe, ed., *Encyclopaedia of the Qur'an*, vol. 1:22-26.

Searle, John R. *Expression and Meaning: Studies in the Theory of Speech Acts*. Cambridge: Cambridge University Press, 1979.

_____. *Intentionality: An Essay in the Philosophy of Mind*. Cambridge: Cambridge University Press, 1983.

Seed, Patricia. *Ceremonies of Possession in Europe's Conquest of the New World, 1492–1640*. Cambridge: Cambridge University Press, 1995.

Sefer ha-Qanah. 1894. Repr. Jerusalem, 1973.

Sen, K. *Identity and Violence: The Illusion of Destiny.* New York: Norton, 2006.

Sheehan, Thomas. "Martin Heidegger." In Craig et al., eds., *Routledge Encyclopedia of Philosophy*, vol. 4:307-323.

Shenk, David W. *Global Gods: Exploring the Role of Religions in Modern Societies*, Scottdale, PA: Herald Press, 1995.

Shiva, Vandana. *Biopiracy: The Plunder of Nature and Knowledge.* Boston, MA: South End, 1997.

_____. *Water Wars: Privatization, Pollution and Profit.* Cambridge, MA: South End, 2002.

Shiva, Vandana et al, eds. *Licence to Kill: How the Holy Trinity—the World Bank, the International Monetary Fund and the World Trade Organisation—Are Killing Livelihoods, Environment and Democracy in India.* New Delhi: Research Foundation for Science, Technology and Ecology, 2000.

Singer, P. W. *Corporate Warriors: The Rise of the Privatized Military Industry.* Ithaca, NY/London: Cornell University Press, 2003.

Sivan, Emmanuel. "The Clash within Islam." *Survival* 45.1 (2003): 25-44.

Sklair, Leslie. *Sociology of the Global System.* 2d ed. Baltimore, MD: The Johns Hopkins University Press, 1995.

_____. *Transnational Capitalist Class.* Oxford/Malden, MA: Blackwell, 2001.

Smart, Ninian. *Worldviews: Crosscultural Explorations of Human Beliefs.* 3d ed. Upper Saddle River, NJ: Prentice-Hall, 2000.

Smith, Christian. *American Evangelicalism: Embattled and Thriving.* With Michael Emerson et al. Chicago/London: University of Chicago Press, 1998.

_____. *Christian America? What Evangelicals Really Want.* Berkeley: University of California Press, 2000.

Smith, Jackie. "Building Political Will after UNCED: EarthAction International." In Lechner and Boli, eds., *The Globalization Reader*, 400-405.

Smith, Jonathan Z. *Imagining Religion.* Chicago: University of Chicago Press, 1982.

Sobel, Dava. *Longitude: The True Story of a Lone Genius Who Solved the Greatest Scientific Problem of His Time.* New York: Walker, 1995.

Sonn, Tamara. "The Islamic Call: Social Justice and Political Realism." In Barazangi, Zaman and Afzal, eds., *Islamic Identity and the Struggle for Justice*, 64-76.

Sontag, Deborah. "As Arafat Embraces Revolt, His Sagging Popularity Rises." *The New York Times* (December 8, 2000): A1, 14.

Soros, George. *The Bubble of American Supremacy: Correcting the Misuse of American Power.* New York, PublicAffairs, 2004.

Soroush, Abdolkarim. *Reason, Freedom and Democracy in Islam: Essential Writings of 'Abdolkarim Soroush.* Translated, Edited, and with a Critical Introduction by Mahmoud Sadri and Ahmad Sadri. Oxford/New York: Oxford University Press, 2000.

Sourdel, D. "*Khalīfa*: (1) The History of the Institution of the Caliphate." In van Donzell et al., eds., *The Encyclopaedia of Islam*, vol. 4:937-47.

Sperber, Jutta. *Christians and Muslims: The Dialogue Activities of the World Council of Churches and their Theological Foundation.* Theologische Bibliothek Topelmann. Berlin: W. de Gruyter, 2000.

Speth, James Gustave. *Red Sky at Morning: America and the Crisis of the Global Environment.* New Haven/London: Yale University Press, 2004.

Sprigge, T. L. S. "Idealism." In Craig et al., eds., *Routledge Encyclopedia of Philosophy*, vol. 4:662-69.

Stackhouse, Max L., Peter L. Berger, Dennis P. McCann and M. Douglas Meeks, eds. *Christian Social Ethics in a Global Era*. Abingdon Press Studies in Christian Ethics and Economic Life. Nashville, TN: Abingdon Press, 1995.

Stanley, Diane K. *For the Record: The United Fruit Company's Sixty Years in Guatemala*. Guatemala City: Centro Impresor Piedra Santa, 1994.

Stassen, Glen H. *Just Peacemaking: Transforming Initiatives for Justice and Peace*. Louisville, KY: Westminster/John Knox Press, 1992.

Stassen, Glen H., and David P. Gushee, *Kingdom Ethics: Following Jesus in Contemporary Context*. Downers Grove, IL: InterVarsity Press, 2004.

Steenbrink, Karel. "Quranic Guidelines for Economy as a Basis for Interreligious Solidarity in Favor of the Poor? Some Reflections on the Indonesian and Dutch Contexts." *Mission Studies* 15.2 (1998): 103-18.

Steppat, Fritz. "God's Deputy: Materials on Islam's Image of Man." *Arabica* 36 (1989): 163-72.

Stiglitz, Joseph E. *Globalization and Its Discontents*. London/New York: W. W. Norton, 2002.

Stoltzfus, Lynn. "Chiapas: Peace and Indigenous Rights." *CPTnet* (March 17, 2001): http://www.cpt.org/archives.php (last accessed August 22, 2005).

Stott, John. *The Lausanne Covenant: An Exposition and Commentary*. Minneapolis, MN: World Wide Publications, 1975.

Streng, Frederick J. *Understanding Religious Life*. Belmont, CA: Dickenson, 1976.

Surty, Muhammad Ibrahim. "The Qur'an in Islamic Scholarship: A Survey of *Tafsīr* Exegesis Literature in Arabic." *Muslim World Book Review* 7.4 (1987): 51-65.

Sweetman, J. Woodrow. *Islam and Christian Theology: A Study of the Interpretation of Theological Ideas in the Two Religions*. 2 vols. London: Lutterworth, 1947.

al-Tabari, Abu Ja'far Muhammad b. Jarir. *The Commentary on the Qur'an* (An Abridged Translation of *Jāmiʿ al-Bayān ʿan Taʾwīl al-Qurʾān*), *with Introduction and Notes by J. Cooper*. New York: Oxford University Press, 1987.

——. *Jāmiʿ al-bayān ʿan taʾwīl āy al-Qurʾān* (The Comprehensive Demonstration of the Qur'an's Exegesis). Edited by Mahmud Muhammad Shakir. 30 vols. Cairo, 1954–68.

——. *Tārīkh al-uman wa-l-mulūk* (History of the Nations and Kings). 11 vols. Beirut: Dār al-Kutub al-ʿIlmiyya, 1987.

al-Tabrisi, Abu Ali al-Fadl. *Jawāmiʿ al-jāmiʿ fī tafsīr al-Qurʾān al-majīd* (All the Elements of a Comprehensive Commentary on the Qur'an). 4 vols. Beirut: Dar al-Adwa, 1985.

Tafsīr Mujāhid. Edited by Abd al-Rahman al-Tahir b. Muhammad al-Surati. 2 vols. Beirut: al-Manshūrāt al-ʿIlmiyya, n.d.

Tafsīr Sufyān al-Thawrī. Edition and with commentary by Imtiyaz Ali Arsh. Rampur, India: Maktabat Riḍā, 1965.

Taha, Mahmud Muhammad. *The Second Message of Islam*. Translated and with an Introduction by Abdullahi Ahmed An-Nai'm. Syracuse, NY: Syracuse University Press, 1987.

Talbi, Mohammad. "Islam and Dialogue: Some Reflections on a Current Topic." In *Christianity and Islam: The Struggling Dialogue*. Edited by Richard W. Rousseau,

S.J. Modern Theological Themes: Selections from the Literature. Scranton, PA: Ridge Row, 1985.

Tamimi, Azzam. *Rachid Ghannouchi: A Democrat within Islamism.* Oxford/New York: Oxford University Press, 2001.

Tennenbaum, Jonathan. "Berlin Dialogue for a New World Order." *Executive Intelligence Review* 32.27 (July 8, 2005): 15-22.

Thiselton, Anthony C. *New Horizons in Hermeneutics: The Theory and Practice of Transforming Biblical Readings.* Grand Rapids, MI: Zondervan, 1992.

Thornton, Russell. *American Indian Holocaust and Survival.* Norman, OK: University of Oklahoma Press, 1987.

Tibi, Bassam. *The Challenge of Fundamentalism: Political Islam and the New World Disorder.* Berkeley: University of California Press, 1998.

Tim, Roger E. "Divine Majesty, Human Vicegerency, and the Fate of the Earth in Early Islam." In Said, ed., *Essays on Islam.*

Tinker, George E. "The Full Circle of Liberation." In Hallman, ed., *Ecotheology*, 218-24.

_____. "The Integrity of Creation: Restoring Trinitarian Balance." In *Constructive Christian Theology in the Worldwide Church.* Edited by William R. Barr. Grand Rapids. MI: Eerdmans. 1997.

_____. *Missionary Conquest: The Gospel and Native American Cultural Genocide.* Minneapolis, MN: Fortress Press, 1993.

Todorov, Tzvetan. *The Conquest of America: The Question of the Other.* Translated by Richard Howard. New York: Harper & Row, 1984.

_____. "Right to Intervene or Duty to Assist?" In Owen, ed., *Human Rights, Human Wrongs*, 28-48.

Toulmin, Stephen. *Cosmopolis: The Hidden Agenda of Modernity.* Chicago: University of Chicago Press, 1992.

Tracy, David. *Plurality and Ambiguity: Hermeneutics, Religion, Hope.* San Francisco: Harper & Row, 1987.

Tuck, Richard. *Natural Rights Theories: Their Origin and Development.* Cambridge: Cambridge University Press, 1979.

Turabi, Hasan. "Principles of Governance, Freedom, and Responsibility in Islam." *The American Journal of Islamic Social Sciences* 4.1 (1987): 1-11.

Turner, Bryan S. *Orientalism, Postmodernism and Globalism.* London: Routledge, 1994.

Turner, Victor. *Revelation and Divination in Ndembu Ritual.* Albany, NY: Cornell University Press, 1975.

Ullmann, Walter. *Medieval Papalism: The Political Theory of the Medieval Canonists.* London: Methuen, 1949.

Vahiduddin, Syed. "Qur'anic Humanism." *Islam and the Modern World* 18.1 (1987): 1-4.

Van Ermen, Raymond. "Intérêts capitalistes et responsabilité planètaire" (Capitalist Interests and Planetary Responsibility). *Le Monde Diplomatique*, Manière de Voir 50 (March–April 2000): 10-12.

Vanhoozer, Kevin J. *Is There Meaning in This Text? The Bible, the Reader, and the Morality of Literary Knowledge.* Grand Rapids, MI: Zondervan, 1998.

_____. *The Trinity in a Pluralistic Age: Theological Essays on Culture and Religion.* Grand Rapids, MI: Eerdmans, 1997.

Van Teeffelen, Toine. "Development Discourse: The Case of Palestine." In *Changing Stories: Postmodernism and the Arab–Islamic World.* Edited by Inge Boer, Annelies Moors and Toine V. Teeffelen, 37-52. Amsterdam: Rodopi, 1996.

Volf, Miroslav. *After Our Likeness: The Church as the Image of the Trinity*. Grand Rapids, MI: Eerdmans, 1998.

Voll, John O. "Fundamentalism in the Sunni Arab World." In *Fundamentalisms Observed*. Edited by Martin E. Marty and R. Scott Appleby, 345-402. Fundamentalism Project 1. Chicago: University of Chicago Press, 1991.

_____. *Islam: Continuity and Change in the Modern World*. 2d ed. Syracuse, NY: Syracuse University Press, 1994.

Voll, John O., and Nehemia Levtzion, eds. *Eighteenth-Century Renewal and Reform in Islam*. Syracuse, NY: Syracuse University Press, 1987.

Vogelaar, Harold. "Open Doors to Dialogue." *The Muslim World* 94.3 (2004): 397-403.

Vroom, Hendrik M., and Jerald D. Gort, eds. *Holy Scriptures and Judaism, Christianity and Islam*: *Hermeneutics, Values and Society*. Currents of Encounter: Studies on the Contact between Christianity and Other Religions, Beliefs and Cultures. Amsterdam: Rodopi, 1997.

Walzer, Michael. *Thick and Thin: Moral Argument at Home and Abroad*. Notre Dame, IN: Notre Dame University Press, 1994.

Wallis, Jim. "The G8 and Global Poverty: God is Acting." *Sojourners Newsletter Online*. http://www.sojo.net (last accessed October 19, 2005).

_____. *God's Politics: Why the Right Gets Its Wrong and the Left Doesn't Get it*. New York: HarperSanFrancisco, 2005.

Watt, W. Montgomery. *The Formative Period of Islamic Thought*. Edinburgh: Edinburgh University Press, 1973.

_____. "God's Caliph: Qur'anic Interpretations and Umayyad Claims." In *Early Islam: Collected Articles*. Edited by W. Montgomery Watt. Edinburgh: Edinburgh University Press, 1990.

_____. *Islam and Christianity Today: A Contribution to Dialogue*. London: Routledge & Kegan Paul, 1983.

Weaver, James H., Michael T. Rock and Kenneth C. Kusterer. *Achieving Broad-Based Sustainable Development: Governance, Environment, and Growth with Equity*. Hartford, CT: Kumerian, 1997.

Webb, Gisela. "Angels." In McAuliffe, ed., *Encyclopaedia of the Qur'an*, vol. 1:84-92.

Weiss, Bernard G. *The Spirit of Islamic Law*. Athens/London: The University of Georgia Press, 1998.

Wenham, Gordon J. *Genesis 1–15*. World Biblical Commentary. Waco, TX: Word, 1982.

Wensick, A. J. *The Muslim Creed: Its Genesis and Historical Development*. Cambridge: Cambridge University Press, 1932.

Wensick, A. J., and J. Mensing, eds. *Concordance et indices de la tradition musulmane*. 8 vols. Leiden: Brill, 1962.

West, Cornel. *Prophesy and Deliverance! An Afro-American Revolutionary Christianity*. Philadelphia: Westminster Press, 1982.

Westphal, Merold. "Phenomenology of Religion." In Craig et al., eds., *Routledge Encyclopedia of Philosophy*, vol. 6:333-43.

Wheeler, Brannon M. *Applying the Canon in Islam: The Authorization and Maintenance of Interpretive Reasoning in Ḥanafī Scholarship*. SUNY Series, Toward a Comparative Philosophy of Religions. Albany, NY: State University of New York Press, 1996.

Wherry, E. M. *A Comprehensive Commentary of the Quran, Comprising Sale's Translation and Preliminary Discourse, With Additional Notes and Emendations.* London: Kegan Paul, Trench, Trubner & Co., 1896.

White, Lynn. "The Historical Roots of Our Ecological Crisis." *Science* 155 (1967): 1203-7.

White, Stephen K. *Sustaining Affirmation: The Strengths of Weak Ontology in Political Theory.* Princeton, NJ/Oxford: Princeton University Press, 2000.

Wildberger, Hans. "Das Abbild Gottes, Gen. 1:26-30." *Theologische Zeitschrift* 21 (1965): 245-59.

Wink, Walter. *Engaging the Powers: Discernment and Resistance in a World of Domination.* Philadelphia: Fortress Press, 1992.

_____. *Unmasking the Powers: The Invisible Forces that Determine Human Existence.* Philadelphia: Fortress Press, 1986.

_____. *When the Powers Fall: Reconciliation in the Healing of Nations.* Minneapolis: Fortress Press, 1998.

Wolfe, Don Marion, ed. *Leveller Manifestos of the Puritan Revolution.* With a Foreword by Charles A. Beard. London/New York: T. Nelson & Sons, 1944.

Woodberry, J. Dudley. "Hasan al-Banna's Articles of Belief." Ph.D. dissertation, Harvard University, 1968.

World Conference on Religion and World Peace. "The Findings of the World Conference on Religion and World Peace, Held in Kyoto, Japan, from October 16 to 21, 1970." *Islam and the Modern Age* 2.1 (1971): 17-29.

World Resources Institute. *World Resources, 2000–2001.* Washington, DC: WRI, 2000.

Wright, N. T. *Jesus and the Victory of God.* Christian Origins and the Question of God 2. London: SPCK. Minneapolis: Fortress Press, 1996.

———. *Simply Christian: Why Christianity Makes Sense.* New York: HarperSanFrancisco, 2006.

Yamani, Ahmed Zaqi. "Foreword." In Watt, *Islam and Christianity Today.*

Yaqub Zaki. "The Qur'an and Revelation." In *Islam in a World of Diverse Faiths.* Edited by Dan Cohn-Sherbok. Library of Philosophy and Religion. Repr. of 1st ed. New York: St. Martin's Press, 1997 [1991].

Younes, Soualhi. "Islamic Legal Hermeneutics: The Context and Adequacy of Interpretation in Modern Islamic Discourse." *Islamic Studies* 41.4 (2002): 585-615.

Zafrani, Haïm, and André Caquot. *La Version arabe de la Bible de Sa'adya Gaon: l'Écclésiaste et son commentaire "Le Livre de l'ascèse."* Collection Judaïsme en Terre d'Islam. Paris: G.-P. Maisonneuve & Larose, 1989.

Al-Zamakhshari, Abu al-Qasim Umar. *Al-Kashshāf ᶜan haqāᵓiq al-tanzīl wa-ᶜuyūn al-aqāwīl fī wujūh al-taᵓwīl* (The Unveiling of the Truths of Revelation and of the Essences of Utterances concerning the Aspects of Exegesis). 4 vols. Beirut: Dār al-Maᶜrifa, 1980.

Zaman, Muhammad Qasim. "God's Caliph: Religious Authority in the Early Centuries of Islam." *Islamic Quarterly* 32.1 (1988): 57-67.

_____. *The Ulama in Contemporary Islam: Custodians of Change.* Princeton, NJ/Oxford: Princeton University Press, 2002.

Zebiri, Kate. *Muslims and Christians Face to Face.* Oxford: Oneworld, 1997.

Zinn, Howard. *A People's History of the United States.* New York: Harper & Row, 1980.

Zucker, M. *Commentary on Genesis.* New York: Jewish Theological Seminary of America, 1984.

INDEXES

INDEX OF BIBLICAL AND QUR'ANIC REFERENCES

INDEX OF AUTHORS

INDEX OF SUBJECTS